Failures and the Law

Failures and the Law

Structural Failure, Product Liability and Technical Insurance 5

Proceedings of the Fifth International Conference on Structural Failure, Product Liability and Technical Insurance

10-12 July 1995

Vienna, Austria

Edited by
H.P. Rossmanith

Technical University of Vienna, Vienna, Austria and
Secretary-General, International Society for Technology, Law and Insurance

Taylor & Francis
Taylor & Francis Group
LONDON AND NEW YORK

Published by Taylor & Francis
2 Park Square, Milton Park, Abingdon, Oxon, OX14 4RN
52 Vanderbilt Avenue, New York, NY 10017

First issued in paperback 2020

First edition 1996

Taylor & Francis is an imprint of Taylor & Francis

A catalogue record for this book is available from the British Library

Publisher's Note This book has been prepared from camera ready copy provided by the individual contributors.

The publisher has gone to great lengths to ensure the quality of this reprint but points out that some imperfections in the original may be apparent

ISBN 13: 978-0-367-57950-0 (pbk)
ISBN 13: 978-0-419-22080-0 (hbk)

Contents

Engineering, Risk, Quality and the Law

Case Studies

Materials and Failure Mechanisms

Preface

The 5th International Conference on **"Structural Failure, Product Liability and Technical Insurance"**, a result of the activities of the newly formed International Society for Technology, Law and Insurance has been organised in response to the ever-increasing need to bring together technologists, legal and insurance experts. The response to a call for contributions from industry, governmental institutions and academia has been exciting. Hence, this volume shows a very large section on case studies where failure cases presented will be studied for their technological importance as well as for their legal and insurance-related aspects.

The contributions in this volume have been grouped into broadly defined topic areas. The first section is devoted to engineering, the risks involved, problems of quality assurance and assessment and legal implications associated with these aspects. The major part of the book covers case studies where technical, legal and insurance-related aspects of some of the most recent failure cases are explained and discussed in detail. The final chapter of the proceedings deals with a variety of problems all related to the materials and the failure mechanisms which finally led to failure of the structure due to fatigue and fracture.

The Editor expresses his deepest appreciation for the technical support provided by all SPT participants and his coworkers at the Institute of Mechanics of the Technical University in Vienna as well as the members of the International Society for Technology, Law and Insurance who have contributed with their expertise to make the SPT conference series an internationally well-received activity. It is hoped that this new Society and its activities shall be attractive to many experts in the fields of engineering, law and insurance and shall serve as an important medium of interprofessional and interdisciplinary co-operation. The timely publication of the proceedings of SPT-5 by the publisher E&FN SPON, an imprint of Chapman & Hall, is kindly appreciated.

It is to be hoped that the work presented and the ideas expressed in this volume will serve as and proof to be a useful and important working basis for everybody in the chain from the designer to the failure assessor and insurance premium adjuster.

H.P.Rossmanith
Vienna, July 1995

INTRODUCTION

H.P. Rossmanith
ISTLI, Institute of Mechanics, Technical University of Vienna, Austria

Recently, several far-reaching political decisions have been made in Europe as well as in Asia and elsewhere, which will fundamentally transform the landscape in the fields of engineering and the law on a global scale. With the extension and formation of the European Union and the introduction of a product liability law in Japan numerous changes in technology and law will become apparent.

Since the First International Conference on "**Structural Failure, Product Liability and Technical Insurance**" in 1983 there has been enormous activity towards establishing a common basis for understanding, co-operation and exchange of expertise among technologists, scientists, lawyers and legal experts and technical insurers. SPT conferences have served as the forerunners of these efforts and it is only now that there is an increased public awareness of the necessity and importance of co-operative interaction between technology, the law and technical insurance within the framework of structural failures.

The main objective of the SPT-Conferences has been to present the State-of-the-Art in Forensic Engineering and Failure Research associated with Structural Failure of plants and machine parts with special emphasis on the technical, legal and insurance related aspects.

The participants at the SPT-Conferences have expressed their interest in establishing an international body related to these issues. Following the SPT-4 Conference, the **International Society for Technology, Law and Insurance (ISTLI)** was founded in Vienna in November 1993 together with a new series of Special Publications (ISP) and a new quarterly, the international, interdisciplinary and interprofessional journal Technology, Law and Insurance.

It is hoped that this new Society and its activities shall be attractive to many experts in the fields of engineering, law and insurance and shall serve as an important medium of interprofessional and interdisciplinary co-operation.

Structural Failure, Product Liability and Technical Insurance, Proceedings of Fifth International Conference. Edited by H.P. Rossmanith. Published in 1996 by Taylor & Francis, 2 Park Square, Milton Park, Abingdon, Oxon, OX14 4RN. ISBN 0 419 22080 1

Engineering, Risk, Quality and the Law

Structural Failure, Product Liability and Technical Insurance, Proceedings of Fifth International Conference. Edited by H.P. Rossmanith. Published in 1996 by Taylor & Francis, 2 Park Square, Milton Park, Abingdon, Oxon, OX14 4RN. ISBN 0 419 22080 1

1 PREVENTION OF HUMAN FAILURE

G.A. PETERS
Peters & Peters Law Firm, Santa Monica, California, USA

Abstract

What are the causes of human error, mistakes, undesirable human performance, and human failure? Why does such error remain, excessive and virtually uncontrolled, as an unwanted attribute of many engineered products, systems and services? There are five general approaches to human error identification, evaluation, cause analysis, fault determination, and the search for adequate remedies. Each has its own bias or perspective that determines outcome. There are five general concepts that have evolved, that serve to enhance or limit the effectiveness of the design techniques and methods that could prevent or correct human error, safeguard against it or control its consequences, and help to communicate any residual risk of injury, damage, or loss. Understanding these concepts is a predicate to any successful attempt to minimize undesirable human performance failures.

Keywords: Human error, human factors, human failure, product liability, risk assessment, safety, warnings.

Structural Failure, Product Liability and Technical Insurance, Proceedings of Fifth International Conference. Edited by H.P. Rossmanith. Published in 1996 by Taylor & Francis, 2 Park Square, Milton Park, Abingdon, Oxon, OX14 4RN.
ISBN 0 419 22080 1

1 Introduction

People make mistakes! In fact, human error associated with engineered systems, products, and processes sometimes seems to be an inevitable certainty. We should be concerned because of its relative magnitude, since some estimates indicate that 50 to 90% of all accidents are caused by human error. Most of us have some personal familiarity with undesirable human performance which might be blamed on carelessness, recklessness, inattention, forgetfulness, or inappropriate risk taking behavior. Our immediate question deals with the prevention of human failure as expressed in terms of undesirable human performance, mistakes, or human error.

Historically, we have reached for and grasped some quick answers in the attempt to prevent or control human failure and its consequences. These answers may be founded on assumptions that we are all human and therefore have developed sufficient understanding from life experiences, that we are endowed with some applicable rational problem-solving capability, and that it is all basically just a matter of common sense. While we need special knowledge in the physical sciences to help solve machine design problems, it may not seem necessary to have specialized behavioral or scientific knowledge to help solve human performance problems. Historically, we have seen many significant product design improvements, but there has been little comparative improvement in needed human performance. In fact, there remains considerable undesirable human error variance associated with user or operator tasks, technical diagnostic troubleshooting and repair tasks, manufacturing and quality assurance tasks, engineering design tasks, and management decision-making.

There have been many attempts to minimize or control unwanted human error, by applying a wide variety of concepts, techniques, methods, and procedures, with mixed results. For example, some promising concepts of accident proneness have had very limited success and, as applied to personnel selection and assignment, may conflict with employment discrimination laws, run counter to labor union contracts, or even interfere with objective assessments of other factors in specific real life fact situations. Thus, some of the answers that seemed to be quite simple and direct have proven to be inadequate for problems that have proven to be complex and confounding. In essence, there have been serious shortcomings in terms of our demonstrated ability to prevent human performance failure.

Because of the magnitude and persistence of the human error problem, it presents a serious challenge as we strive to achieve ever higher quality, safety, environmental, and performance requirements for increasingly sophisticated and complex products, processes, and services. It is a technical challenge in terms of engineering, a moral challenge in terms of our basic value systems as participants in a civilized society, and an ethical challenge in terms of professional group canons stating that the professional's paramount objective is to protect the health and safety of people[1]. Of course, there are legal responsibilities, with both civil and criminal sanctions, for acts of omission or commission, to protect the public health and safety. In other words, there are both incentives and punishments that reflect the technical, professional, legal, and governmental interests ... even for seemingly intractable induced human error.

Assuming human error to be a problem, the basic question is how can we prevent human failure. What have we learned about what is effective and ineffective in terms of the control of human error? This is the objective of this presentation.

2 Philosophical Approaches to Human Error

There are five basic philosophical approaches to any human error evaluation, investigation, reconstruction, cause analysis, or fault determination. Each approach has its own focus and perspective that both limits the scope and biases the results of any human error inquiry. Thus, selection of approach determines the outcome.

2.1 The Passive-Fatalism Approach

One common approach is to fatalistically consider human error as inevitable in an imperfect World. However, this could lead to a relative indifference or "do nothing" attitude in terms of any proactive problem identification and preventive action. So it is a high risk approach from a purely business standpoint and an ethical violation since other canons state "the engineer will have proper regard for the safety, health, and welfare of the public in the performance of his professional duties"[2]. The words of rationalization or excuse include "inherent errors", "harmless errors", "normal accidents", and perhaps, "reasonably expected errors" or "reasonably safe".

A somewhat similar approach consists of simply passively waiting for human error problems to become manifest and, then, to merely classify them into categories having limited or no direct corrective action possibilities. That is, no action because it is someone else's responsibility or of limited action since it can be tolerated. For example, there may be a surveillance of monthly summaries of complaints, deviations, discrepancies, or defects and no corrective action is even considered provided they remain within certain acceptance limits or bands. Action is taken only where it seems needed to reduce sudden excursions having unacceptable quality, cost, or market implications. The associated risks may be low for short service or shelf life products, but high for long life products (particularly, if there could be differing opinions as to what corrective action plans might be technically and economically feasible).

Legal concepts of "no fault", such as workmen's compensation laws, provide little actual incentive to investigate or determine fault for purposes of corrective action.

Therefore, the passive-fatalistic approach with limited or no corrective action is a rather unsophisticated, unaided, and high risk endeavor, that has been found to assure continuance of tolerable, unrecognized. or persistent human error.

2.2 The Behavioral Approach

Another popular approach has been to focus on individual behavior in an attempt to develop "safer people", safer attitudes, create zero defect or error-free performance, and to motivate or encourage workers to act responsibly. This may be facilitated by various types of training for hazard recognition and job skills acquisition, by greater attention to personnel selection and job assignment, more detailed procedures and instructions, by closer workplace supervision, and by better assessment of the psychological and physiological functions required for the desired human performance.

The related legal concepts have been whether there was an exercise of reasonable due care and prudent behavior under the circumstances, was there contributory negligence or reasonable care to protect the person and the equipment, was there good conduct toward others exercising ordinary care, and whether there was an informed voluntary assumption of the risks.

In the case of an industrial accident, the first level supervisor often attempts to shift all of the blame to the worker or other victim, since it usually minimizes his own involvement, simplifies any necessary organizational response, and may minimize liability involvement and damages. Often the people who may be injured are unable to speak or defend themselves, so there may be no conflict or trouble if their individual behavior is determined to be the cause of the accident. Thus, accident investigations tend to be simple and superficial when the blame falls entirely on the proximate individual and how his behavior may be described in a simple blameworthy fashion. Unfortunately, this approach can forestall appropriate preventive action for discoverable human error problems. The consequence is intractable long term human failure problems that stubbornly resist attempts at improvement.

There may be some short term improvements by use of individual motivation, closer supervision, additional training, and job performance evaluations. Attempts may be made to alter some of the individual perceptions of risk vs benefit relationships in a meaningful and personal manner. However, such individual attention is very costly, difficult to maintain over a long period of time, and often results in unacceptable residual excursions of human error.

2.3 The Situational Approach

Still another approach to the prevention of human error is to blame the situation, the work environment, the overall circumstances, group interactions, and the socio-technical aspects of the situation. It is a broader focus that includes workplace and environmental design, the tools and techniques made available to the individual, and workplace management. It includes the psychosocial design of jobs, workstations, and workplaces. It involves organizational design and management, system engineering to provide systems to meet user requirements, and the informed management of people. There is benefit from such a broad perspective as to injury and error causation, but it might be too broad for our present understanding, supporting research, and feasible applications. It may be a necessary productive approach under certain circumstances and, perhaps, a mandatory approach if legal concepts of human rights and social justice continue to grow worldwide.

2.4 The Product Design Approach

During the past half-century, there has been an increasing design engineering focus or emphasis on the interaction between people and engineered products, systems, and processes. There have been pervasive concepts of user-friendly designs, early product concept design consideration of human factors criteria, and a special emphasis on design safety in formal preliminary and detailed design emphasis on design safety in formal preliminary and detailed design reviews. Legal concepts of strict liability for foreseeable human injury has encouraged far better design for producibility or manufacturing, quality systems assurance, installation or set up, maintainability and service, operation, repair, modification, remanufacture, and ultimate disposal.

Some claim that this broadened design safety approach serves to force the design engineer to take more personal responsibility for acts of commision or omission in regard to predictable human error. However, what is predictable depends on the availability of relevant data sources, appropriate design criteria, productive test methods, organized knowledge of the field use or aftermarket problems, as well as what is believed to be encompassed by terms such as human factors engineering and ergonomics as they are understood in various countries. The design focus seems to be one of the least expensive, longest lasting, and most effective approach to human error reduction.

2.5 The Multifactorial Approach

The most recent multifactor approach seems to be the most realistic and productive. It assumes that there may be multiple causation, so the inquiry and remedies are not arbitrarily limited in scope or depth to only one cause. Each cause of injury, damage, loss, harm, or error is considered and explored for possible remedies. Those that are substantial factors or causes are given special attention, as opposed to those that seem overly speculative, theoretical, insignificant, trivial, or negligible in character.

In terms of legal concepts, multiple causation comports with the emergence of pure comparative negligence or fault, concepts of concurrent negligent acts, and the apportionment of damages in proportion to fault.

This multifaceted perspective serves to open-the-door to all causal factors, some of which may have a simple solution while others may be more difficult to resolve. Opening the minds and beliefs of the accident investigators has proven more difficult, since they may have developed the habit of making quick snap opinions as to causation. An early or preconceived opinion only serves to bias perception and provide a gate-keeping function that selectively tailors or accepts facts to fit the initial hypothesis or preconceived opinion. In other words, there may be a tendancy to believe that one person is guilty and the facts are then gathered prove it. Thus, training to recognize sources of bias and the need for full data collection is often necessary for an effective error or injury reduction program.

3 Basic Concepts

In addition to the general philosophical approach to human error or failure, there are some general technical concepts that can markedly influence the results obtained from the application of various methods and techniques for the control of unwanted human variance.

3.1 The Concept of the Average Man

Not too many years ago, designs were prepared for the average man, meaning the 50%-tile male, and often as represented by the designer himself. In terms of human stature or height, some early automobile drivers were tall enough to hit their heads on the roof and others could not reach the pedals with their feet. If the individual operator's head location and eye position could adversely effect the visual lines-of-sight, this might induce human error and accidents. The operator's physical space envelope and arm reach effects the design and placement of hand and foot controls, as well as displays, seats, and access. Thus, to accommodate a greater proportion of people, designers began to use physical dimensions that encompassed 90% of all males, then 95% of males (from small men to large men), and then from the 5%-ile female to the 95%-ile male[3]). This has been labeled "ridiculous" by some experts, because it still eliminates too large a population and the correlations between body segments is modest[4].

Indeed, the physical attributes and kinetics of body segments may be important to consider for injury prevention. Let's assume that a human error results in a rear end automobile collision to another vehicle. The driver who has been struck is slightly mispositioned in his seat and his headrest has been adjusted only slightly above the center of the mass of his head. Hopefully, the seat has been designed to permit relative rearward displacement of the operator's lower torso to minimize initial seat back yield, the head displacement gap, and head rebound, while other design criteria minimize body ramping to reduce neck hyperextension-hyperflexion and abrupt pulses in central nervous system fluids[5[. Perhaps, body self-alignment features, such as side bolsters or automatic wings, and dynamic head rests may correct for mispositioning of the driver. Thus, the human misuse or mispositioning human error has a design correction, the anthropometric body segment kinetic movements have a design injury reduction application, and initial uncorrected but foreseeable human error has prompted defensive safeguards such as cocooning the driver, integrated seat systems, and vehicle body safety cages with controlled crush regions. Therefore, what human error cannot be eliminated can be safeguarded.

In addition, as design engineers have attempted to accommodate a much wider range of users, they have found that it may be more economical to have manual seat adjustment features, automatic correction features, or other creative design solutions to such difficult problems. The design may have to accomodate individuals of varying intelligence, education, experience, cultural backgrounds, language, and even recognize that such factors as child labor are still a reality.

In essence, the basic question is how to design to accommodate or compensate for known human variability, so that we can minimize and control human error, and do this in the most economical, efficient, and effective manner.

3.2 Appropriate Testing

Surprisingly, there has been little formal testing to discover and remedy sources of human error that can materially degrade or cause system failure, malfunction, or damage. The focus usually is to do component testing per specifications and to design, build, modify, or "cut and paste" prototypes until they work satisfactorily, with only peripheral attention given to human factors considerations and almost as an afterthought. A central question might be "who" submits any human error reports, "when", for "what" purpose, and do they receive serious attention from design specialists knowledgeable in an appropriate corrective action specialty?

To accomplish something, an appropriate data base is usually needed. The design can be in terms of detailed standards compliance, then comes the wait for complaints to form a historical data base for future generations of the product. There may be a formal design review process to help utilize the data. Meaningful data may accumulate from field experience in the form of service, maintenance, reliability, or repair reports assuming the generators of the reports have the appropriate training to identify what constitutes human error. Useful data may come from simulator studies and experimental research, provided that it is oriented toward relevant and timely application issues.

Some priority should be given to data sources involving representative user groups rather than company personnel. It is not too costly to gather information from internal and external focus groups. Valuable data can be obtained from product demonstration laboratories using various occupational, educational, and cultural form of direct on-site observation of the product with trained data gatherers.

Whatever human factors techniques, methods, and procedures are utilized, based on what is feasible and the circumstances of the product use or application, the prime objective is to identify and determine what actually happens in real-life situations, so that we can foresee and predict human errors, thus attaining the capability of fashioning effective remedies that can be instituted in a timely fashion. Without such relevant and practical information, design engineering decisions remain speculative regarding the man-machine-environmental interface. Merely guessing about human error and its causes does not constitute a professional or ethical approach to this persistent high risk area.

3.3 Early Engineering Analyses

At an early stage in design, good engineering requires that specific analytic techniques be applied for product design assurance. This should include the timely, detailed, and systematic analysis of human error. That is, an organized and documented effort at the time that product improvements can be best and most easily incorporated into the product design. Since the analysis predicts or foresees human error, it requires something more than mere intuition or general speculative knowledge. Appropriate data is needed as a foundation or basis for predictive estimates. It may come from direct or indirect observation, experimental research, questionnaires, special useability testing, or other data sources.

As an example, direct observational data may be gathered in a variety of ways. In some manufacturing facilities, all work activities are videotaped 24 hours a day so that accidents or errors can be replayed and the circumstances analyzed to determine those remedies that might help to prevent undesirable reoccurrence. Direct human error data may be obtained from trained human factors observers who have designated tasks to be performed before

Direct human error data may be obtained from trained human factors observers who have designated tasks to be performed before them in a maintenance demonstration laboratory. The tasks may be performed on conceptual mockups, pre-production prototypes, first run products, or products that are the subject of field complaints.. Direct observational data on the operation of very large machines or assemblies of machines may be obtained on-site before final acceptance by the customer or as a part of a joint enterprise with the customer. The data reflects what actually happened in terms of human errors, near accidents, and accidents. It includes their frequency, consequences and severity, and such causal product feature descriptors as misleading, confusing, sidetracking, stressful, difficult, and frustrating.

It is usually advisable to do the human error analyses in cooperation with those performing related systematic evaluations; such as, failure mode and effect analysis, fault tree analysis, preliminary hazard analysis, hazard and operability studies, system reliability and maintenance studies, formal design reviews, and other design analysis techniques. This provides an economic approach, more cross-fertilization by disciplines to achieve better systems engineering, more timely access to the design decision-making process, and better fulfillment of document retention requirements under newer directives, standards, regulations, and good practice requirements. Such analyses provide some proof that human factors was considered for every stage of the product use, repair, modification, and disposal.

Some of the safety design features to help prevent human error include:

(a) tamper resistant and fail-to-safe interlocks, switches, and guards to protect against human bypassing, circumvention, or unauthorized adjustment.

(b) safe-stop switches for blind zones to prevent inadvertant human activation and operation.

(c) proximity sensors with brake, speed, and directional control for human injury avoidance or mitigation.

(d) limited movement pulse timers to protect humans who are mispositioned.

(e) activation alarms to permit protective behavioral reactions.

(f) overspeed, overtemperature, overpressure, and leakage limiters and protectors to minimize human injury.

(g) redundancy and other reliability engineering design applications

(h) two hand controls, light curtains, safety mats (edges, boards, and cables), gate monitors, and mechanical guards.

(i) positive guidance for humans and diagnostic troubleshooting aids.

There are many safeguarding devices, on the market or incorporated in products, since human error is common and predictable despite adequate training, supervision, motivation, or even conscientious attempts at safe behavior. Such safeguards should never be an aftermarket option, since human safety should not be a delegated function when it is a primary early design responsibility.

3.4 Risk Assessment and Perception

Risk is the capacity to do harm; for example, will a predicted error be harmless or catastrophic in consequences. The determination of risk has become more objective in character as risk assessment has become popular, advisable, or even mandatory. It usually is a resultant of an estimate of the frequency of errors or incidents, the exposure time or cycles, the severity or magnitude of loss, and sometimes other factors such as the technical and economic feasibility of remedies and their relative effectiveness and the possible introduction of other unwanted or undesirable variance in product performance.

While precise estimates based on quantitative probability are desirable, some qualitative estimates being made are only in terms

10^{-6} level, seem more precise than they actually are and, thus, may not serve as a reliable basis for important decision making. In addition, such estimates are easily biased or bent to achieve standards compliance, contractual specifications, or marketing advantage. However, risk assessment can be a useful and valuable tool, if the problems in risk estimation, classification, and bias are recognized.

If the risk estimates are primarily based on subjective judgment, it may reflect the individual perception of risk of the engineer within a company and his immediate interpersonal or group objectives. Certainly, what may be acceptable risk to him may be quite different than what is acceptable, unacceptable, or intolerable to the customer, user, or general public. If risk estimates are utilized in a risk-benefit study, the perceptions of risk and perceptions of benefits may vary considerably. For example, how safe is safe enough, when the cost of the consequences are shifted to someone else? What if there are immediate benefits to the engineer's organization and the risks become manifest only over a long period of time? The perceptual problems may be exacerbated if the engineer has a strong personal belief, attitude, or conviction that human error is best controlled by conscientious users who should perform as intended without errors, misuse, or abuse of the product. Thus, individual perception can induce bias that could result a failure to adequately control avoidable human error.

The very process of accomplishing a risk assessment may help to identify latent hazards and associated unrecognized risk levels and implications. It permits preventive actions to be focused on a worst-first basis rather than a more haphazard easiest-first approach to problem solving.

In general, the design engineer is and should be first concerned with machine performance and limits, movement and space, time and foreseeable life, fit and wear, durability, and other concerns. The next step is to identify hazards and assess the risks during manufacture, transportation, intallation, operation, servicing, maintenance, repair, modification, resale, and disposal. The final step is to remove the hazards by design to reduce the risk to near zero (such as, using fail-safe techniques and ergonomics design). If the risks cannot be eliminated, safeguards (such as fencing or guards) can be utilized to reduce the risks to a tolerable level. Finally, those who may be exposed to residual risk should be appropriately warned and informed as to their exposure and how to avoid the risk. In all of the design stages, valid risk assessment is a vital predicate to good design.

3.5 Warnings, Instructions, and Communications

Warnings are used only for residual risks because their effectiveness varies as to the appropriateness of their design, testing, and utilization. Many have a communication deficiency, they fail to alert and inform as to the need and methods for protective or avoidance behavior in regard to both hidden and obvious dangers. It has been stated that warnings that fail to communicate their intended message are functionally useless[6]. Legally, there maybe a failure to warn under strict liability or negligence law. The origin of modern warning theory comes from the Nuremberg war trials, more than 40 years ago, which held that each person must consent to his own personal exposure to risk and that such consent must be informed, voluntary, and revocable [7]. How can a person assume risks unless the referent is a specific hazard and he is part of the target group that has been identified and adequately informed?

When warnings are designed, there are neuropsychological considerations that may be important for a particular application. For example, warnings or other information may generate user expectations that serve to suppress competing information presented to the operator by means of visual or auditory displays or by machine condition or status. In terms of attention, what cues does the operator utilize for situational awareness, when cognitive by machine condition or status. In terms of attention, what cues does the operator utilize for situational awareness, when cognitive demands and workloads may lead to certain human error forms and failure pathways? How do warnings effect automatic human information processing, perceptual learning, human memory consolidation and priority, and human risk taking strategies? What happens when we apply the principles and knowledge of warning design to more contemporary psychosocial issues such as the human autonomy inherent in flexible manufacturing, the skilled home office contract worker in an electronic cottage, the impoverished and physically impaired worker, and labor union training for hazard recognition? How do we achieve effective warnings, when the object of the communication may be likely to manifest some degree of genetic personality disorder or other psychopathology, some group environmental behavioral differences, or some individual differences in long or short term memory and processing speed, spatial and verbal reasoning, and general intelligence? Indeed, if this describes the marketplace for products, shouldn't the designer "know the customer" in order to design customer-friendly and customer-acceptable products?

Once again, we are dealing with a fairly complex engineering design issue: how to design warnings, instructions, owner's manuals, recall notices, material safety data sheets, health and safety bulletins, cautionary videotapes, and other communication devices so that they will convey an appropriate and meaningful message that should result in behavior modification of a target audience. This should not be an afterthought to engineering design, but an integral part of the design process. It is an essential part of design, if we are to prevent adverse human failure in the marketplace.

4 Conclusion

The present magnitude of human error and the failure of desired human performance is intolerable in terms of good design practice for many products, processes, services, and engineered systems.

There are known methods and techniques that could and should be used: (1) to predict and prevent human error in new designs, (2) to correct and control human error in existing designs, (3) to protect against and control the consequences of human errors of omission and comission, (4) to supplement and safeguard where there is uncontrolled human error in inadequate designs, and (5) to legally and effectively communicate the residual risks inherent in some designs. This is a problem area in which there are few simple solutions or unsophisticated answers. It is a difficult, but urgently needed area of endeavor that could produce significant technical, liability, social, and professional benefits.

5 References

1. "Engineers shall hold paramount the safety, health and welfare of the public in the performance of their professional duties". Section 1, Suggested Guidelines for Use with the Fundamental Canons of Ethics, Engineer's Council for Professional Development, New York, 1974. Also in the Fundamental Canons, Code of Ethics of Engineers, ECRD, 1974.

2. "The Engineer will have proper regard for the safety, health, and welfare of the public in the performance of his professional duties". Section 2, Code of Ethics for Engineers, NSPE Publication 1102, July 1976, National Society of Professional Engineers, Washington, DC. Also contained in Ethics for Engineers, NSPE Publication 1105, February 1974.

3. USA Male and Female Physical Dimensions for Construction and Industrial Equipment Design, SAE Recommended Practice J833a, Approved 1974. SAE Handbook, Society of Automotive Engineers, 1979.

4. Kennedy, K.W., Christensen J., and Howard, J. Engineering Anthropometry in Vehicle Design, Chapter 2 in Volume 3,

Peters, G.A. and Peters, B.J. (Eds.), Automotive Engineering and Litigation, Garland Publishing Inc., (now John Wiley & Sons) New York, 1990, pages 29-114.

5. Svensson, M.Y. Neck-Injuries in Rear-End Car Collisions: Sites and Biomechanical Causes of Injuries, Tests Methods, and Preventive Measures, Institutionen for personskadeprevention, Chalmers tekniska hogskola, S-412 96 Goteborg, Sweden, 1993.

6. Peters, G.A. Warnings, Notices, and Safety Information. Chapter 17 in Volume 5, Peters, G.A. & Peters, B.J. (Eds). Sourcebook on Asbestos Diseases: Medical, Legal, and Engineering Aspects. Asbestos Abatement. Butterworth, 1991, pages 695-805.

7. Trials of War Criminals Before the Military Tribunals Under Control Council Law No. 10, Vol. 2, October 1946-April 1949, Washington, D.C.: U.S. Government Printing Office, 1947.

2 THE AMBIGUOUS NATURE OF THE TERMS HAZARD, RISK AND DANGER AS USED BY THE LAY PUBLIC, TECHNOLOGISTS AND ATTORNEYS

D. MUSTER
D. Muster & Associates, Inc., Houston, Texas, USA

Abstract
Often, when we use terms in different situations, we change their meanings, sometimes inadvertently, sometimes deliberately and, in most cases, without significant impact. However, the transfer of terms from the lay world, where their meanings are understood with benign ambiguity, to the more demanding worlds of engineering and the courts of law can lead to needless misunderstandings and failures in communication. This can be especially true in the contentious milieu of a court setting where three distinct cultures converge: the attorneys and judges, in their solemn cloaks of magisterial splendor, the jurors, members of the laity whose everyday lives have been interrupted by a call to civic duty, and expert witnesses, whose professional expertise is needed to explicate (in the languages of the laity and the courts) the technical aspects of, say, how and why an industrial product failed, thus causing harm and injury, and how this failure can be related to the manner in which the product was designed, tested, manufactures and marketed. Here a discussion is given of the terms hazard, risk and danger and their definitions and uses in each of the three cultures that converge in a trial setting.
Keywords: Hazard, risk, danger, products liability law, engineering, courts.

1 Introduction

In law and science and technology there are many terms which attorneys and scientists and engineers simply never share with the lay public. The legal implications of the Latin phrases *res ipsa loquitur* and *volenti non fit injuria* and an engineer's interest in the mechanisms of fretting corrosion are simply of no interest to the average citizen --- until, perhaps, the time comes when he may be required to learn their meanings in a courtroom.

Structural Failure, Product Liability and Technical Insurance, Proceedings of Fifth International Conference. Edited by H.P. Rossmanith. Published in 1996 by Taylor & Francis, 2 Park Square, Milton Park, Abingdon, Oxon, OX14 4RN. ISBN 0 419 22080 1

These terms, and others like them, are recognized by everyone as being part of the professional languages of attorneys and engineers. Further, it surprises no one that the practitioners in medicine, psychiatry, economics or the neural sciences use terms peculiar to their fields, terms which are rarely understood by the laity and may not be understood by professionals in other fields. However, when common words, words we find regularly in a newspaper or weekly news magazine, are used in the limited context of the language of a profession, they can take on unexpected and special meanings. This can cause confusion and misunderstanding in the minds of those not in the profession. In general, the problem is not limited to a particular group of terms or cultures, but in this note we limit our interest to the definitions and uses of certain common lay terms which, when used in a courtroom by engineers or attorneys, can take on special meanings.

The lay public uses the terms hazard, risk and danger as synonyms whose common antonym is safety. This lack of precision in the definitions and uses of the terms causes no difficulty in common discourse in our everyday world, but such imprecision may not be acceptable in an engineering designer's office or a courtroom, where the terms must need be defined precisely and unambiguously. In a courtroom, their definitions must be understood in the same way by everyone involved --- the attorneys and judge, the expert witnesses and the jurors. Preferably, their definitions should be phrased in a manner that both highlights and discriminates. Each definition should highlight the essential character of the term and, at the same time, convey a uniqueness of meaning that discriminates one term from another. This is needed especially because the terms hazard, risk and danger deal with different aspects of the subject of safety, a vital concern in the design of an industrial product and the alleged lack of which is often the basis for a court trial. We agree strongly with Abbott [1, p. 58] when he says, "First, we have to establish that 'hazard' and 'risk' are not synonyms, even if the *Oxford English Dictionary* [2] thinks that they are." We would add the term "danger" to Abbott's list.

The languages of the cultures that converge in a court could have a significant effect on the outcome of a trial. Taken as a general statement, this may be a platitudinous truism, but in cases where the central issues involve engineering and scientific principles and practices, a nonobvious trichotomous situation may exist. The terms used by the attorneys and judge and those introduced by the technical experts, may have special meanings in the precise languages of the legal and technical fields involved, but may not be in the normal vocabulary of the jurors. For example, in personal injury cases, the plaintiff's attorney and experts must devise a plausible scenario for the accident in which the plaintiff was injured. In order to explain the scenario to the court, the attorneys and experts may need to use technical terms whose meanings are not understood in the same way by everyone concerned. In this circumstance, these terms must need be explained precisely.

In products liability litigation, the manufacturer or seller of an industrial product, such as, a drug, a vehicle, a machine in a factory, or a cosmetic cream, is in the dock and the issues before the court may include the design, testing, manufacturing and marketing of the product. Here, a hazard may be an identifiable, inherent attribute of the product at issue, the risk may be measurable and the danger associated with using the product a factor which must be taken into account by an "ordinary user" in order that he can use the product safely.

This paper is an expanded and updated version of an earlier paper [3] which was published originally in the Journal of the National Academy of Forensic Engineers. The Academy is a relatively small organization with fewer than three hundred members; thus, it was decided to expand and update the text of the original paper and to seek publication of the revised paper where it would reach a larger audience.

2 Discussion

2.1 The generic and specific meanings of the terms "possibility" and "probability" and "reasonable" and "unreasonable"

Before we discuss the definitions and uses of ***hazard***, ***risk*** and ***danger***, there are two related terms we should define and discuss first, namely, ***possibility*** and ***probability***. Over time, they have acquired specific, contextual meanings in the courts (particularly when used with such adjectival antecedents as "engineering", "medical" and "reasonable") which are significantly more specific than their generic meanings. Moreover, as we show below, the same terms can be used differently in the liability and damage phases of a personal injury suit involving the effects of medical decisions and practices.

The differences between these terms, with and without modifiers, are more than superficial. They are fundamental and, in a sense, serve to highlight a dichotomy between the bases of the expert opinions offered by technologists (such as, engineers) and medical practitioners (such as, surgeons and psychiatrists). Except in the areas of design and marketing, the expert opinions of most technologists are based in the testable, objective criteria of the physical sciences rather than the experience-based, subjective criteria of medicine and psychiatry. Logic and the laws of physics are the usual determinants of whether, under a stated set of circumstances, it is possible (or impossible) for an event to occur. In most cases, an engineer can make such an absolute, go/no-go kind of assessment with relative ease. However, if it is agreed that an event is possible, that is, it can occur, uncertainties arise when we attempt to evaluate the likelihood that it may occur at a given time and under a stated set of circumstances. In careful written usage, ***can*** is used rather than ***may*** when reference is made to capability and certainty (which are akin to the absolutism of the possible/impossible dichotomy) rather than permissiveness and uncertainty (which are associated with the relativism of that which is probable).

The sense of this difference in meaning can becomes important in a trial when an attorney asks an expert, "is it possible for [a specific event] to occur?" The attorney may be ignoring --- perhaps, deliberately --- the more relevant question, "what is the probability that [a specific event] is likely to occur?" An affirmative answer to the first question could lead a jury to believe that, because the event can occur, it is likely to occur, which is often far from the truth. Thus, even though, in the lay world, the terms ***possible*** and ***probable*** have close to the same meaning, in most cases, engineers associate the first term with that which is simply feasible, an event that can occur, and the second with that which is both feasible and likely to occur.

Consider the following examples: we can state, "it is possible that it will rain tomorrow", "it is possible that a certain horse will win a race", or "it is possible that the Dow-Jones industrial average will go up". From an abstract viewpoint, we recognize that each statement simply shows that any or all of these nonrelated events are achievable, that they can occur. The statements are simply a catalog of selected, possible events, not predictive opinions. The likelihood that each event will occur --- even the likelihood that they will all occur on the same day --- can be estimated in terms of quantized measures we call probabilities.

Even in the lay world, we would recognize that a new element has been introduced, if we changed the statements to, "it is probable that it will rain tomorrow", "it is probable that a certain horse will win a race", or "it is probable that the Dow-Jones industrial average will go up". In this new phrasing, each of the statements includes a qualitative sense of the likelihood that the described events will occur. In that context, the second statements are predictive opinions and are stronger and offer more information than the first statements.

Without identifying adjectival antecedents, the terms ***possibility*** and ***probability*** are generic and, usually, have the meanings assigned to them in the discussion above. The first deals with the feasibility of an event, the second with the likelihood that it will occur. However, in the courts and in engineering, the two terms are seldom used in their generic sense. Usually, they are used with antecedents which particularize their meanings, especially with respect to the reasonableness of the events with which the terms are associated and the contextual relationship of the terms with past or future events.

In the courts, the antecedent ***reasonable*** has come to have almost magical powers. It has been held frequently that, "[r]easonable probability . . . is determinable by the substance of the testimony of the expert witness and does not turn on the semantics or on the use by the witness of any particular turn or phrase." [3] For example, a medical expert's opinion preceded by the phrase "reasonable medical probability" can be offered as relating ". . . to the showing that must be made to support an ultimate finding of fact . . .". [4] The weaker phrase, medical probabilities, is deprecated because, "If the witness were permitted to state his opinion only in terms of medical probabilities, the court and jury would have no opportunity to decide the case on the basis of the substance rather than the form of his testimony." [5] Presumably, medical probabilities cited by experts, but unevaluated by them as to their reasonableness, would not be useful to the court and jury in guiding them to an appropriate judgment.

The term ***reasonable*** and its antonym, ***unreasonable***, are used in other legal and engineering contexts as well. For example, as we show later here and have shown elsewhere [6], a properly evaluated "unreasonable danger" can be an essential element of an engineering expert's testimony on design in a products liability suit based on strict tort liability. [7]

In their testimony, engineering experts and medical experts usually use virtually the same defining caveats to indicate their confidence that specific events are likely to occur. A real, but nonobvious, difference in their testimony is that, in most cases, the expert testimony of engineers is used only in the liability phase of a trial to answer questions related to causation and, thus, their testimony deals with events in the past. For example, who or what caused an accident? Only rarely are engineers asked to offer predictive opinions as to the probability of a design defect causing another accident in the future or, if a guard were installed on a machine by the owner or manufacturer, the effect this action would have on the probability of future accidents with the same machine.

In testimony offered to support the likelihood that an event will occur in the future, the standard for expert engineering or medical testimony is the same, that is, it must fall within the realm of reasonable engineering or medical probabilities; however, there is an exception to this rule. When the issue before the court is proof of past, present and future emotional damages or mental anguish, the door can be opened for the admission of testimony that future surgery is a possibility, not a probability. [8] In the context of this special issue, a plaintiff can testify that he has discussed the possibility of surgery with his physician. The fear and anxiety associated with the possibility, not the probability, of this surgery and its consequences can be offered as evidence to support an allegation of the existence of mental anguish. In the language of the Appeals Court, the ". . . testimony of [the] treating physician explaining customer's present and future medical condition and suggesting a strong possibility of [the] need for future surgery amounted to reasonable probability, when taken in context and considered as a whole, and, hence, should not have been excluded as being speculative; . . ." [9] Thus, in this relatively narrow context, a criterion for the equivalence of the terms, medical possibility and medical probability, was established. The decision extends the meaning of the term possibility beyond its generic meaning and could have the effect of introducing an additional relevant matter in a medical malpractice suit. The rule of "reasonable probability" with regard to supporting an ultimate finding of fact and other medical issues related to the probability of future surgery is apparently untouched by this narrowly construed exception.

In the sections that follow, we consider first the generally understood meanings of the terms hazard, risk and danger in the lay world. This is followed by separate discussions of their definitions and uses in the fields of law and technology. Finally, we offer a definition of each term. We trust that the definitions highlight the essential character and uniqueness of meaning of each of the terms, thus, discriminating one term from another.

2.2 The terms hazard, risk and danger in the lay world

Even a casual etymological study of the lay-world uses of the terms hazard, risk and danger reveals the benign ambiguity with which we treat the terms. In *Roget's International Thesaurus* [10] and *Chamber's Thesaurus* [11], each of the three terms has ten or more synonyms, but the lone antonym for each is safety.

In the *Merriam Webster's Collegiate Dictionary*, Tenth Edition [12], the term hazard is defined rather broadly as "**2** a source of danger". Risk is defined (in a generic sense) as "**1** [the] possibility of loss or injury" and, buried in a later definition related to insurance, as "**3 a**: the chance of loss . . . to the subject matter of an insurance contract; also: the degree of probability of such loss". Danger is defined as "**3** exposure or liability to injury, pain, harm or loss", with no mention of hazard and risk.

Oddly and unexpectedly, the *Oxford English Dictionary* [2], is of little help. In essence, it tells us that the three terms are virtually synonyms and, thus, they are defined with a degree of circularity. The *Oxford English Dictionary* contains many definitions and examples of the term hazard, one of which is generic, "**3** risk of loss or harm;" and most of which are related to games of chance, billiards, tennis, golf and even the fact that, in Ireland, a taxi-stand is called a hazard. None deal with the term with sufficient precision to be of more than general use. In the *Oxford English Dictionary*, risk and danger are defined in much the same manner as they are in *Merriam Webster's Collegiate Dictionary*.

Collins Dictionary of the English Language [13] is less general and offers the definition of the term hazard as "**3** a thing likely to cause injury", risk as "**1** the possibility of incurring misfortune or loss;" and danger as "**1** the state of being vulnerable to injury, loss or evil". In *The Random House Dictionary of the English Language*, Second Edition [14], a hazard is defined as "**1** an unavoidable danger or risk, even though often foreseeable:" and "**2** something causing unavoidable danger, peril, risk, or difficulty"; risk (in a generic sense) is defined as "**1** exposure to the chance of injury or loss;" and (in the context of insurance) as "**2 b.** the degree of probability of [an insurance] loss"; danger is defined as the "**1** liability or exposure to harm or injury".

2.2.1 Hazard

In a lay-world sense, excluding the specialized definitions related to games of chance or sports, there does not appear to be an unambiguous, universally accepted definition of the term hazard. The sense of the definitions in the standard sources appears to accept the general concept of a linkage between hazard and danger; that a hazard is a source of danger; that without a hazard, there would be no danger. Some generic definitions suggest that a hazard is an identifiable attribute of a physical entity or situation.

In some definitions, it is stated that a hazard is unavoidable. This concept is contrary to the experience of most engineering designers. In law and in engineering, we accept the fact that there are products which may be dangerous due to inherent hazards associated with their composition or configuration. For example, trinitrotoluene and nitroglycerin are chemical compounds which can be unstable under even the gentlest of handling conditions, some small, light boats (such as canoes) are simply too narrow to stand up in safely, and kitchen knives have sharp, exposed edges. In such cases, the very properties that make the product useful, can also make it dangerous. Usually, we consider the hazards represented by an unstable explosive, a boat that has a high center of gravity when it is loaded, and the cutting edge of a knife as inherent hazards which can render a product inherently dangerous, that is, dangerous *per se*. But a product that is inherently dangerous --- a lay term --- is not necessarily unreasonably dangerous --- a legal term. This distinction is significant to engineers and attorneys, but of little practical interest to a chef who just cut his hand. Most inherent hazards in machinery, such as, pinch points, the exposed shafts, belts and gears in a power transmission system and the cutting edges of power saws, planers and other machine tools can be eliminated or, at least, their effects can be mitigated. In that sense and to that degree, such hazards are avoidable.

2.2.2 Risk

Except in the context of insurance, in the lay world, risk is considered to be the nonquantized chance that an individual might be exposed to harm, injury or loss. As is shown below, this definition is a near paraphrase of the definition used in the field of law as well. But the definition does not include an element of great interest to engineers and designers, namely, the probability (in quantitative terms) that the design of a product, or the way in which it is manufactured, tested and marketed, will expose an individual using the product to harm, injury or loss. Thus, in the lay world and in law, risk is defined in qualitative terms only and not in the quantitative terms needed by engineers in order that they can estimate the probability that a product can cause harm, injury or loss when it is used.

2.2.3 Danger

In the lay-world vocabulary, danger is defined unemotionally, almost in the abstract, as simply "exposure or liability to harm, injury or loss" [12], or "the state of being vulnerable to harm, injury or loss" [13]. Both are short, qualitative statements that do not convey a sense of the menace associated with physical or social situations in which there is an imminent potential for harm, injury or loss arising from an element, or elements inherent in the situations. In a sense, knowledge of the hazard and risk in a potentially dangerous situation permits us to estimate, at least qualitatively, the menace inherent in the situation.

2.2.4 Summary: The terms hazard, risk and danger in the lay world

These lay-world definitions, fall short of our needs in engineering and the law, where we wish to convey a real sense and qualitative measure of the danger inherent in, say, the design of a product. The sense of menace, the sense that a hazard present in the design of a product, or in a physical situation (such as a vehicular accident), when coupled with the probability that this hazard will become operative, should be an element in the definition of danger. The lay-world definitions of danger miss the sense of menace, the real-world potential for imminent harm which, we believe, arise from having knowledge of the hazards and risks inherent in a product or a physical situation. Only when hazard and risk are juxtaposed properly with danger can we appreciate its true meaning.

2.3 The terms hazard, risk and danger in the world of law

The prime source for definitions of the terms hazard, risk and danger as they are used in the world of American law is probably *Black's Law Dictionary*, 6th Edition [15]. In a broad-brush sense, the basic definitions given there represent the current usage of the terms, although case law provides definitions which are more tightly focussed and differ from the basic definitions because of the unique context provided by individual cases. In any event, the subtleties and nuances of the definitions in these cases are generally outside the scope of this note and are not covered here.

2.3.1 Hazard

In the *Black's Law Dictionary* [15], hazard is defined as "a risk or peril, assumed or involved, whether in connection with contract relation, employment, personal relation, sport or gambling. A danger or risk lurking in a situation which by chance or fortuity develops into an active agency of harm." Like the lay-world definitions of hazard, both of the just-cited definitions suffer from the circularity of using danger and risk as synonyms for the term being defined. Nevertheless, the second definition uses a phrasing that couples the concept of a hazard with something ". . . lurking in a situation . . ." which ". . . develops into an active agency of harm." Conceptually, this is close to the notion that a hazard is a source of danger, that a hazard is an attribute of a product or situation which, when it is operable, can develop into "an active agency of harm." Further, the notion that a hazard is "lurking in a situation [or product]" is akin to it being an inherent attribute of the situation or product. Thus, the definition of hazard in *Black's Law Dictionary* and some of the lay-world definitions of the same term cited earlier appear to be conceptually close.

2.3.2 Risk

Excluding the definitions associated with insurance law, in *Black's Law Dictionary* [15] the term risk is defined as ". . . the element of uncertainty in an undertaking; the possibility that actual future returns will deviate from expected returns." As is mentioned earlier, this is a near paraphrase of the lay-world definition of risk, except that the relatively sterile phrase, "the element of uncertainty in an undertaking", does not suggest that the undertaking could expose an individual to harm, injury or loss. However, in the case styled, *Knox Jewelry Co., Inc., v. Cincinnati Ins. Co.* [16], risk is defined by the group of synonymous terms "hazard, danger, peril, exposure to loss, injury, disadvantage or destruction, and comprises all elements of danger." This is not useful as a definition, but it does introduce the concept of exposing oneself to loss, injury, disadvantage or destruction. The legal term "obvious risk: one so plain that it would be instantly recognized by a person of ordinary intelligence", is ambiguous in light of our earlier discussion, because it could be interpreted to mean an "obvious hazard".

2.3.3 Danger

In the vocabulary of the law, the terms "danger" and "dangerous" are used widely with several adjectival antecedents, for example, "apparent danger", "imminent danger", "unavoidable danger", "unreasonable danger", "inherently dangerous", and "unreasonably dangerous". Not all of these compound terms are used in civil law, the most likely venue for an engineer to appear as an expert witness. In *Black's Law Dictionary*, the main definition of danger is ". . . exposure to loss or injury." (*State v. Londe* [17]). This is almost the same as the lay-world definition cited earlier in the *Merriam Webster* [12] and *Collins* [13] dictionaries. Recall our earlier discussion concerning the inadequacy of such an imprecise definition for use by engineers.

2.3.4 Summary: The terms hazard, risk and danger in the world of law

On the record of references cited here, we can conclude that, in general, in the legal and lay worlds, the terms hazard, risk and danger are more likely to be considered synonyms than not. Apparently, in no dictionary are they considered to be related terms with mutually exclusive meanings. This can tend to cause misunderstanding and confusion in situations where we look for clarity and incisiveness.

The sense of the definition of hazard in *Black's Law Dictionary* [15] appears to accept the general concept of a linkage between hazard and danger and that a hazard may be present as a latent attribute in a situation which, when and only when it becomes an active agency for harm, is a source of danger. It follows that without a hazard, an active agency from which harm can derive, there can be no danger. There is no indication in *Black's* that a hazard could be associated with a product.

In the law, apparently the term risk is introduced to provide a qualitative means of indicating that the outcome of a specific event is not certain, that chance may play a role in determining whether a specific outcome can and will be achieved. This sense of uncertainty is not quantized as a probability (that is, it is not expressed as a quantitative measure of the likelihood that a specific outcome will be achieved) but expressed in qualitative terms only.

Apparently, in the law, danger is the dominant element in a class of situations in which a person is exposed to the possibility of loss or injury. In such situations, it is assumed implicitly that an active agency for harm (a hazard) exists and is operable. Further, it is assumed that there is a likelihood (a risk which can be quantized) that the hazard will cause a person to actually suffer loss or injury.

2.4 The terms hazard, risk and danger in engineering

2.4.1 The influence of safety on the world of engineering today

A successful design of an industrial product is based on maintaining reasonable balance among the four principal objectives of the design process, namely, that the product is functional, capable of being manufactured in an assembly line, capable of meeting its market and, to the extent that it can be achieved when the product reaches its ultimate user, it is safe --- not safe in the absolute sense, that it is free of all risk or dangers, but safe in the relative sense, that it is free of unreasonable dangers [6]. It is not required in law or common sense that manufacturers make products that are absolutely safe.

"The concept of strict liability does not impose on the manufacturer a duty to design an accident-proof product." (*Sales and Perdue* [18], p. 14 and *Willis v. Chrysler Corp.*, [19]). Later, the sense of this general rule was particularized by qualifying conditions imposed in several cases. These are summarized by Sales and Perdue ([18], p. 40), "Manufacturers, . . . , are not required or legally obligated to design the safest possible product. (*Weakley v. Fischbach & Moore, Inc.*, [20]); *Henderson v. Ford Motor Co.* [21]). . . . Moreover, the Fifth Circuit has noted: 'It is one thing to show that a defendant might have designed a safer product; quite another to show the product he did design is unreasonably dangerous. The defendant is not obliged to design the safest possible product, or as safe as others make or a safer product than the one he has designed, so long as the one he has designed is reasonably safe'." [that is, it is not unreasonably dangerous]. From another view, Sales and Perdue [22] note that ". . . a manufacturer is not obligated to eliminate or significantly impair the utility of a product to make it absolutely safe." (*Hagans v. Oliver Mach. Co.* [23]; *Boatland of Houston, Inc., v. Bailey* [24]). Finally, in *Hethcoat v. Chevron Oil Co.* [25], ". . . in rejecting, as a matter of law, plaintiff's contention that a design defect existed, the court observed, 'To design a machine incapable of doing injury would require the design of an apparatus that would have no use. The law will not require an absurdity.'" The sense of these case-law examples supports our original statement that it is not required in law or common sense that manufacturers make products that are absolutely safe. However, for all industrial products, we believe it is a reasonable design objective to require that their designs be free of unreasonable dangers [6]. Whether this last can be achieved, may be a matter of debate among engineers and the subject of litigation in the courts, but we believe it should be a design objective in all cases.

2.4.2 The current emphasis on the safety and reliability of our industrial products

Safety in industrial products is hardly a new topic in engineering; however, the current emphasis on safety and reliability is a development of the recent past. At the turn of the century, manufacturers were virtually insulated from responsibility for injuries caused by their products, except when there was a contractual relationship between the manufacturer and the injured party. Since then, the attitudes of the public and our governments have changed dramatically. The safety of the public has become a critical issue of public debate.

In the decade of the sixties, the awareness and grateful acceptance of the benefits of industrial progress by the public was tempered by increasing concern that we must be protected from injury by the products and by-products of technology. In the early seventies, the Committee on Science and Public Policy of the National Academy of Sciences was commissioned to study the subject of safety, publishing the results of their deliberations in 1976 [26]. The pace of our quest to improve the safety of industrial products has quickened since then. Today, the pervasive nature of the public's interest in safety is evidenced across the industrial world. In virtually every available public forum, there is spirited and, sometimes, bitter debate concerning the safety of toys, nuclear power plants, drugs, sports equipment, vehicles of all sorts, and the machinery and equipment that we use in farming, construction and industrial plants.

Engineers find themselves at the center of much of this controversy. They have designed and manufactured the products with which the public is concerned, the airplane that crashes because a system was not designed to fail safe, the automobile that, under certain circumstances, may literally explode, the industrial machine that can maim due to the lack of proper safeguards. Engineers are the designers of industrial plants that produce noxious wastes and pollutants which can affect communities hundreds of kilometers away.

None of this is stated to assign responsibility or blame (if any exists), only to indicate that, today, engineers are involved with open problems of design and manufacture where their technical skills alone are not enough. This leaves many engineers in the uncharted waters where law and technology meet, where the designers of products must learn the engineering significance of legal terms and how to translate this knowledge into designs that can be used safely by an ordinary user.

At this time, in the United States, there is great ferment to change the tort system, the legal means for dispensing compensation for losses most readily available to parties injured by products designed and manufactured by engineers. European engineers are becoming increasingly aware of the new problems ahead of them. The legal basis of the products liability laws of their countries has been changed and harmonized under the *Directive on Liability for Defective Products of the Council of the EEC adopted on 25 July 1985* [27] (see Miller [28], Appendix A, and IMechE [29], Section 7 and Annex A) for effect in 1988. In the United Kingdom, the Directive has been implemented by *Part I, Product Liability, Consumer Protection Act 1987* [30], which is discussed from an engineer's viewpoint in IMechE [29], Section 2, summarized in Abbott ([1], pp. 13-15) and given in full in Abbott ([1], Appendix I).

In these circumstances, the legal implications of these recent events are beginning to penetrate the activities and publications of engineers; however, it is a slow process. For example, the McGraw-Hill Dictionary of Mechanical and Design Engineering [31] and the McGraw-Hill Dictionary of Scientific and Technical Terms [32] do not provide definitions of the terms danger or dangerous, hazard, probability, reliability, risk and safety. In the same context, the authors of the required text for a design course treated products liability litigation and the possible involvement of engineers as defenders of their design in the contentious milieu of a courtroom as an unsuitable subject for inclusion in their book. In their view, "A design book is not the place to dwell on this subject." (Shigley and Mitchell [33]), and, even today, most engineering students are not aware of the importance of designing products that are safe in the sense that they are free of unreasonable dangers. The subject of safety is discussed only rarely in their classrooms. They must wait until they have been graduated to acquire expertise in this facet of design and the acquisition process can be painful, indeed.

Nevertheless, the pace at which safety is affecting the design of industrial products and their manufacture and maintenance is increasing. Design criteria are being changed in order to provide manufacturers with greater assurance that their products are free of unreasonable dangers and, thus, safe for the hypothetical ordinary user referred to in the courts. The lay public is becoming increasingly aware of its power to influence the policies of governments and corporations with regard to safety. Engineers and managers are learning how such terms as negligence and strict liability, marketability and implied warranty, crash worthiness, and design, manufacturing and marketing defects can affect the substance of their decisions and, under some circumstances, even their ability to make them. A detailed discussion of these matters is beyond the scope of this note and is available elsewhere [6]. A brief discussion is given here only in order that this note can be sensibly self-contained.

2.4.3 Summary: The influence of safety on the world of engineering today

Today, the subject of safety in industrial products is high on the agendas of government agencies and virtually every engineering designer and manager. In order to accomplish the goal of making products safe, engineers are conducting extensive studies related to reliability and risk assessment [34] and other aspects of safety [35, 36, 37]. It is almost three decades since the American Law Institute published the *Restatement (Second) of Torts, §402A* [7], which established the legal bases for strict tort liability involving industrial products and which permitted the courts of most states ". . . to forge the rules and concepts which would govern the legal liability of the manufacturers and distributors of products. However, the swiftness with which this body of law developed led to legal anachronisms, conceptual inconsistencies and even conflicting opinions." ([18], p. 1). In keeping with the latter view, the American Law Institute is considering the implications of recent trends in the courts and reviewing the topics of manufacturer responsibility and product liability [38]. Perhaps, in its next Restatement, the Institute will define the terms hazard, risk and danger in an internally consistent manner.

2.5 Definitions of the terms hazard, risk and danger, in the context of their use in the engineering and legal worlds today

With the preceding comments in mind, consider the following terms and their definitions:

> ***Hazard***:
> an attribute of a physical object or arrangement of objects which can, by its simple presence, present a continuing threat of harm, injury, damage or loss to a person or system.
>
> ***Risk***:
> (1) the nonquantized chance that a person or system might be exposed to harm, injury or loss.
> (2) the likelihood, expressed as a quantized probability, that an event will occur.
>
> ***Danger***:
> the combination of hazard and risk that yields a qualitative measure of the

exposure of a person or system to harm, injury, damage or loss.

2.5.1 Discussion

Usually, risk is quantized in one of two equivalent ways:

1. Risk can be expressed as a probability (p) between zero and one (that is, $0 \leq p \leq 1$). This relationship indicates that, if an event cannot happen, the likelihood that it will, its probability, is zero; if an event is certain to happen, its probability is one. The neutral point is one-half; that is, if an event is just as likely to occur as not, its probability is 0.5 and we say, "the odds are even". When the weatherman says, "there is a 50% chance that it will rain tomorrow", we understand him to mean that the odds are even, that it is equally likely that it will, or will not, rain.

2. In reliability and risk assessment studies, the risks are often small (not negligible, just small) and risks are expressed and more readily understood and manipulated as the reciprocals of probabilities. For example, if the number of automobile accidents per year is 50,000,000 and the number of people killed per year due to automobile accidents is 50,000, then it follows that the risk involved can be expressed as 0.001 (or 10^{-3}) deaths per accident or as 1 death per 1000 accidents ([34], p. 9). Often, the latter characterization is given in terms of a convenient base number called the "rate" (for example, see *Accident Facts* [39], an annual publication of the National Safety Council in which it publishes a detailed analysis of accidents, the death rate is given as the deaths per 100,000 population; the mileage death rate as the deaths per 100,000,000 vehicle miles, etc.).

The qualitative nature of danger, our inability to measure it in an accepted, quantitative way is what brings plaintiffs and defendants to the courtroom. We can describe hazards and the potential effects of hazards on persons or systems with a reasonable degree of clarity and precision. We can make acceptable estimates of the risk involved when a hazard is present and operable, but we have no absolute, quantitative measures for determining whether and how a hazard and the risk associated with a specific situation combine to produce an unreasonable danger. It is a judgment call, today, in a designer's office or, tomorrow, in a courtroom.

2.5.2 A hazard with an attendant risk which combine to create a dangerous condition

Consider the following example: The auger in a meat grinder is a *hazard*. Its function is to crush and tear whatever relatively soft material is thrust into it; thus, it has the potential for causing harm, physical damage and injury to objects (say, the fingers and hand of an operator) that may be thrust into it inadvertently. The *risk* involved is the likelihood that such an event will occur. In part, the risk depends upon the openness (readiness of access) of the feed opening into the auger proper. If the feed tube of the grinder is relatively short and has a large cross section, then the risk is great that an operator could be injured by the auger. How great (that is, is it close to 1.0? 0.5? or 0?) is a subjective judgment? The combination of hazard and risk yields a measure of the *danger* of exposing a human operator to the grinder. The danger may sufficiently great that, in legal terms, the grinder could be adjudged "unreasonably dangerous". In contrast, if the feed tube is relatively long with a relatively small cross section and the meat is pushed into the auger with a pestle-like instrument, then we can say that access to the hazard has been denied. In this case, the risk may well be almost zero and, in a literal sense, there may be no danger of exposure to the hazard.

In the examples above, the hazard of the auger exists in both cases. It is there lurking, in the phrasing of a definition given earlier. In the first case, the auger is exposed, readily accessible and the odds (a measure of the risk) that there could be an accident are significantly greater than in the second case, where the designer has reduced the risk to almost zero by the principle of denying the operator access to the hazard. Thus, the danger in the first case could be adjudged unreasonable and, in the second case, negligible.

2.5.3 Risk versus utility: a legal argument, an engineering design choice

Risk versus utility (a nebulous concept associated with the definitions of the legal terms, "design defect" and an "unreasonably dangerous" product), the inability of designers to reduce the risk of a hazard to acceptable levels, the design and use of adequate instructions and warnings for the safe use of a product and other technical questions with legal implications are separate issues not discussed here [6].

2.5.4 Other definitions of hazard, risk and danger

To the best of the author's knowledge, there are no universally accepted definitions of hazard, risk and danger. The three definitions cited here have been synthesized to meet the criteria of being noncircular, highlighting the essential character of the term involved and discriminating in that the phrasing of the definition conveys the uniqueness of its meaning. Hazard, risk and danger are defined differently by Henley and Kumamoto [34], Lowrance [26] and Middendorf [36]. Abbott ([1], p. 58) defines the terms hazard and risk as they are defined in the *British Standard 4778*, Part 1: "Quality Vocabulary", Part 2: "Quality concepts and related definitions", Part 3: "Availability, reliability and maintainability terms" [40]. These standard definitions are near paraphrases of those given here and in the references above. Oddly, Abbott does not define the term danger. Unfortunately, the definitions stated above are not given, even in a paraphrase, in current dictionaries of technical terms (for example, [31] and [32]). We believe the terms and their definitions given here in an earlier section can be used in an engineering office or courtroom to provide a rational basis for evaluating whether a product is unreasonably dangerous or not.

3 Closure

An interest in semantics (the study of the meanings of words) is often deprecated and, sometimes, used in a pejorative exclamation to indicate the triviality of attempting to use words precisely. Nevertheless, all communications originated by engineers and attorneys must need be one-dimensionally clear and unambiguous. In an engineering office or courtroom, the one sending the message wishes to be understood by the receiver of the message in one and only one way. The need for one-dimensional communication is especially important in a courtroom where we have three cultures meeting to reach a decision in which a plaintiff alleges a product designed, manufactured and marketed by a defendant was the cause of his injury or loss. Thus, engineers and attorneys have a clear duty to define the esoterica in their languages in terms that can be understood by jurors. Such terms as hazard, risk and danger, and possibility and probability are often used as synonyms in the lay world from which most jurors come but they take on special meanings in a courtroom setting. We should make it clear that, in a designer's office or a courtroom, when combined, hazard and risk yield a measure of the danger involved with operating or using an industrial product.

4 Acknowledgments

An author may publish alone, but what he writes is influenced by discourse with his friends and colleagues. This author is no different. He wishes to thank those of you who took the time to discuss the substance of this note with him. Special thanks are due Mr. Joe K. Mitchell, Esq., who pointed out errors in the legal context of certain terms and the relatively new meaning and legal relevance of the term "medical possibility" and Mr. James Barker, Esq., for his careful reading of the final draft. Any errors that remain are the author's. This professor is lucky, indeed.

5 References

1. Abbott, H. (1987), *Safer by Design*, The Design Council, London, U.K., pp. 208.
2. *The Compact Oxford English Dictionary*, Second Edition (1991), Oxford University Press, Oxford, U.K., pp. 2386.
3. *Insurance Company of North America v. Myers*, 411 S.W.2d 710, 713 (Tex.1966).
4. *Otis Elevator v. Wood*, 436 S.W.2d 324.
5. *Lenger v. Physicians General Hospital, Inc.*, 455 S.W.2d 703, 707 (Tex.1970).
6. Muster, D. (1990), Safety and unreasonable danger as design criteria for engineers: some effects of products liability law on engineering design. *Proc. Instn. Mech. Engrs.*, Vol. 204, Part B: Jour. Engrg. Manufacture, pp. 185-190.
7. *Restatement (Second) of Torts*, American Law Institute, 1965, §402a.
8. *Thrailkill v. Montgomery Ward and Co., Inc.*, 670 S.W.2d 282 (Tex.App. 1 Dist. 1984).
9. *Thrailkill v. Montgomery Ward*, supra, 383.
10. *Roget's International Thesaurus*, Fifth Edition, (1962), HarperCollins Publishers, New York, NY, U.S.A., pp. 1141
11. *Chamber's Thesaurus* (1986), W. & R. Chambers, Edinburgh, U.K., pp. 750.
12. *Merriam Webster's Collegiate Dictionary*, Tenth Edition (1983), Merriam-Webster, Springfield, MA, U.S.A., pp. 1563.
13. *The Collins Dictionary of the English Language*, Second Edition (1986), Glasgow, U.K., pp. 1771.
14. *The Random House Dictionary of the English Language*, Second Edition (1987), Random House, New York, NY, U.S.A., pp. 2478 + 29.
15. *Black's Law Dictionary*, Sixth Edition (1990), West Publishing Co., St. Paul, MN, U.S.A., pp. 1657.
16. *Knox Jewelry Co., Inc., v. Cincinnati Ins. Co.*, 130 Ga. App. 519, 203 S.E. 739, 740.
17. *State v. Londe*, 345 Mo. 185, 132 S.W.2d 501, 506.

18. Sales, J. B., and J. M. Perdue (1977), *The Law of Strict Tort Liability in Texas*, Houston Law Review, Houston, TX, U.S.A., pp. 298.
19. *Willis v. Chrysler Corp.*, 264 F. Supp. 1010, 1011 (S.D. Tex. 1967).
20. *Weakley v. Fischbach & Moore, Inc.*, 515 F.2d 1260, 1267 (5th Cir. 1975).
21. *Henderson v. Ford Motor Co.*, 519 S.W.2d 87, 93 (Tex. 1974).
22. Sales, J. B., and J. M. Perdue (1981), *Cumulative Supplement*, pp. 174, to Sales, J. B., and J. M. Perdue (1977), *The Law of Strict Tort Liability in Texas*, Houston Law Review, Houston, TX, U.S.A., pp. 298.
23. *Hagans v. Oliver Mach. Co.*, 576 F.2d 97 (5th Cir. 1978).
24. *Boatland of Houston, Inc., v. Bailey*, 609 S.W.2d 743 (Tex. 1980).
25. *Hethcoat v. Chevron Oil Co.*, 364 So.2d 1243 (Fla. Dis. Ct. App. 1978) (per curiam), remanded in part per curiam, 380 So.2d 1035 (Fla. 1980).
26. Lowrance, W. W. (1976), *Of Acceptable Risk - Science and the Determination of Safety*, William Kaufmann, Inc., Los Altos, CA, U.S.A., pp. 180.
27. *Directive on Liability for Defective Products of the Council of the EEC adopted on 25 July 1985*, European Economic Community, Brussels, Belgium.
28. Miller, C. J. (Ed.) (1986), *Comparative product liability*, United Kingdom National Committee for Comparative Law, University of Edinburgh, Edinburgh, U.K., pp. 198.
29. *Product Liability*, Third Edition (1989), IMechE, London, U. K., pp. 56.
30. *Consumer Protection Act 1987, Part I, Product Liability* (1987), H.M.Stationery Office, London, U.K.
31. *McGraw-Hill Dictionary of Mechanical and Design Engineering* (1984), McGraw-Hill, New York, NY, U.S.A., pp. 387.
32. *McGraw-Hill Dictionary of Scientific and Technical Terms*, Second Edition (1978), McGraw-Hill, New York, NY, U.S.A., pp. 1771 + 58.
33. Shigley, J. E., and L. D. Mitchell (1983), *Mechanical Engineering Design*, Fourth Edition, McGraw-Hill, New York, NY, U.S.A., p. 19.
34. Henley, E. J., and H. Kumamoto (1992), *Probabilistic Risk Assessment*, IEEE Press, New York, NY, U.S.A., pp. 568.
35. Thorpe, J. F., and W. H. Middendorf (1979), *What Every Engineer Should Know about Product Liability*, Marcel Dekker, New York, NY, U.S.A., pp. 104.
36. Middendorf, W. H. (1990), *Design of Devices and Systems*, Second Edition, Marcel Dekker, New York, NY, U.S.A., pp. 511.
37. Heideklang, H. R. (1991), *Safe Product Design in Law, Management, and Engineering*, Marcel Dekker, New York, NY, U.S.A., pp. 255.
38. Meyers, T. J. (1992), What's behind all that lawyer-bashing. *Cosmos*, Vol. 2, No. 1, pp. 64-69.
39. *Accident Facts* (1994), National Safety Council, Chicago, IL, U.S.A., pp. 104.
40. *British Standard 4778* (1991): Part 1: "Quality vocabulary"; Part 2: "Quality concepts and related definitions"; Part 3: "Availability, reliability and maintainability terms", British Standards Institution, London, U.K.

3 ADEQUACY OF CONSUMER PRODUCT WARNINGS: A LEGAL REVIEW AND RECOMMENDATIONS

N.C. SILVER
Department of Psychology, University of Nevada, Las Vegas, Nevada, USA

Abstract
In the US, an adequate warning is required to provide the consumer with knowledge of a danger that would not necessarily be obvious. Adequacy of the warning includes consideration of the severity of the danger, the likelihood that the warning will be salient, the intensity and form of the warning, and the cost of improving the warning. The major purpose of this review was to integrate the legal issues and the empirical research concerning saliency of the warning. Specifically, this review examines the issues regarding color. Implications for consumer product warnings are discussed.
Keywords: Warning, Color

1 Introduction

In accordance with the standards set by a number of US state laws, adequacy of the warning is determined by the jury [1]. One aspect of an adequate warning concerns its physical characteristics [2]. These characteristics, such as color, font size and type, and icon, must be sufficiently conspicuous or salient in order to alert a reasonably prudent person of a potentially hazardous situation. Therefore, the purpose of this paper was to synthesize the legal issues with the empirical research of one physical aspect of saliency, namely color.

Structural Failure, Product Liability and Technical Insurance, Proceedings of Fifth International Conference. Edited by H.P. Rossmanith. Published in 1996 by Taylor & Francis, 2 Park Square, Milton Park, Abingdon, Oxon, OX14 4RN. ISBN 0 419 22080 1

2 Color

There have been a number of legal cases in which color of a warning has aided a manufacturer's defense in warding off the plaintiff's charges. For instance, in Hurt v. Coyne Cylinder (1992) [3], the plaintiff was injured by flames resulting from an explosion of an acetylene cylinder. The warning label on the cylinder included the word "FLAMMABLE" with a red diamond surrounding it. Moreover, the signal word DANGER was printed in red followed by the body of the warning explaining that the gas exposed to the air may form an explosive mixture. In Brown v. Sears, Roebuck & Co. (1983) [4], the decedent was electrocuted while he was operating a saw which was connected to two extension cords. A red tag labeled "CAUTION U.S.A. USERS" was provided with a more extensive warning encompassing extension cords. In Andries v. General Motors Corporation (1983) [5], the plaintiff was helping a female motorist who was stuck in the middle of the road. As he looked under the hood, because of insufficient illumination, he lit a cigarette lighter to check the battery fluids. The battery exploded causing serious injuries to the plaintiff. The warning label had colors contrasting with the black battery case. In all of these cases, the color aided in providing a more conspicuous warning.

Conspicuity can be thought of as a reasonable certainty of perception given a short period of time [6]. Empirically, the conspicuity of color was demonstrated by Adams and Lien-tsang (1981) [7]. The researchers systematically manipulated the color, shape, and legend of four traffic signs. Of these three characteristics, manipulations in color produced the highest accuracy for determining correct or incorrect representations of the actual sign. In an examination of determining which features are most salient in an illegible warning, the level of hazard was most dependent on color [8]. Moreover, Young (1991) [9] determined that mean search time decreased when color was added to the warning. However, Cahill and Carter (1976) [10] pointed out that search times increase as more colors are added. This was especially true when there were a plethora of items in the display. Therefore, having a multitude of colors in a warning may reduce cognitive processing capabilities.

There are a number of published standards that designate the appropriate use of color in safety signs and product labels. For example, Westinghouse (1981) [11] and ANSI Z535.4 (1991) [12] paired the colors red, orange, and yellow, with the terms DANGER, WARNING, and CAUTION, respectively. The SAE (1987) [13] standards, however, associated yellow with both WARNING and CAUTION. Therefore, unlike the aforementioned standards [11,12], the color orange is not used to represent potentially hazardous situations.

The ANSI Z535.1 (1991) [14] safety colors include red, orange, brown, yellow, green, blue, purple, white, gray, and black. According to this standard, the word DANGER should be printed in white letters on a red background. The word WARNING should be printed in black letters on an orange background. Finally, the word CAUTION should be printed in black letters on a yellow background. However, what is the information that a color connotes?

Dunlap, Granda, and Kustas (1986) [15] found that red connoted the greatest hazard level followed by orange, yellow, blue, green, and white respectively. It seems feasible that these colors would be paired with the appropriate signal words (e.g., DANGER with red; WARNING with orange; and CAUTION with yellow) in order to produce salient warnings with word-color combinations that would be congruent in perceived hazard level. Unfortunately, this is not always the case. Braun, Kline, and Silver (1995) [16] found that the majority of warnings found on various consumer product labels (e.g., bathroom cleaners, adhesives) were printed in black on backgrounds of different colors. Moreover, on a particular bathroom cleaner, the signal word (WARNING) and the accompanying warning were printed in blue ink on a white background. Therefore, many of these consumer products have warning label colors that are inconsistent with current standards.

Examining how the signal word and color interact, Braun and Silver (in press) [17] found that the perceived hazard level might be compromised if a signal word (e.g., DANGER) is printed in a color other than that recommended by the standard (e.g., RED). For instance, if the word DANGER was printed in blue ink, it would connote the same perceived hazard level as the word STOP in orange ink. Therefore, an "averaging" effect might be taking place between the signal word and color.

Although little research has been conducted concerning the readability of chromatic warning labels, these labels are more readable and connote greater perceived hazard levels than black and white labels [16]. Whereas the difference in perceived hazard level between the chromatic and achromatic stimuli was statistically significant, the effect size was fairly minimal [16]. This result is equivocal, however, because in a study concerning behavioral compliance, [17] found that fictitious products with warnings printed in black were perceived as more salient and readable than those labels printed in green or red.

There is some evidence to indicate that having a chromatic warning label enhances behavioral compliance. Although Rodriguez (1991) [18] found that color alone did not significantly influence compliance, the combination of color and shape, however, increased perceived hazard, retention of label detail, and subject compliance. Wogalter et al. (1987) [19] placed either an achromatic or chromatic warning sign on a drinking fountain indicating the presence of unpotable water. Approximately 88 percent of those individuals who witnessed the achromatic sign drank from the fountain, as compared to only 33 percent exposed to the chromatic version. Braun and Silver (in press) [17] had participants interact with fictitious products in which the warning label was printed in one of three different colors. The participants were to don rubber gloves when using the product. Although there was no significant effect for color with regard to behavioral compliance, a test of proportions revealed that more participants donned the rubber gloves when the warning was printed in red than in green or black.

However, just having a chromatic warning label does not exonerate the manufacturer from liability. In Simonetti v. Rinshed-Mason Co. (1972) [20], the plaintiff was permanently injured after he started his welding machinery after cleaning it with the defendant's product. Although the cleaner had a small red label that contained a warning, the ruling was that the label gave no indication that the product was explosive.

Given the findings of the court and empirical research, there are a number of recommendations with regard to color that might enhance the perceived readability and hazard level of warning labels. First, use safety colors that correspond to the level of hazard associated with the product and the signal word. If this does not occur, then the perceived hazard level will be compromised or "averaged." Second, use red when indicating the presence of an imminent hazard. It is noted that additional issues associated with colors, such as contrast and number of colors used, should also be taken into account in warning label design.

3 References

1. Ferebee v. Chevron Chemical Co. 552 F.Supp 1293 (1982).
2. First National Bank in Albuquerque v. Nor-Am Agricultural Products, Inc., 537 P.2d 682 (Ct. App. 1975).
3. Hurt v. Coyne Cylinder Co., 956 F.2d 1319 (6th Cir. 1992).
4. Brown v. Sears, Roebuck & Co., 667 P.2d 750 (1983).
5. Andries v. General Motors Corporation, Delco Batteries Division, 444 So.2d. 1180 (1983).
6. Peters, G.A. (1984) 15 cardinal principles to ensure effectiveness of warning systems. *Occupational Health & Safety*, May, pp. 76-79.
7. Adams, A.S. and Lien-tsang, H. (1981) The coding of symbol signs, *Hazard Prevention*, March/April, pp. 5-7.
8. Braun, C., Durham, L., Eidson, C., and Silver, N.C. (in press) Interaction of warning label features: Determining the contributions of three warning characteristics. *Proceedings of the Human Factors and Ergonomics Society 39th Annual Meeting.*
9. Young, S.L. (1991) Increasing the noticeability of warnings: Effects on pictorial, color, signal icon, and border, *Proceedings of the Human Factors Society 35th Annual Meeting*, Human Factors Society, Santa Monica, CA, pp. 580-584.
10. Cahill, M.C., and Carter, R.C. (1976) Color code size for searching displays of different density, *Human Factors*, 18, 273-280.
11. Westinghouse Printing Division. (1981) *Westinghouse Product Safety Label Handbook.* Author, Trafford, PA.

12. ANSI (1991) *American National Standard for Product Safety Signs and Labels: Z535.4.* National Electrical Manufacturers Association, New York.
13. Society of Automotive Engineers (1987) *Safety Signs* SAE, Warrendale, PA.
14. ANSI (1991) *Safety Color Code Z535.1.* National Electrical Manufacturers Association, New York.
15. Dunlap, G.L., Granda, R.E., and Kustas, M.S. (1986) *Observer perceptions of implied hazard: Safety signal word and color words.* IBM Technical Report (T 00.3428).
16. Braun, C.C., Kline, P.B., and Silver, N.C. (1995) The influence of color on warning label perceptions. *International Journal of Industrial Ergonomics,* Vol. 15, 179-187.
17. Braun, C.C., and Silver, N.C. (in press) The interaction of signal word and color on warning labels: Differences in perceived hazard and behavioral compliance. *Ergonomics.*
18. Rodriguez, M.A. (1991) What makes a warning label salient? *Proceedings of the Human Factors Society 35th Annual Meeting,* Human Factors Society, Santa Monica, CA, pp. 1029-1033.
19. Wogalter, M.S., Godfrey, S.S., Fontenelle, G.A., Desaulniers, D.R., Rothstein, P.R., and Laughery, K.R. (1987) Effectiveness of warnings, *Human Factors,* Vol 29, pp. 599-612.
20. Simonetti v. Rinshed-Mason Company, 200 N.W.2d 354 (1972).

4 LANGUAGE AND FONT SIZES OF WARNINGS: A REVIEW AND IMPLICATIONS

N.C. SILVER, R. FERRANTE and G. MORRIS
Department of Psychology, University of Nevada, Las Vegas, Nevada, USA

Abstract
By various US state laws, a warning must be placed on a potentially dangerous product to connote a particular level of hazard. If a warning is not provided, then the product might be considered defective. Despite the manufacturers' warnings, numerous injuries still occur. For instance, according to Florida law, one of the criteria in evaluating the adequacy of a warning includes the intensity and form of the warnings given. The major purpose of this review was to synthesize the legal issues and the empirical research concerning two specific factors of intensity and form of the warning, namely language and font size. Implications for warnings are discussed.
Keywords: Font Size, Language, Warning

1 Introduction

Although many consumer products contain warnings and instructions that inform users of the consequences of misuse, many injuries still occur. When injuries occur, however, one must examine the adequacy of the warning. If the warning is judged to be inadequate, then the manufacturer may be liable. According to Florida Law, the adequacy of a warning is determined by a jury [1]. The jury would determine the adequacy of a warning in terms of: (a) dangerousness of the product; (b) the form in which the product is used; (c) the intensity and the form of the warnings given; (d) the burdens imposed by requiring warnings; and (e) the likelihood that the warning will be adequately communicated to those in

Structural Failure, Product Liability and Technical Insurance, Proceedings of Fifth International Conference. Edited by H.P. Rossmanith. Published in 1996 by Taylor & Francis, 2 Park Square, Milton Park, Abingdon, Oxon, OX14 4RN. ISBN 0 419 22080 1

dividuals who will use the product [2]. Therefore, the major thrust of this paper is to provide a synthesis of the legal implications and the empirical research concerning one of the conditions of warning adequacy, namely the intensity and form of the warning.

With regard to the intensity and form, one can call into question the communicativeness and physical aspects of the warning [3]. Two such factors, language and font size, might influence whether an individual would read and understand the warning.

2 Language

In the case of Hubbard-Hall Chemical Co. v. Silverman [4], the employees were from Puerto Rico and had limited knowledge of English. The detailed warning on the pesticide, however, was written in English. According to the court, those individuals who would use such a pesticide are primarily laborers with a limited education. Hence, this warning was declared "inadequate." Furthermore, the California Court of Appeals, Fifth Appellate District ruled that if non-English speaking people are likely to use a product, then the manufacturer may need to provide foreign language warnings [5]. In a similar case, Stanley Industries Inc. v. W.M. Barr & Co Inc. [6], a fire damaged the plaintiff's facility allegedly caused by the spontaneous combustion of rags soaked in an oil. The employees who used the oil were from Nicaragua and their native tongue was Spanish. Hence, these employees had little comprehension of English. Finally, in Campos v. Firestone Tire & Rubber Co. [7], the court concluded that pictorial symbols might be more adequate if a laborer did not speak English.

Unfortunately, the understandability of icons or pictorials is also an equivocal issue. Friedmann (1988) [8] found no significant effect of pictograms on warning compliance and Otsubo (1988) [9] also found no significant differences for pictograms to increase noticeability, reading, compliance, and memory for the message. Young and Wogalter (1990) [10], however, found that memory for the message was enhanced when icons were presented with a conspicuous print. Furthermore, Mayer and Laux (1989) [11] demonstrated that the recognizability of pictograms increased with greater simplicity and specificity. However, there were a number of pictograms (e.g., falling object) that were identified by less than one-third of the participants (all of whom were native English speakers). These findings suggest that if native English speakers have difficulty understanding these pictograms, then understandability of these pictograms might not generalize to other populations.

Therefore, it seems incumbent to determine what signal words (e.g., DANGER, WARNING, CAUTION) used on most warnings would be appropriate for use on consumer products among non-native English speakers. Silver, Gammella, Barlow, and Wogalter (1993) [12] presented a list of potential signal words that were "understood" by at least 80% of the non-native English speakers. Although the common signal words (DANGER, WARNING, and CAUTION) were "understood" by 80-90% of those non-native English speakers sampled, there were other signal words (e.g., LETHAL) that were not known by approximately 50% of the sample. NO or STOP might be better choices as signal words (if appropriate) as they were known by over 95% of the non-native English speaking sample. With regard to meaning across cultures, Tajima, Asao, Hill, and Leonard (1991) [13] pointed out that the connotations of signal words were similar across Japanese and American samples. That is, the word DANGER connoted a similar hazard to the American sample as the word kiken connoted for the Japanese sample. Whether this correlation persists among other languages, however, is equivocal.

Even if the language is "understandable", there is no guarantee that the consumer will read the warning. Hence, this calls into question the issue of saliency. One factor that affects saliency is font size.

3 Font Size

Font size can affect visibility, or how easily a printed character can be identified. Although there are recommendations for the minimum font sizes for positive [14] and negative images [15], there are few guidelines for using specific point sizes. Even the ANSI Z535.4 [16] standard stated that lettering should be of a size that allows a person with normal or corrected vision to read the sign or label at a safe viewing distance. This ambiguity of obtaining the optimum font size has exacerbated the problem of inadequate warnings. For instance, in Huynh v. Ingersoll-Rand [17] the warning was inscribed in "tiny letters" on the trigger of a grinder. The disk on the grinder exploded and fragments struck the plaintiff. In Stanley Industries Inc. v. W.M. Barr & Co. Inc. [6], the warning was written in "small print" on the backside of the label. Finally, in Spruill v. Boyle-Midway, Inc. [18] the signal word (e.g., CAUTION) was written in 9-point type, whereas the body of the warning (e.g., "do not use near fire or flame") was written in 4.5-point type. These examples then beg the question, what is the appropriate font size? To address this issue, we will examine the research in the field.

Much of the research in font size concerns the assessment of reading speed. Paterson and Tinker (1940) [19] found that the optimum reading speed was obtained when the latters were between 9- and 12-points. However, for font sizes above 12-points and below 9-points, reading speed decreased. Vanderplas and Vanderplas (1980) [20] suggested that font sizes for young adults should be between 8- and 11-points. For older adults, however, optimum reading speeds were obtained for 16- and 18-point sizes. In these cases, reading speeds were conducted for larger passages of text. There has been a paucity of research conducted on the perceived readability of warning labels. Silver and Braun (1993) [21] found that when the main body of the warning was printed in 10-point type, there was greater perceived readability than when this text was printed in 8-point type. In a more recent investigation, Silver, Kline, and Braun (1994) [22] found that perceived readability and perceived hazardousness increased as the font size of the signal word and the body of the warning increased. Furthermore, they found an "averaging effect" with regard to the signal word and body of the warning. That is, if the signal word was written in 14-point type and the body of the warning was written in 6-point type, then there would be little difference in perceived readability or perceived hazardousness when compared to a signal word and the body of the warning written in 10-point type. In both cases, the "averaging" effect would be the perceived readability or perceived hazardousness of 10-point type. Hence, the warning is perceived as a whole unit rather than the separate signal word and body of the warning components.

The research on language and font size provide two major implications. First, language should be simplistic yet connote the appropriate hazard level [23]. If the product is marketed in a predominantly non-English speaking community, it is imperative that the warning be conveyed in that language. Pictograms would be helpful provided they are understood by "all" consumers. Otherwise, these pictograms might either be ignored or convey a meaning that is independent or antithetical to the verbal warning.

The second implication of the present review concerns font size. That is, the old adage "bigger is better" would normally apply. To the extent that the signal word is "large" and the subsequent warning text is "small", an "averaging" effect will occur in perceived readability. Moreover, the saliency of the warning declines. This implication, however, is tempered by issues in warning embeddedness, position on the label, and color.

4 References

1. Upjohn Co. v. MacMurdo, 536 So.2d 337, 339 (Fla 4th DCA. 1988).
2. Dougherty v. Hooker Chemical Corp., 540 F.2d 174, 179 (3d Cir. 1976).
3. First National Bank in Albuquerque v. Nor-Am Agricultural Products, Inc., 537 P.2d 682 (Ct. App. 1975).
4. Hubbard-Hall Chemical Co. v. Silverman, 340 F.2d 402 (1st Cir. 1965).
5. Ramirez v. Plough Inc., Calif. Ct. App. No. F01950 (5th Dist., 1992).
6. Stanley Industries, Inc. v. W.M. Barr & Co. Inc., 784 F.Supp. 1570 (S.D. Fla 1992).
7. Campos v. Firestone Tire & Rubber Co., 98 N.J. 198, 485 A.2d 305 (1984).
8. Friedmann, K. (1988) The effect of adding symbols to written warning on user behavior and recall. *Human Factors*, Vol. 30, pp. 507-515.

9. Otsubo, S.M. (1988) A behavioral study of warning labels for consumer products: Perceived danger and use of pictographs. *Proceedings of the Human Factors Society 32nd Annual Meeting,* Human Factors Society, Santa Monica, CA, pp. 536-540.
10. Young, S.L., and Wogalter, M.S. (1990) Comprehension and memory of instruction manual warnings: Conspicuous print and pictorial icons. *Human Factors,* Vol 32, pp. 637-649.
11. Mayer, D.L. and Laux, L.F. (1989) Recognizability and effectiveness of warning symbols and pictorials. *Proceedings of the Human Factors Society 33rd Annual Meeting,* Human Factors Society, Santa Monica, CA, pp. 984-988.
12. Silver, N.C., Gammella, D.S., Barlow, A.S., and Wogalter, M.S. (1993) Connoted strength of signal words by elderly and non-native English speakers. *Proceedings of the Human Factors and Ergonomics Society 37th Annual Meeting,* Human Factors Society, Santa Monica, CA, pp. 516-519.
13. Tajima, N., Asao, K., Hill, G.W., and Leonard, S.D. (1991) Comparison of meanings of warnings in Japan and the USA, in *Advances in Industrial Ergonomics and Safety III* (eds. W. Karwowski and J.W. Yates), Taylor & Francis, Bristol, PA, pp. 739-741.
14. Smith, S.L. (1979) Letter size and legibility. *Human Factors,* Vol 21, pp. 661-670.
15. NDMA. (1990) *Draft Guidelines for Maximizing Label Readability,* Author, Washington, DC.
16. ANSI (1991). *American National Standard for Product Safety Signs and Labels: Z535.4.* National Electrical Manufacturers Association, New York.
17. Huynh v. Ingersoll-Rand, 20 Cal.Rptr. 2d 296 (Cal.App 2 Dist. 1993).
18. Spruill v. Boyle-Midway Inc., 308 F.2d 79 (4th Cir. 1962).
19. Paterson, D.G., and Tinker, M.A. (1940) *How to Make Type Readable.* Harper, New York.
20. Vanderplas, J.M., and Vanderplas, J.H. (1980) Some factors affecting legibility of printed materials for older adults. *Perceptual and Motor Skills,* Vol. 50, pp. 923-932.
21. Silver, N.C., and Braun, C.C. (1993) Perceived readability of warning labels with varied font sizes and styles. *Safety Science,* Vol. 16, pp. 615-625.
22. Silver, N.C., Kline, P.B., and Braun, C.C. (1994) Type form variables: Differences in perceived readability and perceived hazardousness. *Proceedings of the Human Factors and Ergonomics Society 38th Annual Meeting,* Human Factors and Ergonomics Society, Santa Monica, CA, pp. 821-825.
23. Wogalter, M.S., and Silver, N.C. (in press). Warning signal words: Connoted strength and understandability by children, elders, and non-Native English speakers. *Ergonomics.*

5 THE PRODUCT LIABILITY LAW IN JAPAN

T. YAMADA
School of Management, Science University of Tokyo, Japan

Abstract
The Main objective of this report is to discuss and outline the Product Liability Law of Japan. The discussion will center on technical issues concerning the PL Law. The major technical issues defects including defects in products and defects related to handling instructions, defense for development risk and issues related to design. In concluding the report, the author proposes international unification of the PL laws.

Contents

I Introduction

In Japan, a product liability law ("PL law") has been in effect since July 1, 1995. Moves to establish a PL law began to be seen around 1980. Each political party and the Japan Federation of Bar Associations prepared their own bills a number of times, and then it was on June 22, 1994 that a PL law which reflected the opinions of the industrial circles, consumer groups, labor community, legal circles, etc. was adopted in the Diet. The law was promulgated on July 1 of the

Structural Failure, Product Liability and Technical Insurance, Proceedings of Fifth International Conference. Edited by H.P. Rossmanith. Published in 1996 by Taylor & Francis, 2 Park Square, Milton Park, Abingdon, Oxon, OX14 4RN. ISBN 0 419 22080 1

same year, and was effectuated after a period of one year to make its enactment public knowledge. Therefore, there is no precedent of the law's application yet, and it remains to be seen how the PL law will be applied in actual disputes such as litigations.

One major feature of the PL law is that it allows a person affected by a product's defect to claim damages if it is made clear that the damage he sustains resulted from a defect in the product, even if the manufacturer's negligence is not clearly established. This is the concept of liability without negligence which is adopted in PL laws in the United States, Europe, and many other counties, and Japan followed suit after those countries.

The traditional concept had been one of negligence liability. The victim had to prove the negligence of the manufacturer, etc. and it was very difficult for the victim to prove such negligence. For this reason, the PL law is said to be a law to relieve consumers. In Japan, if damage is inflicted on a person or goods because of a defect in a product, the manufacturer, etc. of the product may be held liable under either one of the following three legal theory examples.

(i) Default liability for failure to completely fulfill a contractual obligation;
(ii) Unlawful act liability for causing loss or damage to another person by violating his (her) rights through intention or negligence;
(iii) Warranty liability when there was a fault in the object of sale and purchase or in the object of work.

Although a PL law has come into effect in Japan, its application involves various limitations, and some kinds of loss or damage to which the PL law does not apply can occur. It would be too simplistic, therefore to think that it is not necessary to know the traditional concept now that a PL law has been introduced. Since many regulations of the Ministry of Industry & Trade ("MITI") and other agencies concerning product safety will be eased to a large extent and disputes are going to be settled between the parties concerned as a general rule, it is also possible that the number of disputes concerning product defects, etc. will increase. Another aspect of this matter that should not be neglected is that the introduction of the PL law makes it necessary to learn more about negligence liability which we in the past did not have to pay much attention to. When we think about the PL law, it is important that we have accurate knowledge of the scope of its application.

II Major points of contention when the PL Law was established

The following were the major arguing points:

1. Definition of "defect"
2. Provisions on presumption, and
3. Scope of "product"

"Defect" is a newly introduced concept. The PL law functions as a code for act as well as a code for trial, and therefore, it is desirable to clarify the definition and nature of defect as far as possible. However, if factors for judgment regarding the concept of defect should constitute legal requirements, the victim's burden of proof would be more grave. (1)

There were two ideas about exemplification of judgment factors:

a. they should be limited to the scope of the EC directive at the maximum;
b. the maximum possible number of examples should be cited.(2)

In the end it was decided to judge a defect on the basis of an overall consideration of all circumstances involved in each case, taking into account the victim's burden of proof, the common nature and importance of judgment factors and the like.

Paragraph 2, Article 2 on definition of defect cites:

a. the characteristics of the product;
b. ordinarily foreseeable manner of use; and
c. the time the product is delivered.

These are examples of judgment elements.

A defect is an objective characteristic of a product, and unlike a flaw, it concerns the product's safety. When this is considered, it is reasonable that a product's lack of the safety it normally should have should be regarded as a defect.

As for provisions on presumption:

1. existence of a defect;
2. cause-and-effect relationship between a defect and the loss or damage; and
3. the time the defect existed,

were debated. (3)

The proof of these facts normally relates to many facts such as the nature of each product and each accident. If the law presumed these facts uniformly from predefined facts, it would deviate from the objective of alleviating the victim's burden of proof. For this and other reasons, it was decided not to provide for presumption in the law.

Thus, it was agreed to address the issue of victim's burden of proof through flexible use of factual presumption in the trial, depending on the nature of each case, and also by reorganizing the accident cause investigation organizations of the central government and prefectural government or improving the various private inspection organizations and test and research laboratories.

In this connection, a sentence passed by the Osaka District Court on March 29, 1994, before the bill for the PL law was laid before the Diet in Japan, is worth noting. (4)

The case involved a fire which started in a TV set. Based on the findings of the fire station's on-the-scene investigation, eyewitnesses' testimony when the fire had been detected, and the fact that the power supply cord attached to the TV set may become molten because of overheating rather than a short circuit in the cord, the Osaka District Court rejected the defendant's claim that the fire had occurred because of the plaintiff's inappropriate handling of the power cord. The court presumed that the TV in question had a defect, thus finding that the TV was the source of the fire. This sentence will be very helpful in the application of the PL law. (5)

The PL law defines product as "movable property manufactured or processed," and excludes real estate, and unprocessed agricultural, forest and fishery products. Blood preparations are "products" under this law. The law requires that in the judgment of a defect, due consideration be given to the product's characteristic features. Also, for the relief of victims of blood preparations for transfusion, a special relief organization is going to be established in view of the special nature of the matter. (6)

The extent of damage compensable for product liability is determined on the basis of Paragraphs 1 and 2, Article 416 of the Civil Code of Japan. No upper or lower limits have been set on the amounts of compensation. (7)

The liability period is ten years from the time the manufacturer placed the product into distribution. The short-term statute of limitations is three years. As for accumulated damage, delayed damage and the like, the period is counted from the time the damage occurred in order to provide a relief to the victim commensurate with the nature of the damage.

Defense based on development risk is allowed. The judgment criteria in this case are the levels of the best available science and technology. (8)

With respect to manufacturers of parts, raw materials and supplies, exemption is allowed for the manufacturer of the parts, etc. if the parts, etc. were made according to the design or instructions of the manufacturer of the finished product and if the defect occurred precisely because such design or instructions were followed.

As a general rule, sales agents are not included among entities subject to product liability. The liability of sales agents is to be determined according to the jurisprudence of contractual liability. (9)

III The PL Law of Japan and a brief commentary

III. 1 The PL Law of Japan

The Product Liability Law effectuated in Japanese on July 1, 1995 is as follows.

The Product Liability Law in Japan (Japanese Law No. 85, 1994)

Article 1 [Purpose]

The purpose of this Law is to relieve the injured person by setting forth liability of the manufacturer, etc. for damages when the injury on a life, a body, or property is caused by a defect in the product, and thereby to contribute to the stabilization and improvement of the people's life and to the sound development of the national economy.

Article 2 [Definitions]

(1) As used in this Law, the term "product" means movable property manufactured or processed.

(2) As used in this Law, the term "defect" means lack of safety level that the product ordinarily should provide, taking into account the nature of the product, the ordinarily foreseeable manner of use of the product, the time when the manufacturer, etc. delivered the product, and other circumstances concerning the product.

(3) As used in this Law, the term "manufacturer, etc." means any one of the following:

1. any person who manufactured, processed, or imported the product as business (hereinafter called just "manufacturer.");
2. any person who, by putting his name, trade name, trademark or other feature (hereinafter called "representation of name, etc.") on the product presents himself as its manufacturer, or any person who puts the representation of name, etc. on the product in a manner mistakable for the manufacturer;
3. apart from any person mentioned in the preceding subsections, any person who, by putting the representation of name, etc. on the product, may be recognized as its manufacturer-in-fact, in the light of a manner concerning manufacturing processing, importation or sales, and other circumstances.

Article 3 [Product Liability]

The manufacturer shall be liable for damages caused by the injury, when he injured someone's life, body or property by the defect in his delivered product on which he manufactured, processed, imported or put the representation of name, as described in subsection 2 or 3 of section 3 of Article 2 on. However, the manufacturer is not liable when only the defective product itself is damaged.

Article 4 [Exemptions]

In cases where Article 3 applies, the manufacturer shall not be liable as a result of Article 3 if he proves;

1. that the state of scientific or technical knowledge at the time when the manufacturer delivered the product was not such as to enable the existence of the defect in the product to be discovered; or
2. in the case where the product is used as a component or raw material of another product, that the defect is substantially attributable to compliance with the instruction concerning the specifications given by the manufacturer of the said other product, and that the manufacturer is not negligent on occurrence of the defect.

Article 5 [Time Limitations]

(1) The right for damages provided in Article 3 shall be extinguished by prescription if the injured person or his legal representative does not exercise such right within 3 years from the time when he become aware of the damage and the liable party for the damage. The same shall also apply upon the expiry of a period of 10 years from the time when the manufacturer delivered the product.

(2) The period in the latter sentence of section 1 of this Article shall be calculated from the time when the damage arises, where such damage it caused by the substances which are harmful to human health when they remain or accumulate in the body, or where the symptoms for such damage appear after a certain latent period.

Article 6 [Application of Civil Code]

In so far as this law does not provide otherwise, the liability of the manufacturer, etc. for damages caused by a defect in the product shall be subject to the provisions of the Civil Code (Law No. 89, 1896).

Supplementary Provisions

1. Enforcement Date, etc.
   This Law shall come into force the day after one year from the date of promulgation, and shall apply to the products delivered by the manufacturer, etc. after this Law comes into force.

2. Partial Amendment of the Law on Compensation for Nuclear Damage
   The Law on compensation for Nuclear Damage (Law No. 147, 1961) shall be partially amended as follows:
   In section 3 of Article 4 of that Law, "and the Law relating to the Limitation of the Liability of shipowners (Law No. 94, 1975)" shall be amended as "the Law relating to the Limitation of the Liability of shipowners (Law No. 94, 1975) and the Product Liability law (Law No. 85, 1994)"

III. 2 Brief commentary of the PL Law

Article 1 of the PL law sets forth the purpose of the law. The purpose of the law excludes such a case from the application of this law. In this regard, the extent of damage to be compensated for in accordance with Article 3 is to be judged in each specific case on the basis of the concept of reasonable cause-and-effect relationship (by analogical application of Article 416 of the Civil Code).

Article 4 provides for the cases where the liability of the manufacturer, etc. is exempted. In Subparagraph 1, the law heeds the possible impediment of research and development and technological innovation, and, recognizing the so-called defense for development risk, it stipulates that the manufacturer, etc. shall be exempted if it was not possible, with the scientific or technological expertise available to them when they delivered the product, to perceive the existence of a defect in the product.

Here, the term "expertise" means all knowledge that can affect the judgement of whether a defect exists. It means the whole of knowledge that exists objectively in society, rather than knowledge possessed only by a certain person. It is construed, therefore, that "expertise" cited in defense for development risk is naturally the highest level of knowledge available.

The same article also provides for defense of the so-called parts and raw materials manufacturers. It stipulates that the manufacturer of such a product is exempted from liability if his product was used as a part of raw material for another product, and if a defect occurred precisely because he followed the instructions of the manufacturer of the finished product concerning design. Provided that the parts or raw materials manufacturer himself has to prove concomitantly that there was no negligence on his part to cause the defect to occur, before he can be exempted through this form of defense. Incidentally, the burden of proof in the application of Article 4 lies with the defendant (the manufacturer, etc.) in both cases.

Article 5 provides for time limitations. It says that the right to claim damages under this law shall expire if the victim or his (her) legal representative does not claim damages within three years of the time such party became aware of the damage and of the liable party. The article also sets the so-called disqualification period at ten years from the time the manufacturer, etc. delivered the product, taking into account the need to remove the uncertainty as to the liability of the manufacturer, etc. early or current trends in the U.S. and European countries, and the like. As for accumulated damage, etc., the disqualification period is counted from the time the damage occurred in order to provide a relief to the victim commensurate with the nature of the damage.

Article 6 provides for a relationship between this law and the current Civil Code of Japan (Law No. 89 of 1896). It stipulates that the liability of manufacturers, etc. for defects in products shall be settled according to the provisions of the Civil Code, as well as those of this law. Examples of application of the Civil Code would include comparative negligence (Paragraph 2 of Article 722), relationships between plural liable entities (Article 719), the effect of the exemption clause (Article 90) and the principle of monetary compensation (Paragraph 1 of Article is to provide for liability of the manufacturer, etc. in order to protect the victim in the event a product's defect results in damage to the life or body of a person or a person's property, and thereby contribute to better stability of people's livelihood and sound development of the national economy.

Article 2 defines legally important concepts such as "product" and "defect" to make the law clearer and easier to understand. In Paragraph 1 the law stipulates that "product" means "movable property manufactured or processed," and excludes intangible items and real estate from the application of this law. Also, the so- called unprocessed agricultural, forestry and livestock products are naturally excluded because these are neither manufactured nor processed. [Damage to a house or building resulting from a design fault (e.g., the Osaka District Court's sentence of Nov. 2, 1978 - Hanrei Jiho No. 934, p. 81) and collapse of a breast wall resulting from the absence of drain holes (e.g., the Oct. 29, 1979 sentence of Akashi branch, Kobe District Court - Hanrei Jiho No. 961, p. 107) are excluded from the application of this law because those are not real estate -- the author's note.]

Next, Paragraph 2 of Article 2 stipulates that "defect" means "lack of the safety that the product should ordinarily have," and enumerates "the characteristics of the product," "ordinarily foreseeable manner of use," and "the time the manufacturer delivered the product" as the circumstances to be considered in judging whether a defect really existed. The same paragraph then provides that all other circumstances concerning the product in question should also be taken into account before forming a judgment.

Also, Paragraph 3 of Article 2 says that "manufacturer, etc." means the party or parties who manufactured, processed or imported the product as part of his business operation, as well as any party or parties who represented himself as the manufacturer, etc. or made a representation that may cause him to be mistaken as the manufacturer, etc. or any party or parties who made a representation that allows him to be regarded as de facto manufacturer, etc. of the product based on the way the product is manufactured, processed, imported or sold, or on other like circumstances.

Article 3 provides for product liability which is liability for loss or damage resulting from a product's defect. In other words, Article 3 stipulates that if a defect in a product of a manufacturer, etc. who himself manufactured, processed or imported the product, or delivered the product making the type of representation set forth in Paragraph 3 of Article 2, results in a violation of another person's life, body or property, then, such manufacturer, etc. shall be held liable for the resultant damage, regardless of whether there was negligence on his part. However, the article provides that this stipulation does not apply if the damage occurred in the defective product itself alone. This is because, it the damage is limited to the defective product itself alone, it will not be easy, in actual practice, to distinguish a defect from a fault and this would likely cause many malicious claims to be raised. For this and other reasons, the article, in a proviso,

722; and Article 417).

Incidentally, this law does not exclude warranty liability, default liability or tort liability under the current Civil Code. As long as the requirements for claiming these liabilities are satisfied, the victim can selectively claim damages based on any of these.

This law stipulates in Paragraph 1 of its Supplementary Provisions that this law shall go into effect one year after the day of its promulgation, and that it shall apply to products delivered by their manufacturers, etc. on or after its effective date. With respect to atomic energy damage, it stipulates in Paragraph 2 that the Law Concerning Compensation for Atomic Energy Damage (Law No. 147, 1961) will be revised as required to make it clear that the said law, rather than this law, will apply to atomic energy damage.(10)

IV Technical issues concerning the PL Law (11)

The following are the three major technical issues concerning the PL law:

1. Defect - Defects of the product; defects related to the handling instructions.
2. Defense for development risk;
3. Issues related to design.

IV. 1 Defect

Defects may be classified into two types: the defects of products, and defects related to a handling manual.

(1) Product defects

The burden of proof for a defect of a product lies with the plaintiff (victim). Therefore, to what extent the defendant (manufacturer, etc.) will disclose technical information needed for the proving would become an important issue. Also, even if such technical information is obtained, it may not be possible for a legal expert, such as a lawyer, alone to handle the case because it often takes technical knowledge to fully comprehend the content of the information. In litigations concerning patent rights and the like, a patent attorney directly intervenes in the trial as assistant. It is necessary to take a similar measure also in product-liability litigations. (In Japan, lawyer and patent attorney are separate qualifications.)

Defects of products may be classified into defects arising in the process of manufacture and defects in the design. Generally, a defect resulting from manufacture should be discovered through manufacturing inspection, and an issue arises when such a defect failed to be detected in the inspection.

A design defect is a defect that exists in the product's design itself, and if it exists, all products made according to that design will become defective products. It is believed, however, that it may sometimes be necessary to introduce value judgment regarding the question of whether a design defect should be regarded as a product's defect. Therefore, what criteria such value judgment should be based on remains a most important issue.

(2) Defects related to handling instructions
Defects related to handling instructions are sometimes called defects in warning. Failure to attach a proper warning or instructions to the product will constitute a defect of the instruction manual, even if the product or its design has no defect.

Warning defect is similar in nature to design defect, in the fact that, if the warning or instructions are judged to be inappropriate, all of the products are regarded as having a warning defect, and also that, because value assessment regarding the appropriateness of the warning has to be introduced in determining whether a defect exists.

For example, if warning instructions calling for proper use is attached to a diving suit because its inappropriate wearing may result in the death of the diver, users of the suit may even decrease because the warning may make them more conscious of the danger than when no such warning is given. It may not be advisable to say too much in the warning for, fear of a lawsuit. Thus, it would be necessary also to consider introducing the theory of accepted risk, granting that the user should naturally be prepared for a certain degree of danger.

IV. 2 Defense for development risk

As I have already stated, the concept of development risk means a risk of a product not foreseeable by the level of scientific or technical knowledge existing at the time the product was placed into distribution. This is exactly the same as the concept adopted in the unified European Community directive, and is different from what is called the state of the art in the United States. In the case of a product liability, which in nature is a no-fault liability, whether exemption should be allowed for the manufacture by reason of the manufacturer's inability to foresee the risk by the level of scientific or technical knowledge existing at the time the product was placed into distribution is a political question, rather than a theoretical one. Since PL is a no-fault liability, the availability of such knowledge to the manufacturer is not a matter of question.

Subparagraph 1, Article 4 of the Product Liability Law of Japan stipulates that exemption shall be allowed if the manufacturer, etc. was unable to recognize the existence of the defect at the time of delivery. Basically this can be regarded as all but the same as the European concept of defense for development risk. However, there is the question of what point in time should be regarded as delivery time, and also, the expression, "unable to recognize," may in some cases give the manufacturer, etc. leeway to claim defense for development risk by proving the unavailability of knowledge. Solving these and other like questions will have to await judgments in actual cases in the future.

IV. 3 Issues related to design

Subparagraph 2, Article 4 of the Japanese Product Liability Law means that, in the case where a product was to be made by combining several components, if a defect developed in the finished product when there was no defect in any of the components and the finished product was assembled exactly as the assembly drawings showed, then the manufacturers, etc. of the individual components shall be exempted from liability. In this case, it will be the manufacturer assembling the product who must assume the product liability. And then, the question will be whether there was a defect in any of the drawings, or whether there was a defect in the assembly process.

Although this PL law is based on the viewpoint of the manufacturer vs. the consumer, and most of the commentaries and explanatory notes about it are offered from that point of view, the stipulation of Subparagraph 2 of Article 4 is of special character that, in a way, singles out a certain class of the involved manufacturers as bearer of product liability.

In view of the fact that many products are made under a system whereby each of their components are made in different countries and brought together and assembled in the assembly plant located in one country, it is believed that the application of these provisions of Subparagraph 2 of Article 4 will have an ever higher significance.

Issues concerning design under the PL law boil down after all to the question of "what is design defect?" Therefore, those who are engaged in design work must pursue a new the question of "what is design?" It is essential for the designer himself to try at all times to design safer and easier-to-use products. If he has a firm design philosophy to this effect, and if he designs an article that embodies that philosophy, then he can be confident and declare that the product is free from any design defect, even if a defect is found in the product. Design is always the base of production. Manufacturing is done on the basis of a design. Inspection is also, in a sense, done to see if the product was made exactly as it was designed. Therefore, we must say that it has become more important than ever to recognize afresh that design determines every basic aspect of a product, and try to make reliable designs in terms of safety.

VI. 4 Future issues

A PL law is now enforced in Japan, too. However, specific questions such as to what extent a defect has to be proven and to what extent defense will be allowed will have to be settled by future court judgments.

As I have already stated, many products are made under a system whereby parts made in different countries are brought in and assembled in the assembly plant located in one country. Therefore, the time has come when we should consider international unification of the PL laws.

The United Nations Convention on Contracts for the International Sale of Goods, which was adopted in Vienna on April 10, 1980 and became effective on January 1, 1988, stipulates in its Article 5, as follows: "This Convention does not apply to the liability of the seller for death or personal injury caused by the goods to any person." This provision exists because the Convention strictly provides for the right-and-obligation relationship between the seller and the buyer. (12) Establishing a unified convention regarding product liability is important at least for the purpose of complementing Article 5 of the convention on the sale of goods. Under the current conditions, it seems to be feasible to make a unified PL convention except for the question of the amounts of compensation.

V Closing remarks

As I have explained in detail, a PL law has been enacted in Japan, but there is no precedents for law application. Therefore, we must say that specific questions such as to what extent a defect has to be proven and to what extent defense will be allowed remain to be settled. However, science and technology are developing so rapidly that the number of disputes directly involving technology is expected to keep increasing. Not only product liability issues but patent controversies will also increase, and copyright infringement regarding software programs is another issue in which technology is deeply involved. Thus, it will be important that judges and others who pass legal judgments become more familiar with technical matters, and that means and methods that will help to form proper technical judgments be introduced into the system of settling litigations and other disputes.

Further, it is important for people working in technical fields to learn more about legal liability and acquire a measure of capability to judge the limits of liability.

There are increasingly active moves, worldwide, among experts in the technical, legal and insurance sectors to sit at the same table and discuss issues that concern all such sectors from the above point of view. It is essential that design engineers also take part in these forums actively and try to acquire the ability to think more deeply and more broadly.

VI. References

1) Susumu Hirano: "The Product Liability Law of the U.S. and the PL Law of Japan" - *Horitsu Bunka*, Aug. 1984, from P. 26 on.

"Practical Handbook as Product Liability supervised by Hisashi Honda, from p. 67 on; etc.

2) Yukio Edano: "Background of the Establishment of the Product Liability Law" - *Horitsu Bunka*, Aug. 1994, p. 18.

3) Eiichi Hoshino: "Events Leading to the Establishment of the Product Liability Law" - From the View Point of the Subcommittee on Property Laws, Legal System Research Council" - Jurist, No. 1051, from p. 10 on.

4) Osaka District Court's Sentence, Mar. 29, 1994 - *Hanrei Times*, No. 842, p. 69.

5) Ichiro Kasuga: "Burden of Proof" - *Jurist*, No. 1051, from p. 29 on.

6) Hitoshi Nakai: "Product Liability Law and Issues Involving Blood Preparations" - *Horitsu Bunka*, Aug. 1994, from p. 23 on.

7) Yasuhiro Kawaguchi: "About the Establishment of the Product Liability Law" - *Jurist*, No. 1051, p. 49.

8) "Practical Handbook to the Product Liability Law" (supervised by Hisashi Honda) - from p. 214 on.

9) Tsuneo Yamada: "Problems in Design Involved in the Japanese PL Law" - *Sekkei Kogaku*, Vol. 30 No. 4, from p. 25 on.

10) Yoshio Esaki: "About the Establishment of a Product Liability System" - *Horitsu Bunka*, Aug. 1994, p. 16·p.17.

11) Refer to the above (Tsuneo Yamada)

12) Vgl. Caemmerer/Schechtriem, "Kommentar zum Einheitlihen UN-Kaufrecht", 2 Aufl.

6 ACTIVITY OF JAPANESE SOCIETY OF MECHANICAL ENGINEERS PRODUCT LIABILITY COMMITTEE

Y. MURAKAMI
Department of Mechanical Science and Engineering, Kyushu University, Fukuoka, Japan

Abstract
The Japanese PL laws is supposed to be introduced from July 1, 1995. To discuss the PL-related problems, the Japan Society of Mechanical Engineers started the PL committee in the Materials and Mechanics Division in November, 1994. The meeting is held every 3 months, 8 times in total for two years. The past, present and future activities are outlined.
Keywords: Case studies, Defects, Japanese PL law, JSME committee, PL.

Activities
Facing the introduction of the PL law in Japan from July 1, 1995, the Materials and Mechanics Division, Japan Society of Mechanical Engineers started the PL committee in November, 1994. The objective and activity of the committee are to discuss the PL-related problems which will rise in the near future and to investigate the case studies of failure accidents which happened in the past. The Committee members are composed with 15 persons with the neutral institutions and organizations (9 university professors, 4 national institute researchers, 1 government (MITI) officer and 1 American copy writer) and 45 company engineers.

The chairman, co-chairmen and secretariat are as follows.

Chairman :	Prof. Y. Murakami, Kyushu University
Co-chairmen :	Prof. K. Komai, Kyoto University
	Prof. H. Kobayashi, Tokyo Institute of Technology
Secretariat :	Prof. M. Endo, Fukuoka University

These meetings of the PL committee were already held 3 times in Tokyo. Usually 4 topics are presented by the committee members or invited speakers including foreign

Structural Failure, Product Liability and Technical Insurance, Proceedings of Fifth International Conference. Edited by H.P. Rossmanith. Published in 1996 by Taylor & Francis, 2 Park Square, Milton Park, Abingdon, Oxon, OX14 4RN. ISBN 0 419 22080 1

experts. For every meeting, the proceeding which contains the manuscript prepared by the speakers in advance of the meeting is prepared at the meeting.

The places and the date of the meeting are listed in the following

The 1st meeting : Tokyo, November 30, 1994

The 2nd meeting : Kawasaki, March 3, 1995

The 3rd meeting : Tokyo, May 30, 1995

The 4th meeting is planned to be held on August 21, 1995 in Osaka.

The programs of each meeting follow.

[The 1st meeting]

1. PL Law and Its Relationship with Strength of Materials
   Y. Murakami, Kyushu University
2. U.S. PL Cases
   N. Yasuda, Toyota
3. U.S. PL Case Related to a Fracture Accident and the Result
   K. Hirakawa, Kyushu University
4. Cases Studies of Failure Accidents of Construction Machine Parts and Disk Cutter Chip for Carpenters
   Y. Kitsunai, National Research Institute for Industrial Safety
5. Uncommon Nonsense
   Brian Strawn

[The 2nd meeting]

1. Arrangement of Organizations for Failure Accident Investigation
   K. Morita, MITI, Japanese Government
2. Fatigue under Service Loading
   M. Jono, Osaka University
3. Uncommon Nonsense
   Brian Strawn
4. U.S. PL Cases
   S. Tanaka, Nissan Motors
5. Material and Structural Defects and Accidents of Large Scale Airplanes
   H. Terada, National Institute for Aeronautics and Astronautics.

[The 3rd meeting]

1. Collections and Analysis of Corrosion Failure Cases
   K. Komai, Kyoto University
2. Railway Accidents : Risk, Causes, Reduction and Mitigation
   R. A. Smith, Sheffield University
3. Accidents of Movable Cranes and Safety Assurance
   H. Ito, Nagaoka Technical College
4. Failure and Fracture Causing Leak of High Pressure Equipments
   H. Kobayashi, Tokyo Institute of Technology
5. Uncommon Nonsense
   Brian Strawn

[The 4th meeting (to be held in Osaka, August 21, 1995)]

1. The Strategy for PL in Sport Goods
   Y. Nobuta, Mizuno

2. Failure Analysis : Some Examples
   A. J. McEvily, University of Connecticut
3. Case Study Histories and Future Activities in Private Organizations for Investigation of Fracture Accidents
   (3.1) T. Nishikawa, Sumitomo Metal Technology
   (3.2) T. Ishioka, Kobelco Research Institute
4. Report of SPT-5
   Y. Murakami, Kyushu University
5. Uncommon Nonsense
   Brian Strawn

Japanese PL Law

The tentative English translation of the Japanese PL law is given in the Appendix. The definitions of "defect" in Article 2 is naturally very ambiguous because the term cannot be used to a definite product.

At the 2nd meeting, we investigated the causes of failure accidents experienced by the committee members and classified them to (1) the defects of design, (2) the defects of manufacturing, and (3) the defects of warning and caution. The results were listed up in the proceeding of the 3rd meeting. Of huge number of defects, most common ones follow.

(1) Unusually large material defects.
(2) Wrong materials which were different from designer's indications but used due to mishandling, wrong labels on materials and misindications.
(3) Poor manufacturing, such as notches, scratches, surface roughness, wrong surface treating, cracking and incorrect size.
(4) Poor and wrong heat treatment.

The next task of the committee is to discuss the cause of arising of these defects and to propose the practical methods or know-how to prevent above common 4 defects.

The term "ordinarily" which is used twice in the Law is not defined clearly. The definition and interpretation of this term will be also the source of argument in the court. Although there is no clear definition of "an ordinary person", the definition of the term "ordinary people" in Japan was given by a government committee as "the level of a woman of an average family who received high school education". This definition may require a higher level than that in USA.

The topic "Uncommon Nonsense" given by B. Strawn, an American committee member, reports very "incredible" (it seems so to Japanese) warnings which are seen on many consumer products in USA. (See Fig.1) These warnings indicate the difference in the level of "ordinary people" in Japan and USA. The flood of "Uncommon Nonsense" warnings may be caused by the history and structure characteristic of American society.

The Article 4 [Exemptions] also characterizes the Japanese PL law. Concerning this article the JSME committee is discussing the definition of "the state of scientific or technical knowledge". How can "State of the art" be given equally to engineers or be taught in many universities and colleges ? Among too many fatigue and fracture theories, which one can we authorize to be used for designers ?

(1) Product : ENERGIZER AAA Batteries
Warning : "Keep away from small children. If swallowed, promptly see doctor: have doctor phone (202) 625-3333..."

(2) Product : RAY-O-VAC AAA Batteries
Warning : "If swallowed or lodged in ear or nose, promptly see doctor..."

(3) Product : "GLAD" plastic garbage bags
Warning : "Do not use in cribs, beds, (baby) carriages or playpens."

(4) Product : "SKOAL BANDIT" chewing tobacco
Warning : "This product may cause mouth cancer."

(5) Product : HOTPOINT electric range/oven model RB755
Warning : "Do not use your oven to dry newspapers. If overheated, they can catch fire..."

(6) Product : SEALY Posturpedic Mattress (bed)
Warning : "Never sleep in the same bed as an infant. The infant can become wedged between your body and the mattress and suffocate..."

(7) Product : Q-Tips cotton swabs
Warning : "If used to clean ears, stroke swab gently around the outer surface of the ear, ***without entering the ear canal.***"

Fig. 1. Warnings on American products. (After B. Strawn, Proc. The 1st JSME PL Meeting, Nov. 1994)

Conclusion

The activity period of the JSME PL committee is two year from November, 1994. However, the systematic compilation of case studies will be an importance future activity for the JSME PL committee.

The exchange of information and cooperation with foreign groups will be also necessary.

The support and cooperation of industries are absolutely necessary for such activities.

Appendix
Japanese PL Law

The Product Liability Law (Law No. 85, 1994)

(tentative translation)

Article 1 [Purpose]

The purpose of this Law is to relieve the injured person by setting forth liability of the manufacturer, etc. for damages when the injury on a life, a body, or property is caused by a defect in the product, and thereby to contribute to the stabilization and improvement of the people's life and to the sound development of the national economy.

Article 2 [Definitions]

(1) As used in this Law, the term "product" means movable property manufactured or processed.

(2) As used in this Law, the term "defect" means lack of safety that the product ordinarily should provide, taking into account the nature of the product, the ordinarily foreseeable manner of use of the product, the time when the manufacturer, etc. delivered the product, and other circumstances concerning the product.

(3) As used in this Law, the term "manufacturer, etc." means any one of the following :

1. any person who manufactured, processed, or imported the product as business (hereinafter called just "manufacturer") ;

2. any person who, by putting his name, trade name, trade mark or other feature (hereinafter called "representation of name, etc.") on the product presents himself as its manufacturer, or any person who puts the representation of name, etc. on the product in a manner mistakable for the manufacturer ;

3. apart from any person mentioned in the preceeding subsections, any person who, by putting the representation of name, etc. on the product, may be recognized as its manufacturer-in-fact, in the light of a manner concerning manufacturing, processing, importation or sales, and other circumstances.

Article 3 [Product Liability]

The manufacturer, etc. shall be liable for damages caused by the injury, when he injured somone's life, body or property by the defect in his delivered product which he manufactured, processed, imported or put the representation of name, etc. as described in subsection 2 or 3 of section 3 of Article 2 on. However, the manufacturer, etc. is not liable when only the defective product itself is damaged.

Article 4 [Exemptions]

In cases where Article 3 applies, the manufacturer, etc. shall not be liable as a result of Article 3 if he proves ;

1. that the state of scientific or technical knowledge at the time when the manufacturer, etc. delivered the product was not such as to enable the existance of the defect in the product to be discovered ; or

2. in the case where the product is used as a component or raw material of another product, that the defect is substantially attributable to compliance with the instruction concerning the specifications given by the manufacturer of the said another product, and that the manufacturer, etc. is not negligent on occurrence of the defect.

Article 5 [Time Limitations]

(1) The right for damages provided in Article 3 shall be extinguished by prescription if the injured person or his legal representative does not exercise such right within 3 years from the time when he becomes aware of the damage and the liable party for the damage. The same shall also apply upon the expiry of a period of 10 years from the time when the manufacturer, etc. delivered the product.

(2) The period in the latter sentence of section 1 of this Article shall be calculated from the time when the damage arises, where such damage is caused by the substances which are harmful to human health when they remain or accumulate in the body, or where the symptoms for such damage appear after a certain latent period.

Article 6 [Application of Civil Code]

In so far as this law does not provide otherwise, the liability of the manufacturer, etc. for damages caused by a defect in the product shall be subject to the provisions of the Civil Code (Law No. 89, 1896).

Supplementary Provisions

1. Enforcement Date, etc.

This Law shall come into force the day after one year from the date of promulgation, and shall apply to the products delivered by the manufacturer, etc. after this Law comes into force.

2. Partial Amendment of the Law on Compensation for Nuclear Damage

The Law on Compensation for Nuclear Damage (Law No. 147, 1961) shall be partially amended as follows :

In section 3 of Article 4 of that Law, "and the Law relating to the Limitation of the Liability of shipowners (Law No. 94, 1975)" shall be amended as ", the Law relating to the Limitation of the Liability of shipowners (Law No. 94, 1975) and the Product Liability Law (Law No. 85, 1994)".

7 RELIABILITY, RISK AND INSURANCE: THREE SIDES OF A SPECIAL COIN

P.F. FRUTUOSO E MELO, N.H. HURTADO and P. KALEFF
COPPE/UFRJ, Rio de Janeiro, Brazil

Abstract
It is the task of the insurer to carry out risk studies in order to establish premium ratings for industrial facilities. The discussion between the industrial facility and the insurer concerning the premium ratings could be much improved if sound knowledge of risk analysis and reliability engineering were common practice on both sides. Some features of this knowledge are discussed and practical issues are raised. The role of accident data banks is addressed in this context and the statistical approach to take them into account is discussed.
Keywords: Loss prevention, Maintenance, Premium rating, Reliability engineering, Risk analysis, Risk management, Risk reduction, Statistics.

1 Introduction

Insurance is a means devised by organized societies to distribute the burden of loss. While the insurer accepts to carry the risk and demands a pay for this, the insured is relieved from that burden and pays the requested premium. Since it is the responsibility of the insurer to provide for the payment of the claims when losses happen, premiums will be set by him unilaterally. The criteria on which the insurer relays are based on the knowledge of the statistical behaviour of the occurence of losses in the particular branch of the insured object. When the insured object belongs to a set containing many elements and also a relatively long performance history, it is safe for the insurer to establish premiums more adequate for the insured. Individual objects however, for which no previous experience exists, will pose a significant problem to the insurer and the consequence will be a correspondingly higher premium.

Structural Failure, Product Liability and Technical Insurance, Proceedings of Fifth International Conference. Edited by H.P. Rossmanith. Published in 1996 by Taylor & Francis, 2 Park Square, Milton Park, Abingdon, Oxon, OX14 4RN.
ISBN 0 419 22080 1

For such situations , as in the case of industrial facilities or other engineering objects like off-shore platforms or nuclear power plants, an alternative approach may be provided by using the disciplines of reliability engineering and risk analysis.

For that purpose it is convenient to broaden the concept of loss to the industrial context [1] including the consequences of any event in the industrial environment as, for example, the spillage of hydrocarbons, a manufacturing defect in a production line, a fatality due to an explosion, etc.

Our purpose in this paper is to discuss the issue of establishing premium ratings when a reliability analysis [2] is available or, alternatively, when the plant risk analysis [1] [3] has been performed and losses may be evaluated quantitatively. This point is closely related to the definition of rational safety factors in the context of strucutural safety [4].

2 Reliability Engineering

Reliability engineering may be viewed as the discipline that provides answers to questions like: what is the chance that a pressure vessel will collapse in a given time period?, or, what is the mean time between failures of the new process computer bought by the department? Random events play the dominant role here, so that statistics and probability are fundamental tools for providing answers to these questions.

Central to all this discussion, is the concept of failure. What is a failure? If one considers a bridge, how can failure of it be defined? For example, a bridge might have been designed to withstand a $2x10^2$ vehicle per hour traffic for a life span of 30 years. Suppose, however, that in a few years, the traffic intensity rises to something like 10×10^2 vehicles per hour. What could one say about the 30 year life span? Intuitively, it is to be expected that it should become something less than 30 and perhaps more than 20 years. Does this characterize a failure? One might argue that due to the harsher operation conditions, the expected life span might be reduced due to the aging mechanism caused by the external cyclic loading.

On the other hand, the traffic intensity on the bridge may well be restricted to the 2×10^2 vehicle per hour figure and even so the bridge may collapse before the expected 30 year period ellapses. Its capacity may not have been appropriately taken into account. Material defects might be the very cause of it collapse.

The two situations just discussed concerning the bridge have long been taken into account in the context of what could be called systems reliability [5] [6]. One could say that a component, a structure, etc., undergoes a primary failure when its failure occurs under design and predefined operational conditions. Suppose, for instance, that a pressure vessel is designed so as to withstand a pressure of 15 atmospheres. If it is submitted to a pressure of 20 atmospheres, it might collapse almost instantaneously. If it collapses under a pressure lower than 15 atmospheres, a primary failure is said to occur; on the other hand, for a pressure of 20 atmospheres, a secondary failure has taken place.

Adequate operational practice should be exercized so as to avoid secondary failures, for they might have more serious consequences than primary failures. This will certainly impact insurance calculations.

Reliability engineering deals primarily with the probabilistic modeling of failures and this is performed by means of probability distributions, where the most popular has been the Weibull distribuition, introduced in the 1930s, which has the great advantage of modeling aging mechanisms as well as burn-in effects (dealt with in the design process) and the component useful life, too. It is common practice to employ the concept of failure rate which is related directly to the conditional probability of failure of a component in a small predefined time period, given that it was operational at the beginning of this interval [2]. The identification and modeling of aging mechanisms is very important in an industrial facility for it is the adequate tool for recommending maintenance actions so as to raise the component useful life.

The ability to consider and model failures is not sufficient since a component might exhibit different failure modes. A centrifugal pump, for example, may fail to start on demand or, alternatively, it may cease running, given that is has started successfully. This indicates two failure modes for the pump. Failure data considering different failure modes are very important for the performance of reliability analyses of complex systems, such as protective systems of process plants [7].

Failure models that consider failure modes are termed competing risk models in statistics [8].

It is natural to consider that for a given system, that is, an aggregate of components linked in a predefined logical fashion, the use of such competing risk models is very complicated, since a system may exhibit hundreds of failure modes, the so called minimal cut sets. Different approaches are employed to approach this problem, such as fault tree analysis [2] [9].

The estimation of the reliability, or of the failure rate of a component can be approached by means of reliability testing [8] [10]. Parametric or nonparametric approaches are available.

Of course, expert opinion could be used here and actual evidence taken into account to upgrade it but this approach could take more time than both the plant and the insurance company could afford. This approach called Bayesian analysis, [8] is very important and useful but one should always bear in mind that a plant cannot be viewed as a single component or item and the determination of its failure probability is much more complex.

The point mentioned in the last paragraph is of concern if we consider that past experience must be carefully analysed for particularities of different industrial facilities. Suppose, for instance, that two industrial companies have different safety philosophies and we rely on their past operational experience to establish a premium rating. If we do not take into account these different philosophies, we will certainly be biased in our premium estimate. The management of losses is directly related to what is called today safety engineering [11].

The key problem of considering a component from the structural point of view is the next topic to approach. Some discussion on this subject was performed earlier when primary and secondary failures were commented.

Models are needed to take into account the fact that a failure (for instance, the collapse of a bridge, or the explosion of a pressure vessel) may occur due to loadings that may exceed the capacity.

At this point, the so called systems reliability field may be approached. Many problems to be faced in reliability engineering are typically of this kind. A system may be of concern in many fields. One may have a structural system whose failure is to be evaluated [12]. For example, an offshore structure may be viewed as a system. Loadings and capacities are modeled so as to allow the evaluation of the probability of failure for a given time period. However, a logical model must be considered for some combinations of component structural failures may be responsible for the system failure.

Different connections of components are taken into account for improving the reliability of systems. Design improvements, operational changes, and, most importantly, maintenance considerations are imposed to improve its reliability without impairing costs in an unbearable manner.

It is common practice to adopt redundance in this context, with the so called parallel redundancy. However, a factor that is of concern when considering redundancy is the possibility of occurrence of common cause failures [13], where external causes, like for example, humidity, or fire, can cause the failure of identical redundant components.

Another approach in this sense has long been adopted in the nuclear industry, that of diversity, by which components with different features are employed to reduce the occurrence of common cause failures. For example, if two safety injection pumps are necessary for cooling a reactor in an emergency condition, they should have different characteristics, as, for example, different vendors, different maintenance crews, and so on.

The quantitative impact of common cause failures has been modeled by means of parametric models for which the parameters should be estimated by considering field data [13].

Maintenance is the next step to consider. Traditionally, two maintenance approaches have been considered: that of corrective maintenance and that of preventive maintenance [14] [15].

The first incurs in great losses. The second has proven to lead to lesser losses but, another problem appeared in this context: unbearable plant downtimes may result. In this sense, a third approach has been adopted, that of predictive maintenance, which differs from the latter two in the sense that maintenance actions are performed in such a manner so as not to render components unavailable. This has proven to be an effective approach for example when considering electronic components. [16]. Considerable reliability improvements can be achieved by considering proper maintenance models.

An approach called reliability-centered maintenance [17] has been adopted, in which applicable and effective preventive maintenance strategies are defined and developed. Rather than considering those preventive maintenance tasks that attempt to restore equipment to an ideal condition, it focuses on those which aim at maintaining system function.

The interaction of failure and maintenance events when modeling the reliability of a system is approached by Markovian models when certain conditions are met [2]. Situations like the failure of a pump leading to an additional burden on the redundant pump can be modeled by means of this approach. Extensions are possible in order to take aging into account [18].

The modeling of fatigue crack growth has also been considered by means of Markovian models [19].

The basic models considered above allow for considering the broader area of system safety analysis which is of concern in process plants in the context of loss prevention. A feature that must be taken into account here is the occurrence of human errors which impair reliability considerably. In fact, it is well known that human errors are the most important contributors to system unreliability, as related to maintenance actions and operational actions [20].

The key answer we are interested in this context is the reliability (or unreliability), that is, the probability of failure under given conditions for a specificed time period and under known operational conditions. This figure is very important for it quantifies losses under normal operational conditions in a plant. The knowledge of this figure is very important for insurance discussions.

3 Risk Analysis

A risk analysis of a proposed (designed) or existing industrial facility is a systematic examination aiming at identifying and establishing an opinion on the potentially serious occurrences in it and their respective consequences (deseases, fatalities, etc.) [3] [21] [22]. The quantitative examination of these occurrences is performed by using statistics and probability for one can never be sure whether they will occur and if so when will they occur.

A consideration is due here concerning alternative meanings of the word risk. We can imagine two current meanings for it in the engineering field. The first is the one mentioned in the last paragraph, where risk is understood as a frequency of ocurrence of a focused loss per unit time. The second is related to the structural engineering field where the term risk of failure is understood as a failure probability [4]. Outside the engineering field, it might have different meanings (as in the economics science).

In the nuclear field, it is also common practice to talk about probabilistic safety assessment (PSA) of industrial facilities. Risk studies of this kind have been performed since 1975, when the pioneer study of American nuclear power plants safety by probability models was published.[23].

The risk of an industrial facility may be expressed in terms of fatalities, economic losses, material losses and many other possibilities. The scenarios that may lead to these losses are identified by appropriate techniques, like event trees [21] [22] and their occurrence frequencies are determined. Next, the consequences of each scenario are evaluted in terms of fires, explosions,and the related losses are assessed.

The so called risk curves are obtained, and mean risk figures are calculated from them [21] [22].

Again, the key answer one searches is exactly this risk curve, or its mean value. At this point risk management comes into stage. Risk acceptance criteria can be used to evaluate whether the results are acceptable. If not, risk reduction measures are to be proposed and implemented by the facility, so that the residual risk can be evaluated. If this residual risk meets the acceptance criteria, then insurance discussions can come into play in a quantitative manner.

One word of caution is due here. It concerns the problem of establishing risk acceptance criteria. Point evaluations have been used so far for this purpose. However, they do not contemplate the distinct possibilities of low frequency/high consequence and high frequency/low consequence contributions. On the other hand, a risk curve cannot be easily established as a standard since if it is crossed by the risk curve of a facility, it is very difficult to ascertain whether the risk is acceptable. This feature has been the subject of extensive research.

4 Insurance Considerations

At this point, insurance considerations can be approached properly. It is known that it is the business of the insurer to carry out the assessment of risks [1]. This implies the knowledge of the methods discussed earlier. This, in turn, means that the insurance companies must have risk analysts as employees in order to perform the desired risk analyses and help the company establish premium ratings, as, for example, the Insurance Technical Bureau established in the United Kingdom some twenty years ago.

One might argue that past operational experience can be used to avoid performing risk analyses, as, for example, accident data banks. However, all this data may be of little use if one considers, in the context of process plants, for instance, that many different design and operational conditions are available and, even worse, design concepts are continuously developing so that an accident that occurred 20 years ago would hardly happen today. An appropriate risk analysis can be of much help in this context and the accident data bank is to be used in it, since it may be very important for identifying hazards, [22].

On the other hand, the facility must have its own risk experts (or, if not, it may search for help with consulting companies) in order to perform detailed risk analysis according to established standards [21] [22] so as to be able to argue with the insurance company when establishing premium ratings or renewing them.

Damage insurance can be used as an example. In this context, insurers mention frequently features like the size of the plant, the novelty of technology, process materials, plant layout, building design, fire protection of structures, fire fighting arrangements and housekeeping. These features are fundamental for the purpose of defining premium ratings and probable maximum losses. However, a quantitative result must come out of these features, for qualitative comparisons cannot help optimizing both the figures mentioned above. However, the identification of proper hazards and their indexing has been approached differently [22]. The shortcoming of adopting indexing approaches, like the fire and explosion index, is that subjective weights are assigned to each hazard. On the other hand, a risk analyst relies generally on field data.

The fact is, however, that the ability to perfom risk studies is not as widespread as it should. A risk analyst must have a sound knowledge of reliability engineering so as to be able to give clear quantitative answers. Of course these are probabilistic answers, but they should be evaluated according to the best standards available in order to minimize losses for both the facility and the insurance company.

On the other hand, sound knowledge of disciplines like fluid mechanics, solid mechanics, dispersion of effluents, are fundamental for performing an adequate risk analysis, that is to say, an interdisciplinary team should perform the analysis, both in the insurance company and in the process facility also.

5 Concluding remarks

In order to optimize the evaluation of premium ratings, it is highly recommendable that reliability engineers and risk analysts be aggregated into insurance companies and industrial facilities. However, as discussed, risk analyses require interdisciplinary teams to be performed and the plant personnel plays a dominant role during its development by providing design and operational information. The problem is how insurance companies will solve the problem of validating or performing this same risk analyses for they need to used classified information and this requires negotiation. This is one the costs to be borne when pursuing a lesser premium rating.

6 References

1. Lees, F. P (1986) *Loss Prevention in the Process Industry*, Butterworths, London.
2. Shooman, M. L. (1990) *Probabilistic Reliability: an Engineering Approach*, Robert E. Krieger Publishing Company, Malabar.
3 Rowe, W. D. *An Anatomy of Risk*, John Wiley & Sons, New York.
4 Melchers, R. E. *Structural Reliability Analysis and Prediction*, Ellis Horwood Limited Publishers, Chichester.
5 Fussell, J. B. and Arendt, J. S. (1978) System reliability engineering methodology: a discussion of the state of the art. *Nuclear Safety*. Vol. 20, p. 541.
6 Roberts, H. R., Vesely, W. E., Haasl, D. F. and Goldberg, F. F. *Fault Tree Handbook*, U. S. Nuclear Regulatory Commission, Washington, D. C. NUREG-0492.
7 IEEE Standard (1987) *IEEE Guide to the Collection and Presentation of Electrical, Electronic, Sensing and Mechanical Component Reliability Data*. Institute of Electrical and Electronic Engineers, Piscataway. IEEE-Std-500.
8 Crowder, M. J., Kimber, A. C., Smith, R. L. and Sweeting, T. J. *Statistical Analysis of Reliability Data*, Chapman and Hall, London.
9 Lewis, E. E. (1987) *Introduction to Reliability Engineering*, John Wiley & Sons, New York.
10 Nelson, W. (1982) *Applied Life Data Analysis*, John Wiley & Sons, New York.
11 Blockley, D. (1992) *Engineering Safety*, McGraw-Hill Book Company, London.
12 Ang, A. H-S., and Tang, W. H. (1975) *Probability Concepts in Engineering Planning and Design*, Vol. 1, John Wiley & Sons, New York.
13 Mosleh, A., Fleming, K. N., Parry, G. W., Paula, H. M., Worledge, D. H. and Rasmusson, D. M. (1988) *Procedures for Treating Common Cause Failures in Safety and Reliability Studies*, U. S. Nuclear Regulatory Commission, Washington, D. C.. NUREG/CR-4780.
14 Jardine, A.K.S. (1973) *Maintenance, Replacement, and Reliability*, John Wiley & Sons, New York.
15 Gertsbakh, I. B. (1977) *Models for Preventive Maintenance*, North-Holland Publishing Co., Amsterdam.
16 Guimarães, P.A.M.R.*Statistical Process Control Applied to the Maintenance of Telecommunication Transmission Systems*, M. S. Thesis, COPPE/UFRJ, Rio de Janeiro (in Portuguese).
17 Modarres, M. (1993) *What Every Engineer Should Know About Reliability and Risk Analysis*, Marcel Dekker, Inc., New York.
18 Singh, C. and Billinton, R. (1977) *Reliability Modelling and Evaluation*,

Hutchinson, London.

19 Bustamante, A. and Amendola A. (eds.) (1987) *Reliability Engineering*, Elsevier Publishing Company, Amsterdam.

20 Swain, A. D. and Guttman, H. E. (1983) *Handbook of Human Reliability Analysis with Emphasis on Nuclear Power Applications*, U. S. Nuclear Regulatory Commission, Washington, D. C. NUREG/CR-1278.

21 NUREG/CR-2300 (1983) *PRA Procedures Guide: A Guide to the Performance of Probabilistic Risk Assessment for Nuclear Power Plants*, U. S. Nuclear Regulatory Commission, Washington, D. C.

22 AIChE/CCPS (1989) *Guidelines for Chemical Process Quantitative Risk Analysis*, Center for Chemical Process Safety, American Institute for Chemical Engineers, New York

23 WASH-1400 (1975) *Reactor Safety Study: An Assessment of Accident Risks in U. S. Commercial Nuclear Power Plants* , U. S. Nuclear Regulatory Commission, Washington, D. C.

8 THE FORENSIC ENGINEERING STUDY OF CATASTROPHIC FAILURES IN THE UNITED STATES

M. SABBAGHIAN
Mechanical Engineering Department, Louisiana State University, Baton Rouge, Louisiana, USA
D.E. BECNEL, Jr.
Attorney at Law, Laplace, Louisiana, USA
F.C. DUDENHEFER, Jr.
Cummings et al., Attorneys at Law, New Orleans, Louisiana, USA

ABSTRACT

The authors were involved in litigation which spanned five years in the United States Federal Court. This paper focuses on their experiences in the application of engineering and law, with special emphasis on the interface between the two disciplines. In particular, in:

Section 1: It is demonstrated that as the equipment ages, failures are more likely to occur from a series of isolated events, seemingly unconnected, but ultimately producing a catastrophic failure. The focus of this section is on the United States oil and petrochemical industry with a brief statistical overview of the number of failures and their costs (billions of dollars).

Section 2: The authors discuss a specific catastrophic failure in a catalytic cracking unit, pointing out how a number of isolated events, specifically described, resulted in such a catastrophic failure.

Section 3: The subject of erosion and corrosion in the petrochemical plant is discussed. Attention is drawn to the accelerated rate of equipment deterioration when both mechanisms are present and tandemly active. Unlike the corrosion mechanism that can easily be detected by visual observations, the erosion due to liquid or particle impingement is not always so obvious and detectable through visual examination. The shortcomings of some of the non-destructive techniques in detecting the localized erosion is discussed. In the case of the subject cat cracker explosion, it is shown that the lack of recognition of the liquid erosion potential prevented a rigorous and proper inspection procedure to discover the pending disaster.

Structural Failure, Product Liability and Technical Insurance, Proceedings of Fifth International Conference. Edited by H.P. Rossmanith. Published in 1996 by Taylor & Francis, 2 Park Square, Milton Park, Abingdon, Oxon, OX14 4RN. ISBN 0 419 22080 1

Section 4: The authors discuss how their investigation was controlled and managed by the United States Federal Court.

I. AGE AND FAILURE

Over the last 30 years, the hydrocarbon industry has experienced over 100 large property damage losses[1]. These losses represent approximately $6.38 billion in property damage alone, excluding the costs of business interruption, employee injuries, and third party liability claims[1]. From July 1986 - June 1991, workplace accidents involving explosions, fires, and spills in the U.S. petrochemical industry resulted in 1,129 workplace injuries and 159 workplace fatalities[2]. These accidents, which not only resulted in the loss of life but also affected residents in surrounding communities, have heightened the awareness of the catastrophic potential of the petrochemical industry.

Of substantial concern is the fact that three of the world's worst industrial accidents occurred in the 1980's, all in the chemical and oil industries. On February 25, 1984, 508 people died in San Paulo, Brazil, when a gasoline pipeline ruptured and 700 tons of gasoline were released and later ignited[3].

On November 18, 1984, Mexico City suffered from a catastrophic petroleum accident that occurred as a result of a fire and explosion in a liquified petroleum gas processing plant. More than 550 people were killed, more than 2,000 injured, and another 350,000 left homeless[3].

Two weeks later, the worst disaster in the chemical industry's history occurred at Bhopal, India, when a chemical leak of methyl isocyanide spread into an adjacent town killing 2,500 people and injuring an estimated 10 times that number of people, many of whom suffered permanent disabilities as a result of their injuries[3].

Six more explosions occurred in 1992, nine in 1991, and seven in 1990. The six losses in 1992 totaled $485,100,00 in property damage with an average value of $80,900,000 per loss. The nine losses in 1991 totaled $541,800,000 in property damage with an average value of $60,200,000 per loss, while the seven losses in 1990 totaled $164,900,000 in property damage with an average value of $23,600,000 per loss. The average losses which occurred from 1964 through 1989 was $66,500,000 (stated in January 1, 1993 dollars)[2].

Property damage is not the only consequence of a catastrophic petrochemical plant failure. Loss of production will obviously result in lost revenue adversely affecting profitability. Production interruptions may force customers to seek product elsewhere. If the failure is a result of fault, additional costs are incurred to compensate adjoining property owners for damages caused to their person or property.

The cost, whether to ones own property or in compensation of another's damages, associated with petro-chemical catastrophes are enormous. The reality is that it is not generally the obvious that produces a catastrophic failure, but a series of isolated events occurring over time which ultimately culminate in a catastrophic failure. Equally unfortunate is that many times there were a series of discreet warnings, which if heeded, would have prevented the ultimate catastrophe. The increasing age of petro-chemical equipment and plants in the United States and abroad, and the constant effort to maximize production increase the risk of catastrophic failure. One such failure of a piping system resulted in a combined loss estimated to exceed a billion dollars. This loss was caused by a series of isolated events, seemingly unrelated, which culminated in a catastrophic petro-chemical plant failure.

II. THE FAILURE

The problem of erosion, corrosion and combination of the two is very well known to scientists and engineers, especially those associated with the oil and petrochemical industries. Petrochemical plants replace or repair, annually, a vast number of process equipments that are degraded by the mechanism of corrosion, erosion or both. In those circumstances where only corrosion is the cause of deterioration, the rate of degradation is slowed down by the formation of a protective layer of corroded metal on the surface. This layer of oxide will protect, to some degree, the metal substratum. When both corrosion and erosion attack the surface, the thin corroded layer is removed by the erosion process exposing the fresh metal to the corrosive environment, thus accelerating the process of metal removal from the surface.

The erosion process is the result of the impact of solid particles, or fine liquid droplets, on the surface. The impact of such particles or droplets carried in a gas or vapor stream, creates a large compressive as well as shearing stress. The repeated stress, that may exceed the flow stress for the metal, causes the development of micro-cracks along the weakest grain boundary, and eventually the dislodging of small pieces of metal. The erosion of blades in steam or gas turbines is a constant problem for industry and military. So is the erosion in the slurry pipelines and the two-phase flow pipelines in the oil and petrochemical industry

Several investigators [4,5,6,7,8,9] have discussed the mechanism of wear and erosion process. The rate of erosion is strongly affected by such parameters as the impact velocity, angle of impingement and the hardness of particle as well as the hardness of the surface on which the particles impinge. It is shown [9] that the maximum erosion rate as the result of impingement of aerodynamically entrained solid particles is achieved at impingement angles between 35 and 50 degrees. It is also reported [10] that in the case of liquid impingement erosion, the rate of metal volume loss is proportional to the fifth or sixth power of the relative velocity. This will make the elbows in the process industries uniquely susceptible to such an attack. At the same time, unless special attention is paid to the inspection of such elbows, there is a good chance that the deterioration of the elbow may go on unnoticed. About two decades ago, one of the important, and very common, technique of plant inspection during the major overhaul was for the inspector to tap the elbows with the inspector's hammer and listen to the emitted sound. To an experienced inspector, the sound indicated the state of the elbow, and whether or not it was needed to map more accurately the thickness of elbow with elaborate instruments. Since the advent and wide spread use of Ultra-Sound instruments, the frequency of inspection in some instances has increased, as the instruments can also be used while the plant is on stream. However, the reliability of the data in those erosion prone elbows has not increased. In fact, if the thickness measurements are not periodically taken from several designated and marked spot on the outer surface of elbow, the validity of data can seriously be questioned.

In recent years there have been several accidents initiated by the rupture of an elbow. Such accidents have led to massive release of highly combustible materials and catastrophic explosion resulting in many fatalities, destruction of property and severe environmental pollution. When such cases are investigated, some common basis for the failure of elbows becomes apparent. They include:

1. A two phase flow in the pipe containing solid particles or liquid droplets entrained a in relatively high speed stream.

2. The system suffered from a relaxed and non-methodic approach to the inspection.

In the case study which is the subject of this paper, a Cat-Cracker Unit (CCU) had been in service for more than 30 years. During that time, by adding and upgrading the equipments, the unit's capacity had gradually been increased. In this unit, the overhead from the Depropanizer was being condensed in two rows of heat exchangers each row containing two condensers. To increase the PH of the stream and, thus, to reduce the corrosion rate in the heat exchangers as well as in the associated piping system, a small amount of ammoniated water was being injected into the vapor stream as shown in Figure 1. Here the attention is drawn to the direction of the injection nozzle which is horizontal. As such, the impact velocity of the liquid droplets, normal to the pipe wall, is relatively low (about 1.5 m/sec).

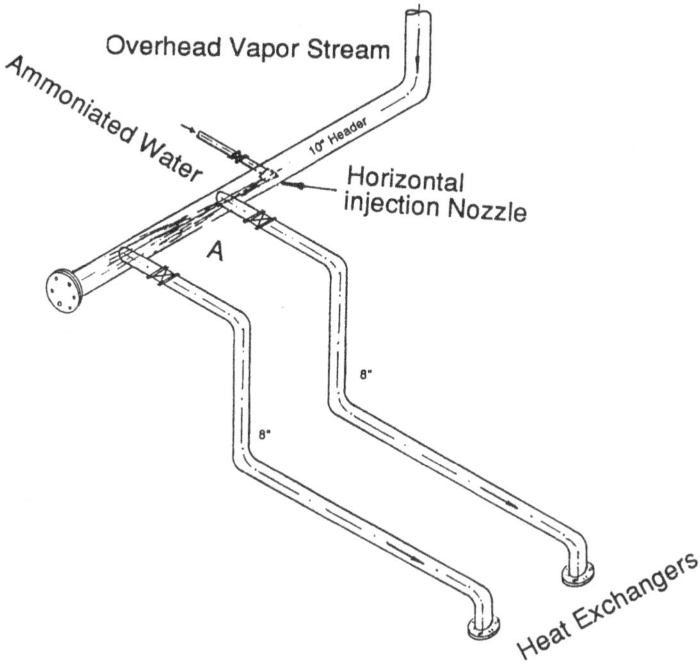

Figure 1. Original system configuration

Although a small amount of erosion could be identified around the liquid impingement zone A, the wall thickness reduction was not significant enough to cause any concern, even after many years of operation. One could attribute the very low rate of erosion at around the impingement point A to:

1. Small impingement velocity.
2. Presence of a pool of liquid at the bottom of the 10" header, further cushioning the impingement velocity.

About five years prior to the catastrophic explosion in one of the turnarounds where the cooling capacity of the exchangers was to be increased, a third row of condensers was added to the previous rows. To meet the required amount of ammoniated water and proper distribution to all three rows of heat exchangers, a second injection nozzle was incorporated into the system. The second injection nozzle, according to some sketches, was also to be placed horizontally. However, the second nozzle was actually placed in the system vertically as shown in Figure 2. The added nozzle was only about 1.5 m from the downstream elbow B, injecting liquid droplets at a relatively high velocity directly onto the subject elbow. One would justifiably think that this is a classic example of an area subjected to corrosion/erosion process. To prevent that, either the nozzle location or orientation could be changed or the elbow could have been lined with an erosion resistant material (gunnite for example).

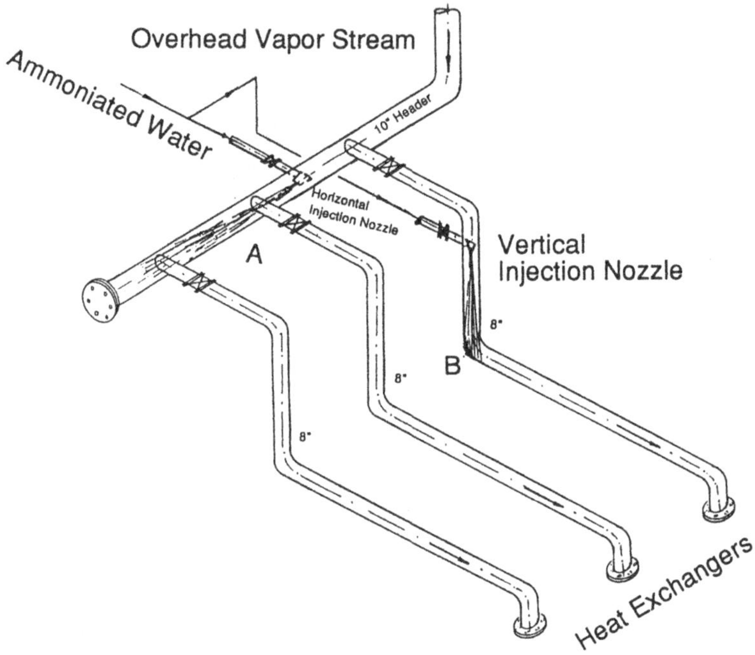

Figure 2. Revised system configuration

The rate of flow through the vertical nozzle, translated into the average flow velocity results in an impingement velocity of approximately 6.4 m/s (21 ft/sec). In a more detailed examination it was found out that the ammoniated-water pump was actually an air driven diaphragm pump. One of the characteristic of diaphragm pump is that the rate of flow is not constant. In fact, as the diaphragm pushes the fluid against the discharge valve the pressure builds up in a pattern that very much resemble a sinosoidal function. Once the pressure inside the pump chamber exceeds the back pressure, the discharge flow commences. That is, only during a fraction of the cycle time the pump discharges, and that the flow of ammoniated water into the vapor stream is pulsating. Therefore, if one determines the nozzle discharge velocity based on average flow rate, the calculated velocity will be less than the actual velocity. In the case of diaphragm pump involved in this particular application, considering the back pressure and the pump's pressure-time characteristic as provided by the manufacturer, one could conclude that the pump was discharging only during one third of its cycle time. The actual velocity, therefore, was about 17% higher than the velocity obtained based on average flow rate. This results in the instantaneous impingement velocity of about 7.31 m/s (24 ft/s). The difference between the instantaneous and average impingement velocity may not seem to be very much, but considering the very strong influence of impingement velocity (the sixth power) on the erosion rate, the increase of velocity from 6.4 m/s to 7.3 m/s amounts to more than doubling the rate of erosion.

The records indicated that the ultrasonic pipe thickness measurements were being taken, but the inspection procedure was erratic in terms of time intervals. In addition, the spots on the elbow where the measurements were being taken at different times were not necessarily the same point. A few scattered thickness data taken from the elbow during the course of five to six years were physically inconsistent. Such inconsistencies were sufficient ground to create a serious doubt about the validity of the data. For instance, attention was not paid to the fact that in almost half the times the wall thickness measured on the subject elbow was greater than the measurement from the previous time (i.e. the pipe wall thickness had increased). Following table is the recording of the wall thickness for the subject elbow.

RECORDED THICKNESS DATA FOR THE ELBOW

Month/Year		5/72	8/73	2/74	2/77	2/80	1/83
Recorded Elbow Wall Thickness:	in	0.300	0.310	0.320	0.270	0.315	0.255
	mm	7.62	7.87	8.13	6.86	8.00	6.48

The post explosion inspection of this elbow revealed a wall thickness that was 7 mm or more at all points except at the outer wall of the elbow and within a few centimeters of the ruptured edge. In that vicinity the wall thickness rapidly decreased and at very near the edge it was about 0.25 mm (0.011 in). Figure 3 shows the subject elbow and the configuration of rupture.

In this case the discrepancies in the thickness measurements and recording simply were dismissed as the inspector's or the ultrasonic instrument's error. A more logical and probable reason for such an erratic reading, one could conclude, would be that the readings at different times were being taken from close but not the same spots on the elbow.

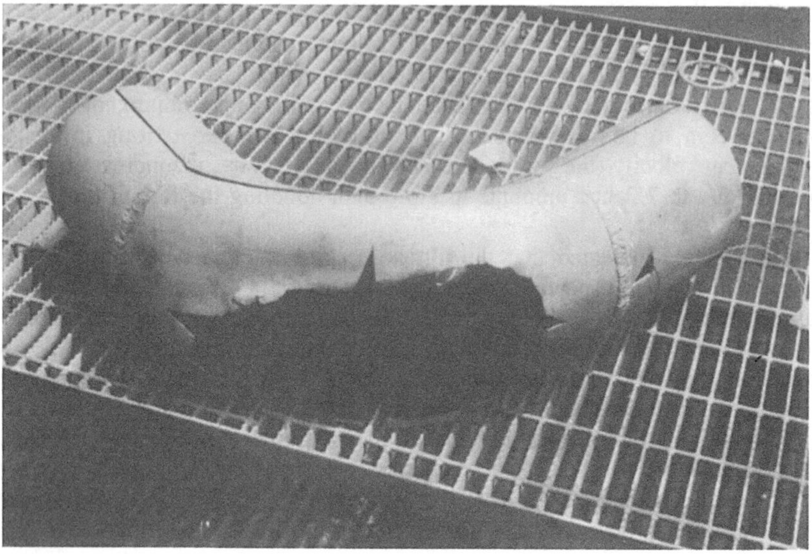

Figure 3. The Ruptured Elbow

When the elbow ruptured, about 10 to 15 tons of hydrocarbon was released into the atmosphere during 30-40 seconds. With the prevailing wind at the time all needed was an ignition source to explode the massive amount of highly volatile mixture. There were several potential sources of ignition including:

1. A large control box located on a platform referred to as the compressor isle. This control panel contained all the instrumentation gears for the control of the drive units (turbines) for the huge air compressor and some of the process gas compressors. Many of the instruments inside this control box were running on 110 volts power, and as any continuous process control requires, many hundreds of circuities were constantly in the state of change, thus creating spark.

2. Two air conditioning units on top of the control room maintaining proper temperature for the instruments inside the control room. It was learned that the drive units for the compressor of these air conditioning units were not sealed and as such were not explosion proof. This was contrary to the NPFA regulation.

3. An excessively hot regenerator skin, resulting from cracked and/or broken inner lining of the regenerator. Although the lining had been repaired earlier, there were still reports of hot regenerator skin after the repair, to the extent that at nights when turning the lights off, one could see the glowing steel, indicating a temperature perhaps in the range of 650-700oC (about 1300oF). Actually the outer insulator had been removed to apply steam and cool the regenerator externally.

4. Presence of a couple heaters in the area with some broken viewing port glasses.

5. Presence of some auxiliary air compressor with hot exhaust pipes.

In almost all incidents involving a plant explosion, it is nearly impossible to definitely identify a single source of ignition. It is always the question of which piece of equipment or component is more probable to have been the source of ignition. After considering all of the above potential sources of ignition in this case, the mode of operation of each of them at the time of accident, the closeness of their location to the area where hydrocarbon was released and the condition that they were found after the accident, one could perhaps identify the instrument box as the more probable source of ignition. This instrument box measuring approximately 3m x 3m x 1m housed numerous control instruments. The outer panels of this instrument box were fabricated from 1/4" stainless steel with stiffener angles. According to the drawings, it had been designed and in fact it was equipped with a purge system. Although it could not be verified, it was supposed to have been under constant purge. Nevertheless, after the accident it was found in a condition that all of its outer walls had severely bulged out while all of the smaller instrument packages inside the box had their walls caved in, strongly indicating the initial explosion within the walls of the instrument panel.

In addition to a number of fatalities and some injuries to persons, this explosion, resulted in many structural damages in the area within approximately 20 Km radius of the explosion epicenter.

III. IDENTIFYING THE FAILURE

In a case study such as the one in question, the key problem is, of course, to be able to determine, after the fact, the causes of the failure. The attorneys and experts attempting to do this rely on three major resources; the physical evidence from the site, the document trail and the testimony of those familiar with the plant. In this section we will show how these tasks are accomplished.

Within hours of the explosion, and while rescue efforts were still ongoing, local, state and federal officials began planning for and organizing an intensive investigation. The owner of the facility was also assembling an investigative team so that their forensic study of this casualty could immediately begin. The first matter of priority for the victims is to organize themselves and thus protect their interest. Shortly thereafter, victims and their families hired counsel who filed suit on their behalf in civil courts on both the state and federal level. These attorneys, after organizing themselves into a committee representing all victims' interest, thereafter retained the services of experts in various disciplines, including process engineering, structural engineering, chemistry, metallurgy, corrosion science, and electrical.

This committee of attorneys, along with their retained experts, then sought the assistance of the court. Under the court's supervision, plaintiffs' counsel and their experts secured an order preserving the explosion site. The Court's order also permitted counsel and experts to enter the plant to conduct an inspection of the site after it had been secured and rescue efforts terminated. This early order was critical as it insured the preservation of all available evidence. The Federal Rules of Civil Procedure contain a provision allowing for the entry onto the land of defendant's for this purpose[11]. In some cases, it may be necessary to gain entry upon the land of someone who is not a party. This is also provided for[12,13].

There are several basic consideration in organizing the inspection, such as, the time of the inspection, the size and composition of the inspection team, the amount of recordation and testing allowed and the duration of the inspection. The inspection must be done as soon as possible. In the case study in question, the investigation was begun as soon as all search and rescue operations were complete and the scene was deemed safe enough to allow an inspection.

The composition of the entry team should consist of experts from a variety of disciplines. First and foremost, their should be someone on hand to provide quality photographs and videotapes of the scene of the failure. In the case study in question, this served to preserve the appearance of important evidence long after it had been discarded or subjected to destructive testing.

Most cases of this type will also require expert testimony relating to chemistry, metallurgy, process engineering. In the case which is the subject of this paper, experts were also retained in the fields of structural engineering, corrosion science and electrical engineering. Detailed measurements of all process lines believed to be involved in the failure were made and recorded.

Following the inspection of the facility, scientific testing has to be done on the components suspected of being the cause of the failure. In the case in question, arrangements were made whereby experts for the plaintiffs attorneys were present when the experts for the owner performed destructive testing on the elbow and other important sections of the process stream.

The next priority after the physical review of evidence, is the marshalling of documentary evidence. Every refinery keeps paperwork in the form of blue prints, piping schematics, diagrams and drawings, log sheets, maintenance reports, daily reports, instrument readings, process manuals and a myriad of others. A copy of these documents should be requested through the discovery process. This is the paper trail through which the functioning of the refinery prior to the explosion can be traced.

In the case at issue here, the defendants produced over four hundred boxes of documents which were housed in a trailer on the plant site. Each of the documents was given a number referred to as a bates number which served to identify it for location purposes. The tasks of the plaintiffs were to search through these hundreds of thousands of documents for the relatively few which would give some indication of what went wrong.

A system had to be set up to review and maintain the documents in a manner which maximized their utility. This included setting up a document depository, reviewing and analyzing the documents and instituting a system for storing the documents and retrieving them when needed. By the time the bulk of the document review gets under way, the experts and lead attorneys should have formulated one or more solid theories of the nature of the failure. In the case in question, these theories were reduced to writing and provided a guide of what to look for in the documents.

In cases involving a large number of documents such as the one in question, the document review process can be broken down into two principal parts. There is a subjective review which should be conducted by the attorneys and experts. This review entails the indexing and description of only those documents which appear to the reviewer to have some relevance to the theories regarding the failure. A description of the level of importance assigned to the document can be included. For example, sometimes the designations "interesting", "important", "hot document", and "smoking gun" are used.

The objective review is usually done by paralegals and other support personnel. This review includes a listing of every document produced by bates number, description, date and the names of the persons listed thereon.

The information gleaned from the objective and subjective reviews is entered into a database designed for the purpose. In the case in question, the database designed could sort the document by bates numbers, dates, document description, author of document, recipients of document and names of persons listed on documents.

After this is done, the information could be recalled for many purposes. For instance, for the deposition of a given employee of the defendant, all documents which he sent, received or which referenced him were pulled.

Likewise, all of the references to a given piece of equipment such as a depropanizer could be pulled as could all log sheets or any other given documents for a specified period of time.

While the use of databases for document utilization has been in use for some time now, one relatively recent development in document management is the utilization of CD/ROM technology. The exact image of literally thousands of documents can be scanned onto one CD/ROM disk. When needed, the documents can be pulled up onto the screen of a personal computer and printed via laser printer. This process produces a quality of document at least as good as most copy machines at a rate of eight or more pages per minute depending on the printer used.

Following the document production and review process, depositions should be scheduled. Those most likely to have information are the employees and manager of the facility in question. In the case under consideration, hundreds of depositions were taken. Through interrogatories, the job title and work descriptions of these employees were discovered. The documents which contained their names were pulled and used to question them.

These employees can serve to complete the picture which has taken shape through the inspection of the plant and the document review process. In the case in question, they testified as to the conditions at the plant prior to the incident in question. They knew which equipment worked well and which gave problems. They testified as to the daily log sheets including such things as the Ph readings taken upstream and downstream from the failed elbow.

In some cases, these employees can also provide eye witness testimony regarding the failure. In the instant case, the people nearest the explosion were tragically killed. However, several employees nearby testified that they heard a loud hissing prior to the explosion which helped to identify the source of the release of twenty thousand, plus pounds of hydrocarbon vapor.

After documents have been reviewed and depositions taken, the pre-trial proceedings then move into what is commonly referred to as expert discovery. Heretofore, the parties have been limited to a discussion of facts. In this stage, facts and opinions are discussed and discovered. Such discovery is lengthy and tedious. Each side probes the others experts to understand his opinions and the forensic and scientific basis for such opinions.

The procedures that have been discussed above, are outlined in the Federal Rules of Civil Procedure which applies to all litigation in the United States Federal Court System. The attorneys on each side have divergent and sometime inapposite views of what is or is not permitted by the Federal Rules. The plaintiffs generally view the Federal Rules as a means to learn as much as it can about as many subjects as it can in an effort to establish fault on the part of the owner. When plaintiffs overstep these boundaries, the owner will likely argue the plaintiff is doing nothing more than embarking on an unpermitted fishing expedition. On the other hand, the owner will typically argue that all request are too broad in an effort to limit inquiry to the narrowest possible field.

It is at this juncture that the Court intervenes and, after argument, fashions an order controlling what information can and cannot be discovered. The role of the Court is not limited to resolving discovery disputes. The Court also oversees and manages the progress of the litigation by establishing schedules for completion of certain tasks necessary to prepare the case for trial. Once discovery is complete, the Court will then entertain motions which will limit and shape the evidence to be presented to the jury. In some cases, the Court may dismiss certain claims or, for that matter all claims that are not legally or factually supportable for trial. The role of the Court is critical in seeing that the litigants follow the Federal Rules of Civil Procedure and that to the extent practicable, the case is made ready for trial in as timely and cost effective manner as is practicable.

IV. CONCLUSION

The forensic investigation as part of the litigation process accomplished two very important goals, it helped to secure compensation for the victims and equally important, drove home the message that process and equipment changes, no matter how slight, can produce catastrophic failure. As a result of such awareness, the owner and many other operators have enacted much more rigorous programs to understand and identify problems associated with planned process and equipment changes. Much more rigorous inspection programs have been adopted. There are perhaps millions of miles of aging pipe in refineries and chemical plants across the world. Strict attention to even the smallest item is necessary to prevent similar catastrophic failures. Industry must always be mindful that the causes of a catastrophic failure are generally isolated, seemingly unconnected events and be avoided only by a most rigorous and organized inspection program, accurately documented in conformity with strict management of change procedures.

REFERENCES

1. Marsh & McLennan, "Large Property Damage Losses in the Hydrocarbon-Chemical Industries: A Thirty-year Review", p. 1 (M & M Protection Consultants, 1993)

2. Wells, Kochan, and Smith, "Managing Workplace Safety and Health: The Case of Contract Labor in the U.S. Petrochemical Industry", p.23 (John Gray Institute/Lamar University, 1991)

3. "Managing Workplace Safety and Health: The case of contract labor in the petro-chemical industry", U.S. Department of Commerce, National Technical Information Service.

4. Finnie, I., "The Mechanism of Erosion of Ductile Metals", Proceedings ofd3 National Congress of Applied Mechanics, American Society of Mechanical Engineers, 1958, pp 527-532.

5. Laitone, J. A., "Aerodynamics Effects in the Erosion Process", Wear, 22 (1979), pp 239-246.

6. Smeltzer, G. E., Gulden, M. E. and Compton, W. A., "Mechanism of Metal Removal by Impacting Dust Particles", Journal of Basic Engineering, Sept. 1970.

7. Finnie, I., "Erosion of Surfaces by Solid Particles", Wear, 3 (1960), pp 87-103.

8. Heymann, F. J., " Toward Quantitative Prediction of Liquid Impact Erosion", STP 474, American Society for Testing and Materials, 1970, pp 212-248.

9. Sabbaghian, M. and Pourmohamadian, H., "Surface Erosion by Impingement of Aerodynamically Entrained Lignite Particles", Journal of Powder and Bulk Solid Technology, 5 (1981), pp 31-36.

10. Hammitt, F. G. and Heymann, F. J., "Liquid Erosion Failures", Metals Handbook, Vol. 10, Failure Analysis and Prevention,

11. Federal Rule of Civil Procedure 34(a).

12. Louisiana Civil Code of Procedure art. 1463

13. Federal Rule of Civil Procedure 34(c), 45.

9 IMPROVED DESIGN THROUGH FAILURE ANALYSIS

I. LE MAY
Metallurgical Consulting Services Ltd, Saskatoon, Canada

Abstract

The paper discusses how analysis of failures can be of benefit to the designer and manufacturer. The problems of improved and "exact" design methods in reducing the safety factor of structures or components when some circumstance has not been taken into account are discussed.
Keywords: Design, failure, failure analysis, safety, safety factors.

1 Introduction

When failures occur it is common for a failure analysis to be made, normally when there is some question of legal liability to be settled, the event has caused a serious accident or a combination of these is present. As a consequence, the idea of a failure analysis being conducted is often a cause of concern, particularly for the manufacturer or seller of the equipment involved.

However, a failure analysis may also be considered to be a very useful and beneficial event, particularly for the manufacturer and designer. It can assist in improvement of the product.

2 Optimum Design

In designing and manufacturing a component or structure, the aim of the designer and manufacturer must be to ensure that the product is sufficiently strong and serviceable that it possesses all the necessary properties to ensure that it does not fail in service

Structural Failure, Product Liability and Technical Insurance, Proceedings of Fifth International Conference. Edited by H.P. Rossmanith. Published in 1996 by Taylor & Francis, 2 Park Square, Milton Park, Abingdon, Oxon, OX14 4RN. ISBN 0 419 22080 1

during what may be considered its design or service life. Premature failure is to be avoided for sound commercial reasons. If this occurs the manufacturer, designer or both may be subject to product liability suits on the basis of the product not meeting the reasonable expectation of the purchaser or, if the failure leads to injury, a costly strict liability suit may result. In any event the reputations of the manufacturer and designer will suffer as a result.

On the other hand, if the component or structure is overdesigned, in that it has an unnecessarily large safety factor, the cost and selling price will be increased and the product may be uneconomic and uncompetitive.

A balance must be struck between the two extremes so that the risk of failure is at a level that is acceptable to society and to the users of the product, so that the product may be deemed "reasonably safe". This does not mean that the risk of failure should be zero. The acceptable level of risk depends on the consequences of failure and on the benefits associated with the risk taken. Thus we may invoke the ALARP (as low as reasonably possible) concept [1].

Improvements in design procedures and in the knowledge of materials behaviour can lead to more refined designs without redundant material and with reduction in cost. This process has accelerated greatly with the introduction of computer methods in the design process. Designs can now be tailored to use a minimum of material because programs (usually of the finite element method (FEM) type) are widely available and higher stresses can be used, together with a reduction in material that carries little stress. Thus, the factor of ignorance or factor of safety, depending on how it is viewed, can be reduced.

This process is all very laudable and to be encouraged, but it has dangers. These occur when the original analysis is flawed and the apparently large factor of safety does not, in fact, exist. Apparent refinement of the design can lead to safety factors that approach unity and can, under specific loading conditions be ≤ 1, leading to failure. To illustrate that this is a historical pattern, Petroski [2], in his recent book, reviews some of the classic errors in design. These include the analysis of Galileo [3] of a cantilever beam in which the fundamental assumption of stress distribution being uniform across the section was implicit. The result was to overpredict the breaking load, something that was not fully resolved during the greater part of the eighteenth century. Other historical errors include the design of the River Dee Bridge by Robert Stephenson in which the formula used for the calculation of the failure load in a cast iron girder did not consider the area of the upper compression flange of the beam nor the beam's actual section. These girders were used in conjunction with eccentric tie-rods to provide prestressing and forming trussed girders, the span of which was increased beyond that of previous bridges. Failure occurred as a result of torsional buckling during passage of a train. Scaling up is a dangerous practice if all the possible failure modes are not considered, some of which were not relevant in the original designs. Another case in point is the failure of the Tacoma Narrows Bridge because of wind-induced vibration and torsional dynamic instability.

That similar problems still arise today is demonstrated by the collapse of a large coal loadout stage in 1982 that the writer investigated, in which the light enclosed cable-stayed structure collapsed on a calm day as a result of prior fatigue cracking caused by wind-induced vibration. This was an elegantly designed structure in which a minimum of material was used, but the possibilities of instability in cross-wind conditions were not checked out thoroughly. Subsequent, and simple, calculations showed the potential for failure and this was later demonstrated in model testing. Design improvement came in this case at a high price and could have been effected without the failure if more imagination had been used at the time of the original design.

Scaling up of existing designs is not the only way in which unexpected failure may arise. When an existing design is used in different conditions (such as a different environment) unforseen problems may arise. Examples discussed previously by the writer include the failure of large storage tanks under low temperature conditions that were not considered in the original design code [4, 5]. Following the 1980 tank failure in Canada [6] and a subsequent one near West Elizabeth, Pennsylvania, USA, in 1988, the American Petroleum Institute (API) published API Standard 653, *"Tank Inspection, Repair, Alteration, and Site Reconstruction"*, in January 1991. This provides a practical fitness for purpose approach to be followed in assessing existing structures or one that may have a change of use or other alteration [5].

3 The Benefits of Failure Analysis

If we do not know the cause of a failure we cannot learn any lessons from it to improve the design and reduce the likelihood of its occurrence in future. Following the failures in the de Haviland Comet aircraft, the very complete investigation and subsequent testing [7, 8] provided valuable knowledge of design against fatigue. In retrospect one may argue that more time should have been spent in building and testing prototypes, but there was a great urgency to produce a commercial jet aircraft. Today we still have failures of a generally similar type from time to time, even though our knowledge of fatigue has increased enormously. They usually relate to something that has been overlooked in the design, or a modification that has taken place later.

The writer has examined a number of storage bins that have collapsed, the most common cause being buckling instability not forseen by the designer. In one particular case the fabricator, who had developed a simple design based on experience, contracted a stress analyst to examine what he thought were the critical areas. These include the region of the support ring, where it acted on the sloping bottom of the hopper section of the bin. FEM analysis allowed the thickness of the bottom to be reduced substantially, but buckling conditions were not considered. Consequently, collapse occurred in the bottom portion of the bins, adjacent to the support ring, a series of such failures taking place.

When life extension studies are made of a plant such as a power station, it is very important to review prior failures that may identify design deficiencies, and some of these may be able to be eliminated in a life extension program. Thus, all such analyses should be recorded carefully and fully at the time.

A crucial problem relates to the provision of information regarding failures back to the design and manufacturer concerned. This does not happen easily and needs to be facilitated by, for example, a copy of the report being supplied to the manufacturer who, in turn, lets the design staff see it.

In the event of a product liability suit, the manufacturer will see the failure reports in due course and should make sure that they are passed on to the designers and acted upon. This latter is not necessarily a simple procedure as companies are sometimes reluctant to make appropriate changes when faced with a legal action. The fear is that this will be an admission of fault on their part, as it might well be in some cases.

The other way in which failures can be useful to the designer and can be communicated to him or her, recently became more prevalent through Conferences such as the SPT series and the ISTLI conferences and journal. It will be pleasant when we hear of specific actions taken as a result of some of these publications.

In closing it is appropriate to quote Robert Stephenson who, in 1856, said [9]:

> *"...he hoped that all the casualties and accidents, which had occurred during their progress, would be noticed in revising the Paper; for nothing was so instructive to the younger Members of the Profession, as records of accidents in large works, and the means employed in repairing the damage. A faithful account of those accidents, and of the means by which the consequences were met, was really more valuable than a description of the most successful works. The older Engineers derived their most useful store of experience from the observations of those casualties which had occurred to their own and to other works, and it was most important that they should be faithfully recorded in the archives of the Institution."*

4 References

1. Health and Safety Executive (1992) *The Tolerability of Risk from Nuclear Power Stations*, HMSO, London.
2. Petroski, H. (1994) *Design Paradigms: Case Studies of Error and Judgement in Engineering*, Cambridge University Press, Cambridge.
3. Galileo (1638) *Dialogues Concerning Two New Sciences*, (trans. by H. Crew and A. de Salvio), (1954) Dover, New York.
4. Le May, I. (1984) Oil tank failures and changing technical specifications, in *Structural Failure, Product Liability and Technical Insurance*, (ed. H.P. Rossmanith), North-Holland, Amsterdam, pp. 115-120.
5. Le May, I. and da Silveira, T.L. (1992) Evaluation of fitness for purpose of some critical structures, in *Materials Performance: Sulphur and Energy* (ed. P.R. Roberge et al.), CIM, Montreal, pp. 209-217.
6. Le May, I. (1982) Failure analysis of a crude oil storage tank. *Metal Progress*, Vol. 122, No. 3, pp. 35-37.
7. Walker, P.B. (1964-1965) The scientific investigation of aircraft accidents. *Proc I Mech E*, Vol. 179, Part 1, pp. 997-1014.
8. Whyte, R.R. (1975) *Engineering Progress Through Trouble*, I Mech E, London, pp. 51-53.
9. Whyte, R.R. (1975) *Engineering Progress Through Trouble*, I Mech E, London, p. v.

10 USING FAILURE MODE AND EFFECTS ANALYSIS TO IMPROVE PRODUCT SAFETY

R.E. McCLAY
Safety Sciences Department, Indiana University of Pennsylvania, Indiana, USA

Abstract
Failure Mode and Effects types of System Safety Analyses have been in use for over 30 years to identify hazards in systems under design/development. The rationale behind the use of these analyses techniques is the identification of hazards associated with the system while there is still the time and resources to effectively control them. They provide an excellent mechanism for exercising legal foreseeability for product hazards. Where used, these analyses have become a viable element in Product Safety Programs and have contributed significantly to product quality and liability mitigation.

This paper addresses the methodology used in Failure Mode and Effects Analysis, and similar analyses. Also covered are the planning and administration of the system safety analysis to include system documentation requirements, the system definition process, establishment of a limit of resolution and the selection of system modes for analysis. The construction of Functional Block Diagrams is explained and the use of matrix methodology to complete the analysis. Finally the paper addresses the formulation of system design changes based on risk information produced by the Failure Mode and Effects Analysis. Each of the topics above is also illustrated through the presentation of an analysis example.
Keywords: System safety analysis, system safety, product safety, product liability, reliability, hazard control.

1 Introduction

In a perfect world, an organization designing and developing a technological system would follow the best possible engineering design protocol. They would utilize the

Structural Failure, Product Liability and Technical Insurance, Proceedings of Fifth International Conference. Edited by H.P. Rossmanith. Published in 1996 by Taylor & Francis, 2 Park Square, Milton Park, Abingdon, Oxon, OX14 4RN.
ISBN 0 419 22080 1

talents of the most knowledgeable engineers. Extensive resources would be dedicated to finding hazards in the proposed system design and in testing the prototype system to verify that all foreseeable hazards had been controlled. Although a great deal of progress has been made in the past 20 years to achieve this ideal, it is quite clear to safety scientists everywhere that we do not yet live in a perfect world.

Willie Hammer in his outstanding text on product safety, has noted that with regard to the legal concept of foreseeability, a system designer...

> "may be held liable for actions that result in injury or damage only when he was able to foresee dangers and risks that could reasonably be anticipated." [1]

As technology advances and dramatically improves the engineer's ability to access information about hazards and how to control them, society will expect an improved ability to reasonably anticipate system hazards. This means that those hazards that should be foreseen will multiply accordingly and that we will experience a greater level of safety in our technological systems. One must consider, however, whether or not the actual design process will take full advantage of our improved knowledge of hazards.

In his book about the fallibility of engineers and the design process, Petroski maintains the inevitability of system failures by observing...

> "Although the design engineer does learn from experience, each truly new design necessarily involves an element of uncertainty. The engineer will always know more what not to do than what to do. In this way the designers' job is one of prescience as much as one of experience. Engineers increase their ability to predict the behavior of their untried designs by understanding the engineering successes and failures of history. The failures are especially instructive because they give clues to what has and can go wrong with the next design. They provide counter examples." [2]

And so in a difficult design process with imposing knowledge limitations, it becomes important to examine the education and professional preparation of engineers to determine how well they can respond to this challenge. But here too there are concerns. C.O. Miller, distinguished authority and pioneer in aviation system safety, has consistently maintained that the education of engineers in the U.S. has conditioned them to see and think about how a system works not how a system can fail. What is needed then is a device or process by which engineers together with safety scientists can exercise their responsibilities for foreseeability by seeking out in a rigorous way the various mechanisms by which any system element can fail. Fortunately such a process exists within the discipline of system safety.

System safety is a discipline recognized at least in the U.S. for applying scientific, engineering and management principles to optimize the safety of systems. [3] The guiding principle in system safety is that hazards must be recognized and controlled in the system design. Tagging-on safety features to an existing hazardous operating system is regarded as an undesirable outcome resulting when system safety principles are ignored. System safety analysis techniques are methods used to identify hazards in systems which are in the early stages of their life cycles. These techniques are very rigorous in that they force an analysis team to examine every element within the system or every step in a human procedure or every logical possibility for the occurrence of a catastrophic event. Other techniques called hazard analysis, though less rigorous, can also be very helpful in the early identification of system hazards. All of these techniques are well known in the safety science literature and are available as essential aids in the foreseeability process.

2 Text

2.1 Failure mode types of analyses

One set of important techniques which can be used with great effectiveness for the identification of hazards in product systems is the failure mode type of analysis. This type includes three (3) separate techniques which are discussed in this paper:

- Failure mode and effects analysis
- Failure mode effects and criticality analysis
- Fault hazard analysis

These techniques are all of the same genre in that they direct attention to the possible malfunction of non-human elements in the system. The failure mode type of analysis was developed within the discipline of reliability engineering over 30 years ago and has been used with effectiveness during this period. Though the subject of some inconsequential debate over whether it should be called a reliability analysis or a system safety analysis, it clearly meets the criteria for use in both applications. [4]

2.2 Important definitions

The approach taken in the failure mode type of analysis is to look at each system element and determine what consequences will result if this element fails to perform as intended in the system design. Important terminology utilized in this approach is as follows:

- Critical function - a mission to be performed by a system element which is critical to the functioning and survival of the system.

Nearly every element of a given system is designed to perform one or often more than one critical function. These functions affect the performance of other system elements and frequently influence elements outside the given system.

- Failure - occurs whenever a given system element fails to carry out a critical function assigned to it.
- Failure mode - the manner in which a failure occurs.

System element failures constitute important hazards which need to be considered by the design engineer as a reasonably foreseeable event. The possible consequences need to be identified for each failure mode.

Example: If Gear A was assigned a critical function to transfer torque to Gear B, then a failure occurs whenever torque is not transferred according to system specifications. Three failure modes can be envisioned as shown in Figure 1:

- excessive torque is transferred
- insufficient torque is transferred
- torque is transferred in the wrong direction

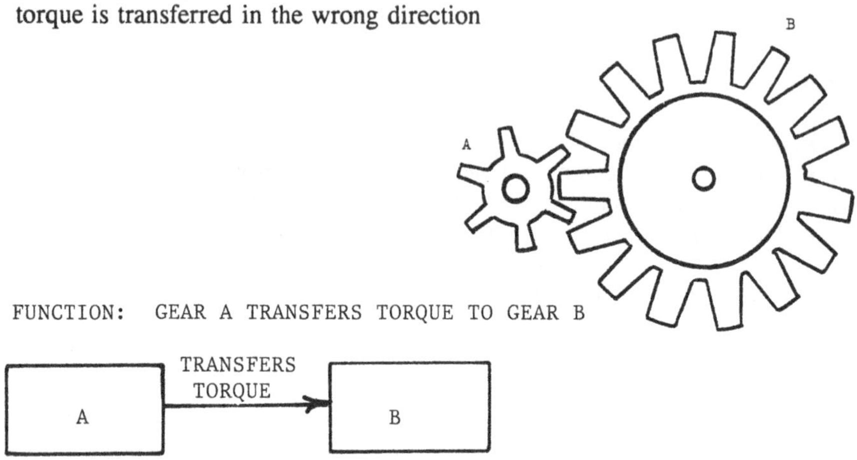

FAILURE: TORQUE NOT TRANSFERRED ACCORDING TO SPECIFICATIONS

FAILURE MODE #1: EXCESSIVE TORQUE TRANSFERRED
FAILURE MODE #2: INSUFFICIENT TORQUE TRANSFERRED
FAILURE MODE #3: TORQUE TRANSFERRED IN WRONG DIRECTION

Fig. 1. Critical functions, failures and failure modes

Three types of failures are recognized by reliability engineers:

- Primary failure - occurs when an element fails while operating under conditions it was designed to tolerate.
- Secondary failure - occurs when an element fails while operating under conditions it was not designed to tolerate.
- Command fault - occurs when an element fails to perform its design function because it receives either no signal or an improper input signal.

All three (3) types of failure are encountered when performing a failure mode type of analysis.

When preparing to perform any system safety analysis, the analysis team must decide upon the scope and depth of the analysis. The scope is determined by selecting specific system modes for inclusion in the analysis.

- System mode - the state of the system within a given life stage.

Example: Systems may exist in several or many operational modes during their operational life stage. For example, an automobile might be seen to have the following system modes; parked, rearward movement, forward movement, passenger loading and unloading, refueling, maintenance, cargo loading and unloading, etc. Hazards which are non-existent in some system modes might be quite serious in others.

The depth of the analysis is established by choosing a limit of resolution.

- Limit of Resolution - the level in the system element hierarchy below which the analysis will not progress.

The levels of elements in the system element hierarchy are as shown in Figure 2 below and include the system itself (as a single element), subsystems, components, and parts. These levels are identified by the importance of the element's function. So if we subdivide a subsystem into elements with the next greatest functional importance, we will have components. Major components have a more important function by definition than do minor components. In reliability engineering, these hierarchial levels are referred to as "indenture levels". [6]

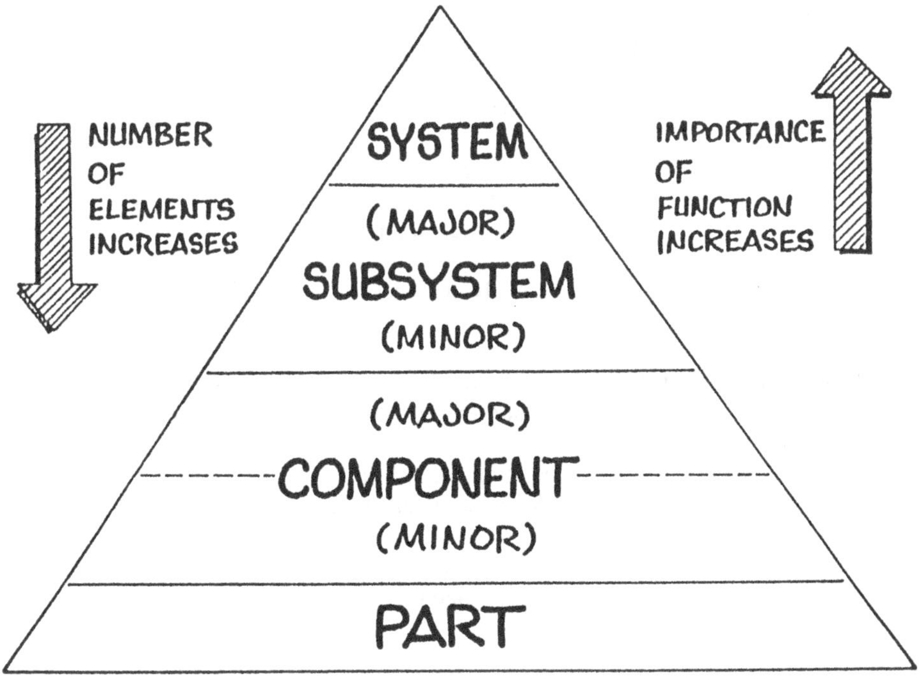

Fig. 2. System Hierarchy

An analysis performed at a lower level in the system element hierarchy will require more information, take longer, but should produce more information about hazards. In any case, a decision must be made about the limit of resolution before the analysis begins. Frequently the amount of information available on the new system and/or the time available for the analysis will determine the hierarchial level at which the analysis is performed. Since every system element at the chosen level is to be analyzed, the analysis is thorough and systematic no matter which level is selected.

2.3 System definition

It is really quite impossible to perform any sort of meaningful system safety analysis on any system without first defining that system. When a system is properly defined for analysis, it will be clear what the system's overall objective is, one will see which elements are considered to make up the system and to know what each of their functions are. The general manner in which the system operates also becomes apparent. This information is so essential for an effective system safety analysis (SSA), a SSA is defined to be an analysis technique for which a system definition is necessary. Hazard analyses on the other hand have no such requirement.

The steps followed in defining a system for analysis will normally include the following:

- Establishing the system's objective.
- Bounding the system by identifying those elements within the system and those elements considered as being in the system's environment.
- Identifying the system's general functions. These are those things which must be accomplished within the system in order for it to achieve the system objective. For example the general functions for a coffee cup would be:

  *contain the coffee
  *provide a means for lifting and carrying the cup
  *insulate against the conduction of heat

The general functions of a system always become the functions assigned to the subsystems.

- Constructing the system element hierarchy. With the functions of the subsystems known, it becomes possible to add the components and parts that are needed to accomplish those functions.
- Developing a brief system operating procedure.
- Listing any assumptions which are necessary because of a lack of information.

When the system has been completely defined, a limit of resolution established and system modes selected for analysis, only one more step remains before a failure mode type of analysis can be started.

2.4 Functional Block Diagram

A functional block diagram (FBD) is a depiction of the functional relationship of the system elements at the limit of resolution (LOR). One block is created for each element at the LOR, arrows are drawn between the blocks to represent the critical functions. The arrows are drawn from the element performing the critical function to the element or elements affected by that function. An element is affected by a critical function if it undergoes some intended change when the function is performed. Only direct effects are considered.

If a system contains three (3) elements, a coffee cup, the coffee and a spoon, critical function arrows would be drawn as follows:

- One arrow from the cup to the coffee because it is intended that the cup contain the coffee.
- Another arrow from the cup to the coffee because it is intended that the cup keep the coffee warm.
- One arrow from the spoon to the coffee because it is intended that the spoon stir and mix the coffee.
- No arrows would be drawn from the coffee to the spoon or to the cup because it is not intended that the coffee affect these in any way (although this may occur because we can't prevent it).
- An arrow could be drawn from the coffee to a point outside of the system because a change may be desired in the person drinking the coffee.

Critical functions are very concise and are written on each of the arrows drawn. The functional block diagram when it is complete allows a failure mode type of analysis to be performed because it identifies all of the critical functions performed by system elements at the LOR. When the analysis team has documented these critical functions they can then begin to study the effects that failures in these functions would cause if these failures were to occur.

2.5 The FMEA matrix format

The actual analysis begins when the analysis team starts to record information on the matrix sheets. System safety analysis techniques utilize one of two (2) formats for recording the analysis information; logic tree or matrix format. Information is not recorded vertically in the columns as one might intuitively believe. The analysis team records the details about a particular failure in a horizontal fashion from left to right across the page. Each line treats an individual failure and constitutes one entry. Entries are numbered for reference purposes.

The functional block diagram serves as a reference document for the analysis. Elements at the system's limit of resolution as shown in blocks on the FBD are considered individually. The order in which the system elements are analyzed is not important. All system elements shown on the FBD are analyzed; actually it is the critical functions performed by the system elements which are analyzed. Again, the order for selecting these is not important. One critical function is selected from the FBD at a time and the analysis team presupposes that a failure has occurred in this function. A particular failure mode is identified for the failure under study. Next the team attempts to determine what event may have the capability to cause this failure mode. Causes themselves may be primary or secondary failures or command faults. There is really no limitation placed on what causes per entry should be recorded.

After the cause for the failure mode is listed, the analysts predict the worst effects that might occur as a result of the failure mode under study. The U.S. Department of Defense has established a classification system to categorize the severity of the worst credible effects that may arise out of the failure sequence being documented. This classification system is shown in Table 1.

Table 1. Hazard classification system

Hazard Class	Worst Credible Effects
I. (catastrophic)	Death or system destruction
II. (critical)	Severe injury, illness or major system damage
III. (marginal)	Minor injury, illness or minor system damage
IV. (negligible)	No injury, illness or system damage

In the FMEA the final column on the right is for the analysts to record any preventive measures that might either eliminate the cause for the failure or lessen the severity of the effects. Normally this column is only completed for those entries where Class I or II effects are predicted. The analysts should list here any and all ideas for better controlling the hazard they have identified.

One of the strongest features of the matrix format and the most often misunderstood feature is what happens after a complete entry is made on the matrix. From the rightmost column in the sequence just completed, the analyst returns one column at a time from right to left to see if some variation of this entry can be found. Since the hazard classification is based on the worst credible effects listed in this entry, no other variation is possible in this column. Since the effects are the worst credible effects arising out of the failure mode, no other worst effects can logically be listed. However, in moving further to the left to the cause column, the analyst may find another cause which should be considered for the particular failure mode. If so, another entry can be created starting with the cause column and again being completed from left to right. This process is repeated over and over until all possible causes have been considered.

When no other cause for the predicted failure mode can be identified, the analyst moves to the next column on the left and considers any other failure mode which might apply to the failure under study. Most failures will have two (2) or more failure modes. Again, as soon as another entry is started, it is completed by making column entries from left to right. This left to right analysis with right to left searching for additional variations gives this technique its systematic nature. This process is depicted in Figure 3.

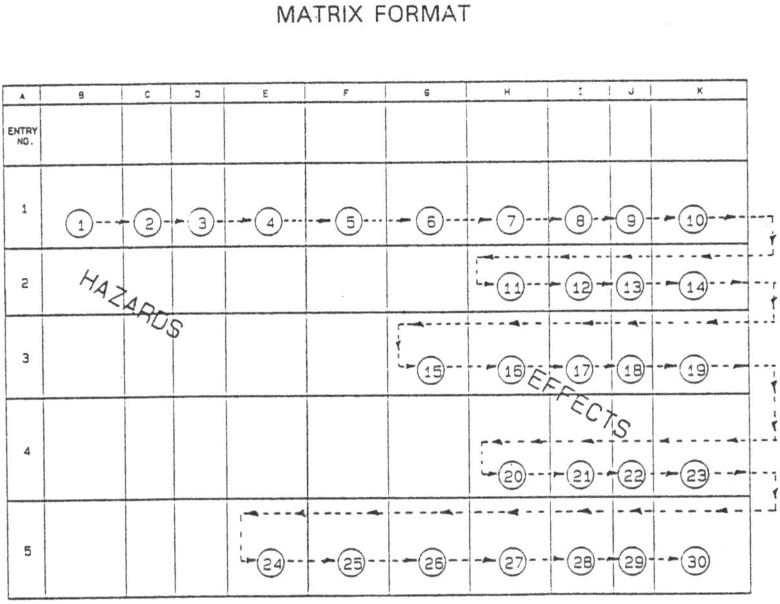

Fig. 3. Completing entries in a Matrix Format

When all possible failure modes for a selected critical function have been explored, the analysts are ready to examine any other critical function that may be performed by the system element being analyzed. These are taken from the FBD. The analysis continues in this way until every possible cause has been analyzed for each failure mode in each critical function for all system elements at the limit of resolution.

As a review let us summarize again the column headings which appear on the FMEA matrix:

- Entry number - this is nothing more than a line number.
- System element at the limit of resolution - selected from the functional block diagram, this could be a subsystem, component or part.
- Function - the critical functions depicted by arrows on the FBD.
- Failure mode - the manner in which the failure has been predicted to occur.

- Cause - any event which has the capability of producing the predicted failure mode in the critical function under study. Component failures are the most frequent cause for subsystem failures and part failures are the most likely cause of a component failure.
- Failure effects (local) - the worst effects that the particular failure mode can have upon neighboring or interfacing elements at the limit of resolution.
- Failure effects (system) - the worst credible effects that the failure mode can have on the overall system and its environment.
- Hazard Classification - the category shown in Table 1 from Mil-Std 882 based upon the worst credible system effects.
- Preventive measures - ideas for eliminating the cause of the failure or reducing the severity of its effects.

Occasionally FMEA matrices will include a column titled "Failure Detection". Here is noted the means by which operating personnel can detect the predicted failure. This often can be an important hazard control feature in any system. [8]

The FMEA, FMECA and FHA all utilize what is called inductive logic to complete the failure sequences on the matrix sheets. This means that the analysis team is using their collective experiences and knowledge of the literature to identify failure sequence possibilities. There are no logical rules to follow as exist with fault tree analysis. Judgements are based on analogies drawn from similar systems or different systems with similar operating principles.

2.6 The FMECA and the FHA

Whereas the basic FMEA is almost totally qualitative, considering only the severity of the effects from a given failure, the FMECA attempts to better indicate the risk associated with each hazard.

Risk = f (frequency of hazard; severity of the effects produced by the hazard)

The FMECA introduces the concept of criticality to represent the risk associated with a given failure. Criticality as reflected in the criticality index computed for each entry, takes into account not only the severity of the worst credible effects but also the failure probability.

Criticality Index = Failure Probability x Hazard Classification Weighting Factor

The failure probability in reality represents the likelihood of the failure cause shown in the cause column on the FMECA matrix. This probability can be computed from the failure rate determined by reliability testing and published in handbooks similar to reference [9]. The time period over which the failure probability would be calculated could be either taken as unity since the criticality index is only a relative indicator or the actual operating period of the respective system element. For additional precision, the failure probability can be multiplied by the failure mode ratio since not all failure modes of the element would be likely to produce the worst credible effects.

- Failure mode ratio - the fraction of the element failure rate occurring within the failure mode under consideration. (These values are also available in Reliability Data handbooks.) [10]

The hazard classification weighting factor is a value assigned to each hazard classification only for purposes of criticality computation.

Table 2. Hazard Classification Weighting Factors

Hazard Class	Weighting Factors
I	10,000
II	1,000
III	10
IV	1

The criticality index from the FMECA allows the analysis team to prioritize the failure entries on the basis of relative risk and this can be very helpful in achieving a cost-effective resolution to identified hazards.

The fault hazard analysis (FHA) has as its primary focus the identification of factors which may create the cause of the failure. The cause, failure mode and effects are identified as with the basic FMEA matrix but two (2) additional columns are provided:

- Secondary factors - abnormal and out-of-tolerance conditions (generally environmental) capable of leading to the cause of the failure mode under study.
- Upstream events - situations, errors or failures capable of leading to the cause of the failure mode under study. [11]

Through the use of these two (2) columns, the FHA is able to provide the analysts with additional information about secondary failures and command faults that may constitute important hazards in the system.

2.7 Capabilities and limitations of the failure mode type of analyses

Each system safety analysis technique has its own capabilities and limitations. The capabilities of the failure mode type of analysis are as follows:

- Identifies single failures which are capable of producing critical or catastrophic effects.
- Determines if failure detection mechanisms are adequate in the system design.
- Establishes what levels of redundancy exist in complex subsystems.
- Provides a basis for evaluating system displays and controls. [12]

Limitations of the failure mode and effects analysis also need to be noted.

- Is not able to evaluate combinations of failures at the limit of resolution. Only fault tree analysis is able to achieve this goal.
- Is not the best analysis technique to use in order to evaluate the effects of human error.

To illustrate the use of the FMECA technique an example has been provided of an analysis performed on a portable combustible gas indicator instrument. System definition, a functional block diagram, a schematic diagram and FMECA matrix sheets are shown in figures as indicated.

Raheja predicts in his excellent reliability textbook that, "the future role of FMECA will be bigger than it is today." [13] This is because of its flexibility and because of the greater need today to establish a rigorous foreseeability mechanism in the system design process.

In summary, the failure mode types of analysis while venerable and time-tested have not been universally used by design engineers and safety professionals to identify non-human hazards in new system designs. As engineers receive better classroom education in safety science and liability increases for those in the design function, these techniques should experience much wider usage.

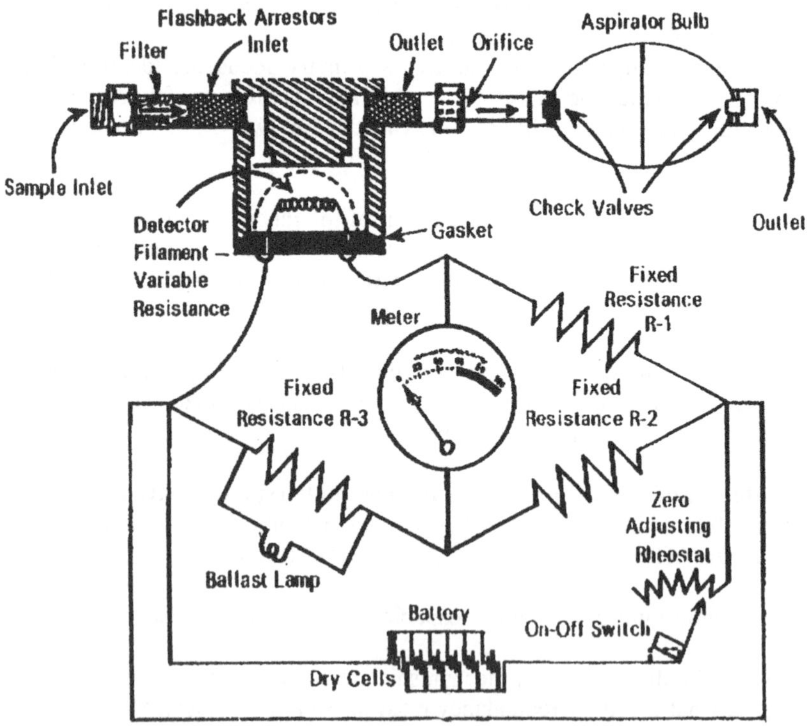

Fig. 4. Combustible Gas Indicator Schematic Diagram (Simplified)

Fig. 5. Combustible Gas Indicator instrument

SYSTEM DEFINITION FOR THE COMBUSTIBLE GAS INDICATOR (CGI) SYSTEM

SYSTEM OBJECTIVE: To measure the concentration of a fuel gas in a given location.

BOUNDING:

SYSTEM	SYSTEM'S ENVIRONMENT	TRANSIENTS
The instrument case and everything it contains	The instrument operator	Electrical current
The carrying harness	All other objects and persons in the area	Gases that enter and leave the instrument
Sampling line		
Sampling probe		

Fig. 6. System Definition for CGI System

GENERAL FUNCTIONS:

1. Draw gas into the instrument and discharge it
2. Contain and burn gas while holding the heat inside the combustion chamber
3. Measure the change in resistance of the detector filament
4. Contain and protect the elements of the system and hold them together
5. Balance the Wheatstone Bridge before sampling
6. Allow the instrument to be carried from place to place

SYSTEM ELEMENT HIERARCHY: *

Subsystem A: (Flow)
  Function - Draw gas into the instrument and discharge it

  Component A-1 (Aspirator)
    Function - Creates a pressure difference which causes flow into the instrument
  Component A-2 (Sample Inlet)
    Function - Admits gas into the instrument
  Component A-3 (Sample Outlet)
    Function - Permits gas to leave the instrument
  Component A-4 (Filter)
    Function - Prevents solid material from entering the instrument
  Component A-5 (Flashback Arrestors)
    Function - Prevents a flamefront from igniting gases inside the instrument

OPERATING PROCEDURE:

1. Open lid
2. Aspirate fresh air into the combustion chamber
3. Turn selector switch to "check"
4. Pull up voltage adjustment knob and turn it all the way to the left
5. Turn on switch marked "SW" to energize the instrument
6. Aspirate more fresh air through the instrument
7. Use voltage adjustment knob and turn it in the direction to move meter pointer to "check" on the dial
8. Turn the selector switch to "%LEL"
9. Pull up zero adjust knob and adjust meter pointer to zero
10. Turn selector switch to "10%LEL"
11. Pull up zero adjust knob and adjust meter pointer to zero
12. Turn selector switch to "%LEL"
13. Wearing SCBA and following hazardous site entry procedures, carry the instrument to the site of the atmospheric test
14. Using an appropriate instrument, test for the presence of elevated Oxygen levels
15. Verify that no contaminants are present that would disable the CGI system
16. Place the end of the sampling probe in the area to be tested
17. Use the aspirator to draw a sample into the chamber
18. Read the % concentration from the indicator dial

ASSUMPTIONS: None

* For purposes of brevity, the structure of only one subsystem has been shown here.

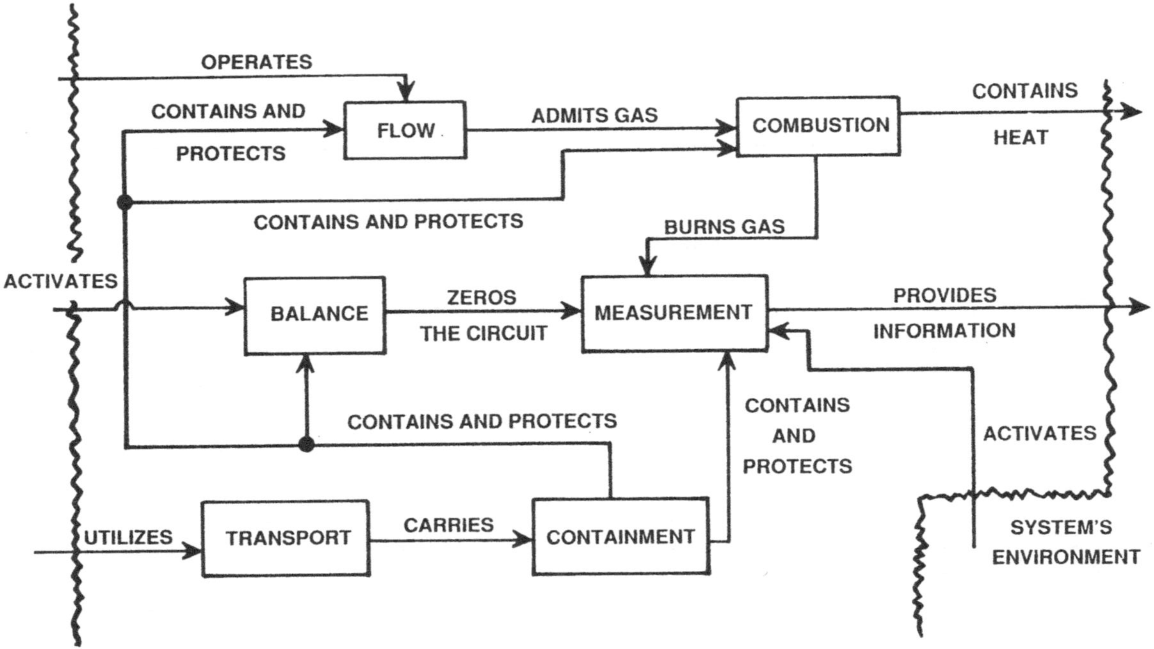

Fig. 7. Functional Block Diagram for CGI System

SYSTEM MODE(S) GAS DETECTION

MATRIX - FAILURE MODE, EFFECT AND CRITICALITY ANALYSIS

1	2	3	4	5	6		7			8
ENTRY NUMBER	SYSTEM ELEMENT AT LIMIT OF RESOLUTION	CRITICAL FUNCTION	FAILURE MODE	POSSIBLE CAUSE	FAILURE EFFECT		CRITICALITY			PREVENTIVE MEASURES
					6A LOCAL	6B SYSTEM	7A FAILURE PROBABILITY	7B HAZARD CLASS	INDEX 7C	
1	COMBUSTION SUBSYSTEM	CONTAINS HEAT	HEAT ESCAPES THROUGH INLET	INLET FLASHBACK ARRESTOR FAILS	HEAT ESCAPES FROM SYSTEM	POSSIBLE EXPLOSION	5×10^{-8} PER HR.	I	5×10^{-4}	REDUNDANT ARRESTORS
2			HEAT ESCAPES THROUGH OUTLET	OUTLET FLASHBACK ARRESTOR FAILS	HEAT ESCAPES FROM SYSTEM	POSSIBLE EXPLOSION	5×10^{-8} PER HR.	I	5×10^{-4}	ENSURE INHERENT SAFETY OF CIRCUITRY
3			HEAT ESCAPES THROUGH GASKET	GASKET FAILS	HEAT ESCAPES FROM SYSTEM	POSSIBLE EXPLOSION	2×10^{-7} PER HR.	I	2×10^{-3}	TESTING OF SYSTEM IN GASEOUS ATMOSPHERE
4		BURNS GAS	GAS BURNS WITH EXCESSIVE HEAT	HIGH O_2 CONCENTRATIONS PRESENT	HEAT ESCAPES FROM SYSTEM	POSSIBLE EXPLOSION	1×10^{-4} (est)	I	1.0	TEST FOR O_2 PRIOR TO VAPOR TESTING
5										HIGH O_2 WARNING LIGHT

Fig. 8. Sample FMECA Matrix Sheet

5 References

1. Hammer, W. (1993) *Product Safety Management and Engineering*, American Society of Safety Engineers, Des Plaines, IL, p. 4.
2. Petroski, H. (1992) *To Engineer is Human*, Vintage Books, New York, NY, p. 105.
3. McClay, R.E. (1990) *Readings in System Safety Analysis*, 6th Edition (unpublished text), Safety Sciences Department, IUP, Indiana, PA
4. Benner, L. and J. A. McDevitt (1981) *"System Safety Methodology"* , White Paper No. I, *Proceedings of System Safety Conference*, Washington D.C., SSS Chapter, Arlington, VA, pp. 8 - 11.
5. Henley, E.J. and H. Kumamoto (1981) *Reliability Engineering and Risk Assessment*, Prentice Hall, Englewood Cliffs, NJ, pp. 59 - 60.
6. Military Standard 1629 A, (1980) *Procedures for Performing a Failure Mode Effects and Criticality Analysis*, U.S. Department of Defense, Washington, D.C., p. 4.
7. Military Standard 882 C (1993) *System Safety Program Requirements*, U.S. Department of Defense, Washington, D.C., p. 11.
8. Sundararajan, C. (1991) *Guide to Reliability Engineering*, Van Nostrand Reinhold, New York, NY, pp. 149 - 150.
9. *Nonelectronic Parts Reliability Data*, (1991) Reliability Analysis Center, Rome, NY.
10. Sadlon, R. J. (1993), *Mechanical Applications in Reliability Engineering*, Reliability Analysis Center, Rome, NY, p. 147
11. Ericson, C.A. (1972) *System Safety Analysis Technology - Fault Hazard Analysis, Report* No. D2-113072-3, Boeing Company, Seattle WA.
12. Arnzen, H.E. (1966) "Failure Mode and Effect Analysis: A Powerful Engineering Tool for Component and System Optimization", *Proceedings of the Fifth Annual Reliability and Maintainability Conference*", New York, NY.
13. Raheja, D.G. (1991) *Assurance Technologies; Principles and Practices*, McGraw Hill Book Co., New York, NY, p. 61.

5 References

11 QUALITY ASSURANCE CONTRACTS: NECESSARY DETAILS AND LIMITS OF RESPONSIBILITY

C.O. BAUER
Wuppertal, Germany

Abstract

Quality assurance contracts as organizatorial regulations between industrial manufacturers and their suppliers have been developed to define duties and requirements between all parties involved. Under the headline of quality assurance they contain clauses in nearly all cases which are not at all necessary to support quality but change legal requirements and transform duties to the smaller, weaker part, the supplier.

All commercial contracts and quality contracts too have to obey legal requirements of national law. Most times this is not realized with the consequence that nearly all quality contracts are legally unvalid because they do not follow necessary legal rules.

In cases of damages to be regulated under product liability procedures they proove their unvalidity to a remarcable extent: Over decades they have never been presented to insurance companies or at courts, because the authors felt or feared that the unvalidity might be publicly known by the following decisions.

There are multiple duties between supplier and manufacturer, as diffent as the products involved. Legal requirements always have to be observed carefully,as carefully as technical details even by people without legal training.
Keywords: quality assurance contracts, legal requirements, limits of validity, practical value of product liability, consequences of audits.

Structural Failure, Product Liability and Technical Insurance, Proceedings of Fifth International Conference. Edited by H.P. Rossmanith. Published in 1996 by Taylor & Francis, 2 Park Square, Milton Park, Abingdon, Oxon, OX14 4RN. ISBN 0 419 22080 1

1 Historical development

The first quality assurance standards were originated by the automotive industries in the United States to discipline their suppliers. The most famous example constitutes the Ford Q 101, which exists in various issues from the 1970th on with new content every three to five years. Ford Q 101 as one of the best known company standards reflects the varying organizatorial ideas valid at the time of the special issue about duties of manufacturers and their noumerous suppliers purely derived from American practices and thinking with no interference from Europe. [1]

This quality assurance contract - or what varying title is used - was transformed by strict binding concern-directive to Europe, regardless of the diffences between the USA and Europe with regard to industrial organization, functions of suppliers and legal national framework, which differ in the various European countries from those in the United States. The Q 101 does not take notice of the legally binding structure of generally accepted technical rules as the legally authorized substitute to fulfill legal requirements expressed in undefined legal terms, neither of the overall function of the state of art nor the references given by different legal obligations for all parties involved. This quality assurance company standard tried to transform many original responsibilities from the automotive manufacturer to the supplier, regardless of their own legal obligations and the limits of legal delegation as stated by national law and High Court decisions.

An investigation into the clauses of the 1985 issue of the Ford Q 101 therefore could proove more than two dozens of important violations of German national law and High Court decisions, but as a strict concern directive nobody was authorized in whole Ford of Europe to alter even a word, not to say anything about procedures or technical details. The Q 101 made history, followed by many other quality assurance contracts which were established by other companies under the well sounding headline of promoting quality or similar slogans, but in reality prooved to be a new mean of suppression of suppliers not regarding own binding legal responsibilities. This tendency can be prooved to exist by the overwhelming majority of all textes of quality assurance contracts up to today whatever declaration may be issued to obscure this fact.

2 Varying contents

The Ford Q 101 and its many followers delegated as many responsibilities from the manufacturer to the supplier and constituted many additional responsibilities for them far above their functions and the limits of controle for the supplier's companies. These contracts were drawn up mostly by quality assurance people with no legal training or sufficient knowledges, not even the simplest knowledges of the legal base for commercial contract duties between parties could be prooved much to often. With the buying power of important clients they forced suppliers to accept these contracts regardless of legal, technical or organizatorial objections against principles and many details if they intended to stay in business. Fair trading as headline disguised a poor power acting.

With the installation of special supplier orientated quality inspectors or auditors the buying companies reserved themselves the right inspect or to interfere into processes and internal activities, normally under the sole responsiblity of the independant supplier at the same time decreaded by clauses of the same contract their total stipulation of exemption from liability for any such action or demands established by their own personal.

This created curious situations by comparing the various issues of the Ford Q 101 which hailed some methods in one issue, e.g. in 1982 which were thoroughly condamned in the following one. So to follow up the content of the various issues of the Ford Q 101 documents an interesting variation of the ideas and the development of procedures which were transformed by this quality assurance standards upon their many suppliers regardless of any theoretical findings or practical experiences as well. "Product observation" as a general duty by strict product liability jurisdiction in the US and in Europe was and is completely neglected.

The best example for this is the US-version of the Failure Mode and Effect Analysis (FMEA) as expressed in Ford Q 101. It does even neglect the mathematical law of failure tolerance propagation, undisputed in science since more than 160 years. What an amount of time and money was vasted in multiple supplier's companies to establish such bureaucratic procedures to follow this customer demand even for parts with no main functions. The technical and economical results are far less than their missionaries promissed! It would be worthwile to compare the thereby wasted millions of dollars with those which were levished by the famous Ford Edsel, the most remarkable fiasco in the US-american history of automotive trade.

A German standard DIN 25 448 exists as a scintific solid procedure and is used as the base of security checks of all German atomic power plants but with no chance of even been recognized or implemented into the Ford standard as a convincing example of the famous NHI-effect documented on many other places within Q 101 also (NHI = not here invented).

3 Legal framework

Analyzing the legal validity of quality assurance contracts at first everybody has to investigate what constitutes the legal base and framework to be observed. As a contract between independant companies it has to follow the national laws of commercial contracts. Special national laws about conditions of sales have to be observed in addition which in the 1960th and 1970 were issued from nearly all states to controle the socalled "Small printed conditions" at the reverse-side of the orders. (Gesetz zur Regelung des Rechts der Allgemeinen Geschäftsbedingungen, Germany 1976). This legal subject is regulated in addition by a lot of court decisions, alone in Gemany there exist over 3.000. They filled the many open questions and problems of the law and give a detailed framework binding lower courts and therewhith the companies. One of the most decisive ruling in Germany outrules categorically the delegation of cardinal responsibilities from one company to its suppliers for actions out of their practical control. What is regarded as a cardinal responsibility has to be decided in single cases, knowing their special conditions.

One example is the main quality assurance clause as such:

"Responsibility for the total quality of the supplied parts rests solely by the supplier"

regardless of the fact that the **construction** of the part was done **solely** by the purchaser who in addition **assembles** them - a process known to be full of causes for failures and therewith faulty products. [2]

One of the principle decisions of the German High Court categorically outrules any transformation of the responsibility for incoming goods-inspection (Wareneingangsprüfung) upon the supplier himself. It does not need any long legal theory to explain that this would be the transformation of an obligation upon a party to controle itself and act as its own inspector, a complete reverse of the necessary division of responsibilities.

Another example is the undue prolongation of warranty far above legal terms. A further example is the negation of the duty to test the goods immediately after receiving them and to claim any defect of major characteristics according to the state of the art.

Without going too far into legal details there exists a collection of 256 quality assurance contracts in Germany, collected since more than 20 years. In analysing their contents only 6 would legally be acceptable by the courts, all others have more than one clause which violates legal binding principles.

In Germany a federal central agency exists as an anti-trust agency (Bundeskartellamt) which has the duty to gather all court decisions on commercial clauses. This agency has the obligation to controle recommandations given by commercial organizations to their members upon their legal content before they can be published. Only in three cases commercial organizations have asked for such legal tests, and later published their recommandations in the state journal (Bundesanzeiger). Their recommandations constitute a good framework even for managers, engineers and quality assurance people with no direct legal training. What contradicts those recommandations cannot become a legally valid contract.

4 Valuable content

The legal obligations of buyer and supplier are splitted up within various general and special laws, not to forget duties defined, specified and limited by High Court decisions. As a collection to describe all necessary duties and obligations of all parties involved, quality assurance contracts are helpful.

Most of the goals and limits given by law are expressed in undefined legal terms which have to be filled by court decisions taking into consideration all conditions of the single cases. Collecting those court decisions this gives a network which allows to mark the limits which have to be observed.

This starts e.g. with the definition of the defect - in a purely legal sense - which differs from such definitions in technical standards. ISO has recognized this situation and gives general definitions in ISO 8402 which are quite helpful and conform to the legal ones (picture 1 and 2). The complicated relations between legal definiti-

Nonconformity

The non-fulfilment of a specified requirement.

Notes:

1. In some situations **specified requirements coincide** with **customer usage requirements**

   In other situations they may not coincide, being either more or less stringent, or the exact relationship between the two is not fully known or understood.

2. Nonconformities will generally be classified according to their degree of seriousness. The number of classes and the assignement to a class should be appropriate to the quality requirements of the specific situation. Usually class A will include those nonconformities for the highest concern, requiring the most stringent acceptance criteria.

Nonconformity

Definition ISO 3534-2, 1993 (1.5.6)

Defect

The non-fulfilment of an intended usage requirement

Notes

1. The term "defect" is appropriate for use when a **quality characteristic of a product, process or service is evaluated in terms of usage** (as contrasted to conformance to specifications).

2. Since the term "defect" now has definite meaning within the law, it should **not be used as general term.**

Defect

Definition ISO 3534-2 1993 (1.5.8)

ons and technical situations are correctly explained by ISO/DIS 3534-2 (1.5.8) see picture 3.

Because of the illegitimity of the stipulation of exemption from liability as set by the EEC directive on product liability such contracts should try to apply the extreme accuracy in defining and marking the limits of liability of all parties involved in strict conformance with law and legal decisions [3].

Another example: if an audit is undertaken by a purchaser, he cannot stipulate any exemption from liability for facts,processes and installations which the auditor has seen and evaluated at the supplier's plant and premises. By legal evaluation everything any auditor as a member of the workforce of a client has seen is regarded as known to all experts of that auditing company and accepted as in con formance with the state of the art [4]. The company itself has to establish an internal information system so that all such informations are distributed within its organization to everybody qualified to evaluate them and not only be collected in the files of the quality assurance department. In many cases such knowledge has exempted suppliers from liability because the defect causing process and its conditions were fully known to the buying company, not criticized, therefore regarded by them as technically sound, the thereupon arising defects being regarded as outside supplier's responsibility. Defects and damages arising from such processes are alone to the responsibility of the client, no quality assurance contract can alter this even if this is the silent intension of many quality assurance contracts.

Such risks of audits are rarely known and evaluated by most companies believing in the value of their own auditing for special products or processes.

5. Product liability and quality assurance contracts

With great vigour some authors of quality assurance contracts emphazise the importance of quality assurance contracts for product liability cases. This actual importance has been stated as a fact in many articles all over Europe and in even more oral presentations. But no proof ha been published so far with verifiable details.

To investigate the real practical proof in damage cases of quality assurance contracts a survey was carried out by one of the leading German industrial insurance companies within their decentralized organization and in cooperation with a leading independant legal attourney with some depandances in different cities.

The surprising result:

Neither in negociations of product liability cases of this insurance company any of the different parties involved nor in any legal process at courts of appeal any such quality assurance contract had ever been presented as a legal valid base for claims or for the rejection or transformation of responsibilities. Not one company presented its quality assurance contract to the court which then would be obliged to check at first their legal validity before they could take them into consideration. This happended even in cases where such quality assurance contracts did exist, but were not presented by the issuing party.

In ISO/DIS 3534/2 defect (1.5.8) is defined as

"The nonfulfillment of an intentive usage requirement"

with the following notes:

(1) "The term "defect" is appropriate for use when a quality characteristic of a product, process or service is evaluated in terms of usage (as contrasted to conformance to specifications)".

(2) "Since the term "defect" now has definite meaning within the law, it

should not be used as a generell term".

Nonconformity should not be equated with defect.

To classify nonconformance as a defect, specific knowledge of magnitude of the variation from specification limits and of the specific consequences are requiered. The reality is that many nonconformances are not defects, which is particularly important for the legal considerations. This is consistent with the definition for various classifications of defects in ISO 3534.

On the other hand a product conforming to specifications could be defective."

Nonconformance - Defect

- ISO/CD 10 205 Dez. 1991

The consequences:
The practical test of validity and importance of quality assurance contracts within product liability cases showed the unique result, that all the outspoken and claimed intentions formulated in quality assurance contracts were not dared to put to the legal test presenting them to any court as independent institution only qualified to check their validity against law and jurisdiction. This can only be interpreted that the originators and users of such quality assurance contracts are aware of the unlegitimited content and that they did not dare to present them to a qualified court whose decisions would confirme this fact in public and therewith destroy this tool for further use.

6. Legal and organizatorial better solutions

This total desaster of the legal validity of quality assurance contracts in critical situations as consequence of legally unvalid clauses neglecting legal responsibility and legal fundaments - or by willingly neglecting legally binding principles and High Court decisions for obvious reasons. Legal conditions between manufacturer and supplier are governed by different legislation. They differ sometimes to a varying extent just which law is applied. Therefore general schemes cannot be drawn up to demonstrate correctly the connections for all states under all legally possible conditions.

A comparison between the legally set obligations of companies according to German High Court decisions and the content of ISO 9004 and 9001 is given in table 1. Any such comparison upon the base of other national legislation in comparison to the clauses of the ISO standards would be very helpful but is not known to be published so far.

The necessary consequence of such a comparison is quite obvious, certificates of third parties stating only the conformance of company's organizatorial setup and processes according to the minimum clauses of the incomplete international standard ISO 9001 have no any importance or legal relevance for a resonable selection of suppliers and fulfilling the necessary obligations for an effective integration of supplier's knowledges into manufacturer's organization. Picture 4 shows the relation between technology and legal principles with the decisive connecting link to proove specific explicive reference to actual conditions of use. All procedures and processes which can not proove this technically reliable link are irrelevant for any legal evaluation. This old prooven fact caused the commission of the European Community to state very simple and understandable: [5]

> Test programs, supervision and **certification** may decrease risks and therewith the probability of damages, but have **no effect upon the liability of a producer"**.

7. Conclusions

General evaluation:
No quality assurance system certificate releases any company from the strict duty to select its suppliers with the outmost caution and tho-

	Criteria	ISO 9004 Section-Nr.	ISO 9001 Section-Nr.	Importance High Court Decisions
1.	Product Safety/ Liability	19	- -	* * *
2.	Responsibility for Organization	(3.3)	- -	* * *
3.	Observation of Product Behaviour	(7.3;16.3)	- -	* * *
4.	Responsiblity for Construction	- -	- -	* * *
5.	Informationssystem external/internal	- -	- -	* * *
6.	Evaluation of State of The Art, Generally Accepted technical Rules	- -	- -	* * *
7.	Responsibility for User Instructions	(16.2.3)	- -	* * *
8.	Risk Analysis	(0.4.2;8.4)	- -	* * *
9.	Controll of Completeness of Technical Specifications	- -	- -	* * *
10.	Separation of Development from Construction	- -	- -	* * *
11.	External Damage Analysis	- -	- -	* * *
12.	Decisions about production Processes	(10.2)	(4.9.6)	* * *
13.	Purchasing	(9.7)	(4.6.1)	* * *
		(12.1)	(4.10.1.2)	* * *
14.	Sales/Marketing	(7)	(4.3)	* * *
15.	Qualification of Personel	(18.1.2)	- -	* * *
16.	Economically Justifiable Costs	- -	- -	* * *
17.	Total cost of Quality	(0.4.3.2) (6)		* *
18.	Classification of Characteristics/Defects	- -	- -	* *

19.	Reliability of Test Procedures to Conditions of Use	(8.3)	- -	* *
20.	Documentation of Reliable Quality Figures	(17)	(4.16)	* *
21.	Contract Control on Completeness for Conditions of Use	(7.1c) (7.2)	(4.3) incomplete	* *
22.	Limit of Validity of Quality Elements	(8.3)	- -	* *
23.	^Notice of Incompleteness of Quality Elements	(1) (5.1.1)	(4.2)	*
24.	Supervision of Test Procedures at Subsuppliers	(13.3)	- -	*
25.	Confidence/Secret	- -	- -	- -

* * * = extremely important, * * = Important, * = necessary

( ) = Section number which needs addition and/or changes and/or specialization to conform with legal requirements

- - Section missing to transform legal requirements into company organization

Legal Requirements compared with the Content of ISO Standards

Legal Principles

Requirements

- Carefulness
- Unsubstanciated legal concepts for technical products and processes

Criteria for

- Case sumption
- Complete evaluation of actual process
- Proportion of responsibility of all parties involved

Final judgement for all parties involved

Technology

- Reliable functions
- Reproducible processes
- Predictability
- Precaution
- Organization
- Steering
- Reliability
- Safety
- Test-procedures
  *Part test under actual loads
  * Machine fatigue test under actual conditions of use

Product specific referred to actual conditions of use

Literature: Haftpflichtverband der Deutschen Industrie V.a.G.
Qualitätssicherungssysteme, Normen und Zertifikate
Quality Assurance Systems, Standards and Certificates (H III-10/89)

Correlation between Technical Reliability and Legal Importance

roughness as to possible damages which might be caused if the special products he is intended to supply may fail under foreseeable conditions of use. This obligation has to be performed always regarding the individual product and its intended forseeable usage requirements - no formal procedure or even standard can sufficiently realise it. This needs to apply the full state of the art to design, manufacture, test and assembling individually. The attempt to regulate this obligation bureaucratic only by paper work and formal checking is contradictional to law and High Court decisions and will not release the companies from their binding obligations. [6]

A lot of money is mislet in many companies not only all over Europe because they believe in legal effects of such much too simple methods which by their construction cannot fulfill the propagated aims. These procedures only help to install a bureaucratic system with a lot of unnecessary paperwork and expenditures for third party organizations and councellors to earn money which definitively will give no dividends in a foreseeable period. [7]

Each action as well as test procedure has to be technically explicit evaluable for the foreseen usage requirements to get any legal value. Mandatory scale will always be the full state of the art, not only formally documentated procedures or figures.

The consequences:
Only strict and thorough engineering work and sufficient organizatorial knowledges regarding national legally binding conditions can fulfill company obligations. Anything else has to be regarded as secterial missionary work which much too often wastes too much time and money which can better be used for upgrading solid engineering work.

REFERENCES

[1] Ford Motor Company Q 101 Quality system recommandation, Cologne 5/95

[2] Bauer, C.O. Gefahren und Schwierigkeiten für Qualitätssicherungs-Vereinbarungen (Dangers and Problems with Quality Assurance contracts) Giesserei 81 (1994) Nr. 13.

[3] EEC Directive of 25. July 1985 on the approximation of the laws, regulations and administration provisions of the member States concening liability for defective products. Official Journal of the European Communities Nr. L 210/7.8.85

[4] Official Journal of the European Communities C 267/12, 19.10.1989

[5] Bauer, C.O. Der Lieferantenaudit, Ziele, Methoden, Aussagefähigkeit (Supplier audit - goals, methods, reliability) Giesserei 80 /1993) Nr. 20, S. 687-692)

[6] Westphalen, F. von, Bauer, C.O. Just-in-Time-Lieferverträge, Qualitätssicherungsvereinbarungen, (Just-in-time contracts, Quality assurance contracts) RWS-Script 264, Köln 1993

[7] Sprenger, R. Der große Bluff (The great bluff), Manager Magazin August 1995 S. 128/131

12 CAN PATENTING PREVENT FAILURE?

P. REVY von BELVARD
Wil, Switzerland

Abstract

This contribution unveils how the present patent system encourages and forces improvements even when financial interests may speak against this because failure of machine parts due to wear assists in selling the parts. Moreover, it is shown that the approaches to solve similar problems may be quite different.

Key Words: Patent, mill, pelletizing, grinding, die, life time, service.

1 Introduction

When in the late 15th century the venerable City Elders of Venice decided to introduce the first modern Patent Act, the intention was to motivate inventors into publishing their ideas in order to advance technical progress.

The concept worked - without doubts - and still works in many cases. In other cases, however, there is a counter-motive: Do not alter things too gravely. Such a case occurs with parts which are subjected to conditions which will lead to deterioration of the product. As is well known, products have a certain life span and progressively become worn out and need to be replaced. In such a case, two human factors combine and dominate the scene:

- the knowledge that *all earthly things are perishable*, a fact which is perfectly natural and corresponds to our daily experience in life, and
- the consideration that any *prolongation of the life span* of a product would also reduce the sales figures of such parts.

Structural Failure, Product Liability and Technical Insurance, Proceedings of Fifth International Conference. Edited by H.P. Rossmanith. Published in 1996 by Taylor & Francis, 2 Park Square, Milton Park, Abingdon, Oxon, OX14 4RN. ISBN 0 419 22080 1

2 Pellet Mill

A typical example for this competition of arguments is given by the pellet mill which (Figure 1) which is used to produce food pellets for pets and other animals. The function of the pellet mill is similar to a meat grinder. Food to be pelletized passes through a metering unit "m" into a mixer "M" beyond which there is an entrance channel "EC" and a gate "d" through which the mixed ingredients are fed into a perforated die for pelletization.

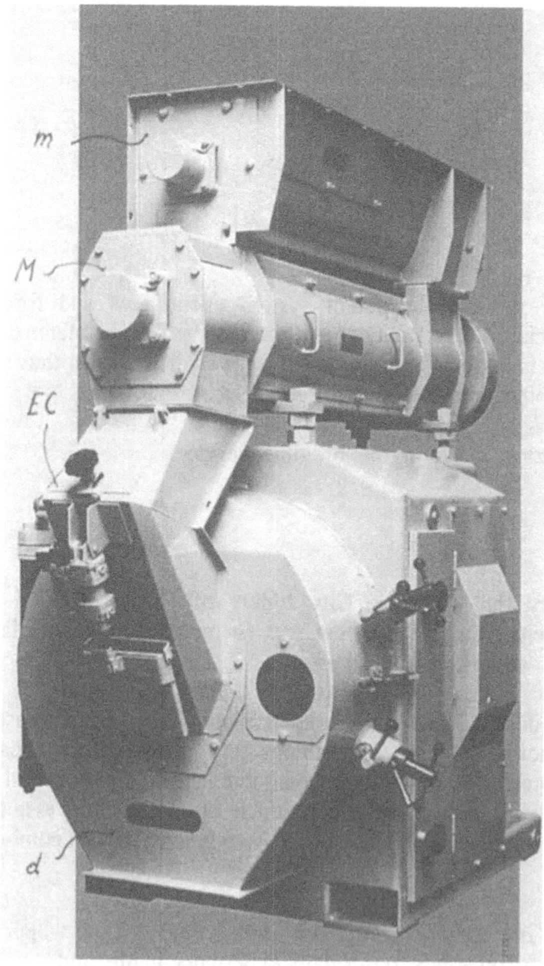

Figure 1 Pellet mill (Photo: Bühler AG, Uzwil, Switzerland)

As shown in Figure 2 in contrast to the household meat grinder (where the meat is axially pressed through the machine) here the raw material (food "F") is radially pressed out through the die "a" by means of press rolls "b". Blade edges "e" shear off the pellets "g" for further collection, processing and packaging. Of course, there is no meat the fat content of which would lubricate the die. On the contrary, most of the ingredients are subjected to abrasion. It is, therefore, not surprising that the dies,

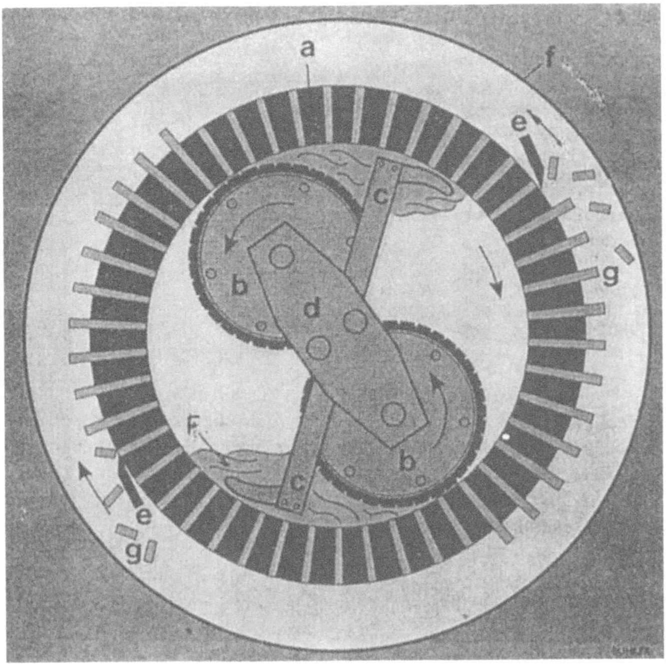

Figure 2 Schematic of a rotary pellet mill

after a certain service life time, show signs of serious wear similar to the ones represented in Figures 3a and 3b. The dies as well as the press rollers will be worn out during the pelletization process. In many cases the dies may even break as shown in Figure 4.

In addition, there are other factors which promote wear of machine parts. When the dies rotate due to the movement of the machine, the press rollers squeeze the raw material into a small gap thus exerting pressure on the die. Although the die is perfectly cylindrical, it will attain an oval shape with a longer diameter which continuously changes its position during the rotation of the die. This means that strains and stresses in the die also change continually and this situation will eventually result in failure of the part. With fractures certainly to occur after about 10,000 hours of service, manufacturers of dies of this kind would do good business at present conditions and terms. This "business and design" situation prevailed until an inventor felt the need for improvement and filed a patent for an improved die shown schematically in an exploded view in Figure 6. The improved was designed with an inner lining of wear-resistant material. The lining consists of several segments "13" which are mounted in *exact alignment* with the perforations "11" of the outer ring "10". The crux of this particular design was lying in the alignment. Because of the enormous pressure acting on the ring the segments would never align exactly. This results in difficulties in maintaining sanitary conditions because there was always some

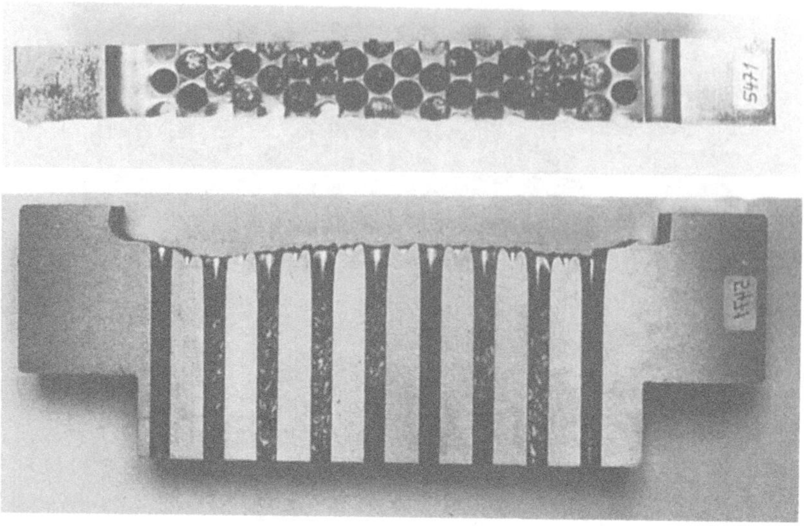

Figure 3 Wear damaged dies : a) top view , b) cross-sectional view

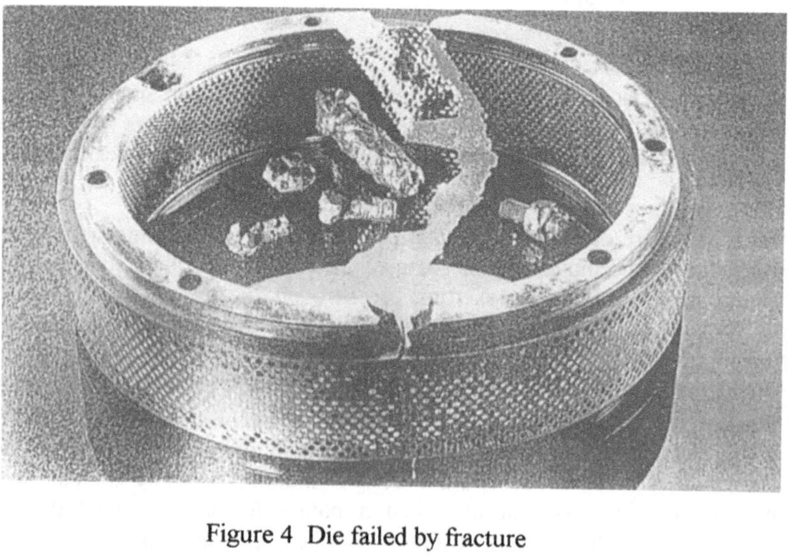

Figure 4 Die failed by fracture

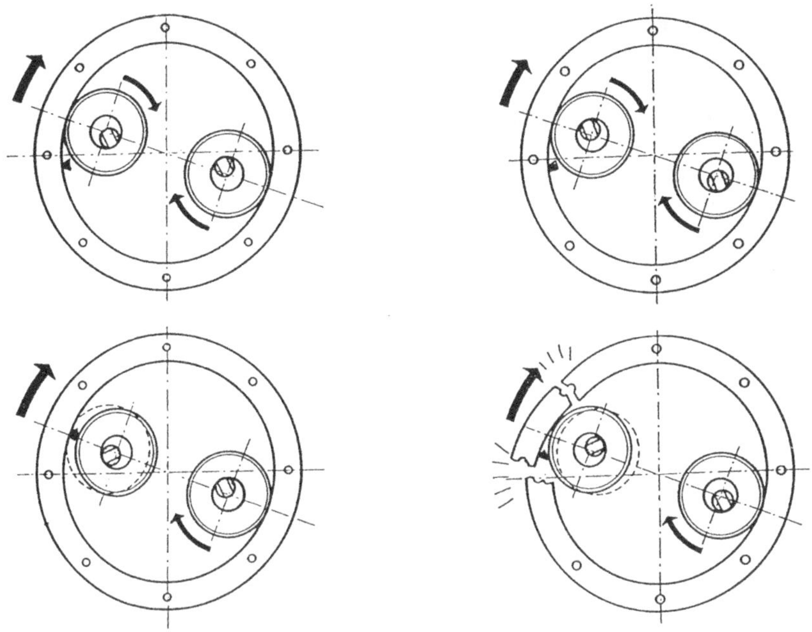

Figure 5 Schematic showing the process of fatigue failure of a die

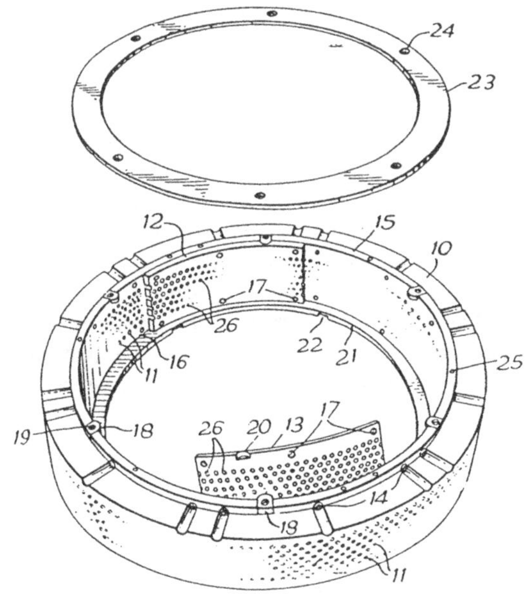

Figure 6 Die with an inner lining of wear-resistant material [1]

food material trapped in the slots and gaps between the lining and the outer ring and exposed to mould and bacteria.

3 The sectorial die and the alignment problem

The first attempt at improvement of the die was quickly followed by another inventor's creation. "If alignment of the perforations is such a problem", the second inventor thought, "then the solution might be to design the entire die to consist of individual sectors". This can be seen from still another British invention, British Patent No 992,352, given in Figure 7.

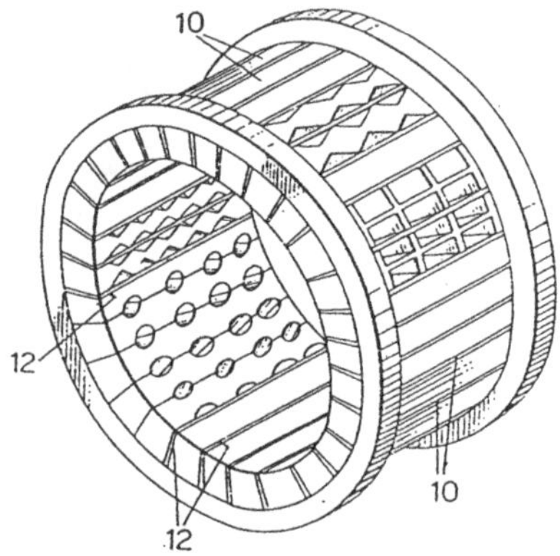

Figure 7 Die consisting of a set of individual sectors [2]

Now there was no longer any need to align perforations, but of course, the fact remained that there were slots and gaps between the individual sectors and with them the sanitary problem did not disappear. Therefore, this idea was also abandoned and all manufacturers who saw their business already gravely disturbed thanked god and sighs of relief could not go unnoticed. Some time later another inventor finally concluded: "If the improvement of the design of those parts of the die which wear out so fast is such a problem then why not improve the outer surface of the press rollers which are similarly subjected to wear and abrasion. Figure 8 shows the design according to European Patent No 4,992 where a press roller is equipped with a wear resistant outer lining "1" and a tough core "2". One can see, that the inventor returned to the initial idea of lining and he put an outer lining of wear-resistant material around an inner core of softer metal.

This was a suggestion made by a professional who not only knew what he was talking about but also caused a great deal of disturbance on the market as customers were very interested in the longer life time of the new design. This caused the competitors to tackle the matter from another viewpoint. The most expensive part which will be worn out during service is the die and therefore, a solution regarding the die had to be found. In order to avoid the alignment problem a composite die consisting of three axially successive parts as shown in Figure 9 taken from the US-Patent No. 4,496,298 was suggested. With this design the problem of wear on the inner surface of the die

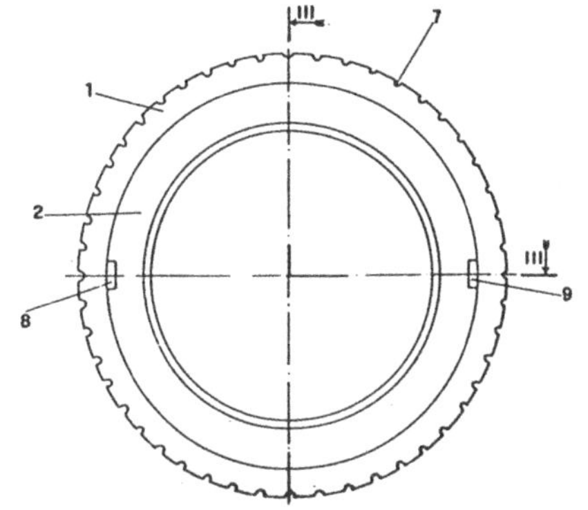

Figure 8 Press roller with a wear resistant outer lining and tough core [3]

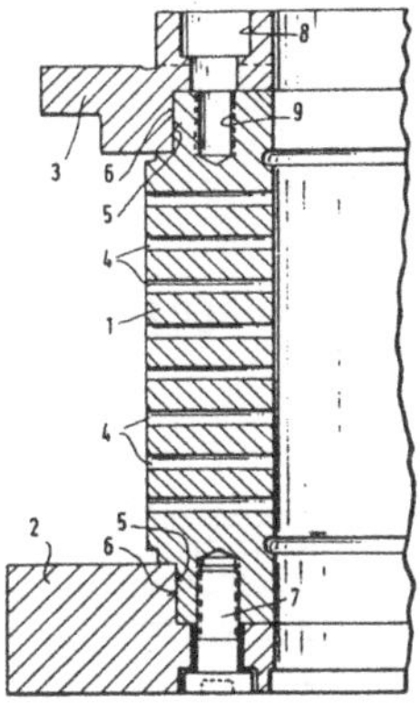

Figure 9 Schematic of a composite die [4]

still remained. By now all manufacturers were on alert. There was a great pressure and urgency generated by the first successful attempts to produce a significant improvement so as not to fall into second place.

After enormous research activity all difficulties with composite dies and their gaps and slots have been resolved by finding a new alloy which will extend the life time of the product by a factor of 6. This alloy is harder than the alloys employed previously. The structural component as being in use today is shown in Figure 10.

4 Drum mill

However, having failure prevention in mind, it may seem straight forward to look for a harder material. However, there are situations where the selection of a harder material leads to a contrary result. This situation typically occurred in the problem of reducing wear of drum mills. In a drum mill or ball mill there are milling media in the

Figure 10 Die design based on new alloy

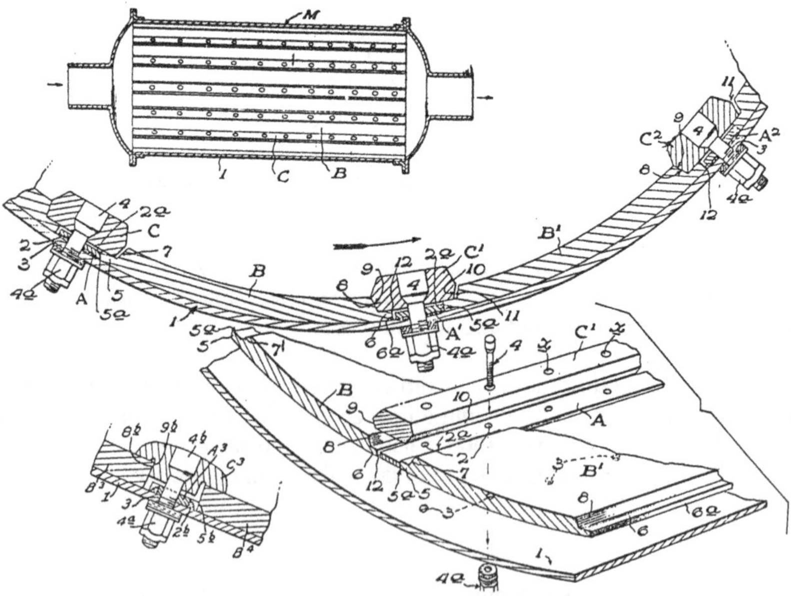

Figure 11 Schematic of a drum mill [5]

form of balls and the material to be ground is filled into the drum and agitated by rotating the drum. The balls are lifted by the motion of the drum, fall back into the material and grind this to a fairly find level. But as the balls cannot differentiate between the material and the drum, the latter will also be ground and is therefore subjected to wear and tear.

As stated earlier it is relatively easy to look for a harder and more wear-resistant drum material and this was the situation for some time. US Patent No 2,611,546 shown in Figure 11 shows a drum lined with a special wear-resistant material which can easily be replaced when worn out.

A typical material which is processed in drum mills is municipal and household waste. An inventor thought that if the inner surface of the drum is unnecessarily ground while the refuse should, in fat, be comminuted, then why not make the lining from waste material? This idea led to the German Patent Spec. No. 1,297,120 shown in Figure 12 where the ribs "10" suitably arranged around the drum periphery retain the waste material "11" in-between them forming a soft and cheap lining for the drum "1".

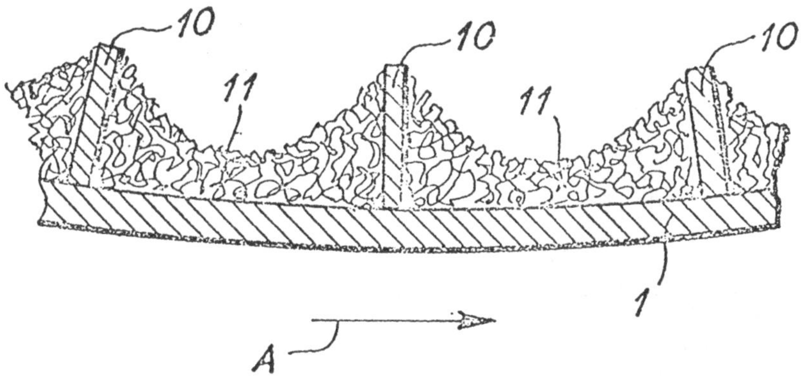

Figure 12 Rib system protective lining design [6]

It is, though, not always a question of hardness and wear-resistance, or, on the other hand, of softness of the material when wear and tear occurs which may lead to failure. While some constructive solutions can be found in an environment of privacy in the designer's office, in other cases one needs to know the practical performance and behavior of a machine.

5 Agitator mill

A typical case is an agitator mill which is a similar type of mill. An agitator mill is normally used to grind a dispersion of inks to such a fine level of size that the individual particles are no longer perceptible when used in service. Figure 13 shows the agitator mill according to German Spec. 1,949,221 with a bottom inlet orifice where the dispersion to be ground is fed in and an outlet "20". Sandwiched between is a milling chamber "1" with hard material to be ground "4" which is agitated by a

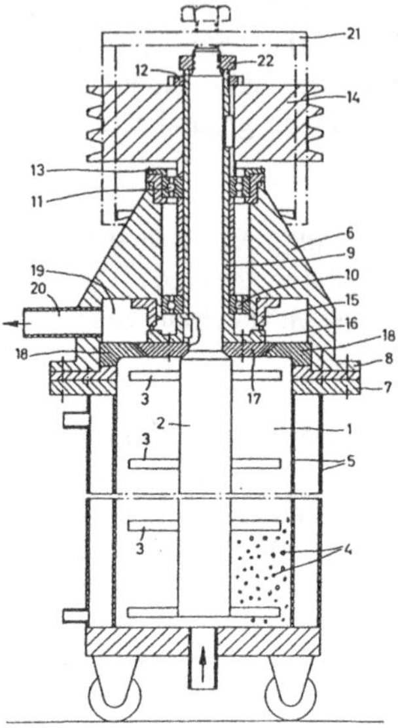

Figure 13 Schematic of an agitator mill [7]

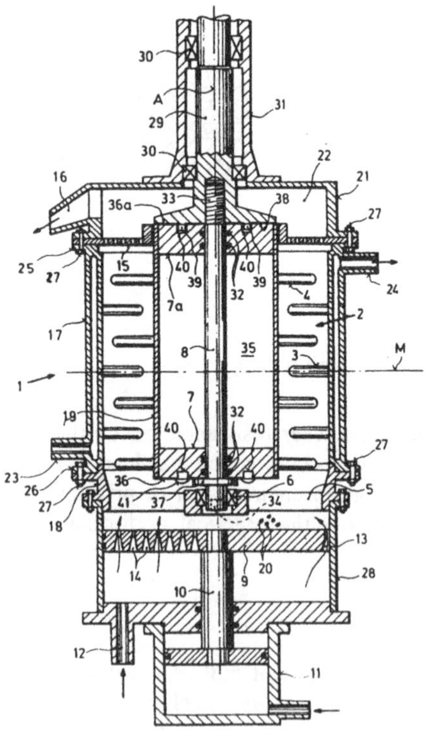

Figure 14 Schematic of an agitator mill with detachable rotor [8]

whisk-like rotor "2" moving the media "4" such that the push and rub of the particles against another produces the grinding and comminution of the particles of the ink dispersion to micron size.

Normally one would expect that the media to be ground to cover more or less the bottom of the milling chamber. But now imagine a stream of ink flowing upwards through a gap (between "17" and "18") which is narrower than the media "4" used and thus prevents the media from escaping. The higher the amount of liquid which flows through the milling chamber the larger the rate of flow and the larger the drag effect of the streaming ink.

When servicing these mills after a certain duration of operation the maintenance staff found that whereas the upper region of the rotor and its pins were worn out the lower region could still be used in service. Of course, in principle, there was no need to have anything altered, as it is tacitly accepted that there are components which will age by wear. But in this case the information came to the attention of the designer who was in charge of the agitator mills in a particular company manufacturing these mills. This designer's new design (German Spec. No 3,61,721) is shown in Figure 14. A rotor "2" is detachable from its shaft "29" by a long tightening screw "8" onto which the main body of the rotor "2" is fixed. The body is symmetrically shaped with respect to a horizontal center plane "M" such that the rotor "2" having suffered from wear at its upper end can be turned upside down and in this way be re-used.

6 Conclusions

This present situation supports the request made by Professor Reh at the ISTLI Conference in Wil in Switzerland in 1994 [9] for a closer liaison between maintenance and service experts and designers and experts in the R&D departments for an improved learning from failures which occur outside the company. This feedback to the designers and researchers will then enable them to direct their efforts to those issues most urgently pressing in practice.

References

[1] British Patent No 870 597
[2] British Patent No 992 352
[3] European Patent No 4 992
[4] US-Patent No 4 496 298
[5] US-Patent No 2 611 546
[6] German Patent No 1 297 120
[7] German Patent No 1 949 221
[8] German Patent No 3 614 721
[9] L.Reh: ISTLI Symposium " Repair and Liability" Wil, Switzerland, Nov.1994.

Case Studies

Structural Failure, Product Liability and Technical Insurance, Proceedings of Fifth International Conference. Edited by H.P. Rossmanith. Published in 1996 by Taylor & Francis, 2 Park Square, Milton Park, Abingdon, Oxon, OX14 4RN. ISBN 0 419 22080 1

13 MULTIPLE FATALITIES CAUSED BY LOSS OF A TRACTOR–TRAILER'S LOAD

A.A. JOHNSON
University of Louisville, Louisville, Kentucky, USA

Abstract
A flatbed tractor-trailer carrying fifteen junked and flattened cars was traveling through downtown Louisville, Kentucky, on an interstate highway when the driver lost control of the vehicle. It grazed a concrete median and then ran into it. The fifteen cars, which had been in two stacks each retained by three steel wire ropes, slid off the flatbed. Twelve of them went over the median into oncoming traffic causing three fatalities and two serious injuries.

Reconstruction of the accident showed that two of the three ropes retaining the forward stack of cars broke long before the vehicle reached the accident site. They failed by friction and wear because they were not tightened sufficiently. The vehicle entered a curve in the highway at about 60 m.p.h. (96.6 km/hr). This caused the load to shift and the right side of the trailer to leave the ground. The consequent loss of control caused the six cable tighteners on the left side of the trailer to grind against the concrete median generating a cloud of concrete dust. The heat generated softened the wire ropes sufficiently to cause tensile overload failures. Loss of the load followed immediately.
Keywords: Accident reconstruction, flatbed tractor-trailer, friction and wear, tensile overload, wire rope.

1 Introduction

On the morning of August 10th, 1989, Mr. Gerald St. Peter took possession of a 1981 International flatbed tractor-trailer at a restaurant known as The Red Rooster near the city of Bardstown, Kentucky. The trailer was loaded with fifteen junked and flattened cars in two stacks each of which was secured with three wire ropes.

Structural Failure, Product Liability and Technical Insurance, Proceedings of Fifth International Conference. Edited by H.P. Rossmanith. Published in 1996 by Taylor & Francis, 2 Park Square, Milton Park, Abingdon, Oxon, OX14 4RN. ISBN 0 419 22080 1

Fig. 1. An aerial photograph of the accident taken shortly after the accident (Photograph courtesy of The Courier-Journal, Louisville)

He drove the vehicle towards the city of Louisville and eventually entered the northbound lanes of Interstate 65 which is the main north-south artery through Louisville. In downtown Louisville he entered a curve to the right. The vehicle went out of control, grazed the concrete median, and then collided with it. The load came off the flatbed and twelve of the fifteen cars went over the concrete median into the oncoming traffic.

One of the junked cars which went over the median separated from the rest and landed on a Nissan Sentra. The two young women in it were killed. The other eleven junked cars which went over the median stayed together. They collided with a 1984 Chevrolet Cavalier and a 1984 Chevrolet C-20 truck with a small trailer. A nurse who was driving the Cavalier was killed and the two occupants of the truck were both seriously injured. Two other vehicles hit by the junked cars were damaged but their occupants were not injured.

The Louisville Division of Police thoroughly photographed the accident scene; documented and mapped tire marks, scrapes and gouges; took witnesses' statements; and impounded the accident vehicles. The Office of the Jefferson County Commonwealth Attorney's Office asked the writer to use this material to reconstruct the accident.

2 The configuration after the accident

Figure (1) is an aerial photograph taken shortly after the accident by a staff photographer of the Courier-Journal, the leading newspaper in Louisville. The flatbed tractor-trailer may be seen at "A". Two of the junked cars are sandwiched between it and the concrete median barrier. Another junked car lies in the roadway a few yards behind it. The eleven junked cars which stayed together are shown at "B". They have been sprayed with a white foam by the fire truck which may be seen at the top right of the photograph. The car in which the nurse was killed is shown at "C", and the truck whose two occupants were injured at "D". To the right of the photograph at "E" is the Nissan Sentra in which the two young women were killed.

3 Characteristics of the accident site

The road surface at the accident site was of concrete and, at the time of the accident, it was dry and visibility was excellent. The radius of curvature at the accident site was 863 ft (263 meters) and the superelevation 6%. The curve as a whole had a radius of 955 ft (291 meters).

Using a drag coefficient of 0.65 to 0.85 the speed for side slip comes out to be about 94 to 107 m.p.h. (151 to 172 km/hr). The lateral acceleration for a speed of 60 m.p.h. (96.6 km/hr) comes out to be 0.21 g. The speed limit at the accident site was 55 m.p.h.(88.5 km/hr).

4 Witness statements

Statements were taken from fourteen witnesses and some of these were useful in developing the reconstruction. Several witnesses gave estimates of Mr. St. Peter's speed as he approached the accident site. These, together with the knowledge that his vehicle had a governor set at 65 m.p.h. (104.6 km/hr), suggested that he was traveling at a speed in the range 60 to 65 m.p.h. (96.6 to 104.6 km/hr). One witness was driving alongside Mr. St. Peter's vehicle when the accident occurred and he stated that both vehicles were traveling at about 60 m.p.h. Accepting this value we can conclude that the accident was not triggered by side-slip.

Three witnesses stated that they saw the tractor-trailer's load shift as it was approaching the accident site. One truck driver reported that at least one wire rope was loose and flailing about dangerously. Two witnesses driving close to the tractor-trailer just as the accident occurred saw it tilt. One of them reported seeing the bed of the trailer tilt to the left twice. The other reported seeing the rear wheels of the tractor lift 6 to 8 inches (15-20 cm) off the ground. Mr. St. Peter himself reported that he felt the load shift at the time of the accident.

Two witnesses saw the accident from a building overlooking the accident site. They reported seeing the right side of the trailer lift off the ground. The trailer then rode on the concrete median and the wire ropes broke. After most of the junked cars slid off the flatbed and across the median the vehicle righted itself. The three remaining junked cars then slid off but did not go over the median.

5 Examination of the tractor-trailer

Examination of the tractor-trailer revealed that there had been six wire ropes securing the load. Three had secured a stack of junked cars at the front end of the flatbed and the other three a stack at the rear. Since there were fifteen junked cars it was assumed that there were eight in one stack and seven in the other. Each rope was secured to a short chain anchored to the right side of the flatbed. On the left edge of the flatbed were six tighteners which protruded from the side of the vehicle. It was immediately obvious that these had ground against the concrete median barrier when the vehicle went out of control.

For convenience let us number the ropes "Rope #1", "Rope #2", etc., starting at the front of the vehicle. Figure (2) shows the tightener for Rope #5. The outer edges of the two plates perpendicular to the side of the vehicle in which the tightener is mounted were heavily abraded. Indeed, it became apparent, by comparing this tightener with others near the front which were less heavily abraded, that about 0.5 inch (1.27 cm) of the plates had been ground away. The degree of abrasion varied from one tightener to another but, in general, those towards the end of the vehicle were most heavily abraded. It was noted that the tightener for Rope #6 had suffered damage in a prior incident and was not functional. Ropes #1 and #2 had failed high up in the front stack. Ropes #3, #4, #5, and #6 had failed at their tighteners. Apart from the failure of the ropes and the abrasion damages to their tighteners the tractor-trailer was only superficially damaged.

Wire samples, each including a fracture surface, were cut from the core of each

Fig. 2. One of the wire rope tighteners on the flatbed.

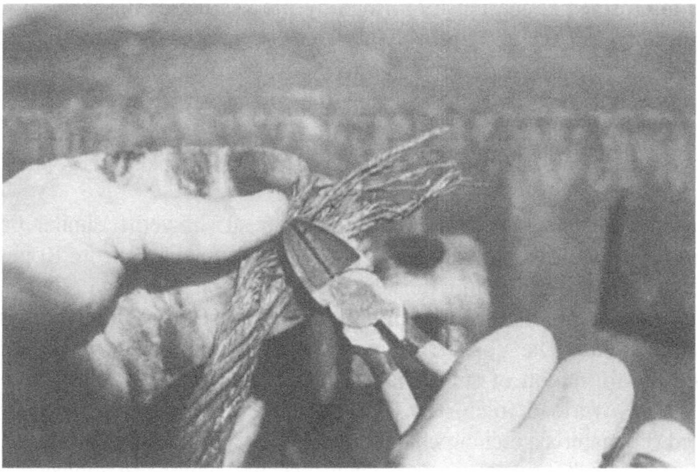

Fig. 3. The cutting of wire samples from one of the broken wire ropes.

rope. The cutting of a specimen from Rope #6 is shown in Fig. 3. Each wire specimen was placed in a separate carefully labeled plastic container for future laboratory study.

6 Laboratory work on the wire ropes

Fracture surfaces of wires from each rope were examined in a scanning electron microscope equipped for semi-quantitative energy dispersive analysis by X-rays (EDAX). Samples were also prepared for metallographic examination using standard techniques. The etchant used was 2% nital.

Figure (4) shows a fracture surface from a wire in Rope #1 at a magnification of about X98. It is not a readily identifiable type of fracture surface. An EDAX trace from the center of the fracture surface showed the presence of about 65% iron, 26% aluminum, 8% silicon and smaller amounts of zinc and calcium. An EDAX trace from the surface of the wire a short distance from the fracture gave 96% aluminum and 4% silicon. Thus, the iron wires were coated with an aluminum-silicon alloy. This is apparently a common method of corrosion protection. The silicon suppresses the formation of brittle intermetallic compounds at the iron-aluminum interface [1]. The results for a sample from Rope #2 were somewhat similar.

Figure (5) shows a metallographic section through the fracture of a wire from Rope #2. The white areas contiguous to the fracture surface, and on the lower edge of the section, are interpreted as martensite formed by the heating of the steel to above its A_1, temperature (~723°C), and probably above its A_3 temperature, followed by rapid cooling. The wire was evidently subjected to vigorous friction and wear which raised its surface temperature sufficiently to cause these microstructural changes. A wire from Rope #1 showed similar effects. Evidently Ropes #1 and #2 were not tightened and this gave rise to the relative motion of the ropes and junked cars which caused the friction and wear.

Wire specimens from Ropes #3, #4, #5 and #6 all showed a similar pattern of behavior but it differed substantially from that observed for wires from Ropes #1 and #2. Figure (6) shows the fracture surface of a wire from Rope #5. It has necked down almost to a point. Figure (7) shows a metallographic section through a wire from the same rope. The upper surface shows gross abrasion which has caused shearing and the formation of sub-surface voids. The necking is characteristic of a high temperature overload fracture. There are ill-defined microstructional changes which altered the materials etching characteristics near the fracture surface. Thus, we conclude that Ropes #3, #4, #5 and #6 failed because they were subjected to gross abrasion and heating. The heating softened the steel sufficiently to cause failure by overload.

The unaltered microstructure of the wires is characteristic of a heavily cold-drawn steel. Microhardness measurements were made on one polished and etched specimen from each rope. Three measurements were made on each wire using a Buehler Micromet instrument with a 200 gm load. When the results were converted to Rockwell hardesses they came out to be in the range R_c 46 to 51. This corresponds to a range of ultimate strengths from 221,000 to 264,000 p.s.i. (155 to 185 kg/mm^2). Thus, the wire ropes were of adequate strength.

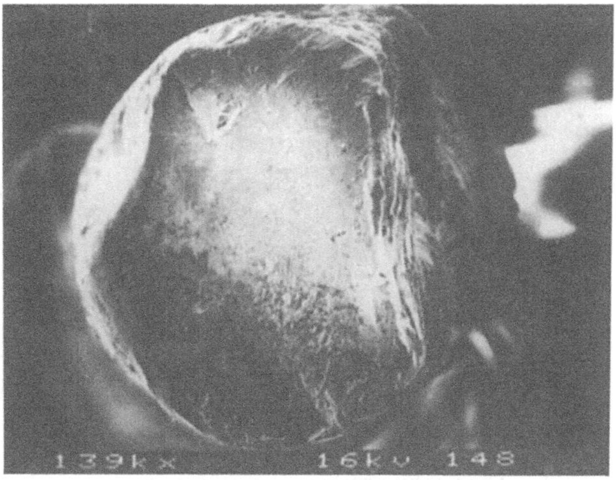

Fig. 4 The fracture surface of a wire from Rope #1 (x~98).

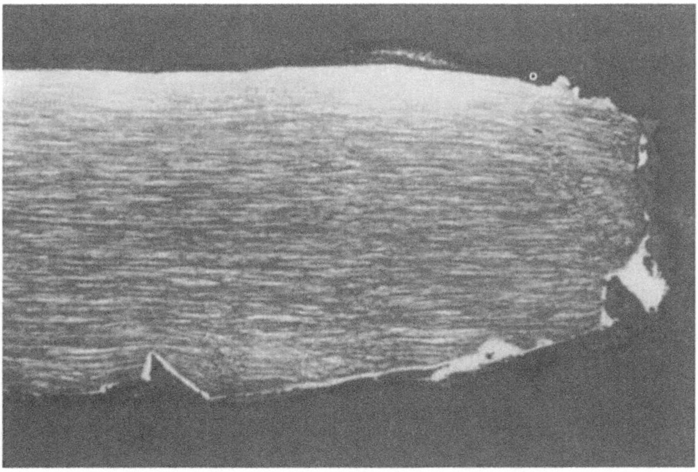

Fig. 5 A polished and etched section through a fracture surface from Rope #2(x~40).

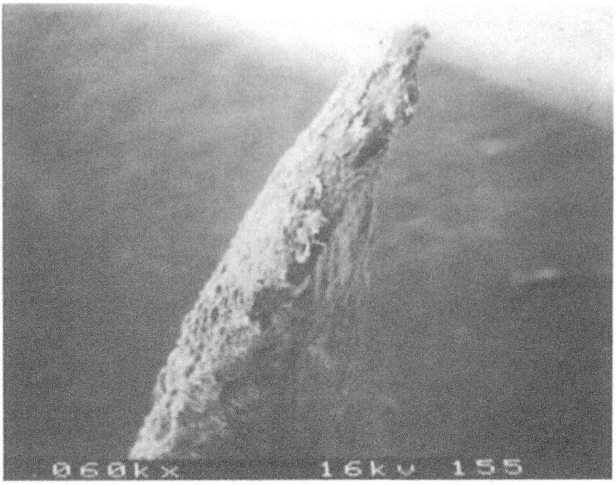

Fig. 6. The fracture surface of a wire from Rope #5 (x~43).

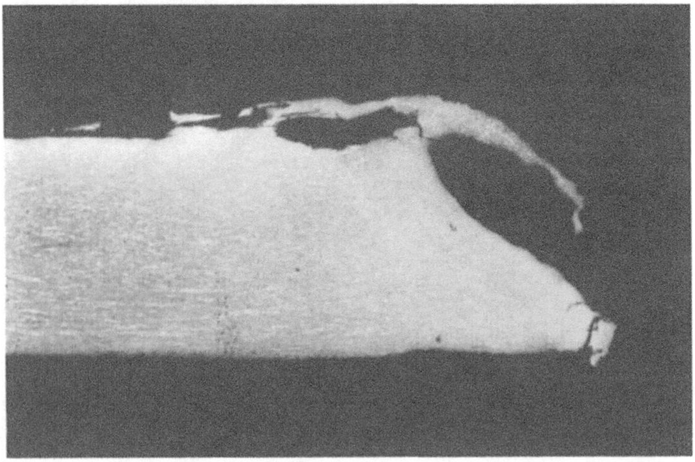

Fig. 7. A polished and etched section through a fracture surface from Rope #5 (x~32).

Aluminum-silicon alloy was found on all of the fracture surfaces examined. Aluminum and silicon form a eutectic which melts at 580°C. We can therefore conclude that, in the vicinity of the fractures, all of the ropes reached at least this temperature.

7 The reconstruction

In a paper of this length it is not possible to describe all of the details of the evidence which went into the reconstruction. For example, the tire marks at the accident site and the damage to the concrete median have not been described. Neither is there space to discuss all of the technical literature consulted. Of particular importance, however, is a study of the rollover of loaded trucks carried out at the Transport and Road Research Laboratory in England [2]. This work, which used nine different loaded articulated vehicles, showed that rollover occurred at lateral accelerations of from 0.20 to 0.32 g. Thus, Mr. St. Peter's vehicle, which we showed above to have experienced a lateral acceleration of 0.21g, was, at the very least, close to rollover even without a load shift.

All of the details which went into the reconstruction have been presented in a 62 page report [3]. The reconstruction which emerged may be summarized as follows:

(i) On the morning of the accident Mr. St. Peter went about securing the load on his truck. The load consisted of two stacks of junked cars with seven or eight cars in each stack.

(ii) On the left-hand side of the vehicle were six wire rope tighteners. The steel rope from each one connected to a chain which was anchored on the right-hand side of the vehicle. The rearmost rope tightener was inoperable because it had been damaged previously.

(iii) Mr. St. Peter secured the ropes on the rear stack of junked cars but was able to tighten only the two with operable strand pullers. The rearmost rope could only be tightened to the extent that adjustment of its chain made possible.

(iv) The forward stack of junked cars was secured by the third rope, but Mr. St. Peter omitted to tighten the two forward ones. Thus, the forward stack was inadequately secured at the outset of the journey.

(v) On the journey from the Red Rooster to Louisville the bouncing of the forward stack of junked cars caused such intense wear of the two forward steel ropes that they failed. Thus, approaching Louisville, the forward stack was secured by only one rope.

(vi) As the vehicle entered the curve where the accident occurred the centrifugal force acting on the forward stack caused it to shift to the left. It probably also rotated because it was restrained only by the third strand.

(vii) Application of the brakes by Mr. St. Peter caused the vehicle to exhibit the phenomena known as "trailer swing". The trailer swung outwards until it brushed against the concrete barrier. It bounced off the barrier several times.

(viii) The shift in the center of gravity of the forward stack reduced the lateral acceleration needed for rollover and, as a result, the right-hand trailer wheels and tractor drive wheels left the ground.

(ix) As the movement of the trailer to the left caused the trailer to brush against the concrete barrier, the two steel mounts of each of the last four strand pullers suffered abrasion by the concrete and substantial heat generation occurred. The rope in the vicinity of each tightener was heated sufficiently to melt its aluminum-silicon alloy coating and soften the steel.

(x) The front of the tractor hit the concrete barrier, and the impact triggered the failure of the four remaining heat softened strands. With no restraints, the two stacks of junked cars slid off the flatbed into the oncoming southbound traffic.

8 Prosecution of the driver

Mr. St. Peter was taken before the Jefferson County, Kentucky, Grand Jury. He was indicted on the following charges:

(i) 3 counts of Manslaughter II	(Class C felony, 5 to 10 years)
(ii) 2 counts of Assault II	(Class C felony, 5 to 10 years)
(iii) 2 counts of Wanton Endangerment I	(Class D felony, 1 to 5 years)
(iv) 5 counts of Criminal Mischief I	(Class D felony, 1 to 5 years)
(v) 1 count of Reckless Driving	(Fine of $20 - $100)

The case was set for trial by jury in Jefferson Circuit Court in Louisville. During the trial a plea bargaining agreement was reached and Mr. St. Peter pleaded guilty to the following charges and agreed to the sentences and fine shown:

(i) 3 counts of Reckless Homicide	(1 year each, consecutive)
(ii) 2 counts of Assault IV	(1 year each, concurrent)
(iii) 2 counts of Wanton Endangerment II	(1 year each, concurrent)
(iv) 5 counts of Criminal Mischief II	(1 year each, concurrent)
(v) 1 count of Reckless Driving	($100 fine)

The judge sentenced Mr. St. Peter to three years in prison and the $100.00 fine but suspended both. He was given five years of probation and was required to pay $500 to Mothers Against Drunk Driving at the rate of $25.00 per month.

9 References

1. ASM Metals Handbook (1982), Ninth Edition, Vol. 5: Surface Cleaning, Finishing,
   and Coating, Materials Park, Ohio.
2. Kemp, R.N., Chin, B.P., and Brook, G. (1978), "Articulated vehicle roll stability, methods of assessment and effects of vehicle characteristics, Laboratory Report 788, Transport and Road Research Laboratory, Crowthorne, England.
3. Johnson, A.A. (1990), An evaluation of the physical evidence in Commonwealth vs St. Peter, University of Louisville, Kentucky, U.S.A.

Acknowledgments

The author is indebted to Patricia Lumley for typing the manuscript and Barbara Johnson, for proofreading it.

14 FATIGUE IN TANK VEHICLES

T. SVENSSON and H. TORSTENSSON
Swedish National Testing and Research Institute, Borås, Sweden

Abstract

Tanks for the transport of hazardous materials are usually designed only for static overloads. As a basis for the development of design criteria regarding fatigue damage a proposed method for calculation is presented, based on performed field measurements. The method is founded on the assumption that only events such as road imperfections contribute to the fatigue damage. The results indicate that the fatigue damage can be quantified with the proposed method, but further investigations are needed to implement the method in design.

1 Introduction

An investigation of tank vehicles for dangerous goods was carried out in 1990 by the Swedish Plant Inspectorate. There was a surprisingly high percentage of tanks for the transport of gasoline that were damaged. In particular in tanks of aluminium fatigue cracks were found in 46 % of the tanks. Often such tanks are built on longitudinal chassis beams, and the cracks usually originate at the weld between tank shell and beam.

The present design rules for tanks for dangerous goods (ADR, 1995) do not address the fatigue problem. Additional design criteria need to be developed to reduce the fatigue problem. Such design rules are not easy to develop since many uncertainties are connected to the fatigue phenomenon. One such uncertainty is the characterization of the road severity. Traditionally transport vibrations are characterized by means of there power spectral density or simply by their overall RMS-values (Root Mean Square). Such methods are based on the assumption of stationary signals, and consequently rare events like for instance pot-hole responses are averaged out. However, these rare events do have a considerable influence on fatigue damage and should be taken into account. Therefore methods combining stationary vibrations with

Structural Failure, Product Liability and Technical Insurance, Proceedings of Fifth International Conference. Edited by H.P. Rossmanith. Published in 1996 by Taylor & Francis, 2 Park Square, Milton Park, Abingdon, Oxon, OX14 4RN. ISBN 0 419 22080 1

non-stationary events have been proposed with different degrees of sophistication (see Richards and Hibbert (1993) for a survey). Here we propose a simpler method, based only on the non-stationary events, but taking the vehicle dynamic characteristics into account.

The method is developed as a result of an investigation of the vibrational response of a specific tank vehicle (see Figure 1 and Olofsson and Svensson (1992)).

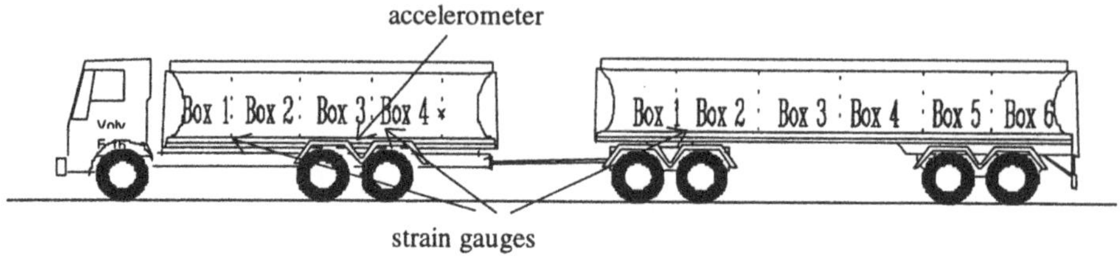

Figure 1. Tank vehicle equipped with accelerometer and strain gauges (from Olofsson and Svensson (1992)).

2 Method

In engineering practice fatigue design is usually based on the Wöhler curve for the material in question,

$$N = \alpha(\Delta\sigma)^{-\beta} \tag{1}$$

where N is the number of cycles to failure, α and β are material dependent constants and $\Delta\sigma$ is the stress range. When the dimensioning applied stress is of variable amplitude the linear Palmgren-Miner summation rule is used,

$$D = \sum_i \frac{n_i}{N_i} = \sum_i \frac{n_i}{\alpha(\Delta\sigma_i)^{-\beta}} = \frac{1}{\alpha}\sum_i n_i(\Delta\sigma_i)^{\beta} \tag{2}$$

where D is the cumulative damage caused by n_i stress cycles of ranges σ_i (i = 1, 2, 3, ..., M). The rule says that failure occurs when D exceeds one.

In order to use the Wöhler curve and Palmgren-Miner rule we need

- material properties in the form of parameters α and β
- characterization of the assumed stress history ($n_i, \Delta\sigma_i$)

2.1 Material properties

To carry out the above analysis we need to know the fatigue properties of the material used. In particular fatigue data for welded material are needed. If a complete Wöhler curve is not available, we can make approximations. The exponent in the Wöhler curve, β, generally lies between 3 and 5 for welded materials, and it applies to steel as well as to aluminium alloys. Once β is fixed, the constant in the Wöhler curve, α, can be determined if one point of the curve is known. One such point that often is available is the fatigue limit. For aluminium alloys it has also been demonstrated that the fatigue limit (for example the ten-million limit) is rather similar for very different alloys, cf. Figure 2. However, one must be cautious about corrosion effects. A striking property, which can be seen in Figure 2, and which may explain the short service lives in Sweden is the much reduced fatigue properties of aluminium alloys in a salt water environment.

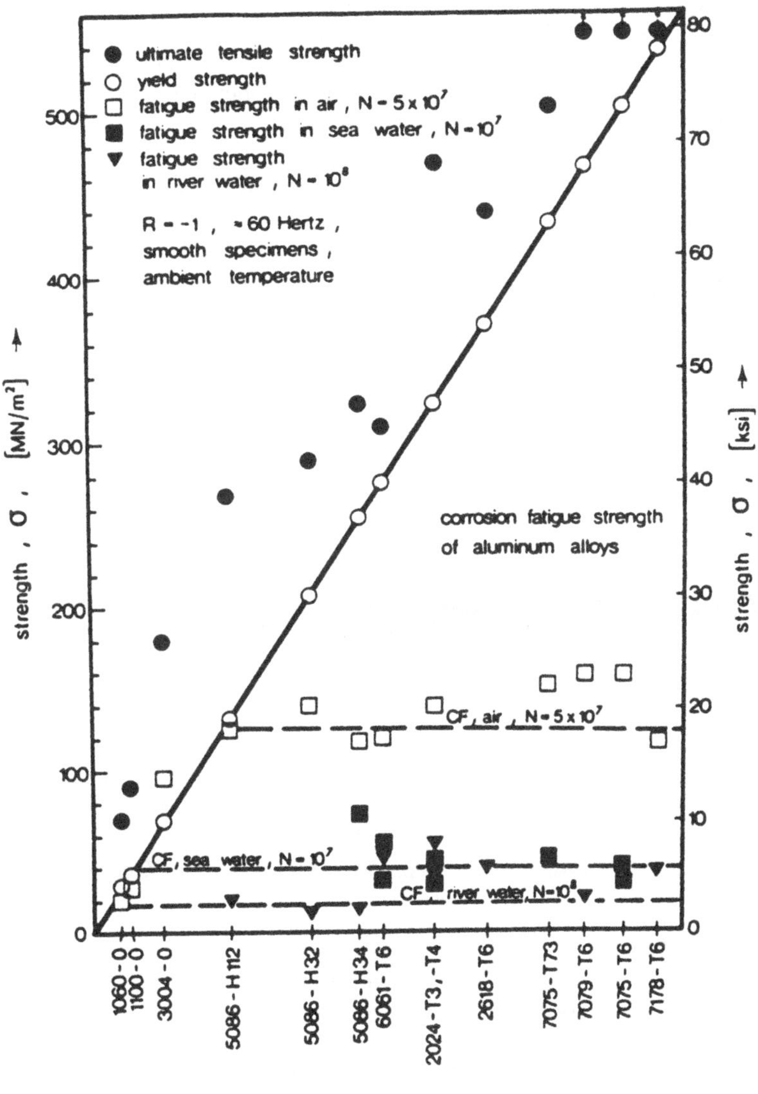

Figure 2. Fatigue properties for various aluminium alloys (from Boyer (1986)).

2.2 Stress history characterization

The vibrations that cause fatigue on road vehicles originate both from road irregularities and internal vehicle vibrations. The internal vibrations consist of rather high frequencies and we here claim that small high frequency vibrations can be neglected in fatigue calculations. To motivate this we make a rough estimate of the order of magnitude of the loads that cause damage. Since the analysis regards dimensioning against fatigue, primarily load levels which recur less than 10 - 20 million times during the life of the tank vehicle should be interesting. Let us assume that more high-frequent loads are small enough to fall below the practical fatigue limit. If the service life of the vehicle is, say, 10 years and the vehicle is used 60 hours per week during 40 weeks per year, a total life of 24,000 hours or 86 million seconds is obtained. A fatigue limit is usually defined at about a few million cycles, so frequencies above 0.1 Hz should not be interesting.

Based on this discussion about relevant frequencies it can be concluded that focusing should be on low-frequent events caused by, say, the vehicle passing uneven spots on the road. Figure 3 illustrates the idea, showing a pot-hole response in a high frequency vibrational environment. Included in the vehicle response on these events is of course also the high frequency vibrations which here consequently must be accounted for.

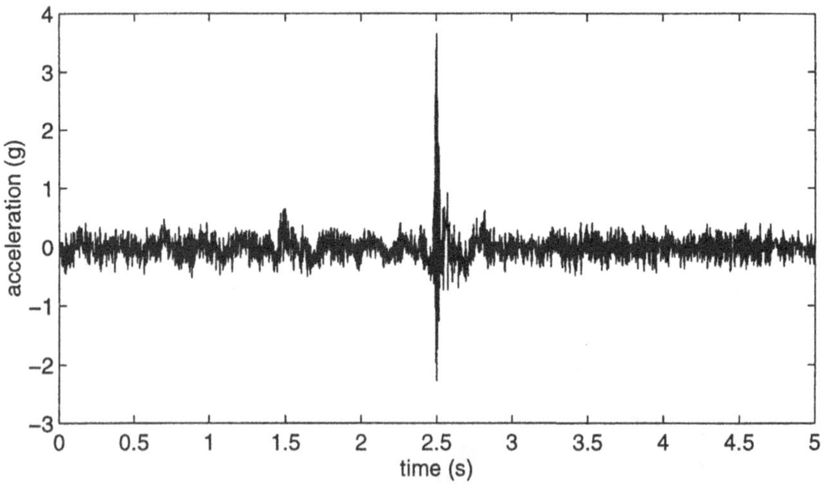

Figure 3. Pot-hole response (from a Round-Robin investigation (Richards, 1993))

2.3 Acceleration - stress

The Wöhler curve is based on mechanical stress, but road irregularities are often represented as accelerations at some specific position on the vehicle. There is consequently a need to analyse the transfer functions from accelerations at one position to stresses at the most critical positions in the tank. The dynamics of the tank and its connections to the vehicle have an influence here, but an analysis based on these dynamics would be very complicated. However, following the assumptions made above, only low frequent events need to be taken into account in the fatigue analysis. In the low frequency area the tank acts as a stiff object and a static analysis will be sufficient. In Figure 4 the acceleration and strain responses from one event are plotted and it can be seen how the two signals follow each other at the dominating low frequency.

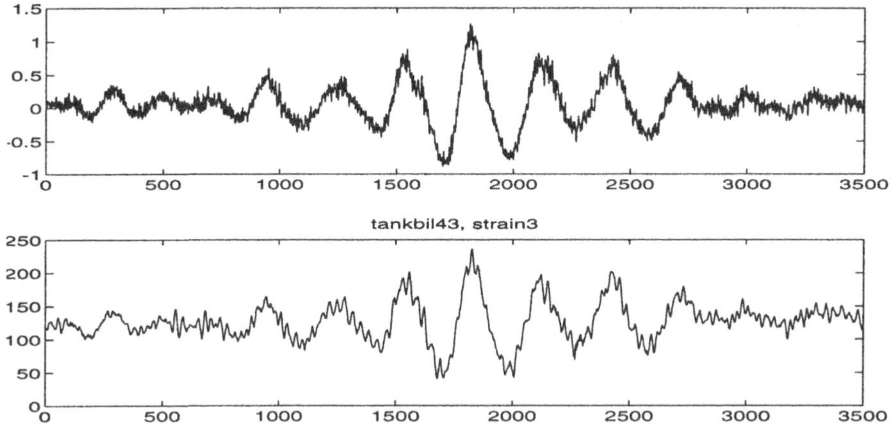

Figure 4. Time history for acceleration and strain respectively, from Olofsson and Svensson (1992)).

Thus, the strain can be regarded as a linear function of the vertical acceleration, $\varepsilon = \varepsilon_0 + ca$, where ε is the strain, ε_0 is the mean level of the strain, c is a proportionality constant and a is the vertical acceleration. Primarily we are interested in welded joints, which usually contain high residual stresses, so the mean level will not be very important. We disregard the mean level and consider only the stress ranges, which means we refrain from estimating ε_0.

The stress can now be calculated from the vertical acceleration (a), the proportionality constant (c), the elasticity modulus of the material (E), and the stress concentration factor (K_x)

$$\sigma = c \cdot E \cdot a \cdot K_x \tag{3}$$

For determining the different parameters in formula (3), the proportionality constant c can be calculated by a static analysis using FEM (Finite Element Method) or analytical methods. The modulus of elasticity is rather well-defined for most materials (aluminium alloys: 70 GPa, steel: 200 - 210 GPa). The stress concentration factor K_x depends both on the geometry and on the weld class and has to be studied in detail for each type of construction. Guidance for the determination of K_x is found in handbooks for steel or aluminium structures.

2.4 Damage accumulation

If we assume that the spring/damper system of the vehicle behaves as a second order linear system, the distribution of the number of cycles caused by an event can be calculated (cf. Fig. 5 for a typical damped response of this kind). One event gives approximately the cycles

$$\sigma_t, \quad \sigma_t e^{-2\pi\varsigma}, \quad \sigma_t e^{-4\pi\varsigma}, \quad \sigma_t e^{-6\pi\varsigma}, \quad \ldots \tag{4}$$

where ζ represents the relative damping, originating from the spring/damper system of the vehicle, and σ_t is the peak load at the event in question. The relative damping is normally of the order of magnitude of 5 %, which gives a rather rapid damping, and already after two or three cycles the amplitude will be halved.

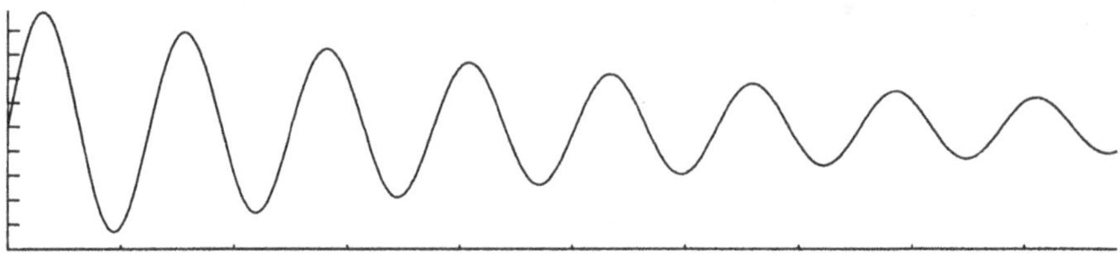

Figure 5. Ideal damped response

To calculate the damage caused by the response of one event with its maximum at σ_t we use (1) and (2),

$$D(\sigma_t) = \frac{1}{\alpha}\sum_{k=0}^{n}\left(\sigma_t e^{-2\pi\zeta k}\right)^{\beta} \tag{5}$$

where n represents the number of load cycles in the damped response that cause damage. A conservative estimate is obtained by letting n approach infinity, which gives a simple analytical expression,

$$D(\sigma_t) = \frac{1}{\alpha}\sum_{k=0}^{n}\left(\sigma_t e^{-2\pi\zeta k}\right)^{\beta} \leq \frac{\sigma_t^{\beta}}{\alpha}\sum_{k=0}^{\infty} e^{-2\pi\zeta k\beta} = \frac{\sigma_t^{\beta}}{\alpha\left(1 - e^{-2\pi\zeta\beta}\right)} \tag{6}$$

The damage caused by those cycles following the maximum one for each event is given by the factor $\dfrac{1}{\left(1 - e^{-2\pi\zeta\beta}\right)}$ which for the damping 2 - 6 % and $3 < \beta < 4$ is between 1.3 and 2.3.

To characterize the spectrum completely, the distribution for the maximum load of the events (σ_t) must be determined. The total damage is then given by

$$D = \sum_i n_i D(\sigma_{t,i}) = \frac{1}{\alpha\left(1 - e^{-2\pi\zeta\beta}\right)}\sum_i n_i \sigma_{t,i}^{\beta} \tag{7}$$

where n_i is the number of events with the maximum load $\sigma_{t,i}$. One complication is the deviation from an ideal damped oscillation. Several small superimposed cycles are added in the real load sequence (see Figure 4). This has in several tests given a more severe damage effect than what can be predicted by cycle counting methods. From for instance Dowling (1972) it can be concluded that the real service life is about one half of the predicted one in such cases. For design of tank vehicles the predicted damage must be doubled to compensate for this error.

3 Application

In the investigation Olofsson, Svensson (1992) measurements were made on a specific tank vehicle, driven on various roads representing normal use of such vehicle. Analysis of the measurements gave values of the proportionality constant c at different locations and a distribution of acceleration events during two hours drive. The maximum value of c was $c_{max} = 100\ \mu/g$ (microstrain per gravity acc.) and a count of the events during two hours gave results according to Table 1. For comparison a corresponding count is given in the table for a lorry drive from the Round-Robin exercise summarized in Richards (1992).

Table 1. Counted acceleration events during 2 hours' drives

Relative acceleration level, a_t/a_{max}	1.0	0.9	0.8	0.7	0.6	0.5	0.4
Number of events, tank drive	2	4	5	15	38	92	252
Number of events, lorry drive	2.3	2.3	2.3	2.3	0	32	831

The relative damping was estimated from the response (represented in Figure 4) to $\zeta = 5\%$. The material in the actual tank is aluminium and its fatigue properties for weldments is estimated from ECCS (1992),

$$\beta = 4.32 \quad , \qquad \alpha = 6.6 \cdot 10^{13}$$

where the stress is measured in MPa. The modulus of elasticity for aluminium is set equal to 70 GPa.

3.1 Design for twenty years' service life

We intend to find the maximum allowed acceleration level of the vehicle carrying the tank. Assume that the tank vehicle is driven for 60 hours per week and 50 weeks per year. The dimensioning 20-year life is then 60,000 hours. We first calculate the maximum stress at which $D_{60,000}$ equals unity. Using (7) and the parameter values chosen above we obtain,

$$1 = D_{60,000} = 30000 D_{2\,\text{hours}} = \frac{30000}{\alpha\left(1 - e^{-2\pi\zeta\beta}\right)} \sum_i n_i \sigma_{t,i}^{\beta} =$$

$$= \frac{30000\sigma_{max}^{\beta}}{\alpha\left(1 - e^{-2\pi\zeta\beta}\right)} \left[2 + 4(0.9)^{\beta} + 5(0.8)^{\beta} + 15(0.7)^{\beta} + 38(0.6)^{\beta} + 92(0.5)^{\beta} + 252(0.4)^{\beta}\right] \quad (8)$$

Solving this equation for the maximum allowed stress gives the value $\sigma_{max} = 66$ MPa.

Next we calculate the maximum allowed acceleration level from (3),

$$a_{max} K_x = \frac{\sigma_{max}}{cE} = \frac{66 \cdot 10^6}{100 \cdot 10^{-6} \cdot 70 \cdot 10^9} \approx 9.4 \quad (9)$$

With a K_x-factor equal to 3 and a safety factor of 2 the maximum allowed acceleration level will be about 1.9 g. This means that the vehicle spring/damper system must be soft enough to give an acceleration level (at the dimensioning position) less than 1.9 g, when subjected to a maximal road irregularity.

4 Conclusions

A methodology is proposed for fatigue damage prediction on tank vehicles. The predictions are based on fatigue properties of the material, the distribution of coarse road irregularities and the dynamic properties of the vehicle carrying the tank. The method is simple enough to be useful for design purposes. There are however important uncertainty factors associated with the fatigue damage assessment:

* corrosion effects can have a considerable influence,
* the connection between acceleration and maximum stress is not always easy to obtain,
* multi-axial effects may be important,
* the presented linear summation technique is an approximation

A combination of calculation methods and testing will be required to increase the knowledge in these fields. However, the method can be a helpful tool in the design process, in particular for comparison of different road types and of different tank carrying vehicles.

Acknowledgements

The research reported in this paper was carried out in co-operation with SAQ Inspection AB and was financed by the Swedish Rescue Services Agency.

References

ADR (1995) European Agreement concerning the carriage of dangerous goods by road.

Richards, D. P. and Hibbert, B. E. (1993) A round robin exercise on road transportation data, Journal of the Institute of Environmental Sciences, July/August 1993.

Olofsson, U. and Svensson, T. (1992) Shock response spectra and fatigue damage response spectra for a tank vehicle. Report SP-AR 1992:53, SP Borås.

ECCS (1992) European Recommendations for Aluminium Alloy Structures, Fatigue Design. ECCS - Technical Committee 2, Aluminium Alloy Structures.

Dowling, N. E. (1972) Fatigue Failure Predictions for Complicated Stress-Strain Histories, Journal of Materials, Vol. 7, No. 1, pp 71-87.

Boyer, H. E. (editor) (1986) Atlas of Fatigue Curves, American Society for Metals.

15 USE OF RIGID BODY DYNAMICS THEORY TO RESOLVE A COMPLICATED PATENT LITIGATION CLAIM

B. ROSS and J. KOKARAKIS
Failure Analysis Associates, Inc., Menlo Park, California, USA

Abstract

A patent dispute concerning the disparate non-positive rotating counterweight drive mechanisms employed in mechanical grape harvesters was resolved using kinematics and rigid body dynamics. In particular, contention that the allegedly infringing patent violated the single pivot axis claim of the predecessor patent was proved by analysis to be incorrect because the suspension system at issue was constituted as a four-bar linkage rotating about an infinite number of instantaneous centers which could not be construed to represent a single pivot axis. Analytical development of the governing equations for free body motion and the fundamental means of their interpretation from standpoint of inertia forces, restoring spring forces and Coriolis damping forces are discussed in the paper

Keywords
patent claim, mechanical grape harvester, shaking motion, single pivot axis, four-bar linkage, instantaneous center, Coriolis damping

Structural Failure, Product Liability and Technical Insurance, Proceedings of Fifth International Conference. Edited by H.P. Rossmanith. Published in 1996 by Taylor & Francis, 2 Park Square, Milton Park, Abingdon, Oxon, OX14 4RN. ISBN 0 419 22080 1

1 Introduction

This paper concerns the technical arguments pursued in an ugly patent litigation lawsuit. The underlying disparate claims were readily formulated and distinguished through engineering mechanics analyses founded on rigid body dynamics and kinematics theory. The case involved a self-propelled mechanical grape harvester machine that was conceived and placed on the global viticulture market with great success by the Up-Right Company. The original Up-Right models employed a clumsy heavy picking head driven in reciprocating lateral translation by crankshafts and mechanical linkages (Figure 1). As expected, vibration levels were high, and machine service life for the picker head was only around 400 hours. Apropos, the FMC Corporation, a conglomerate manufacturer of agricultural and food processing machines, recognized a prime business opportunity whereby they developed a completely new type of picking head which oscillated in pure rotation under non-positive centrifugal driving forces produced by a pair of counter rotating eccentric masses (discs) in the vertical plane (Figure 2). This picking head was envisioned and designed for direct installation in the Up-Right machine as a replacement embodiment; thus, FMC gained great marketing advantage from the huge fleet of existing Up-Right machines already operating worldwide. The FMC picking apparatus greatly reduced vibrations transmitted to both machine and operator plus extended the unit service life about five fold.

Up-Right experienced major losses both in sales of picker head spares and even new machines which were being ordered bare without the OEM reciprocating harvester unit. Thus, the company's engineers were obliged, as a matter of corporate survival, to come up with a comparable design that incorporated purely rotating elements and at same time would not be in conflict with claims of the FMC patent. Their means of solution, called a Rotary Pulsator, devised a compact picking head mounted in the machine

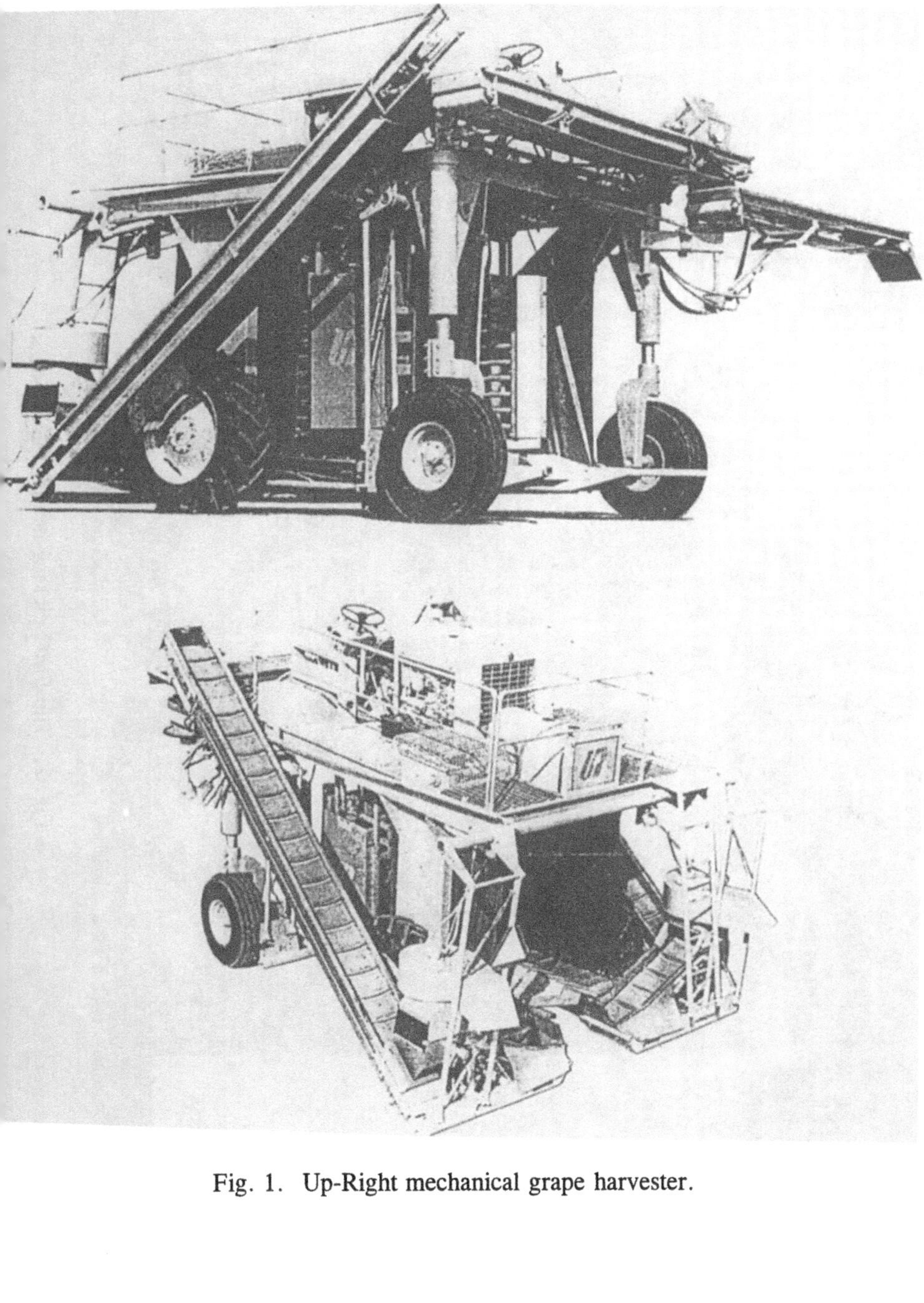

Fig. 1. Up-Right mechanical grape harvester.

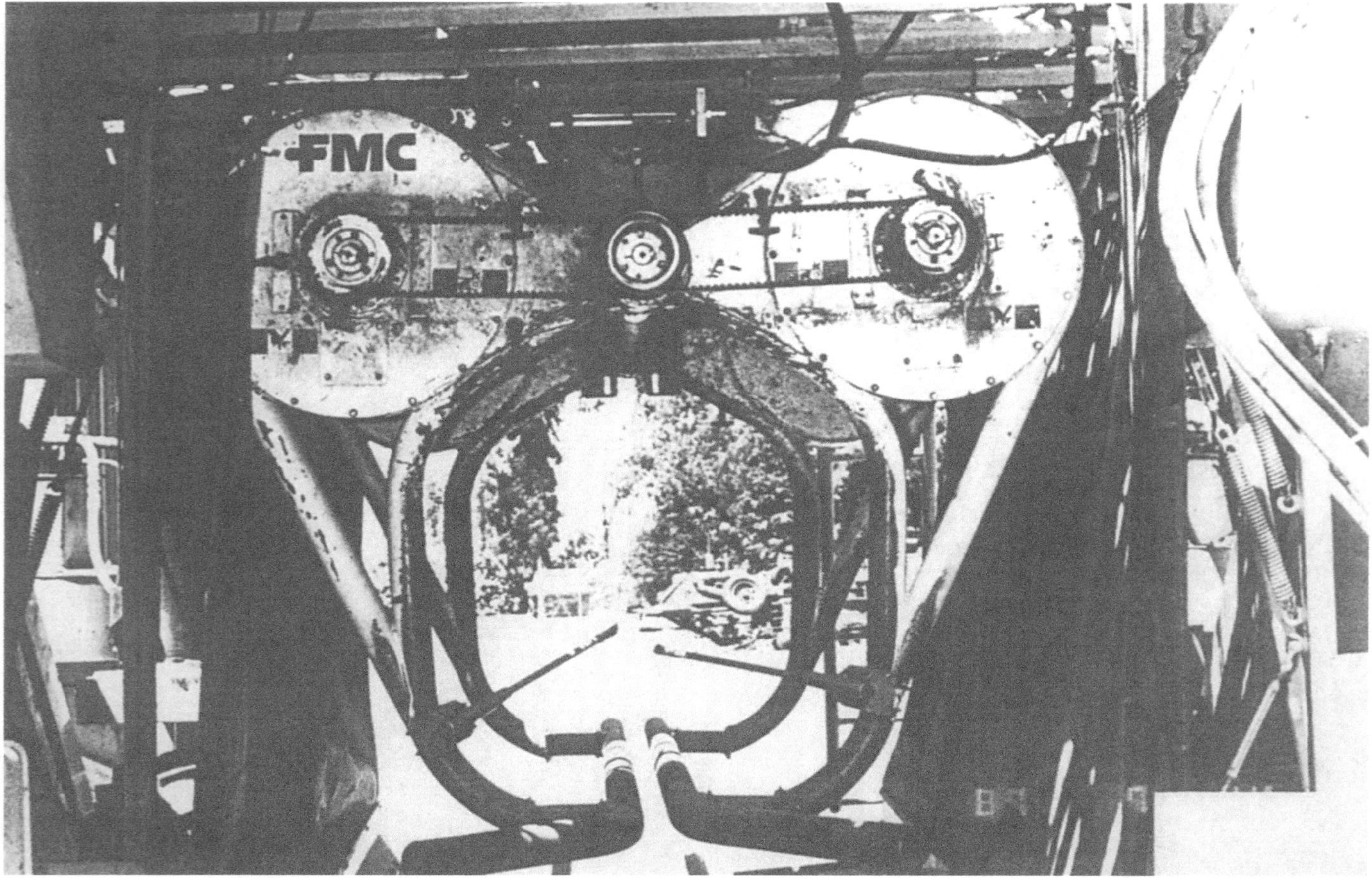

Fig. 2. The FMC picking head.

Fig. 3. The Up-Right compact picking head (Rotary Pulsator).

frame through a four bar mechanical linkage (Figure 3). The driving assembly contained dual, fore and aft, counter rotating disk packages. Each package incorporated two different size half circle eccentric masses turning about the same vertical axis with the masses stacked horizontally opposed, one above the other (Figure 4).

A fundamental claim element of the FMC patent, required to avoid conflict with a prior art British patent for a towable berry picker (Hally, 1,304.356) was that their picking head configuration rotated, implicitly with stable motion, about a single fixed pivot axis (Abstract, Figure 5). Thus, it had to be proved to a lay jury that the Rotary Pulsator did not violate the single pivot axis claim. In reality, this assertion was valid because the Up-Right assembly actually pivoted about an infinite set or locus of instantaneous centers. Moreover, the Rotary Pulsator in contrast to the FMC arrangement embodied a composite center of gravity that was at all times above the instantaneous center or transient pivot axis (Figure 6). Thus, the Up-Right system was actually statically unstable yet dynamically stable.

Both qualitative and quantitative arguments were developed to establish that the Up-Right machine did not violate the FMC single pivot axis design. In this context, the governing homogeneous equations for rigid body free vibration motion were developed for each system and an analytical representation of torque, unbalanced force, and partition of energies between the two configurations extracted from the analysis. The equation of motion for the FMC system consisted simply of an inertia torque term and a restoring torque term and was in essence identical to the large displacement equation for a classic pendulum. The equation of motion for the Rotary Pulsator included far more complicated expressions for the inertia torque and restoring torque but also introduced a non-linear term in rotational angular velocity squared corresponding to a Coriolis damping torque.

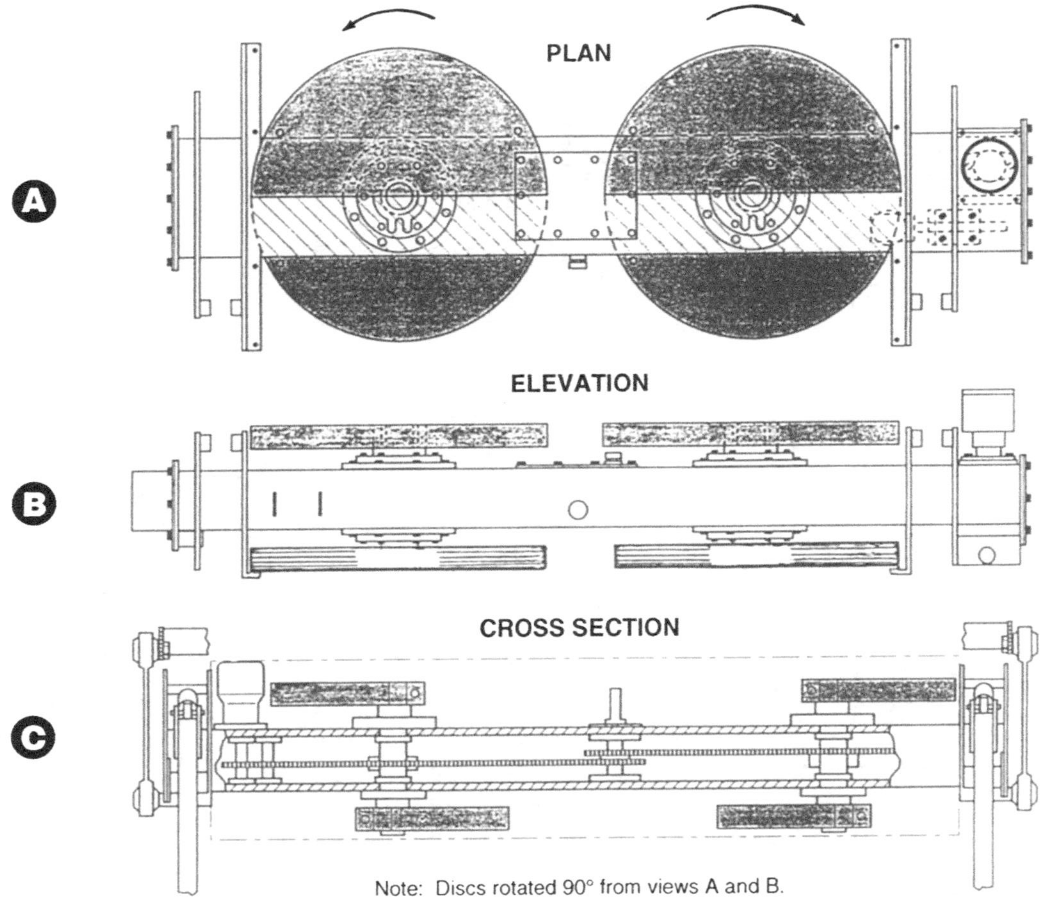

Fig. 4. Rotating counterweight arrangement in the Up-Right grape harvester.

56-330 SR

9/1/81 OR 4,286,426

United States Patent [19]

Orlando et al.

[11] 4,286,426

[45] Sep. 1, 1981

[54] VIBRATORY FRUIT HARVESTER

[75] Inventors: Franklin P. Orlando, Morgan Hill, Richard W. Fitzmaurice, Santa Clara, both of Calif.

[73] Assignee: FMC Corporation, San Jose, Calif.

[21] Appl. No.: 95,822

[22] Filed: Nov. 19, 1979

[51] Int. Cl.[3] A01D 46/26; A01D 46/28

[52] U.S. Cl. .... 56/330; 56/328 TS

[58] Field of Search .... 56/330, 328 TS

[56] References Cited

U.S. PATENT DOCUMENTS

3,184,908	5/1965	Rust	56/330
3,187,493	6/1965	Harrett	56/330
3,229,453	1/1966	Harrett	56/330
3,344,591	10/1967	Christie et al.	56/330
3,396,521	8/1968	McKibben et al.	56/330
3,413,789	12/1968	Studer	56/330
3,439,478	4/1969	Orton	56/330
3,439,482	4/1969	Orton	56/330
3,485,027	12/1969	Ganger	56/330
3,492,801	2/1970	Olmo et al.	56/330
3,535,856	10/1970	Olmo et al.	56/330
3,611,689	10/1971	Patzlaff	56/330
3,636,688	1/1972	Fontan et al.	56/330
3,703,072	11/1972	Patzlaff	56/330
4,016,711	4/1977	Claxton	56/330
4,063,406	12/1977	Burton	56/330
4,085,572	4/1978	Bruel	56/330
4,170,100	10/1979	Hood, Jr. et al.	56/328 TS
4,172,352	10/1979	McCarthy et al.	56/330
4,179,871	12/1979	Claxton	56/330
4,207,727	6/1980	Poytress	56/330

FOREIGN PATENT DOCUMENTS

2403734	5/1979	France	56/330
2419004	11/1979	France	56/330

Primary Examiner—Paul J. Hirsch
Attorney, Agent, or Firm—R. S. Kelly

[57] ABSTRACT

A harvesting apparatus for plants grown in rows, such as grapes, includes a vehicle having a support frame of an inverted U-shape and a shaker frame that also has an inverted U-shape and that is pivotally mounted to the vehicle support frame about a single axis that is parallel to the direction of travel of the vehicle. Striker bars are rigidly attached in spaced, opposing and parallel relation at the lower end of the shaker frame and at a height above ground level to engage the trunk of the plants. A non-positive oscillatory drive mechanism is attached to the upper end of the shaker frame to rock the frame about the single pivot axis to cause the striker bars to alternately impact opposite sides of the trunks. The drive mechanism consists of two eccentric weights rotatably mounted to the shaker frame above and equidistantly spaced from the pivot axis so that the center of gravity of the shaker frame, weights and other elements attached to the shaker frame is approximately coincidental with the pivot axis. The shaker frame includes frame sections adapted to scissor about the pivot axis so that the distance between the striker bars is greater than the diameters of the trunks to permit continuous harvesting, and a grape collection and conveyance system is also provided on the vehicle below the shaker frame.

23 Claims, 6 Drawing Figures

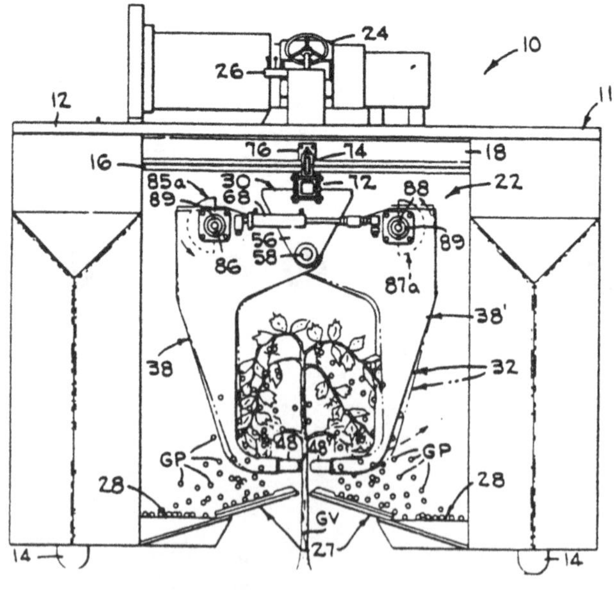

Fig. 5. Abstract of FMC patent.

2 Up-Right Grape Harvester

A front elevation schematic drawing of the Up-Right grape harvester is shown in Figure 6. The structural system (Figures 3 and 6) consists of four support arms, two pairs front and back, each arm shaped in the form of a Crescent (C) section. The C arm pairs which resemble grapples support two longitudinal horizontal shaker rails at their lowest ends (#56, #57)* which alternatively impact the grape vines in lateral translation. An added convenience feature enables the separate grapple arms to be spread by a horizontal hydraulic cylinder (i.e. ram) and thereby create a variable shaker rail width spacing to accommodate different growth varieties of grapevines. The C-arms are bolted at their upper parts to a longitudinal rectangular box section (Figure 3, #26). The box section is hung on its fore and aft ends by two pivoting convergent links (#29, #30) which form a classic four-bar system. The box section and attached C-arms are oscillated by a harmonic driving torque generated through the assemblage of opposite rotating half circle counterweights. The eccentric mass pairs which turn about vertical axles are mounted asymmetrically above and below the box section (Figures 4 and 6). Drive chains that power the rotating counterweights are submerged within the box section in an oil bath. A hydraulic motor runs the chains through a shaft mounted sprocket gear. The pair of lower counterweights has smaller mass than the upper ones. Consequently, as shown in Figure 6, net resultant horizontal forces on the C- arms induce both translational and rotational motions, due to unbalanced forces and torsional moments, respectively.

In the idealized model drawing (Figure 7), the box section assembly is suspended from the harvester's rigid horizontal framework at points D and C via the bars DA and CB. It is apparent that the upper rigid frame (DC), the grapple support box section (AB) and

---

* Number designations correspond to the Description of Preferred Embodiments; Patent Number 4.621.488.

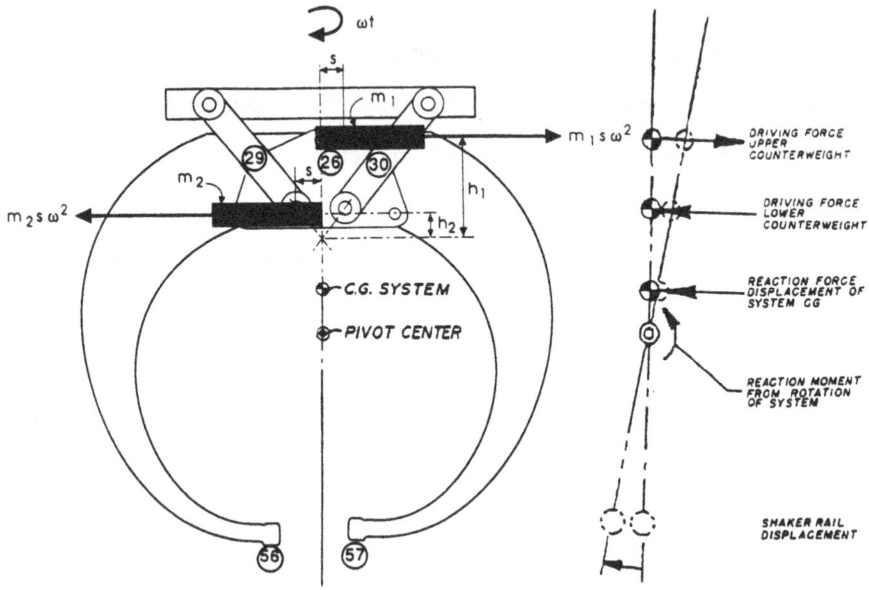

$$\text{Torque} = (h_1 m_1 - h_2 m_2) s\omega^2 \sin \omega t$$

$$\text{Unbalanced Force} = (m_1 - m_2)\, s\omega^2 \sin \omega t$$

Fig. 6. Schematic drawing of the drive means of the Up-Right Rotary Pulsator.

the suspension links (DA, CB) constitute a four bar linkage mechanism. The mass and time varying polar moment of inertia of link AB which effectively represents the entire box section assembly (i.e. grapples, rotating eccentric weights, chains, sprockets, shafts, hydraulic drive motor, etc., etc.) can be computed from physical measurements and geometrical data. The composite center of gravity, G, of the complex oscillating four link mechanism is also determined and located with respect to link AB (Figure 7).

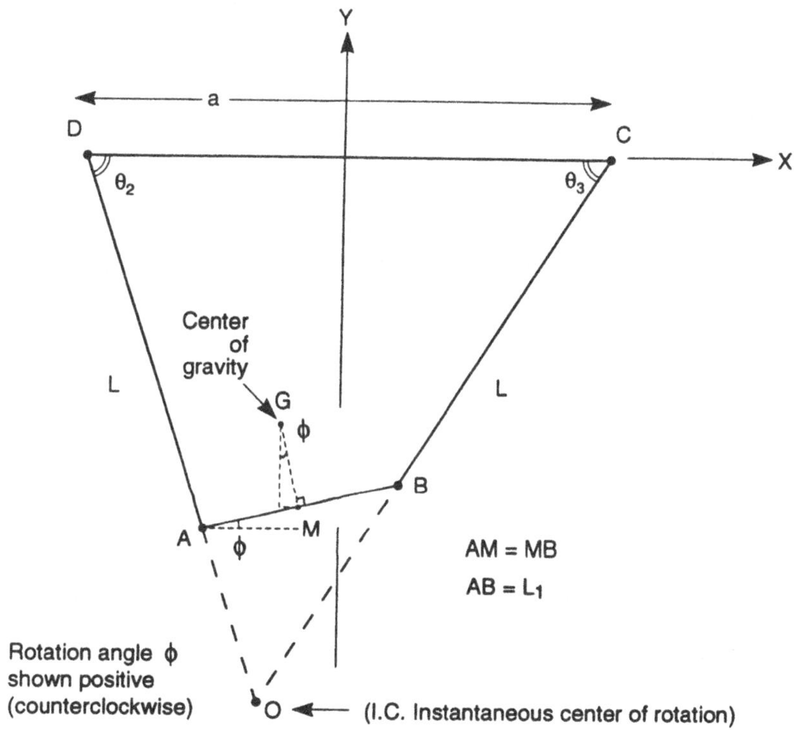

Fig. 7. Idealization of the four bar linkage of the Up-Right harvester, Rotary Pulsator drive system.

3 Equations of Motion

Development of the governing equations of motion for the mechanical oscillating system was based on the Lagrange formulation. The four-link mechanism was treated as a one-degree of freedom system where the angular rotation, ϕ, of the link AB was chosen as the independent variable, necessary to define the compound motion.

For the given system, the Lagrange equation is:

$$\frac{d}{dt}\left[\frac{\partial E}{\partial \dot{\phi}}\right] - \frac{\partial E}{\partial \phi} + \frac{\partial U}{\partial \phi} = 0 \qquad (1)$$

where E and U are the kinetic and potential energies respectively. The following operating conditions are assumed:

(1) Viscous damping neglected (can be added as an additional velocity dependent term in the final equation.)

(2) Free vibration case, complementary solution (particular solution can be obtained by adding a driving torque or forcing function term in the final equation).

The potential energy U is:

$$U = mgy_G \qquad (2)$$

where m is mass of the mechanical system undergoing compound motion (essentially the box section assembly, link AB), g the acceleration of gravity and y_G upward vertical

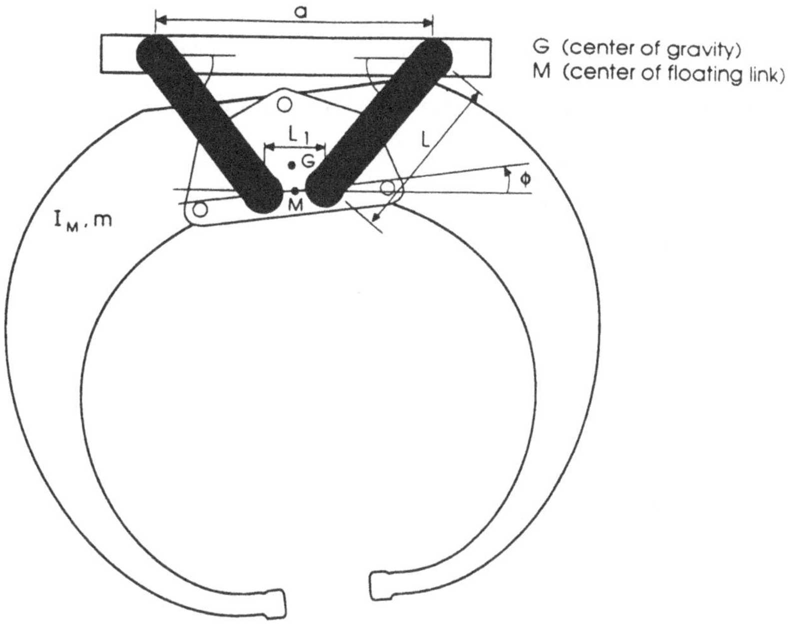

Fig. 8. Means of support of the Up-Right Rotary Pulsator.

displacement of the center of gravity G, given by (Figure 7):

$$y_G = -L\sin\theta_2 + \frac{L_1}{2}\sin\phi + GM\cos\phi \qquad (3)$$

where L and L_1 are lengths of links AD, AB respectively; GM is distance of the center of gravity from the midpoint of link AB.

Following some algebraic manipulations and differentiation with respect to ϕ, Equation (2) becomes:

$$\frac{\partial U}{\partial \phi} = mg\left[L_1\frac{\sin(\phi - \theta_3)\cos\theta_2}{\sin(\theta_2 + \theta_3)} + \frac{L_1}{2}\cos\phi - GM\sin\phi\right] \qquad (4)$$

It can be shown that the term defined by Equation (4) is a restoring moment equal to mg (x_G - x_O), where x_G, x_O are horizontal translations of the center of gravity G and the instantaneous center of rotation O, respectively. Equivalently, this term represents the contribution of a torsional spring.

The kinetic energy, E, is:

$$E = \frac{1}{2} I_M \dot{\phi}^2 + \frac{1}{2} m \left(\dot{x}_G^2 + \dot{y}_G^2 \right) \tag{5}$$

where I_M is mass moment of inertia of the mechanical system with respect to the Z axis, passing through the center of gravity G and $\dot{x}_G$ and $\dot{y}_G$ are horizontal and vertical velocity components, respectively, of the center of gravity motion.

Following some lengthy algebraic manipulations, the kinetic energy contribution is computed as:

$$\begin{aligned} \frac{d}{dt}\left[\frac{\partial E}{\partial \dot{\phi}}\right] - \frac{\partial E}{\partial \phi} = & \left\{ I_M + m\left(\frac{L_1^2}{4} + GM^2\right) + 2mLf\left[Lf + L_1 \cos(\phi + \theta_2) - 2GM \sin(\phi + \theta_2)\right]\right\} \ddot{\phi} \\ & - mL\dot{\phi}^2 \left\{ \cos(\phi + \theta_2)\left[3GM(f^2 + f) - L_1 \eta - \frac{rL_1}{2}\right] + \sin(\phi + \theta_2)\left[\frac{3L_1}{2}(f^2 + f) + 2GM\eta + GMr\right] \right. \\ & \left. + Lf(2\eta + r) \right\} \end{aligned} \tag{6}$$

where

$$f = \frac{\dot{\theta}_2}{\dot{\phi}} = \frac{-L_1}{L} \frac{\sin(\phi - \theta_2)}{\sin(\theta_2 + \theta_3)} \tag{7}$$

$$r = \frac{d\dot{\theta}_2}{\dot{\phi}\, d\phi} = \frac{-L_1 \cos(\phi - \theta_3)(f - k) - L\cos(\theta_2 + \theta_3)(f^2 + kf)}{L \sin(\theta_2 + \theta_3)} \tag{8}$$

$$k = \frac{\dot{\theta}_3}{\dot{\phi}} = -\frac{L_1}{L}\frac{\sin(\phi - \theta_3)}{\sin(\theta_2 + \theta_3)} \tag{9}$$

and

$$\eta = -k + \frac{f^2}{\tan(\theta_2 + \theta_3)} + \frac{k^2}{\sin(\theta_2 + \theta_3)} \tag{10}$$

The kinetic energy expression can be distinguished by three different groups of terms. Specifically: terms stemming from movement of the center of gravity, G; terms due to movement of the instantaneous center, IC; and finally cross coupling terms due to combined movement of both the center of gravity and the instantaneous center. The resultant homogeneous equation of motion for the free vibration case can be recast in simpler form and is given by:

$$\underbrace{\left(I_M + \frac{mL_1^2}{4} + mLL_1\, c_1\right)\ddot{\phi}}_{\text{Inertia Torque}} + \underbrace{mLL_1\, c_2\dot{\phi}^2}_{\text{Coriolis Torque}}$$

$$\underbrace{mg\left[L_1\frac{\sin(\phi - \theta_3)\cos\theta_2}{\sin(\theta_2 + \theta_3)} + \frac{L_1}{2}\cos\phi - GM\sin\phi\right]}_{\text{Restoring Torque}} = 0 \tag{11}$$

where c_1 and c_2 are time dependent coefficients deduced from equation (6).

The instantaneous center of rotation for the four bar linkage case coincides with the mutual intersection point of center line projections for links DA and CB (Figure 7). Obviously, the instantaneous center is a fictitious point which moves spatially as the grape harvester grapple (link AB) oscillates laterally with both rotation and back and forth translation in the cross section plane. Consequently, the grapple actually rotates around an infinite number of axes, rather than a single pivot axis fixed in space. This situation is depicted in Figure 9. Horizontal and vertical displacements for the center of gravity and the instantaneous center of rotation are also found in this plot over a range

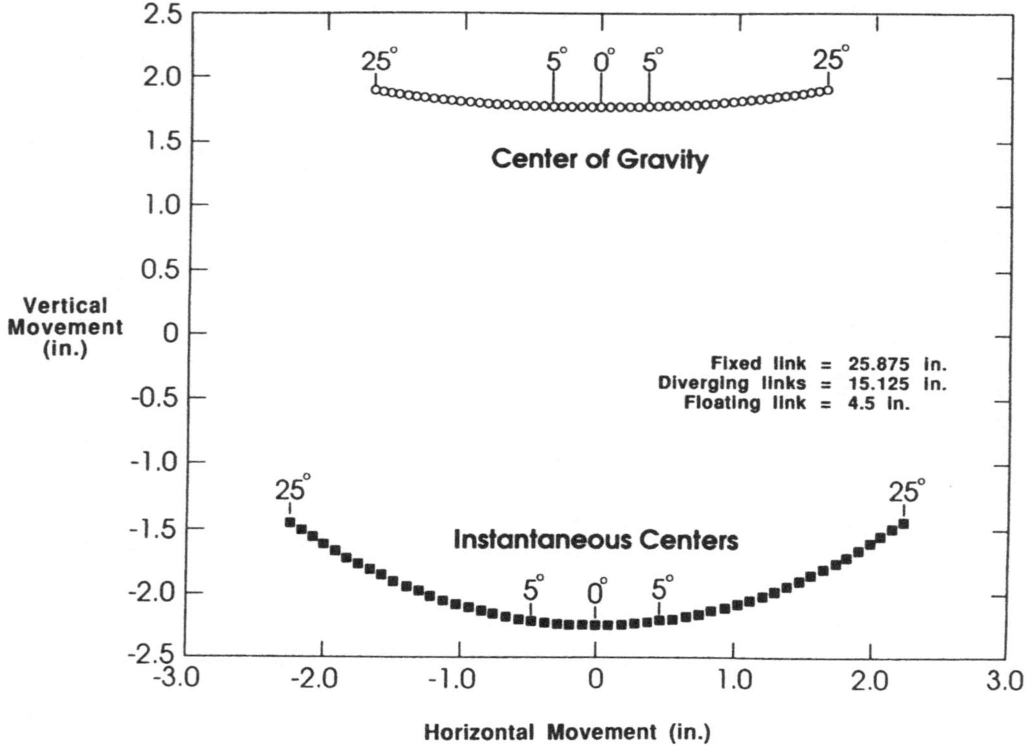

Fig. 9. Loci of center of gravity and instantaneous center of Up-Right Rotary Pulsator.

of rotations, ϕ, between zero and ± 25 degrees. The differential horizontal distance between the instantaneous center and the center of gravity, which represents the lever arm for the dynamic restoring moment, can be obtained readily as a function of rotation. It is noted that:

a) Horizontal displacements for both the center of gravity and the instantaneous center are far more significant than vertical displacements.

b) Vertical displacement of the center of gravity is an order of magnitude less than that of the instantaneous center.

c) Horizontal distance between the instantaneous center and the center of gravity is always such that the gravity moment due to the weight force tends to restore the zero slope condition. Moreover, at relatively small rotation angles, say ±10°, the restoring moment; that is, the effective torsional spring, exhibits linear stiffness or spring rate.

4 FMC Grape Harvester

The FMC grape harvester mechanical system design is portrayed in Figure 10. The configuration consists of front and rear portal frames suspended from a live central shaft running longitudinally down the vehicle mid-line. The suspension element is thus a single pivot axis fixed to the harvester chassis. This stationary pivot acts as a center of oscillation for the shaker. A non-positive rocking couple in the vertical cross-section plane is produced via two equal size eccentric masses rotating in the same direction but 180° out of phase as shown in Figure 11.

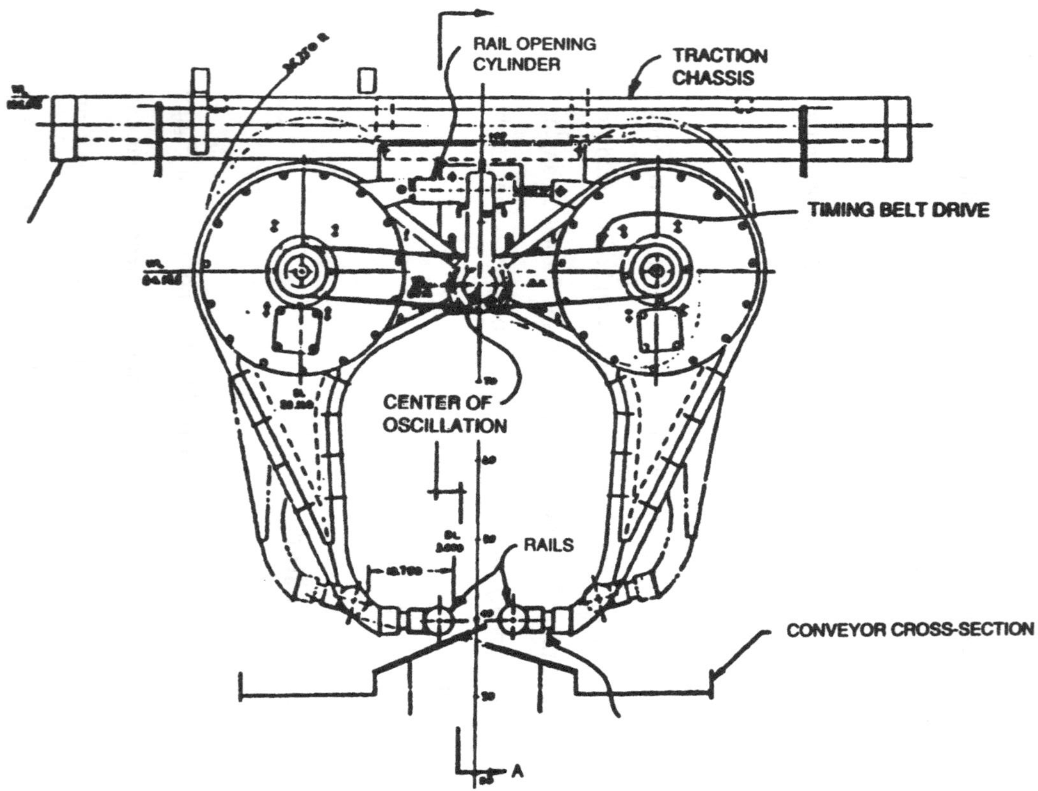

Fig. 10. Cross-section of the mechanical system of the FMC grape harvester.

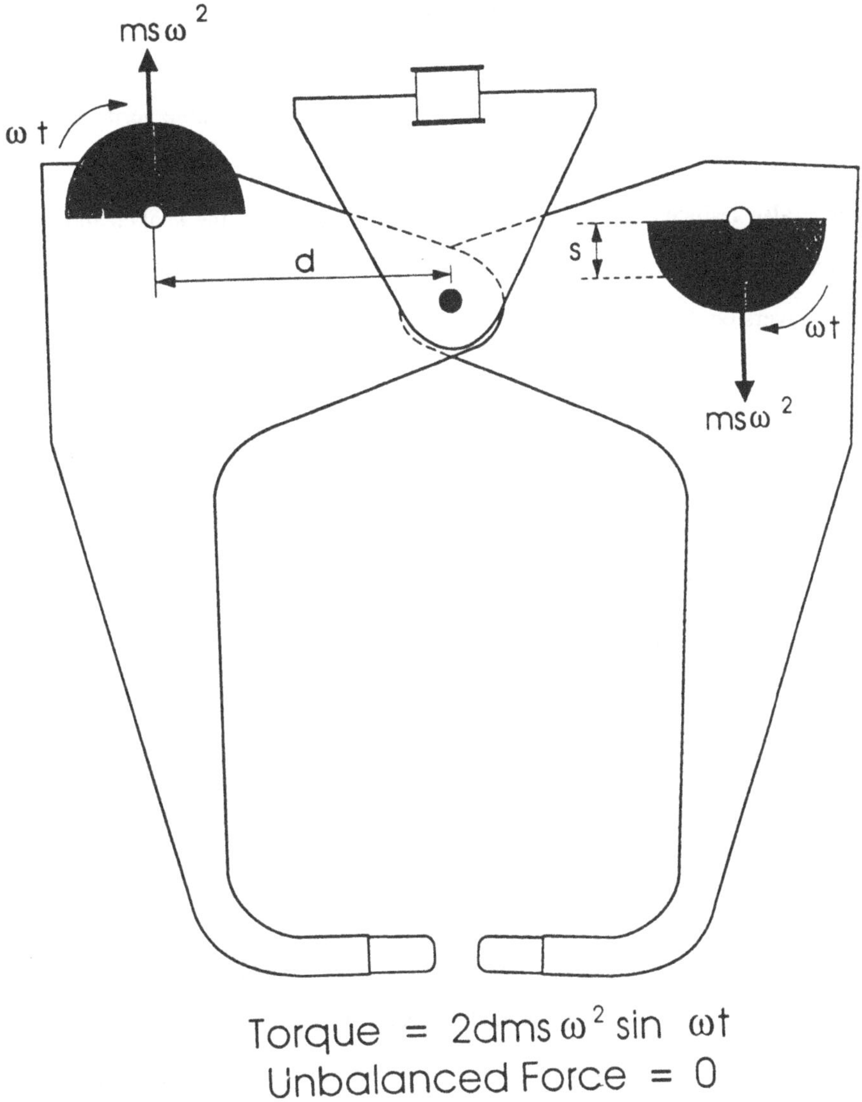

Fig. 11. Drive means of the FMC grape harvester, FMC Patent 4,286,426.

The FMC design philosophy attempted to minimize the vertical distance between the center of gravity and the center of oscillation to its lowest possible value. This approach was undertaken in order to reduce the mass moment of inertia of the shaker about the pivot axis and also to lower the magnitude of undesirable vertical vibrations transmitted to the vehicle frame and operator. On the other hand, compressing this distance also reduces the restoring torsional spring rate of the shaker and can lead towards an undesirable state of neutral stability with possibility of portal frame (i.e. shaker) inversion.

The FMC support means is presented in Figure 12. Kinematics of the FMC mechanism are fully defined by two rotational degrees of freedom; rotation of the eccentric masses ωt (Figure 11) and rotation of the shaker, ϕ, (Figure 12). From the governing equations, it can be shown that if the driving and driven belt pulleys have equal radii (Figure 2), the mechanism degenerates to a single degree of freedom system. Then the equation of free vibrations is:

$$\underbrace{I_0\ddot{\phi}}_{\substack{\text{Inertia}\\ \text{Torque}}} + \underbrace{mg\,(OG)\sin\phi}_{\substack{\text{Restoring}\\ \text{Torque}}} = 0 \tag{12}$$

where I_0 is mass moment of polar inertia for the shaker and tynes assembly with respect to the center of oscillation O and OG is distance between the center of gravity and center of oscillation. As recognized, the system represented by Equation (11) is a pendulum oscillating with large displacements about the spatially fixed point, O.

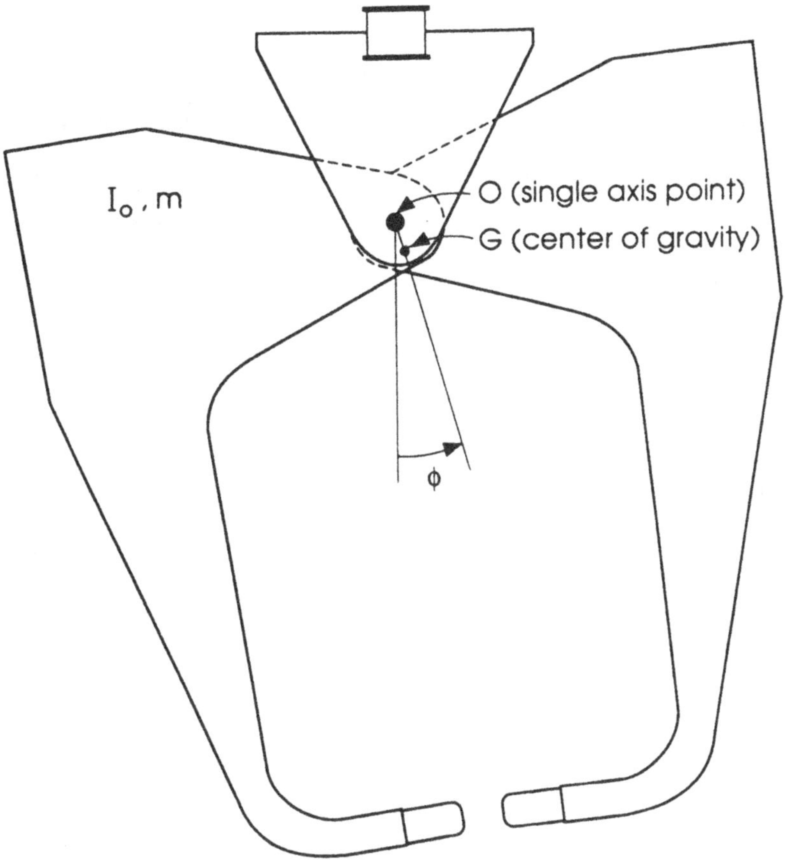

Fig. 12. Support means of the FMC grape harvester, FMC Patent 4,286,426.

5 Comparisons between the Up-Right and FMC Grape Harvesters

The essential kinematic difference between the two harvester designs is that the FMC machine is constituted by a classic simple pendulum system, whereas the Up-Right design behaves as classic four-bar linkage kinematic mechanism. This observation is readily proved by substituting $L_1 = 0$ in the governing equations of motion (i.e. (4) and (6))for the Up-Right design. As a result, M coincides with O and there obtains:

$$I_0\ddot{\phi} - mg\,(OG)\sin\phi = 0 \qquad (13)$$

The difference in sign between Equations (12) and (13) is due to fact that, effectively, point O lies below point G in the Up-Right design.

The FMC shaker rotates around a dedicated fixed point in space (Figure 13), whereas in the Up-Right design the phantom pivot axis moves spatially as an "instantaneous center of rotation" with an instantaneous velocity of zero. Moreover, there are an infinite number of instantaneous center locations created during each full cycle transit of the shaker assembly (Figure 9). Thus, by definition, fundamental to the theory of rigid body dynamics, the Up-Right grape harvester apparatus does not rotate about a single pivot axis.

The driving power in both the FMC and Up-Right machines is provided by pairs of eccentric mass counterweights rotating out of phase by 180°. These seemingly equivalent "engines" have four fundamental differences.

(a) In the FMC design the counterweights rotate in a vertical plane aligned with (i.e., parallel) to the plane of rotation of the shaker. In the Up-Right design, the

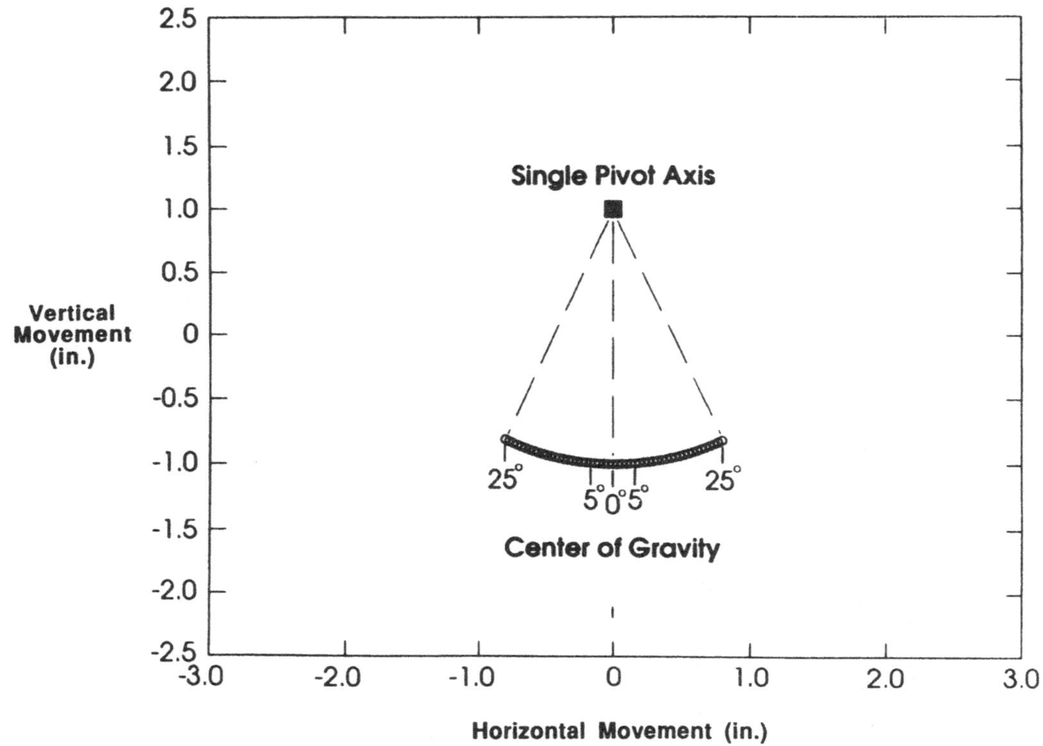

Fig. 13. Loci of center of gravity and single pivot axis of FMC grape harvester, FMC Patent 4,286,426.

counterweights rotate in upper and lower horizontal planes normal to the plane of rotation of the shaker (C-arms).

(b) The counterweights of the FMC design have equal mass, whereas the upper counterweights for the Up-Right design are heavier than the lower ones. This differential between upper and lower masses introduces a net horizontal unbalanced force (Figure 6), causing translation, in addition to torque or rotational moment in the vertical plane (Figures 6 and 8).

(c) The offset sidewise positions of rotating counterweights in the FMC design relative to the centerline pivot axis yield a driving torque which is dependent on the global rotation, ϕ, of the shaker assembly (Figure 12). Specifically:

$$T = 2dms\omega^2 \sin \theta(t) \tag{14}$$

where the geometric quantities are indicated in Figure 11. The time dependent angle, $\theta(t)$, is given by:

$$\theta(t) = \omega t - \phi \tag{15}$$

Thus, the operating torque is a time dependent asymmetric function with respect to the rotational belt drive direction of the counterweight pulleys. This behaviour adversely reduces snapping action of the shaker rails on the grape vines during the anti half cycle swing with respect to the direction of belt drive rotation.

(d) The arrangement of horizontal half circular discs in the Up-Right design, yields a non negligible time varying mass moment of inertia due to fact that distance of the composite center of gravity for the counterweight pairs from the vertical

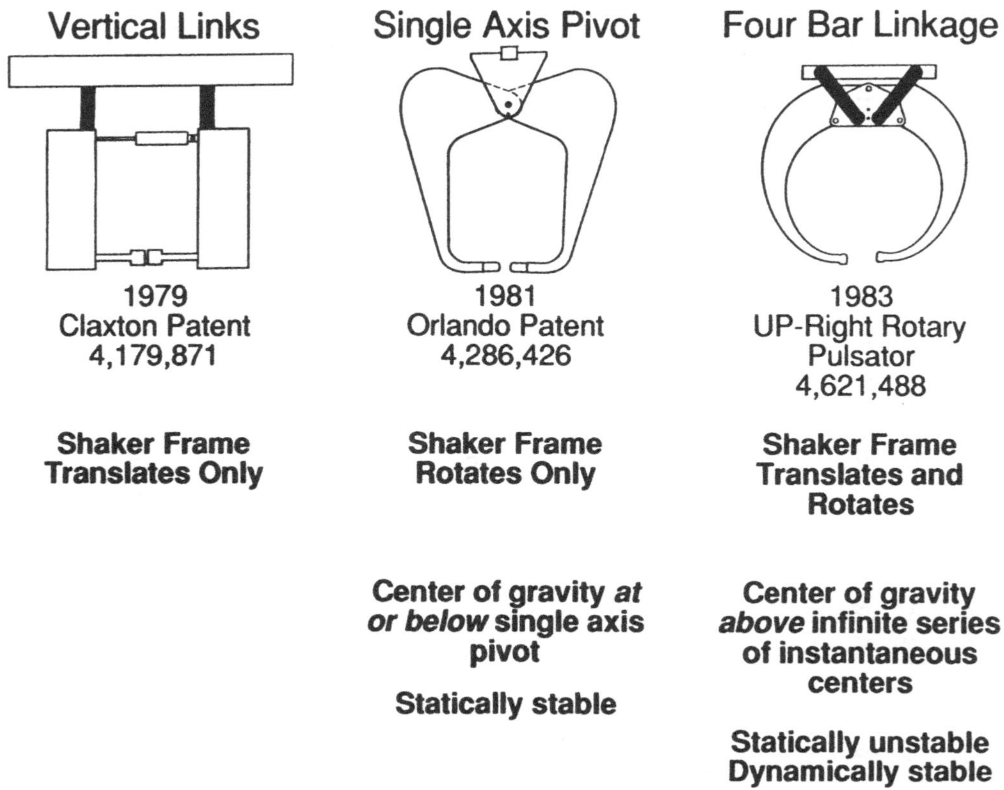

Fig. 14. Support means - comparison with prior art.

center plane varies with counterweight rotation angle. This phenomenon results in a band of natural frequencies for the Up-Right machine rather than a distinct value and is comparable to the time varying mass moment of inertia for the connecting rod assembly of a diesel engine (Draminsky effect).

Blatant disparities between the two grape harvesting machines are summarized in Figures 14. As readily perceived, individual characteristics of the basic modes for both shaker support and shaker drive systems render the two designs fully distinguishable and unequivocally non infringing.

References

"Inertial Grape Shaker Development," Orland F., Wiederrich J., Mortensen D., 1982 Summer Meeting, AMERICAN SOCIETY OF AGRICULTURAL ENGINEERS

U.S. Patent Specification 1304356, Fruit Gathering Machine, Filed 19 August 1969, Hally, William Whytock.

U.S. Patent Specification 4,179,871, Suspension System for Harvester Shaker Member, Filed 25 December 1979, Claxton, Gerald L.

U.S. Patent Specification 4,286,426, Vibratory Fruit Harvester, Filed 1 September 1981, Orlando, Franklin P., and Fitzmaurice, Richard W.

U.S. Patent Specification 4,336,682, Continuous Harvester for Plants Grown in Rows, Filed 29 June 1982, Orland, Franklin P.

U.S. Patent Specification 4,621,488, Oscillatory Shaker Rail Harvester, Filed 11 November 1986, Claxton, Gerald L.

16 A FATAL ACCIDENT CAUSED BY A SPINDLE/KNUCKLE SEPARATION

K.L. SALES
Law Firm Segal et al.
A.A. JOHNSON
University of Louisville, Kentucky, USA

Abstract
The driver of a 1984 Ford Ranger truck lost control of his vehicle on a two-lane road and crossed the center line into the path of a 1972 Ford pickup truck. The driver of the Ranger was killed and the driver of the pickup seriously injured. Reconstruction of the accident showed that it had been caused by a spindle/knuckle separation in the left front wheel assembly. The spindle stem is designed to be retained in a hole in the knuckle by a shrink fit. Because of dimensional errors in this particular assembly the stem had been in contact with the surface of the hole over only a small fraction of its length. This enabled the stem to work loose causing separation of the left front wheel assembly from the vehicle with consequent loss of control.

In ensuing litigation, in which the case was tried by a jury twice, the two drivers were awarded a total of $2.3 million in compensatory damages and $4 million in punitive damages against Ford. This award has been affirmed by the Kentucky Court of Appeals, but Ford has asked for a discretionary review by the Kentucky Supreme Court.
Keywords: Compensatory damages, Ford pickup, Ford Ranger, jury instructions, jury trial, knuckle, punitive damages, spindle separation.

1 Introduction

At about 3:45 a.m. on June 2nd 1985 Troy Cilone, a 17 year old male, was driving his 1984 Ford Ranger truck northbound on the road joining the cities of Bardstown and Louisville, both of which are in Kentucky. He was returning from a fishing trip and was towing a trailer bearing a small aluminum boat. In a rural area a few miles

Structural Failure, Product Liability and Technical Insurance, Proceedings of Fifth International Conference. Edited by H.P. Rossmanith. Published in 1996 by Taylor & Francis, 2 Park Square, Milton Park, Abingdon, Oxon, OX14 4RN. ISBN 0 419 22080 1

south of Louisville his truck went out of control and collided with a Ford pickup truck traveling southbound driven by William Fulkerson. As a result of the collision Cilone was killed and Fulkerson was badly injured.

The initial investigation was handled by the Jefferson County, Kentucky, Police Department. Its personnel carefully documented the accident site photographically and mapped out the positions of the point of impact, significant pieces of debris, gouge marks, and the positions at which the two vehicles came to rest. This information was compiled into the Police Department's report of the accident.

Shortly after the accident the Kentucky Medical Examiner, Dr. George Nichols, put together an interdisciplinary team to conduct a more thorough investigation of the accident. This consisted of Detective Joseph Ash (accident reconstructionist), Mr. (now Dr.) William Smock (anatomist and accident reconstructionist), Mr. Philip Schaad III (accident reconstructionist) and Dr. Alan A. Johnson (metallurgist and accident reconstructionist).

In this paper we first present an outline of the results of this more detailed investigation and of the accident reconstruction which emerged from it. We then summarize the very different conclusions reached by Ford personnel. Finally we describe the litigation in which these opposing views were presented in court [1]. This litigation is still unfinished.

2 The accident site

The accident occurred on a straight and level stretch of road in clear almost dry conditions. The road surface was of asphalt with an aggregate. Figure 1 is a scale drawing of the accident site showing the point of impact, as determined from the distribution of debris, and the rest positions of the two vehicles, the trailer and the boat. The trailer remained attached to the Ford Ranger but the boat came off the trailer. The collision occurred when the Ranger crossed the center line and invaded the southbound lane.

The investigating police officers noted gouge marks on the roadway starting 142 feet (43.3 meters) south of the point of impact and continuing intermittently to the resting place of the Ranger. From these the path of the Ranger could be reconstructed. These gouge marks are shown in Figure (1). There were also tire marks at the site of the accident, but these did appear to have been created by the two vehicles involved in the collision.

The left front wheel came off the Ranger and was found south of the point of impact. Because a wheel can easily roll a considerable distance after separation its resting place is not considered significant. Brake shoes were found 198 feet (60.3 meters) south of the point of impact, pieces of tire rim at 152 feet 4 inches (46.4 meters) and 132 feet (40.2 meters), and a piece of brake caliper at 47 feet 7 inches (14.5 meters). All of these pieces of evidence are marked in Figure (1).

Figures (2) and (3) respectively show the Ford pickup and Ford Ranger after the accident. It is obvious from the patterns of damage that the left front of the pickup engaged the left front of the Ranger. Since no skid marks were observed there is no way of estimating the speeds of the vehicles at impact. It is obvious from the severity of the damage to both vehicles, however, that it was a relatively high speed collision.

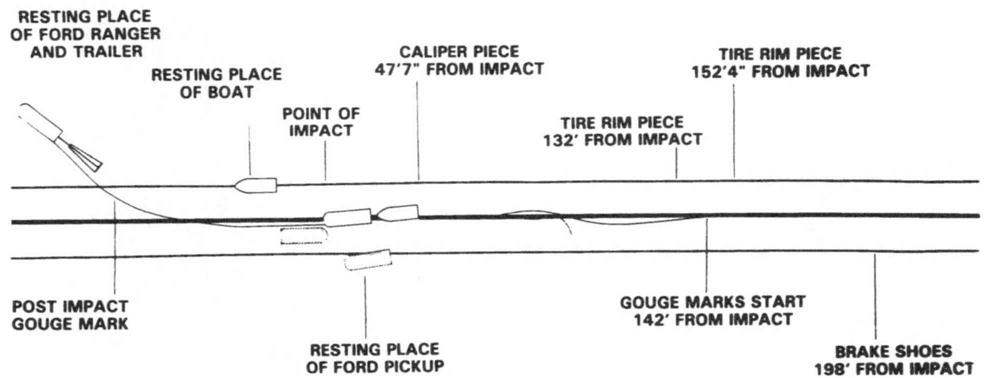

Fig. 1 A schematic diagram of the accident site.

Fig. 2 The Ford pickup truck after the accident.
(Photograph by Philip H. Schaad III)

Fig. 3 The Ford Ranger after the accident.
(Photograph by Philip H. Schaad III)

3 Examination of the Ford Ranger

Early in the investigation it was noticed that the spindle on which the bearings of the left front wheel were mounted had separated from the steering knuckle. It remained with the wheel when the wheel separated from the vehicle. This attracted attention because the spindle stem is locked into its mating hole in the knuckle by a shrink fit as the truck is manufactured. The joint is not intended to come apart under any circumstances. The spindle design for the 1984 Ranger is such that retention of the spindle stem in its mating hole depends entirely on the shrink fit. There is no other means of retention.

The spindle was removed from the wheel and cleaned. It is shown adjacent to its mating hole in Figure (4). It was found to slide into the hole quite easily but it would not go all of the way. The stem exhibited pitting corrosion with dark red-brown corrosion products along about 80% of its length. This is shown in Figure (5). Evidently there existed a gap between the stem and its mating hole over about 80% of the length of the joint. The shrink fit was operating over only about 20% of the joint.

A series of measurements were made on both the spindle stem and its mating hole. Initially an impression of the hole was made using a precision dental casting stone. Systematic measurements were made on the stem and the stone model of the hole using vernier calipers. All of the measurements on the spindle stem came within the range 3.31 ± 0.02 cm. There were therefore no major variations in diameter along the spindle stem. The measurements on the stone model of the hole showed that there was significant flaring at the outboard end of the hole. The maximum diameter measured was 3.45 cm and this was along a diameter which was roughly parallel to the ground when the knuckle was in place on the vehicle.

These approximate measurements enabled us to map out the main dimensional features of the spindle stem and its mating hole. These were followed by precise measurements using a Cordax 1808 machine with an MP3 Smart Terminal at the Vermont American laboratories in Louisville, This machine will measure both internal and external diameters to about ±0.0002 in (±0.005 mm). Figure (6) shows a Cordax computer printout showing the dimensional relationship between the spindle stem and its hole along the horizontal diameter. This makes it obvious why the stem will go about 80% of the intended distance into the hole and no more. Evidently there could never be an adequate shrink fit between these two components.

Examination of the wheel revealed that the rim on the inboard side was damaged and about a quarter of it had broken off. The tire was torn and deflated. On the outboard side the wheel was essentially undamaged. This pattern of damage was considered unusual because a collision usually results in damage to the outboard rather than inboard parts of the wheel.

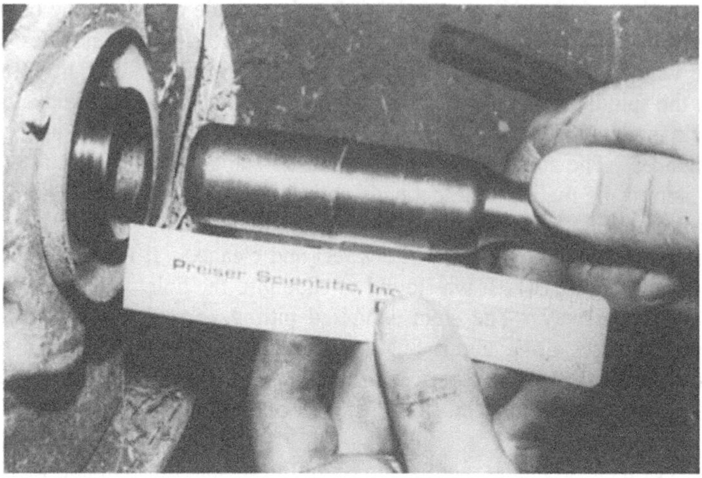

Fig. 4 The spindle and the hole in the steering knuckle with which it mates. (Photograph by William Smock)

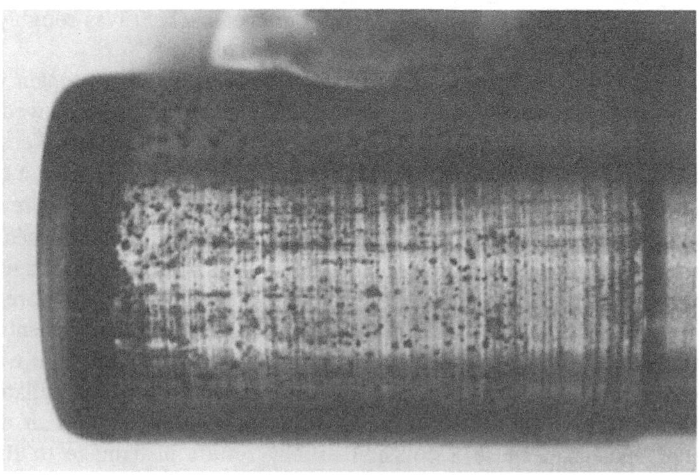

Fig. 5 Pitting corrosion on the spindle stem. (Photograph by Alan Johnson)

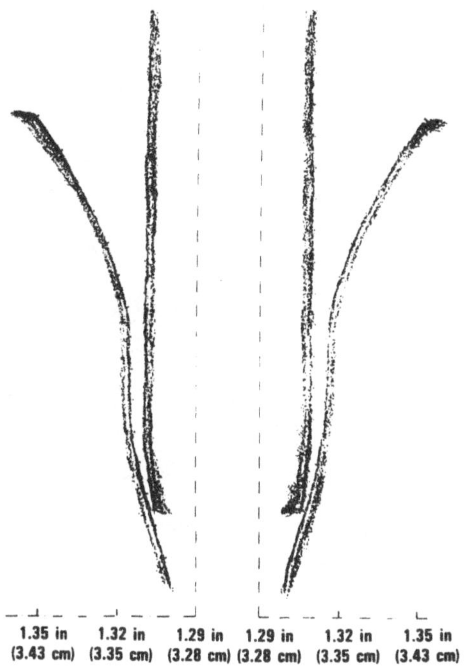

Fig. 6 Measurement of the diameter of the spindle stem and of the internal diameter of its mating hole made using a Cordax machine.

The tapered bearings were taken out of the wheel and were examined for Brinell marks, i.e. marks caused by the driving of the bearing's rollers into the races. Such marks are usually found when a wheel is in place during a collision and is impacted. Brinell marks were found on three adjacent rollers (Fig. 7) on the inner bearing. Matching marks were found at the inboard edge of the outer race (Fig. 8). Thus, the rollers and the race were not in their normal relationship when the Brinell marks were formed. No Brinell marks were found on the rollers of the outer bearing but there were faint marks on the outer race, one corresponding to each roller. These may have been caused by "false brinelling" [2].

Examination of the filaments of the brake lights showed that they had undergone "hot shock", i.e. that the coiled filaments had become distended as a result of the collision. This can only occur if the filaments are incandescent at the instant the collision occurs. Filaments which are not lit are brittle and usually break up into pieces [3].

Many other observations were made during the examination of the vehicle. We have presented those which played important roles in the development of our reconstruction of the accident.

Fig. 7 Brinell marks on the three adjacent rollers in the inner left front wheel bearing of the Ranger. (Photograph by Detective Joseph Ash)

Fig. 8 Brinell marks on the outer race of the inner bearing corresponding to those shown in Fig. 7. (Photograph by Detective Joseph Ash)

4 The reconstruction

As the reconstruction team discussed and digested the evidence a consensus emerged that the accident had been caused by the defective shrink fit between the left front spindle and knuckle. Complete unanimity as to the exact sequence of events which triggered the accident was not achieved but, in outline, it was concluded that the sequence was roughly as follows.

As the truck was used in normal service the defective joint gradually loosened. The spindle shaft gradually moved out of its mating hole. This caused relaxation of the preload on the wheel bearing causing the wheel and its attached brake rotor to wobble. As the spindle stem came further out of its hole the geometrical relationship between the rotor, which was then wobbling, to the brake caliper assembly, which is mounted on the knuckle, changed. Exactly what happened at this point is not clear but probably the spindle became fully disengaged from its hole and the spindle, rotor and wheel assembly interacted with the spindle and brake caliper assembly. Pieces of the assembly such as the brake shoes and part of a caliper were scattered along the road surface. The wheel was damaged, perhaps on being forced under the front of the truck after separation, and pieces broken off the rim were scattered on the road surface.

The Brinell marks on the bearing rollers and races were created at some point after the spindle came loose as evidenced by the fact that the marks show that the rollers and the race with the mating marks were not in their correct geometrical relationship. They may have been created as the wheel and spindle stem separated.

As the disintegration occurred the driver applied his brakes but the vehicle could no longer be steered. It was out of control with the dust shroud and front left suspension in contact with the ground. The vehicle crossed the center line and collided with the oncoming Ford pickup truck.

Ford's team of experts came up with a very different reconstruction. They denied that the spindle stem/knuckle assembly was defective, claiming that it could not have been because each assembly is tested with a 23,625 lb. pushout test at the manufacturing plant. Witness marks indicating that the test was indeed carried out can be seen on the end face of the spindle. They also used the existence of Brinell marks on the bearing rollers and races to claim that the wheel must have been in place when the collision occurred. Using these arguments and others they made the case for a simple scenario where a sleep-deprived young man fell asleep at the wheel causing his vehicle to cross the center line of the road and collide with the oncoming pickup truck.

5 The first jury trial

The first trial of this action took place in June, 1987, in a court in Louisville, Kentucky. Crucial to the first trial was the legal question of whether the 1984 Ford Ranger spindle/knuckle assembly was in a defective condition unreasonably dangerous at the time of the collision between the two vehicles. The basic factual question was whether the wheel and tire assembly separated prior to collision or upon collision. The applicable legal standard in Kentucky in a products liability claim is governed by a national model standard called the Restatement, Second, Torts, §402A, which holds the seller of a product is liable for injury to a user or consumer if the product is sold without substantial change in condition in a defective condition unreasonably dangerous.

Pursuant to this Rule, the Kentucky Supreme Court has consistently held that considerations in a products liability case, such as feasibility of making a safer product, patency of the danger, warnings and instructions, subsequent maintenance and repair, misuse, and the product's inherently unsafe characteristics, while they have a bearing on the question as to whether the product was manufactured in a defective condition unreasonably dangerous, are all factors bearing on the principle question rather than the separate legal questions [4, 5]. Accordingly, the plaintiff's purpose in the first trial was to merely show that the product was defective and that it would not, if designed and manufactured properly, separate prior to collision with the Fulkerson vehicle. Ford's goal was to show that the wheel and tire assembly had separated after collision.

The plaintiffs retained their own reconstruction team, which included John Stilson, a former Ford engineer who had worked on the design of components for the Ford Ranger, and Jerry Wallingford, an engineer and accident reconstructionist who was part of a larger testing facility doing independent tests for manufacturers of vehicles. Stilson concentrated on defects in manufacture and design of the spindle stem, whereas Wallingford presented an expert reconstruction of the collision. Stilson's opinion corroborated that of the state medical examiner's reconstruction team to the extent that he found that the lack of a secondary retention feature on the spindle stem constituted a design defect, which in this case proved fatal upon a failure of the shrink-fit. Dr. Alan Johnson provided the metallurgical analysis of the failure of the shrink-fit. Wallingford concluded in his reconstruction that indeed the wheel had separated prior to collision which, lacking evidence of other trauma to the wheel, could only have been caused by a defect in design, manufacture or both.

Ford's primary defense in the first trial was that the presence of Brinell marks on the race and bearings was clear evidence that the spindle stem had not separated until collision between the two vehicles. Stilson's opinion, as a counterpoint, was that the impact of the bearings upon the race when the wheel and tire assembly separated prior to collision caused the markings and that the off-centered nature of the markings indicated that the bearings were not properly seated upon collision. It was his opinion that, had the impact been the cause of the separation of the wheel and tire assembly, the Brinell marks would have been centered upon the race rather than offset (Figure 8).

Ford presented the testimony of James Varin, a Ford engineer and metallurgist; Dennis Guenther, an automotive engineer and Professor at the Ohio State University; and Richard Eisenberg, a metallurgist. Varin and Eisenberg concentrated on the spindle stem assembly and based their conclusion that it was not defective on what they characterized as brinelling of the race and bearing. This indicated to them that the impact of the two vehicles had separated the wheel and tire. Guenther is an accident reconstructionist and, while he was unable to explain the gouges and scrapes preceding the collision, rendered the opinion that the wheel and tire had separated upon collision, relying primarily on post-collision scrapes, gouges and debris.

Significant evidence not presented to the jury were changes Ford had made in the design of the spindle stem assembly of the Ranger after its release. The Plaintiffs sought to introduce evidence of a pattern of design change, which included the swaging or flaring of the in-board head of the spindle assembly for retention purposes as evidence of the defective nature of the shrink-fit method. The trial court disallowed use of changes in design under a theory known as "subsequent remedial measures" which is a public policy based legal principle. Essentially, this principle means that if a manufacturer corrects a product defect it should not be punished by having that fact used against it in a trial.

The trial court's instruction to the jury on the law prior to their deliberations undermined the strict liability concept advanced by the plaintiffs by requiring them to prove that the product was defective either on the basis of a manufacturing or design defect regardless of whether the wheel and tire assembly separated prior to collision. The jury could determine that the wheel and tire had separated prior to collision but, if it had not determined that the plaintiffs had proved design or manufacturing defects, still find against them.

The result of the first trial was that the jury found in favor of Ford and determined that the driver of the Ranger was at fault. In discussions with jurors after the verdict, plaintiffs' attorneys found that they were quite distressed because, while they felt that the wheel and tire had separated prior to collision, the court's instructions had limited their actions.

The Supreme Court of Kentucky reversed the Trial Court holding that the issue of strict liability was not properly presented to the jury and, of critical importance in the second trial, altered the law of Kentucky concerning the introduction of design changes. The Court instructed the trial court on retrial to admit the design changes on the grounds that they were relevant in proving the defect [6].

6 The second jury trial

The second jury trial altered the dynamics of the situation in its entirety. It was clear that this time the trial court would properly instruct the jury on the evidentiary standard in a strict liability case and that evidence of subsequent design changes to the spindle stem on the Ranger were admissible as evidence. However, several events which occurred between the two trials altered the course of the second trial dramatically.

The first event was the entrance of Larry Bihlmeyer, another former Ford engineer, who had originally been involved in the development of the shrink-fit assembly designed for the Ford F-Series vehicle and later used on the 1984 Ford Ranger. At the time of the original trial, Bihlmeyer was still employed by Ford Motor Co., and therefore unavailable to the plaintiffs. By the time of the second trial, however, Bihlmeyer had left Ford and entered into a private consulting business. Plaintiffs' original expert on design defect, John Stilson, was a colleague of Bihlmeyer's and recommended his use in the case due to his detailed knowledge of the design of the shrink-fit.

To say that veritable "flood gates" opened upon Bihlmeyer's entry into the case would be an understatement. Plaintiffs' counsel developed an entirely different strategy for the trial of the case with the addition of Bihlmeyer and gained access to information through Bihlmeyer's knowledge of Ford Motor Co. not previously available, even though in pre-trial preparations they had requested the information from Ford. For instance, Ford had claimed in the original trial that it had not designed the swaged or flared end on the spindle stem for the Ranger as a secondary retention feature, but rather had designed it as a cost saving manufacturing process. Bihlmeyer assisted counsel for the plaintiffs in requesting specifically numbered product change requests from Ford Motor Co. which stated that the change to the spindle stem was a retention feature and not a cost saving feature. This was demonstrative of the defect and showed the necessity for the admission of evidence of design changes subsequently to the identification of a defect.

Additionally, Bihlmeyer was keenly familiar with the original design of the spindle stem for the Ford Ranger which was essentially identical to the F-Series spindle stem. This design included a head and shoulder on the in-board side of the spindle shaft which made it impossible for it to separate from the vehicle should a manufacturing defect cause the shrink-fit to fail. As an engineer at Ford, Bihlmeyer had been advised by his superiors that he should never design a spindle stem based on the shrink-fit without a secondary retention feature such as the shoulder or head that appeared on the original Ranger design. All of this was unavailable at the original trial.

Plaintiffs' strategy in the second trial required a shuffling of the players. Bihlmeyer was used as the design and manufacturing engineer with in-depth knowledge of the product. Stilson had extensive background in accident reconstruction and therefore became the privately secured accident reconstructionist. A scale model of the scene including its topography was constructed and used as a graphic display to reconstruct the collision. In addition, a large display was used with each witness to help the jury retain their recollection of the location of critical pieces of the vehicles, scrapes, and gouge marks on the highway.

The testimony of the original group of experts assembled by the Kentucky Medical Examiner was given first and the independence of their work was stressed. The privately retained experts then used this testimony as a spring-board from which to add their own special knowledge of the design of the vehicle and the workings of the Ford Motor Co.

Not only did this prove effective in presenting the case to the jury, but in the second trial the plaintiffs sought punitive damages for the gross negligence of Ford Motor Co. in permitting this design on the market when it clearly understood it to be unsafe at the outset. In the first trial, plaintiffs withdrew the punitive damage claim at the close of the evidence because they simply did not have available to them the product change requests, and the other information now provided by Bihlmeyer, demonstrating that Ford realized significant cost savings by the introduction of this design. Additionally, there was a pattern throughout the history of the Ford Ranger of trying to design around the defect instead of withdrawing or recalling the product. The cost of such a recall would have been substantial. The piece de resistance in the evidence of design change was the implementation in 1992 of a new spindle stem which included the shoulder and head that was the foundation of the original design for the Ford Ranger, and which, in the final analysis, would have prevented the injuries to Fulkerson and the death of Cilone.

Finally, the cross examination of Ford's expert witnesses was more effectively carried out in the second trial than the original due to the greater depth of information available and the change in tactics. In the original trial, the accident reconstructionist for Ford Motor Co. had been extremely effective in his presentation but, in the second trial, was somewhat neutralized due to his failure to test the Ranger for hot-shock deformity in the brake light filaments. An examination of these filaments by plaintiffs' experts had indicated an engagement of the brakes by the driver of that vehicle prior to collision. This evidence led to the conclusion that some event had taken place, such as separation of the wheel prior to collision, causing the driver of that vehicle to brake.

Additionally, a fortuitous event took place in the second trial. On the bumper of the 1972 Fulkerson pick-up truck was a substance transference, black in color, which appeared to come from one of several sources including the front left tire of the 1984 Ford Ranger. This substance had not been precisely identified. In the first trial, Ford's experts had contended that this was a tire transference showing that the left front tire had been in place upon collision. Some of the photographs seem to confirm this, and it was quite an effective tool with the jury. During the course of the second trial, Ford took a calculated risk and requested the trial court, in the middle of trial, to have this substance tested by the Kentucky State Crime Laboratory which request the court granted. To Ford's dismay, the results showed that the substance was not from a tire and, while the crime laboratory could not exactly identify this substance, its composition was consistent with the plastic liner in the wheel well of the left front tire. This bolstered the plaintiffs' claim that the tire was not in place upon collision with the left front bumper of the Fulkerson vehicle.

In researching the case it had been found that two of Ford's experts, Varin and Eisenberg, had testified in a related case in Alabama, in which a 1985 Ranger went out of control and hit an embankment killing the driver. Ford's position in that case had been that the impact with the embankment had caused the wheel to separate. Interestingly, the 1985 vehicle included the swaged or flared in-board spindle stem design, which apparently proved ineffective. Varin and Eisenberg provided testimony in that case diametrically opposed to what was provided in this action. Both individuals had concluded in that case that it was clear to them that the wheel and tire had not separated prior to the individual's collision with the dirt embankment because the roadway did not bear evidence of the characteristic scrapes and gouges that one would expect to find if the wheel and tire had separated prior to impact. On the other hand, in the Kentucky action they had testified that the scrapes and gouges were irrelevant and not associated with the wheel and tire separating prior to collision. Plaintiffs' attorney had obtained copies of their testimony in consultations with the attorney handling the case in Alabama and used it in cross examining them at trial thereby undermining their credibility with the jury. This was effective in developing the case for punitive damages.

The evidence also showed that Ford recognized that the spindle/knuckle joint on the 1984 Ranger was defective immediately after the vehicle was introduced. Discussions between Ford and the spindle manufacturer aimed at developing a positive retention feature were initiated within months of the introduction of the vehicle.

The final result of the trial was the verdict for the Plaintiffs for $2,300,000 in compensatory damages and $4,000,000 in punitive damages. The case is presently on appeal and the Kentucky Court of Appeals has affirmed the judgment of the trial court. Ford has requested the Supreme Court of Kentucky to review the case.

6 References

1. Herrensmith, M. (Plaintiff) and Fulkerson, W.F and P. (Intervening Plaintiffs) vs Ford Motor Company Co. (Defendant) and Herrensmith, M. (Intervening Defendant), Twelfth Division, Jefferson Circuit Court, Kentucky, File No. 85-CI-06869.
2. Anon (1980) Bearing Failure Prevention Guide, Fafair Bearing Division of Textron Inc., New Britian, Conn.
3. Harmeling, D.E. and Johnson, A.A. (1993) Metallurgical theory underlying post-impact lamp filament examinations, J. Accident Reconstruction Vol. 5, issue 5, pp. 52-56.
4. Nichols vs Union Underwear Co., Inc. (1980), Ky., 602 S.W. 2d 429.
5. Montgomery Elevator Co. vs McCullough (1984), Ky., 676 S.W. 2d 776.
6. Ford Motor Co. vs Fulkerson et al. (1991), Ky.,812 S.W. 2d 119.

7 Acknowledgments

The authors are indebted to Randall Storey who drew Figure (1), Patricia Lumley who typed the manuscript and Barbara Johnson who proofread it.

17 CONCRETE DETERIORATION IN AN OLD OIL REFINERY IN THE ARABIAN GULF - CASE STUDY

O.K. DAOUD
Dar Al-Handasah Consultants, Egypt
I.A. IBRAHIM
Department of Civil Engineering, Cairo University, Cairo, Egypt

Abstract
The high degree of cracking and corrosion occurred to concrete footings in a 45 years old oil refinery in the Arabian Gulf is assessed. A combined investigation using visual inspection and laboratory testing revealed that the concrete above ground, exposed to the environment had suffered from high concentration of chloride and sulfate contents, higher carbonation, lower cement content and pH value than buried concrete. Also, lower compressive strength and density, higher voids content and absorption were obtained for exposed concrete. This investigation shows that the major design criterion for concrete in industrial facilities such as oil refineries should be resistance to harsh environment.
Keywords: Carbonation, cement content, chloride attack, concrete deterioration, concrete properties, industrial facilities, oil refineries, sulfate attack

1 Introduction

Concrete is one of the most durable man-made materials currently in use. Advanced technology in materials and concreting techniques has improved the performance of concrete in the past few years to unbelievable limits of high strength and physical properties. However, these advances are not applied to everyday concrete works, only the few elite receive such special treatment under very tight control. The majority of concrete structures are made of conventional concrete components in order to maintain their cost at competitive levels with other building materials. This is particularly true in projects built in the Arabian Gulf states during the construction boom in the 50s and up till the 80s. Several studies reported the deterioration in industrial projects [1,2] and office buildings [3] in the Gulf states. However, these studies included limited

Structural Failure, Product Liability and Technical Insurance, Proceedings of Fifth International Conference. Edited by H.P. Rossmanith. Published in 1996 by Taylor & Francis, 2 Park Square, Milton Park, Abingdon, Oxon, OX14 4RN. ISBN 0 419 22080 1

information on the field performance of concrete as related to the prevailing environment due to the limited number of samples involved.

This paper shows a case study for the performance of conventional concrete used some 40 years ago in the harsh environment of the Arabian Gulf and under severe operating conditions in one of the oldest oil refineries in the area. In order to have sound statistical results, a large number of concrete cores were cut from the footings supporting different equipment in the refinery . The study shows that many lessons are to be learnt from the failure of the concrete under these conditions. In the following sections a brief description of the research study is given together with a summary of the results obtained.

2 Description of the project

The project under study is an oil refinery located on the Gulf shore line, some 80 km south of Kuwait City. The refinery was commissioned in 1949 with a limited number of production units. Many additions have been made since then, but many of the old production units are still in operation. Some have been put out of operation due to aging equipment or the introduction of new technology. All the units, totalling 63, used concrete footings to support the refinery equipment which varies in shape, size and function. The footings in some areas suffered serious damage regardless of age or location, see Figs. 1 and 2.

Fig. 1. Badly cracked footing supporting a steel frame.

Fig. 2. Highly spalled footing supporting a storage tank.

The cost of the repair work escalated in the past few years to a very high level which warranted the need to analyse the causes of failure and the associated lessons for future extensions. A pilot investigation was commissioned to study 7 units, some of which were among the oldest units in the refinery.

The following units were included in the study with their age as of 1990:

- Platformer aged 32 years (out of operation now).
- Crude Distillation Unit No.3 aged 32 years.
- Fractionating Unit No.1 aged 27 years.
- Fractionating Unit No.2 aged 24 years.
- Washery aged 40 years.
- Production Loading Pump house 32 years (out of operation now).
- Bitumen Plant aged 11 years.

3 Research programme

The 7 units contained over 1000 concrete footings. The data available on the design requirements and construction records were quite limited. However, a thorough review of these data was necessary to diagnose the causes of failure and the sources of distresses. The footings were plotted on dimensioned plans using the as-built drawings, and were numbered according to their location, on the plan and on the site. A data collection sheet was developed to record the footing shape, dimensions, condition, and other physical information during the visual survey. A computer spread-sheet was used to store these data for quick retrieval of information.

Training of the team conducting the visual survey was important to unify the bases of the assessment and to eliminate the human factor. During the survey, remarks were recorded for possible locations for core cutting. This depended on the types of distress observed or the degree of exposure to aggressive environment. Later on, the team plotted the locations of footings to be sampled on the plan to ensure that they were well distributed in the unit.

Some 102 concrete cores, 100 mm in diameter, were cut under meticulous safety conditions to avoid any sparks. The cores were evenly distributed between concrete exposed to external environment (above ground) and buried concrete (under ground), between good, medium and poor condition footings and between high, medium and low importance footings.

The cores were examined visually upon arrival at the laboratory and were photographed and labelled. Figs. 3-5 show examples of highly deteriorated cores with evidence of poor concreting practice. Each core was divided into 3 portions to measure its chemical and physical properties. The surface layer was used to determine the profile of sulfate, chloride, cement, $CaCO_3$ contents and pH value. The middle portion was used, for compressive strength, E modules, ultra pulse velocity and density; and the last portion was used for voids ratio, absorption and specific gravity

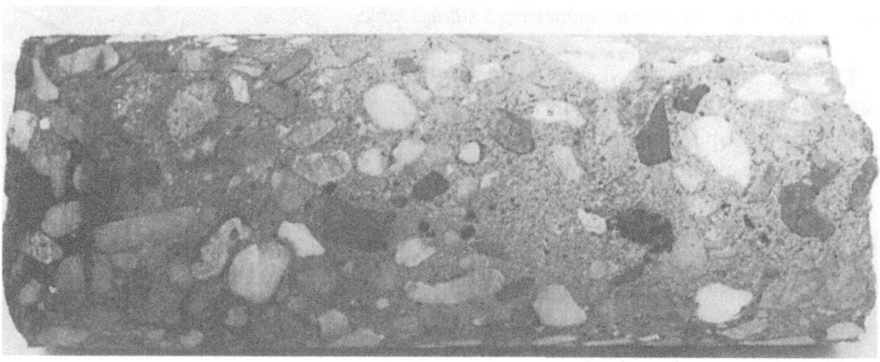

Fig. 3. Large voids in a core indicating high water/cement content and poor compaction.

4 Results and discussion

4.1 Chemical analysis

The results indicated that the chemical composition of the outer skin of concrete varied widely between buried and exposed concrete. Exposed concrete contained on average, 4-5 times more chloride. This aggravated the corrosion of steel reinforcement and produced an alarming degree of cracking. The source of the chloride was the daily wash of the concrete floor using sea water from the fire hydrant system. The average chloride content in exposed concrete at the steel level was 0.75% of the cement weight. While buried concrete contained 0.15% only, which is the maximum allowed by most design codes, individual results as high as 3.7 % were obtained in some areas.

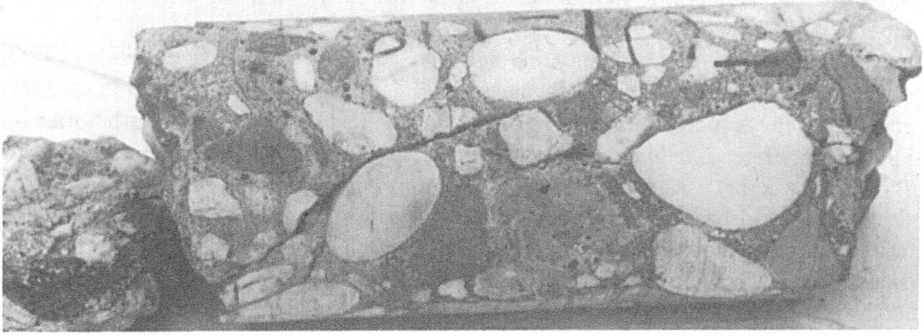

Fig. 4. Badly cracked core showing large rounded, unevenly-distributed aggregate and corroded steel bar.

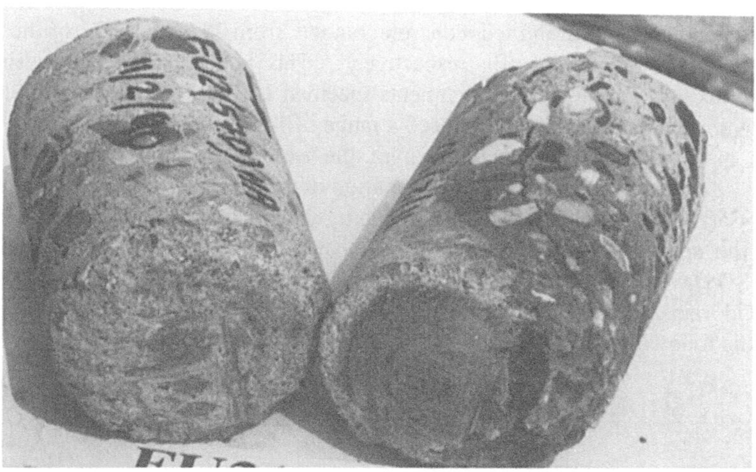

Fig. 5. Two piece core indicating the existence of repair work (the left-hand portion) and cracked original concrete (the right-hand portion). Notice the piece of wood separating the two portions indicating poor quality control during the repair work.

The average sulfate content in the outer 10 mm of concrete was 3.7 and 3.2% of cement weight for exposed and buried concrete, respectively. Individual results as high as 8 and 5.2% were obtained for exposed and buried concrete, respectively. The sulfate source could either be from the basic components of the concrete (cement and sand), or from the H_2S and SO_2 gases released during the refinery operation. These gases when combined with water or humidity form acid rain which attacks the concrete forming ettringite. Ettringite is highly water absorbent and expands to 5-6 times its volume when exposed to water, causing internal cracking in concrete. Ettringite, however, is highly soluble in chloride-rich water. Thus the daily wash by sea water may have moderated the effect of sulfate attack.

Carbonation reduces alkalinity in concrete and hence increases possibility of steel corrosion. Carbonation extent is measured by the $CaCO_3$ content at the steel level (5cm from the concrete external surface). The results showed that both buried and exposed concrete contained the same average level of $CaCO_3$ at 50 mm depth (26.5% of cement weight). At shallower depth (10 mm) this was increased to 65 and 93% for buried and exposed concrete, respectively. At 50 mm depth individual readings of $CaCO_3$ content for exposed and buried concrete ranged up to 60% and 40%, respectively. The above results show that steel corrosion observed in the footings is mostly attributed to chloride attack rather than carbonation. These results compared well with the pH readings obtained at the same level. The average pH value of exposed concrete at the outer skin (0-10 mm) was 10.5 with individual values as low as 9.1, while buried concrete produced an average and minimum values of 11.4 and 11, respectively. At steel level, the pH average values were 11.8 for both buried and exposed concrete. Normally in well preserved concrete the pH value is between 12 and 12.5.

The cement content of buried concrete ranged from 28.7 to 31.7% of the mortar weight at 10 and 50 mm depth, respectively. This is the expected value since the concrete mix given in the design documents specified 1:2:4 mix proportion for cement: sand: gravel. Exposed concrete produced a range of 13.2 to 29.3% for the same levels. The loss in cement near the outer skin of the exposed concrete is attributed to the leaching. Leaching is caused by the dissolution of $Ca(OH_2)$ by water containing CO_2 forming $Ca(HCO_3)_2$ which is leached out due to water attack. Sulfur dioxide escaping through the smoke stacks combines with atmospheric moisture and forms a sulfurous acid (H_2SO_3) which gradually reacts with oxygen to form sulfuric acid (H_2SO_4). Both create acid rain which is aggressive to steel and concrete alike. This is explained through the following equations:

$$CaCO_3 + \xrightarrow{SO_2 \cdot H_2O} CaSO_3 \cdot \frac{1}{2} H_2O + CO_2 \xrightarrow{O_2 \cdot H_2O} CaSO_4 \cdot 2H_2O \quad (1)$$

and

$$CaCO_3 + H_2SO_4 + H_2O \rightarrow CaSO_4 \cdot 2H_2O + CO_2 \quad (2)$$

The above component in Equ. (2) will react with cement components and produce ettringite, known also as calcium sulfoaluminate, following this reaction:

$$3(CaSO_4 \cdot 2H_2O) + 3CaO\, Al_2 \cdot O_3 \cdot 6H_2O + 20H_2O \rightarrow 3CaO.Al_2O_3 \cdot 3CaSO_4 \cdot 23H_2O \quad (3)$$

Ettringite is easily soluble in chloride and is washed away during the daily wash by sea water. The high voids ratio observed at the surface, as will be discussed later, may have contributed to the high sulfate attack on the concrete. These voids could be the result of poor curing during the construction phase or high water/cement ratio [4]. It could also result from the continuous leaching of calcium sulfate or gypsum. Aggression due to SO_2 attack is further increased if the ambient temperature is higher than 25°C, which is the case in Kuwait.

4.2 Physical properties

The exposure of concrete to the refinery environment was found to degrade noticeably its physical properties. Table 1 shows that the compressive strength was reduced from an average 41.6 MPa for buried concrete to 30.9 MPa for exposed concrete *i.e.* 25% reduction. Voids ratio and absorption increased by 30%. Similar results were obtained for E modules and UPV, where a 31% and 12% reduction were obtained, respectively. Bulk and dry densities were reduced by 0.9 and 2.52%, respectively.

Cohen and Mather [5] reported the failure criteria for the loss of mass and strength drop resulting from sulfate attack. Previous research quoted in ref. 5 defined loss of mass limits of 5% and 2.5% for beams and cubes, respectively. The loss of mass for a beam is double that for a cube because of the higher surface-to-volume ratio. Strength drop limit was defined at 25%. Accordingly, assuming that the buried concrete properties represent the original material, exposed concrete in this refinery has performed unsatisfactorily.

Table 1. Mean values for properties of buried and exposed concrete

Property		below ground	above ground
Absorption	(%)	5.34	6.98
Voids ratio	(%)	12.74	16.45
Bulk density	(t/m^3)	2.36	2.34
Dry density	(t/m^3)	2.26	2.21
UPV	(m/s)	4471	3931
Compressive strength	(MPa)	41.6	30.9
E - Modulus	(MPa)	15893	10925

These reductions in the physical properties are attributed to two main factors:

1. Leaching due to the carbonic and sulfuric acids attacks which is explained above.
2. Internal cracking due to cyclic heating and wetting and drying. The arid climate of the Arabian Gulf causes a thermal range exceeding 30 C° daily in summer. Also, in the refinery heat is dissipated from high towers, stacks, boilers and other equipment. Exposed concrete subjected to these conditions will experience thermal movements creating internal cracking; whereas buried concrete is protected by the surrounding soil and thus not exposed to such harsh treatment. Wetting and drying by sea water used in the daily wash of the footing cause continuous build up of these salts in the

concrete. These salts crystallize when water evaporates with increase in volume which causes increased internal pressure and cracking.

Furthermore, the exposure to the environment was found to alter the relationships between the physical properties of concrete [6]. This was attributed to the loss of the binding effect of cement in the exposed portion of concrete due to the sulfate attack and other detrimental factors.

5 Lessons Learnt

- Under severe environmental or operational conditions in industrial facilities the prime design criterion for concrete mix should be durability. Strength may not be as important when compared to durability in these plants. The shutting down of such facilities to repair damaged concrete may be far more expensive than the initial investment of producing high performance concrete. High cement content, low water/cement ratio and good curing will guarantee dense concrete and will provide protection to the concrete from ingress of harmful gases and liquids.
- Insulating concrete surface by a protective layer is a must in oil refineries. Good concrete alone may not be sufficient to maintain the concrete in operational condition throughout the lifetime of these facilities. It is more important to maintain this layer periodically. Thus the maintenance staff should be aware of the expected lifetime of the protective layer. This life is to be considered in the light of the harsh environment to which the material is exposed and not the life mentioned in the manufacturer's brochure which is normally applicable only to temperate climates.
- The need to develop a systematic method to assess concrete damage for various structures cannot be over-emphasized. More work in this field will provide good tools for the maintenance staff in important facilities to evaluate damage in concrete structures. A guide for sampling and testing may be helpful in diagnosing failures at the early stages when repair cost will be minimum.
- The analysis of concrete deterioration due to the combined attack of different factors needs a team of specialists in wet and analytical chemistry, geology, materials and metallurgical engineering. With such a team, the cause of failure can be diagnosed accurately and hence the proper selection of the most appropriate repair measures is possible.
- It is very important for designers to study carefully the macro-environment and operational conditions to which the structure they design wil be exposed during its life span. Designers used to tamperate climates should consider the prevailing hot weather in the Middle East and the lack of periodical maintenance in this area and make the necessary modifications in their designs to coop with these conditions.

6 Acknowledgment

The authors are grateful to Dar Al-Handasah Consultants for providing facilities to produce and present this paper, to the Kuwait Institute for scientific Research (KISR) for financial support and to Mr. G. Al Qazweeni and Mr. K. Hijab, of the Civil Engineering and Building Dept., KISR for their help and assistance.

7 References

1 Shalaby, H.M. and Daoud, O.K.(1990) Case studies of deterioration of coastal concrete structures in two oil refineries in the Arabian Gulf Region. *Int. J. of Cement and Concrete Research,* Vol. 20, pp. 975-985.

2 O'Connor, J.P. (1994) Middle Eastern Concrete Deterioration: Unusual Case History *J. of Performance of Constructed Facilities,* Vol. 8, No. 3, pp. 201-212.

3 Qazweeni, J.A. and Daoud, O.K. (1991) Concrete deterioration in a 20-year-old structure in Kuwait. *Int. J. of Cement and Concrete Research,* Vol. 21, pp.1155-1164.

4 Knöfel, D. and Röttger, K. G., Contribution on the Influence of SO_2 on Cement Mortar and Concrete, Mortar and Concrete, Conf. Proc. University Siegen, Germany.

5 Cohen, M.D. and Mather, B. (1991) Sulfate Attack on concrete-Research Needs, *ACI Materials Journal,* V. 88, No. 1, Jan-Feb, pp. 62-69.

6 Daoud, O.K. and Ibrahim, I.A. (Under publication) Impact of environmental and operating conditions in oil refineries on concrete properties. *ACI Materials Journal.*

18 ENGINEERING ANALYSIS OF THE EXXON VALDEZ TRACKLINE PATH AND GROUNDING

R. KADLEC and E. BERES
Failure Analysis Associates, Los Angeles, California, USA
D. HAYES
Failure Analysis Associates, Houston, Texas, USA

Abstract

Shortly after midnight on March 24, 1989, the *Exxon Valdez* went aground in Prince William Sound. This paper presents the reconstruction of the ship's path and control inputs for the half hour preceding its grounding. Additionally, this paper presents engineering analyses of the grounding which quantify vessel damage, and interpret this damage to determine how the vessel encountered the reef.

The quantitative analysis of the ship's path and control inputs includes an analysis of the ship position, time of position, speed, heading, and rudder setting. The analysis is based on interpretation of evidence such as reef survey data, printouts from the vessel's instrumentation, and witness statements, and includes a numerical simulation using a hydrodynamic model for the *Exxon Valdez*, calibrated from sea trials. The ship track analysis predicted the rudder position required to produce the observed vessel responses.

The engineering analysis of the grounding began with a quantitative inspection of the vessel in dry dock, prior to any repairs. Analyses of inspection data showed that vessel damage occurred in three separate phases, only the first two of which contributed to energy dissipation which brought the vessel to rest; the third phase was caused by tidal variation and wave action.
Keywords: course recorder, crude carrier, *Exxon Valdez*, fathometer, hydrodynamic model, trackline, sea trials

Structural Failure, Product Liability and Technical Insurance, Proceedings of Fifth International Conference. Edited by H.P. Rossmanith. Published in 1996 by Taylor & Francis, 2 Park Square, Milton Park, Abingdon, Oxon, OX14 4RN. ISBN 0 419 22080 1

1 Introduction

In the late evening hours of March 23, 1989, the Tanker *Exxon Valdez* left the Valdez, Alaska crude oil marine terminal bound for Long Beach, California, with a full cargo of North Slope crude oil. Several hours after departure from Valdez, the ship ran aground on Bligh Reef in Prince William Sound. After grounding, the damaged vessel released approximately 11,000,000 gallons of crude oil into Prince William Sound.

In the aftermath many important questions were raised about the causes for the grounding and who should be ultimately responsible. Critical to many of these questions was establishing a clear understanding of the sequence of events which led to the grounding and the detailed mechanical nature of the grounding impact sequence. This paper presents a detailed engineering analysis of the reconstruction of the ship's path and the vessel control inputs required to produce the vessel track for the approximate half hour preceding its grounding. Additionally, this paper presents engineering analyses of the grounding which quantify vessel damage, and interpret this damage to determine how the vessel encountered the reef, and the sequence and severity of damage.

2 Background

The *Exxon Valdez*, pictured in Figure 1, is a modern, very large crude carrier (VLCC) of all-welded structural steel construction. The vessel measures 987 feet long overall, 166 feet wide, and 88 feet deep from the main deck to the flat keel. The tanker has a maximum draft (fully laden) of 64.5 feet, with a corresponding dead weight of approximately 210,000 long tons. At maximum draft, the vessel can transport approximately 1.48 million barrels of crude oil per voyage.

Fig. 1. *Exxon Valdez.*

The vessel cargo space, forward of the pump house, consists of eighteen cargo, ballast, and slop tanks (see Figure 2). The space is divided into five tank sections, numbered 1 through 5, beginning at the bow, with a port wing tank, a center tank, and a starboard wing tank in each section. The segregated ballast tanks consist of the fore peak, No. 2 and No. 4 port and starboard wing tanks. The cargo oil tanks include No. 1 across, center No. 2, No. 3 across, center No. 4, and No. 5 across.

The *Exxon Valdez* was delivered to its owner on December 11, 1986 and was, at the time, the largest ship ever built on the U.S. West Coast. It is the first of two Alaska-class vessels designed and built for the Exxon Shipping Company (Exxon) by the National Steel and Shipbuilding Company (NASSCO) in its San Diego, California shipyard from plans approved by the U.S. Coast Guard and the American Bureau of Shipping. The vessel was designed to meet the standards of the International Convention for the Prevention of Pollution from Ships, 1973, as modified by the Protocol of 1978 (MARPOL 73/78). These standards provided for, among other items, protectively-located segregated ballast tanks, allowable tank compartment length, and damage stability. The vessel was certified by the U.S. Coast Guard for the carriage of crude oil products and flammable or combustible liquids Grade B or lower. The vessel was designed with a single bottom and met all applicable regulations of the day.

3 Analytical methodology

The engineering methodology applied to this study is consistent with our knowledge and experience of the analysis of large engineering structural failures and marine disasters. The analysis relies on failure analysis principles which include a quantitative documentation of the post-grounding vessel structural damage, a quantitative survey of the grounding site, full-scale vessel testing, and examination of available information and documentation describing the events, facts, and circumstances surrounding the grounding. This study employed contemporary engineering methods from disciplines including structural, mechanical, marine, and materials engineering. These disciplines were integrated in a synergistic manner to obtain the quantitative information required to describe the vessel motion, vessel control inputs and vessel grounding sequence.

As is often the case in failure analysis studies, the trackline and grounding sequence of the *Exxon Valdez* had to be established on the basis of partial known information regarding the vessel damage, the grounding site, and recorded information about the vessel's movement. The components of the complete analysis are depicted in Figure 3 and are composed of the following tasks:

- Vessel damage inspection and analysis
- Grounding site inspection and reef bathymetry survey
- Vessel sea trials and performance tests
- Course recorder data analysis
- Vessel trackline analysis, including development of an analytical hydrodynamic model

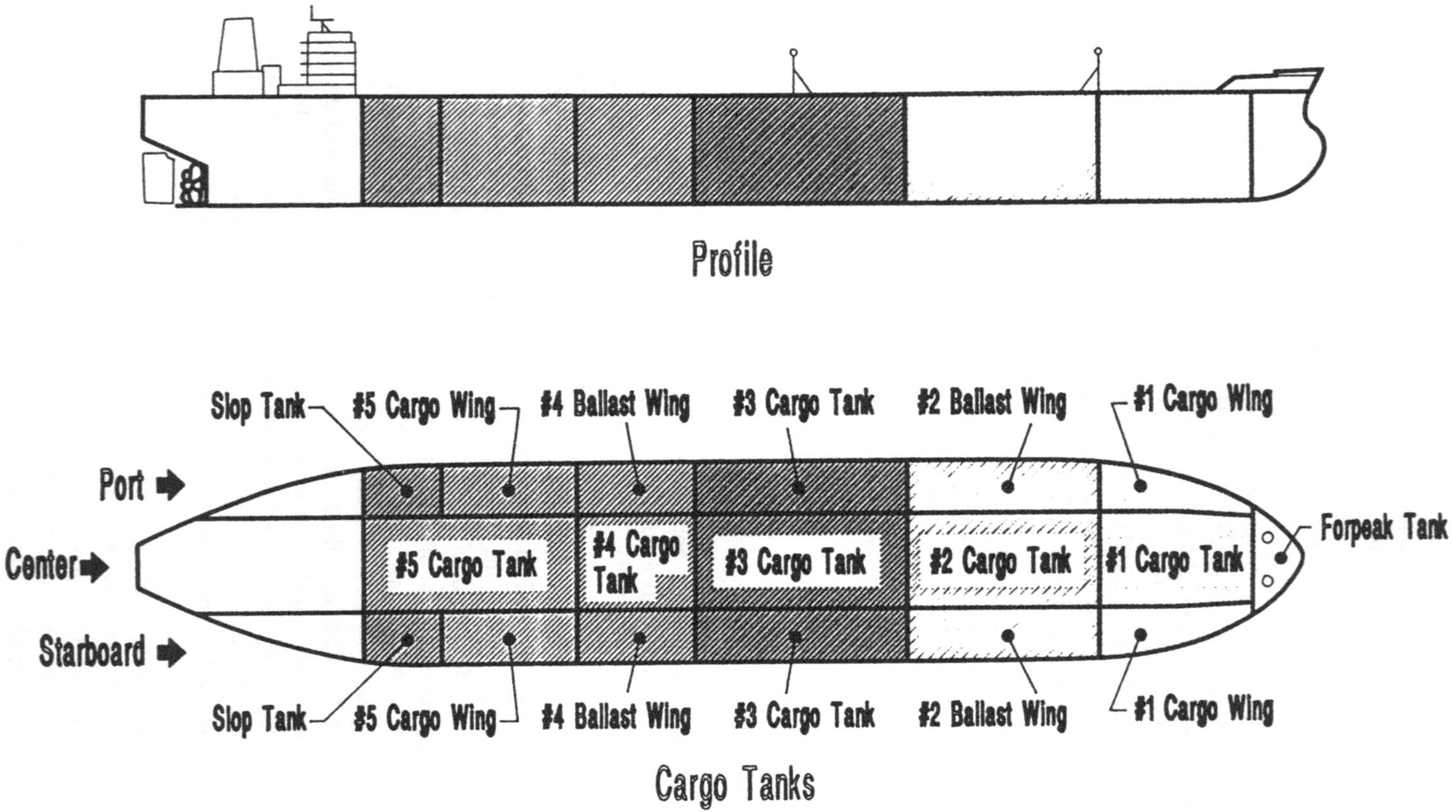

Fig. 2. Plan and elevation views of the *Exxon Valdez*: tank layout.

- Fathometer data analysis
- Final synthesis of results regarding trackline, timing, and control inputs

4 Vessel damage inspection and analysis

Introduction-scope

An engineering study and evaluation of vessel damage requires quantitative documentation of the vessel subsequent to the grounding. This task was accomplished by conducting a vessel inspection at San Diego, California, while the *Exxon Valdez* was in dry-dock and prior to any repairs. This information was supplemented by reports of changes to the vessel damage in transit from Bligh Reef to the NASSCO shipyards at San Diego, including documentation obtained while the ship was at Naked Island, Alaska and San Clemente Island, California.

Description of vessel inspection

Figure 4 depicts the damaged *Exxon Valdez* shortly after it arrived in dry dock. We conducted a stem-to-stern quantitative inspection of the *Exxon Valdez* from September 8 to September 14, 1989 at this setting with a 30-person engineering team. A detailed inspection protocol was developed prior to the inspection which served to facilitate execution of inspection tasks and maximize the opportunity for complete and accurate data collection. Examples of some of the inspection data sheets used during the inspection are depicted in Figures 5, 6, and 7, were included in the protocol. Major objectives of the inspection were divided into three interrelated phases: global hull survey, detailed internal and external hull deformation survey, and specification conformance verification.

Fig. 4. *Exxon Valdez* in dry dock.

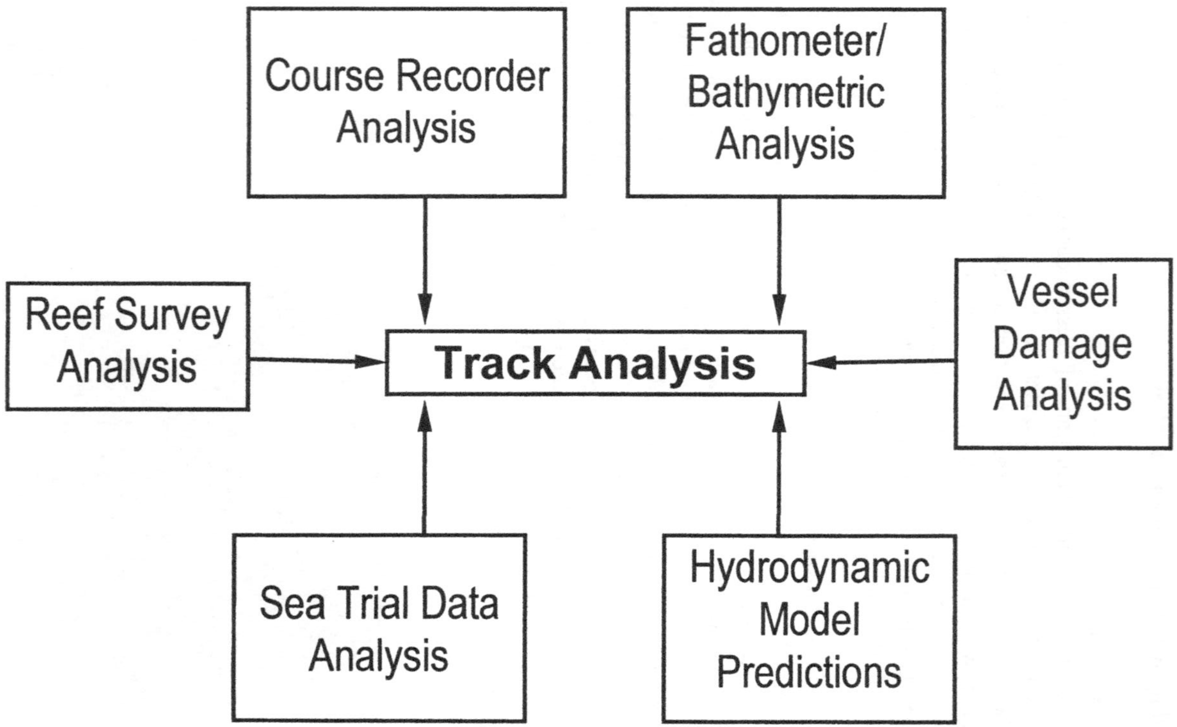

Fig. 3. Track analysis flow chart.

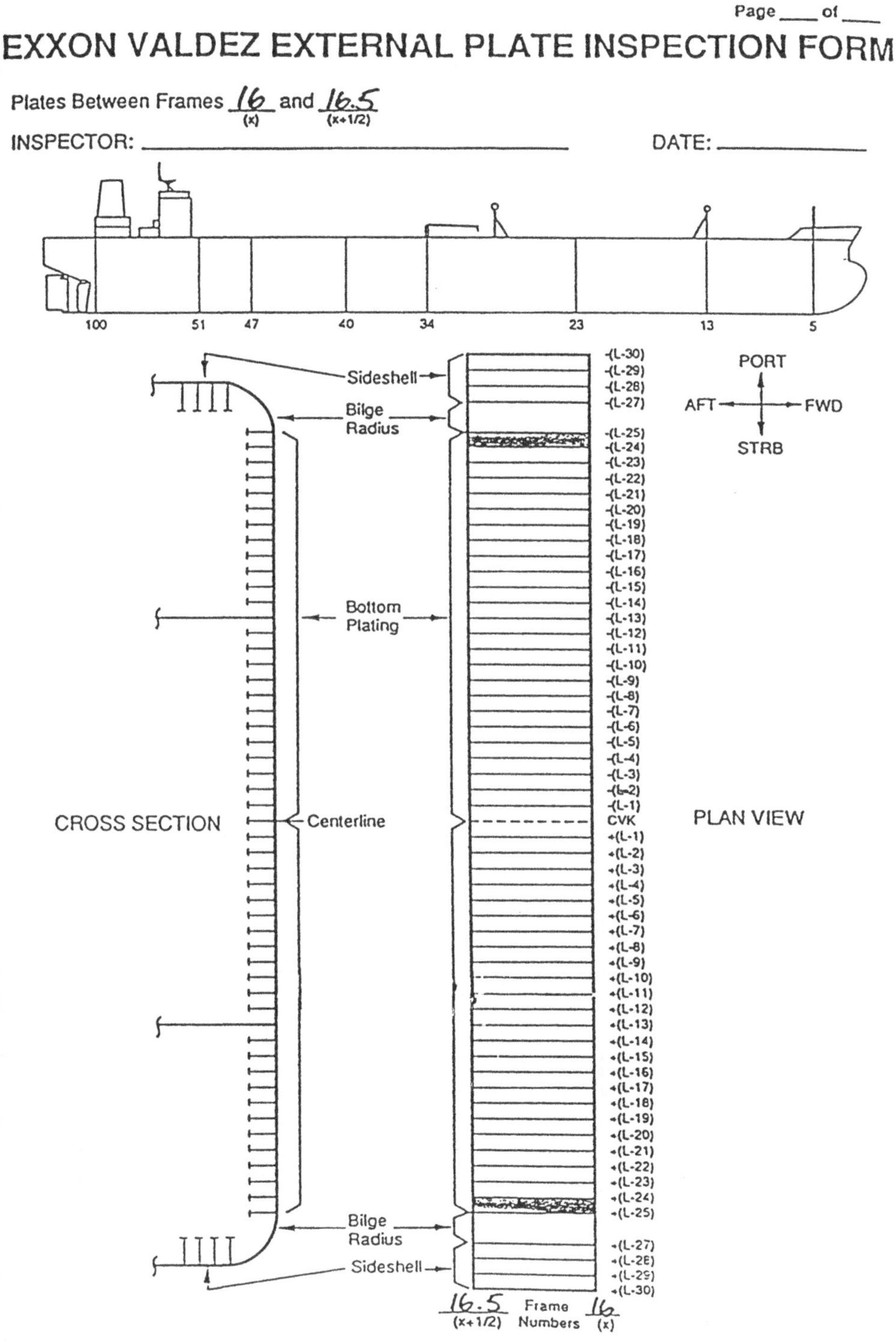

Fig. 5. Typical external plate inspection form.

Page ____ of ____

EXTERNAL PLATE INSPECTION FORM

Plates Between Frames 16 (x) and 16.5 (x+1/2)

CENTER

INSPECTORS: ______________________________

Member Type = P

Ship Frame Reference = 16.25 (x+1/4)

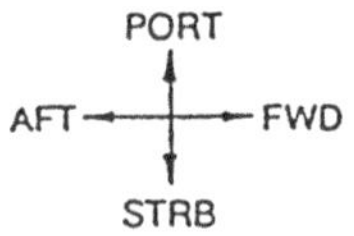

Longitudinal BHD, L-12, L-11, L-10, L-9, L-8, L-7, L-6, L-5, L-4, L-3, L-2, L-1, CVK, L-1, L-2, L-3, L-4, L-5, L-6, L-7, L-8, L-9, L-10, L-11, L-12, Longitudinal BHD

Reference Number	Inspection Status Code (I=Inspected, M=Missing, O=Obstructed)	Fracture Damage Code (F=Fracture, O=No Fracture)	Scrape Damage Code (S=Scrape, O=No Scrape)	Dent Damage Code (D=Dent, X=No Dent)	Hole Damge Code (H=Hole, O=No Hole)	Torch Damage Code (T=Torch Cut, O=No Cut)	Photographed (P=Yes, O=No)	Sketched (S=Yes, O=No)	Photograph (Roll Number - Picture Number(s))	Sketch Page Numbers(s)
-12.5										
-11.5										
-10.5										
-9.5										
-8.5										
-7.5										
-6.5										
-5.5										
-4.5										
-3.5										
-2.5										
-1.5										
-0.5										
+0.5										
+1.5										
+2.5										
+3.5										
+4.5										
+5.5										
+6.5										
+7.5										
+8.5										
+9.5										
+10.5										
+11.5										
+12.5										

16.5 (x+1/2) 16 (x)

Frame Number

Fig. 6. Typical external plate inspection data form.

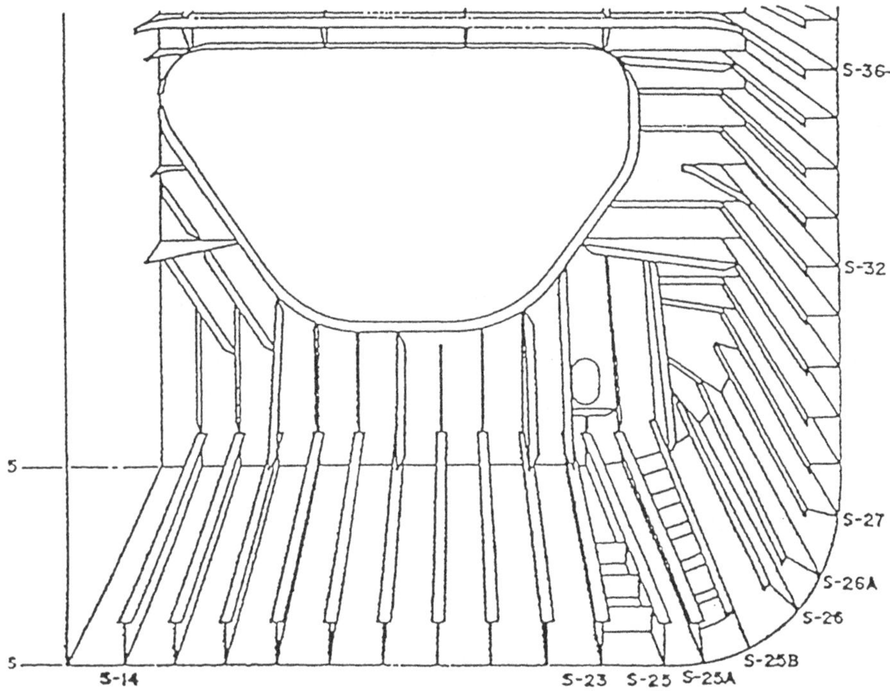

Wing Tank 2S - LKG. FWD. at Fr. 15 Plate 15

Fig. 7. Typical internal tank inspection form.

Quantitative description of vessel damage

Data collected during the San Diego dry-dock inspection were categorized as follows: (1) outer hull deformed shape, (2) plate damage, and (3) internal structural damage. The first category corresponds to the maximum outer hull vertical displacement, whether hull plate was present or not. The second category includes only damage to the exterior plate or outer hull skin. Figures 8a and 8b show two photographs of some typical external vessel underside grounding damage documented during the inspection. Finally, the internal structural damage includes deformation to the longitudinal stiffeners, transverse web frames and bulkheads, and longitudinal bulkheads. A typical internal ship tank structure where internal damage was documented is shown in Figure 9.

Analysis of structural damage data collected during the dry-dock inspection was performed to quantify the extent of damage in the three categories listed. This information indicates how and where some of the vessel kinetic energy was dissipated through structural damage at the time of grounding. Finally, comparisons were made between San Diego dry-dock inspection data and that obtained by divers while the vessel was positioned at Bligh Reef, Naked Island, and San Clemente Island, since the condition of the outer skin while positioned on the reef could not be determined solely from dry-dock inspections.

Fig. 8a. Damage to vessel bottom.

Fig. 8b. Damage to vessel bottom.

Outer hull damage shape

The shape of the deformed hull in dry-dock was documented by laying out a 3-by-8 foot grid pattern of reference points, corresponding to intersections of frames (and midframes) and longitudinal stiffeners, and recording measurements of the vertical offset to the bottom of the vessel. In most cases, this meant measuring up to the hull plate. However, where the plate was missing, the vertical offset was measured to the bottom of a longitudinal stiffener or frame element. These measurements characterize the extent of penetration into the hull. The spacing of measurement points was decreased where necessary to capture complex deformation patterns, and increased in areas of the port side where there was no damage. More than 3000 measurement points were made in this survey.

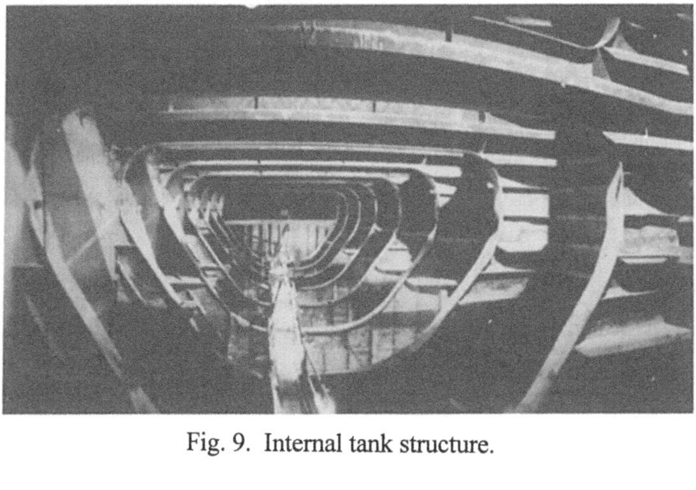

Fig. 9. Internal tank structure.

Figure 10 shows a digital mesh representation of the deformed vessel bottom. This illustration reveals damage patterns which correspond to the three distinct damage shapes. These damage patterns are conveniently characterized as: the long gouge, which is the long narrow feature that extends along the middle of the vessel, curving toward the starboard side, and ending past frame 48; the tunnel, which represents deformation on the starboard side, outboard of the long gouge, between frames 4 and 25; and the cavern, or the area between frames 22 and 27 on the starboard side where the vessel ultimately came to rest.

Plate damage

Drawings of all damaged external plates recorded during the dry-dock inspection were digitized and combined to create a composite damage map. To develop an accurate assessment of plate damage, information recorded in the composite damage map included missing plate material, locations of plate fractures and torch cuts, plate denting and scraping, as well as locations of crack arrest holes.

This composite damage map allowed quantification of the various types of structural damage sustained during the grounding event. Also, plate fracture locations identify where tanks were breached and the size of openings through which cargo leaked.

Analysis of the composite damage map reveals that approximately 10,000 square feet. of external plating was missing or had been removed by the time the vessel reached San Diego. This amounts to approximately 150 tons (short tons) of material. All subsequent weights are listed in short tons. Combining this with the total area of fractured, dented, and scraped plates gives approximately 560 tons of plate damaged.

The status of missing plates is known to have changed during transit of the vessel to San Diego. Therefore, additional data which reflect the plate status at Naked Island were digitized, including a damage map developed by divers, and sketches made based on video tape recordings while the vessel was positioned at Naked Island and Bligh Reef. Approximately 8,000 square feet of hanging plate were removed from the vessel when it was near San Clemente Island.

Table 1 Summarizes areas and weights of damaged plate observed on the *Exxon Valdez*.

Table 1. Plate deformation summary

Category	Description
Total Area of Missing Plate	10,000 sq. ft.
Total Area Dented	22,000 sq. ft.
Total Area Scraped	26,000 sq. ft.
Total Area Deformed	36,000 sq. ft.
Total Weight of Damaged Plate	560 S.-tons

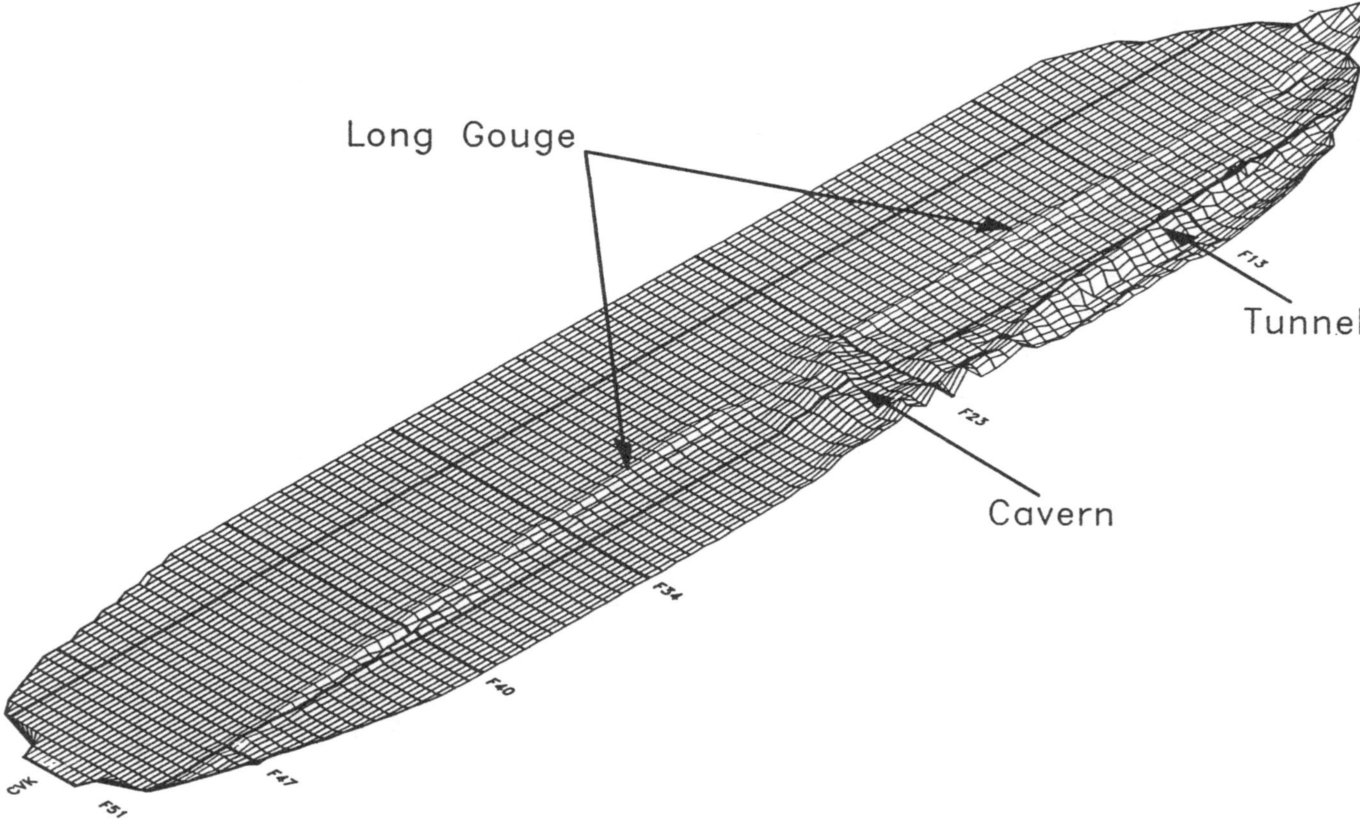

Fig. 10. Digital mesh representation of deformed vessel bottom.

Internal structural damage

The extent of internal structural damage is characterized in terms of maximum height of deformed structural steel above the undeformed bottom for bulkheads, frames, and longitudinal stiffeners, and total weight of damaged steel. This information was derived from internal tank inspections. The total weight of deformed structural steel was determined separately for the long gouge, tunnel, and cavern. These results are combined with the plate damage information to evaluate the total weight of structural steel damaged during the grounding process.

Due to the high strength, rigidity, and ductility of the transverse and longitudinal bulkheads, bulkhead deformation is in the form of gouging. The zone of damage above penetration varies and can extend more than 20 feet above the vessel bottom. The weight of damaged bulkhead material was calculated to be approximately 10 tons.

Deformation characteristics of open web frames are similar to bulkheads due to their strength, rigidity, and ductility. However, transverse profiles through frames indicate greater vertical deformation than at the bulkheads. Because of increased stiffness, both transverse and longitudinal bulkheads tend to ride-over the obstacle more than the frames do. This total weight of deformed frame material is approximately 34 tons.

The structural damage to longitudinal stiffeners was recorded during the internal tank inspections in terms of the following parameters:

- Deformed depth of longitudinal stiffeners
- Maximum separation between longitudinal stiffeners and outer hull
- Depth of maximum penetration
- Number of plastically-deformed longitudinal stiffeners
- Number of fractured longitudinal stiffeners
- Number of longitudinal stiffeners with connection failures

The total weight of deformed longitudinal stiffeners is approximately 440 tons.

Engineering interpretation of hull damage

The observed hull damage indicates three distinct phases of vessel damage; the first grounding, which resulted in a long gouge; the second and final grounding, which formed the tunnel; and additional deformation caused by tidal variation and wave action, which resulted in the cavern. Detailed descriptions of these features follow. (See Figure 10.)

First grounding: the long gouge

Damage associated with the first grounding has been characterized as a long gouge starting just starboard of the CVK, and extending from frame 4 to frame 48.5, with a total length of approximately 700 feet. The cross-sectional shape remains relatively constant, approximately 15 feet in width. This long gouge reflects impact with two separate points, as revealed in profiles of bottom deformation along the length of the long gouge. The total plan area affected by the first impact is approximately 11,000 square feet Plate fractures are observed between frames 4 and 47, a length of 682 feet. The total weight of structural steel deformed during this event is about 300 tons. The overall length and consistency in shape of the long gouge indicates that the vessel passed over and continued moving beyond the site of the first impact.

Tanks damaged during the first grounding include the Fore-Peak, 1C, 2C, 3C, 4S, 4C, 5S, 5C and the starboard slop tank. Each of these tanks suffered external plate ruptures. The total area of damaged plate in the first grounding is approximately 11,000 square feet The maximum height of penetration into each tank for both groundings is shown in Table 2. Also listed in this table is the maximum height of damaged longitudinal stiffeners. Numbers in the table show how high into the vessel structural steel was deformed at locations between frames and bulkheads. Maximum heights of deformation at frames and bulkheads range from several feet to over 20 feet.

Table 2. Penetration of tanks for both groundings

Tank	Maximum Bottom Penetration (in.)	Maximum Height of Longitudinal Stiffener Damage Above Undeformed Bottom (in.)
First Grounding		
Fore Peak	12	40
1C	28	58
2C	17	64
3C	16	(*)
4C,4S	16	43
5C,5S	18	46
Second Grounding		
1C,1S	73	88
2C,2S	91	106
3C,3S	110	(*)

(*) Obscured by deformation caused by tidal variation and wave action.

A plan view of the long gouge reveals that it is not straight, but curves from the CVK to starboard, which indicates that the vessel was turning to starboard during this grounding. Impact damage sustained during the first grounding resulted in a substantial kinetic energy loss that slowed the vessel.

Second and final grounding: the tunnel

The tunnel outboard of the starboard longitudinal bulkhead was caused by the second and final grounding. This tunnel was approximately 20 feet wide and 6 feet high. The cross sectional shape of the tunnel remains relatively constant along most of its length but finally tapers somewhat near frame 23, which is a transverse bulkhead. This change in tunnel shape is consistent with the position of the bulkhead located at frame 23. The bulkhead provided substantial structural resistance to the remaining momentum and energy.

The discrete wire frame representation of the vessel bottom in Figure 10 distinctly shows the tunnel, which starts at frame 7, continues past frame 23, and ends within the cavern. The maximum tunnel penetration height before reaching the cavern is approximately 91 inches. This occurs at the intersection of frame 22.5 and longitudinal stiffener No. 24 starboard. The second grounding resulted in tanks 1C, 1S, 2C, 2S, 3C, and 3S being breached.

Profiles of transverse bulkheads and frames show that bulkheads deform less than frames. Smaller deformation at the bulkheads, relative to the adjacent web frames, shows that the vessel had a tendency to ride over at these locations. This condition is consistent with an interplay between kinetic and potential energy, and affects the total stopping distance.

During this second and final grounding, approximately 500 tons of steel were deformed, or about 50% of the total weight of damaged steel. Table 2 shows the maximum height of penetration and maximum damage height. These tunnel damage characteristics are consistent with a high-energy impact where a substantial amount of vessel kinetic energy was dissipated.

The curved shape of the tunnel is consistent with a starboard turn, and is attributed to increased right rudder, the higher rate of turn, and increased resistance on the starboard side.

Deformation due to tidal variations and wave action: the cavern

Cross sectional profiles of the vessel bottom show a dome-shaped cavity, called the cavern, with a maximum vertical offset of 9.2 feet at frame location 23.5. The maximum height of structural steel damage is approximately 14 feet above the undeformed bottom in the cavern region.

The cavern shape is different from the long gouge and tunnel. A significant amount of structural deformation occurred as a result of tidal variation and wave action. This deformation extends from about frame 23 to frame 27, and from the CVK to the turn-of-the-bilge on the starboard side. The damage in this area does not exhibit the directional and energy dissipating shapes similar to long gouge and tunnel deformation, but instead is dominated by vertical deformation caused by the vessel settling down on the reef as the tide went out. Waves, wind, and tidal currents acting on the vessel contributed to deformation in this area, as indicated by the physical evidence in dry-dock. The total weight of deformed structural steel caused by tidal variation and wave action is approximately 230 tons. This is significant compared to the approximately 300 tons damaged during the first impact, and approximately 500 tons damaged in the second impact. Also, since the cavern is aft of the tunnel it played no appreciable role in stopping the *Exxon Valdez* during grounding.

Vessel lightering and freeing from reef
It is well known that a grounded vessel can be "broken" or capsized by lightering cargo without regard to the hull integrity or stability. The vessel was surveyed to determine it's damaged condition and the cargo was lightered from the super tanker while stresses in the vessel structure were kept at safe levels, and vessel stability was maintained. The result was that all of the remaining cargo (about one million barrels or approximately 80% of the original load) was successfully saved and the vessel was safely refloated and towed to the NASSCO ship yards for repairs.

5 Grounding site inspection and reef bathymetry survey

Obtaining information about the grounding location, shape of the reef area where the grounding occurred, and the wreckage distribution of vessel fragments left behind was also an important piece of the puzzle necessary for these trackline and vessel impact analyses. This survey was quite challenging since Bligh reef, at the grounding location, was submerged under approximately 55 feet of salt water.

We carried out grounding site and reef surveys with the support of an underwater construction company from Alaska. They provided logistical support and divers to map the reef, surveying services, and a 28-foot aluminum survey boat. Other equipment and services consisted of tidal gauges, current meters, side-scan sonar, a magnetometer and an underwater positioning system. We also had the use of a 130-foot work vessel with a landing-craft-type bow door/ramp. This work vessel was well suited for the dive operations and wreckage search activities.

At the time of the reef survey, we had little accurate information regarding the position of the *Exxon Valdez* when it grounded on Bligh Reef. The only information provided were bearings to several nearby artifacts taken by ship's officers at the time of grounding. This information could only locate the grounding site within 500 feet. Further, when we arrived at the reef, the underwater visibility was less than 2 feet due to glacial silt in the water. Other environmental conditions persisting at the time of the survey in early September were 38° F water temperatures, and iceberg flows from Columbia Glacier, just a few miles across Prince William Sound from the grounding site.

In order to locate the actual grounding site from these approximate parameters we first set up a short baseline positioning system. This was accomplished with a Mini-Ranger system and land survey techniques which used three previously-surveyed locations. Further a the remotely operated diving vehicle (ROV) was launched and a search pattern begun. Because the underwater visibility was less than 2 feet, this effort proved futile during the early search efforts. Knowing that some steel hull plate had been lost on the reef during impact, we employed a magnetometer search using the bathymetry survey vessel. This method located a metal target which proved to be a hull fragment from the grounding. Once this first sighting was accomplished, the full reef survey then commenced.

An area approximately 0.5 nautical miles by 1.0 nautical miles was mapped using the bathymetry survey and side-scan sonar. Ultimately, detailed contours of this entire region were defined, and a set of 48 unique targets were identified by ROV and/or divers. Of those 48, 34 were identified as either impact damage to the rock or sand reef bottom. The other fourteen targets were steel plate from the vessel. This survey also established that the vessel impacted Bligh Reef at two different ridges of the reef approximately 1140 feet apart. The depths at these two impact locations were consistent with the vessel draft.

Figure 11 shows the resultant contour map that was developed for Bligh Reef on the basis of these measurements and analyses. In addition to showing contour details of the reef that were not previously available, it also locates the first and final grounding positions. The detailed nature of these findings were quite important to accurately establishing the final trackline and vessel grounding sequence.

By logically comparing these grounding positions, vessel wreckage debris, and damage to the reef with the general characteristics of vessel motion and with details of the long gouge and tunnel, it is clear that the long gouge was created during first impact with the reef, while the tunnel damage was produced at the final grounding.

6 Vessel sea trials

The quantitative analysis of the ship's path and control inputs requires data on the performance characteristics of the vessel. To obtain this data, we performed a series of sea trials with the vessel once it had been repaired and recommissioned for service. These trials were performed from August 31 to September 5, 1990, off the coast of San Diego at various cargo loads (draft) and speeds conditions. The objectives of these tests were to ascertain and compare the vessel's pre- and post-incident performance characteristics, to determine the ship's maneuvering characteristics under the same loading conditions and vessel speed as the day of the incident, to perform a series of vessel maneuvers for hydrodynamic analysis, and to determine the advance, transfer, and diameter of the vessel.

Vessel performance data were gathered by connecting electronic data acquisition equipment to the ship's gyrocompass, speed log, shaft speed indicator, rate-of-turn indicator, and rudder angle indicator. Additionally, a global positioning system (GPS) was installed on this vessel for these tests, so that an accurate record of the ship's position during each test could be obtained. A microwave antenna was installed on the port wing of the bridge as part of this system. The accuracy of this satellite tracking system was claimed to be within approximately three meters over long times or distances, but with relative accuracy to within perhaps one meter over short times or distances. This accuracy was adequate to characterize the turning performance of the vessel.

Tests were performed to establish the ship's turning characteristics under loading and speed conditions which existed on the day of the incident. Series of turns in both the starboard and port directions were made with constant rudder inputs, starting at 5°, and increasing in 5° increments for each succeeding test. A 180° turn was performed for each maneuvering test.

Although there were turning performance data for the *Exxon Valdez* in its as-built condition, there had been a minor modification to the rudder between the time it was built and the grounding. Hence, we also tested the ship not only for its turning

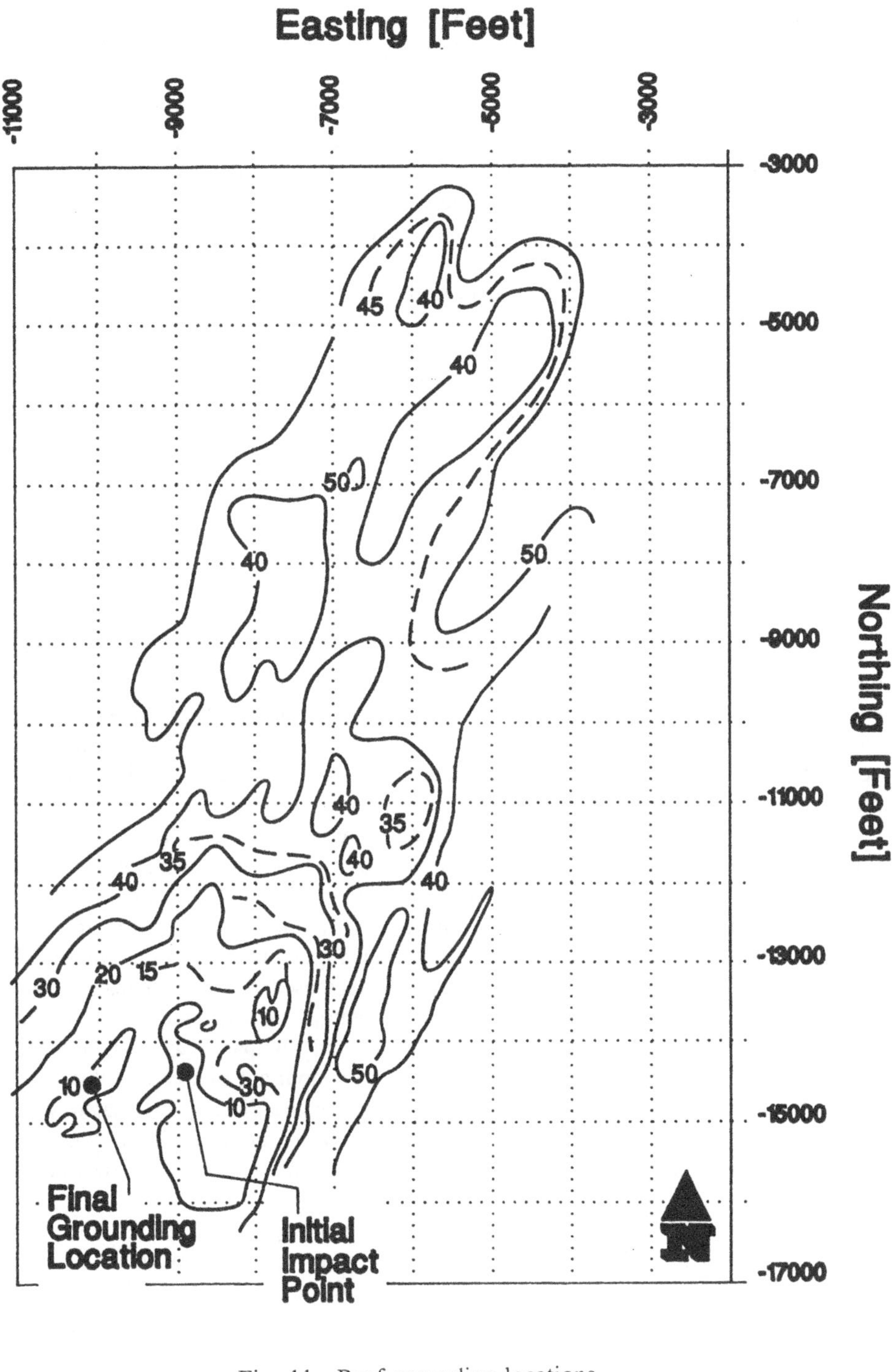

Fig. 11. Reef grounding locations.

characteristics, but also for its speed-time performance with constant heading under the computer control system called "Load Program Up" (LPU) used to slowly ramp up to sea speed.

The collected data were analyzed for environmental and other factors and reduced to tables and plots of position, speed, and heading versus time and other data. These measurements, made while testing the actual vessel under identical load and relevant speed conditions existing on the night of the *Exxon Valdez* grounding, enabled us to establish the ship's turning characteristics on the night of the grounding.

Figure 12 is an example of data developed from sea trials where trackline paths are shown for different rudder control inputs as a parameter. Because these trackline data were taken in the open ocean, adjustments were made to correct for current effects by conducting the same sea trial in opposite directions and then removing that motion effect due to current. These current corrections are also shown in Figure 12.

7 Course recorder data analysis

A copy of the vessel's course recorder trace is shown in Figure 13. This information also provided an important data record from the incident. The course recorder is actually a gyro repeater electrical signal, and thus the course recorder plots the vessel heading versus time. Analysis of the incident course recorder trace consisted of first producing photographic enlargements of it. These images were then digitally scanned and computer enhanced to provide high resolution of heading versus time. The time scales of these scans were then expanded to allow greater resolution of the heading record.

Figure 14 shows the digitized incident course recorder trace from the point in time where the heading began to change from 180° to larger headings, that is from the point in time when the vessel began a right turn. This portion of the course recorder trace was most critical for analysis of vessel motion prior to grounding. In particular, the portion shown within the square on Figure 14 is a region where the heading rate of change versus time suddenly decreased for approximately one minute. An enlargement of this portion of the course recorder trace is shown in Figure 15. The actual digitized trace is shown and compared with an average continuous curve for the same digitized curve. It is important to note that this rate-of-change decrease for approximately one minute can be directly correlated with first impact of the vessel with Bligh Reef. This impact caused the large gouge in the vessel. As noted in Figure 13, this decrease in rate of change only shows up as a small "kink" in the heading curve.

We studied and analyzed the course recorder equipment to determine inherent equipment biases introduced into the chart recording. A comparison of the sea trials course recorder digitized data and sea trials digital data resulted in the formulation of a heading bias calibration. A final calibrated incident heading was then determined using these results. Figure 16 depicts the final reduced data for the

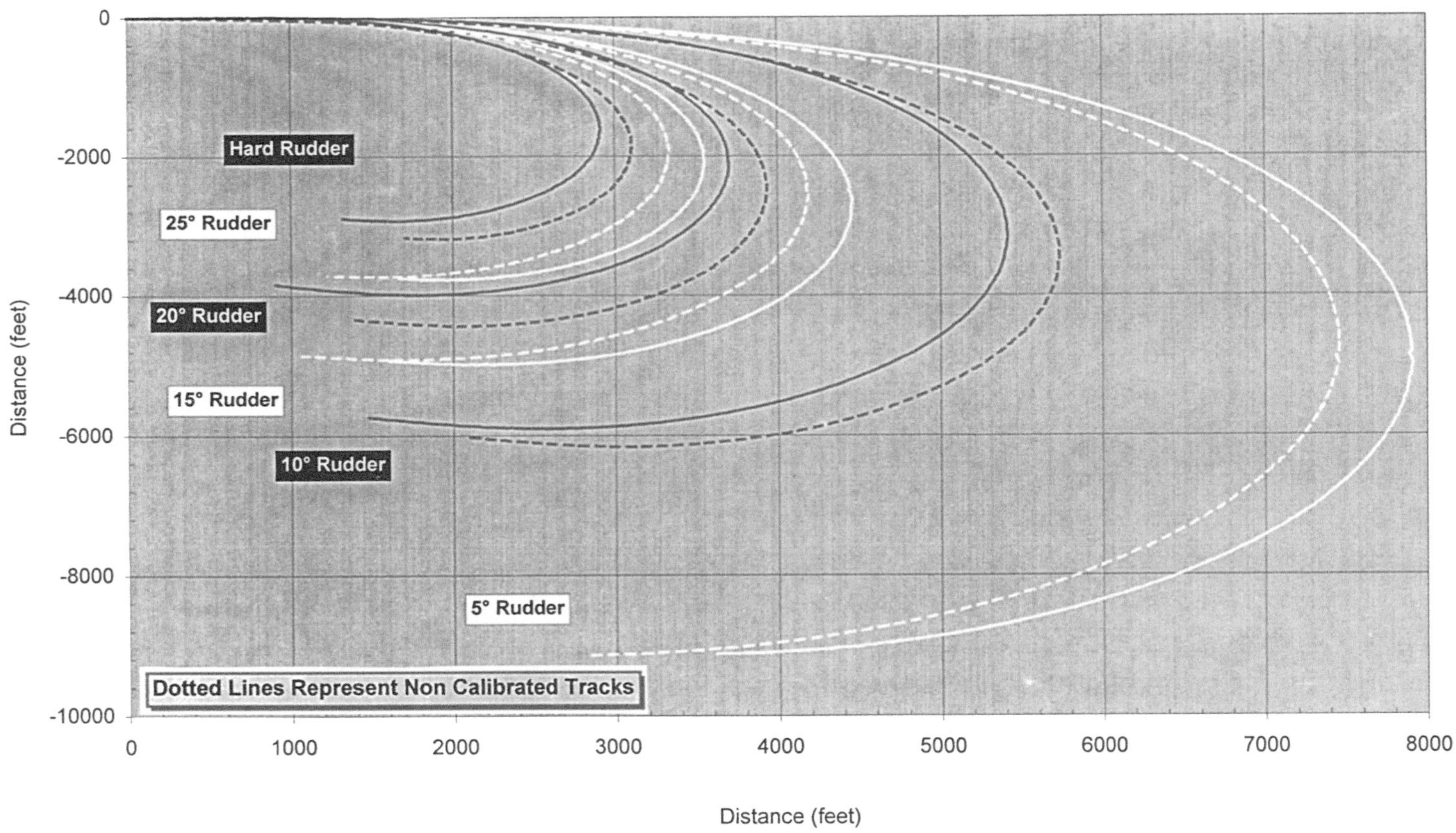

Fig. 12. Effects of current calibration to sea trials track data.

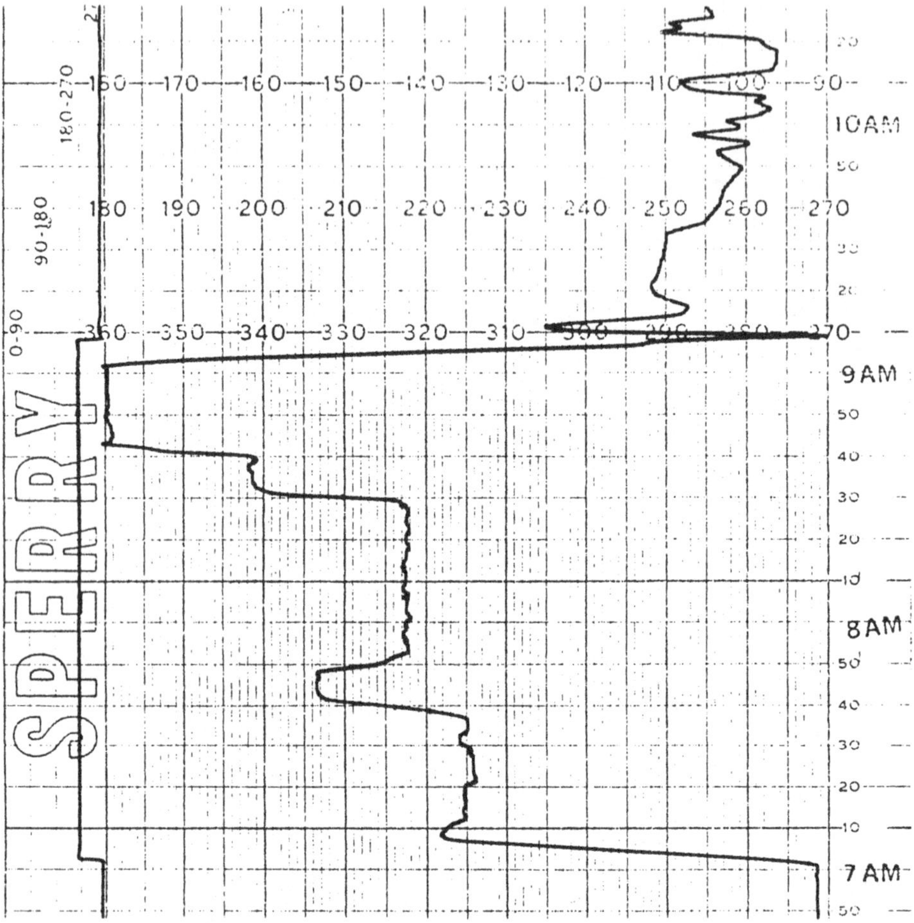

Fig. 13. Incident course recorder trace.

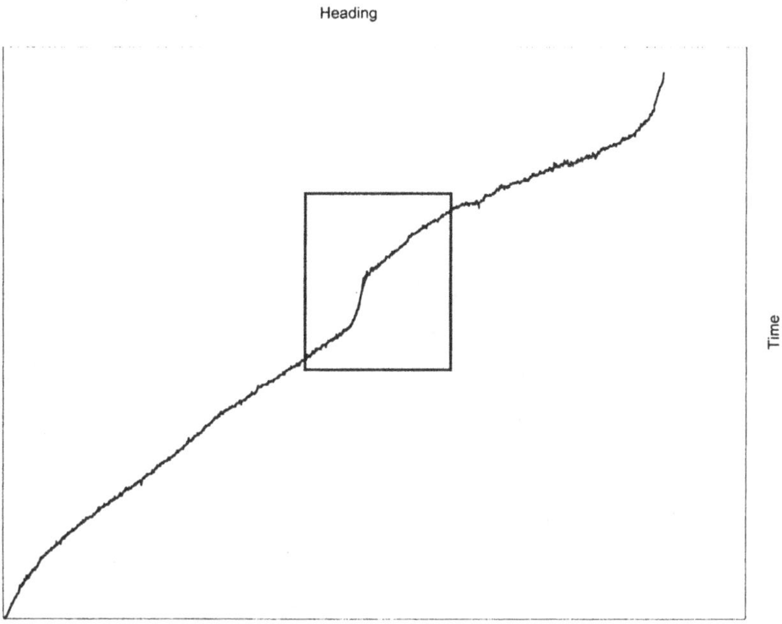

Fig. 14. Computer enhancement of course recorder trace.

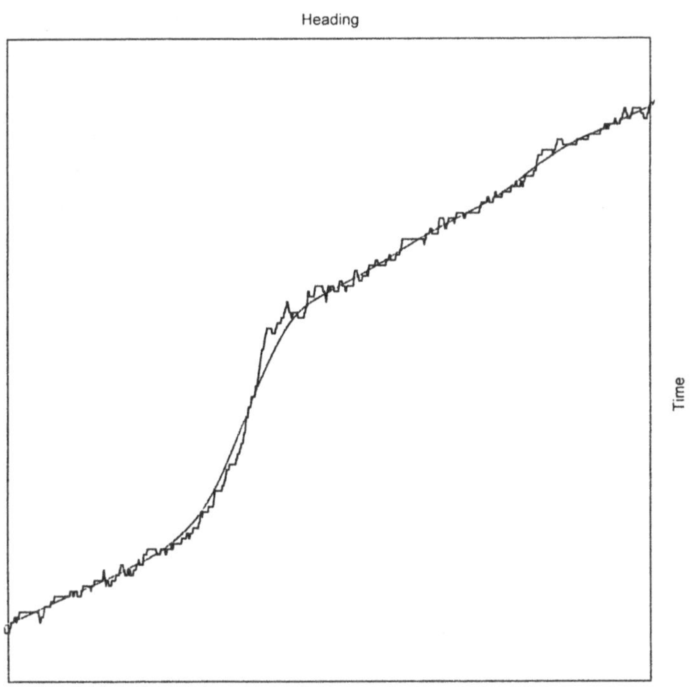

Fig. 15. Enlarged view: computer enhancement of course recorder trace.

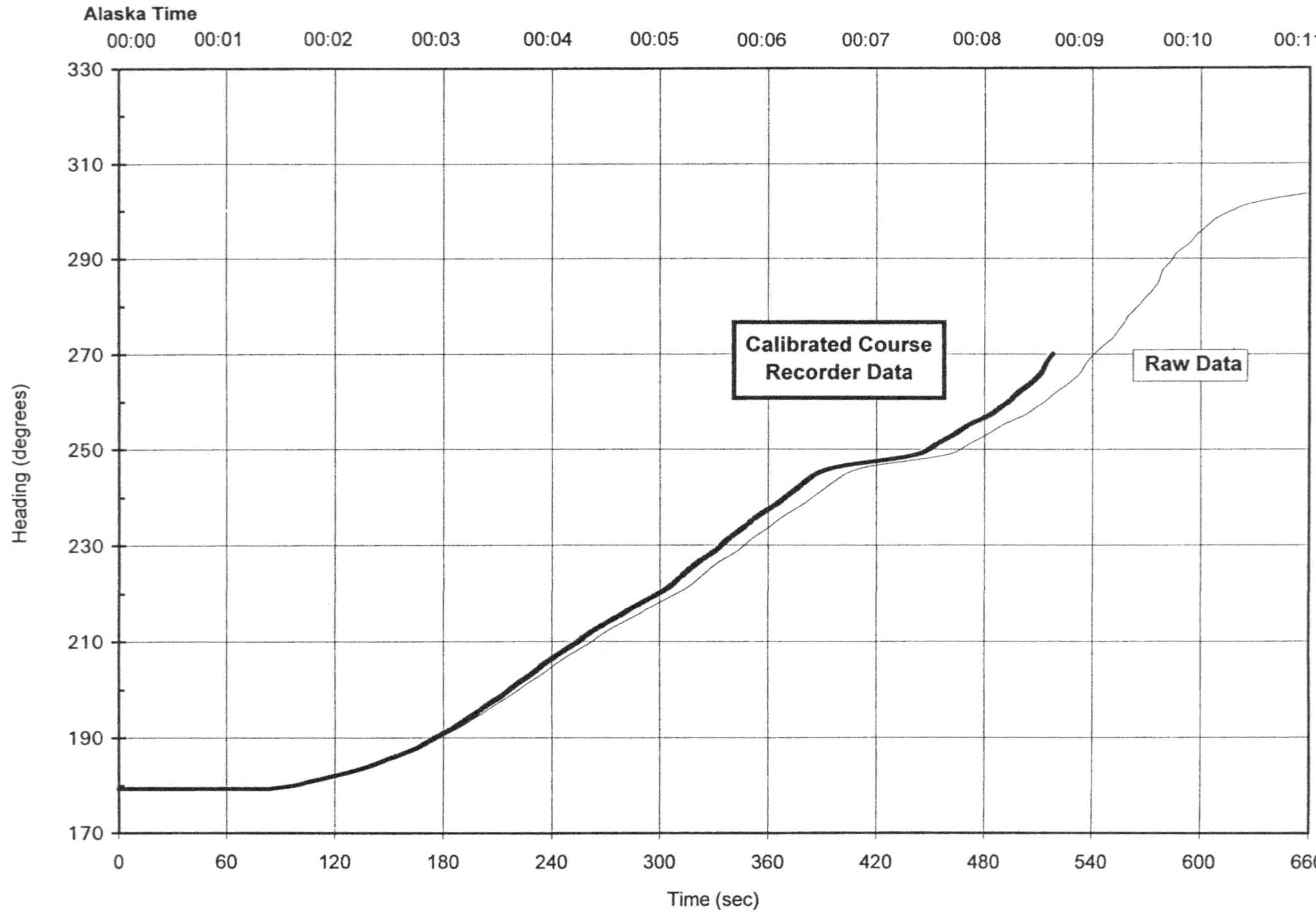

Fig. 16. Course recorder calibration.

heading versus time of the *Exxon Valdez* in the approximate 10 minutes prior to grounding. The calibrated and uncalibrated traces are labeled as shown.

8 Hydrodynamic model

The hydrodynamic model formulated for this study computes the lateral forces and yaw moments for a given velocity and rudder input to determine the trajectory of the ship. In the model's hydrodynamic equations, the ship track and attitude are defined by six coefficients that characterize the dynamic properties of the vessel in water: three derivatives for the lateral forces, and three derivatives for the yaw moments. Based on these six coefficients, the physical properties of the ship, the rudder input, and the velocity profile from the sea trials, the lateral acceleration and yaw rotational acceleration are calculated. These accelerations are then used in the kinematic equations of motion to determine the ship's heading, turning rate, and position.

A matrix optimization was utilized in order to determine the ship coefficients that best characterized the *Exxon Valdez* based on a systematic comparison of predicted positions with data collected during sea trials. For inputs of rudder position and engine shaft speed, the model predicts vessel position, speed, heading, and water depth below the hull as functions of time. These topics are more thoroughly covered in Reference [1].

9 Fathometer data analysis

Another important piece of information was the incident fathometer trace, shown in Figure 17. The fathometer trace was digitally scanned and computer enhanced to provide high resolution of water depth versus time. The digitized fathometer data is shown in Figure 18. Once in this form, the fathometer data could be used to check the validity of the vessel track by comparing the actual fathometer trace with traces made from hypothetical vessel tracklines.

Using the tracks generated by the hydrodynamic analytical model for different vessel rudder control, given in terms of rudder angle, a series of depth-below-hull versus time hypothetical plots were generated. These depths were then correlated and compared with the vessel fathometer recorded on the night of the grounding.

Careful analysis of the water depths obtained from our bathymetry survey and published navigational charts for Prince William Sound showed that the actual fathometer trace from the *Exxon Valdez* most closely matched a track 1.05 nautical miles from Busby light as the ship steamed at a heading of 180° in the minutes before the turn. As closely as it could be measured, the vessel's crew had placed a navigational fix on the chart at this location.

Comparison of the vessel's fathometer trace with depth data for a set of hypothetical right turns (different rudder angles) using a least squares fit technique showed that the vessel had made a right turn produced by a 10° rudder angle. Figure 19 shows the trackline for the vessel superimposed on the contour chart for

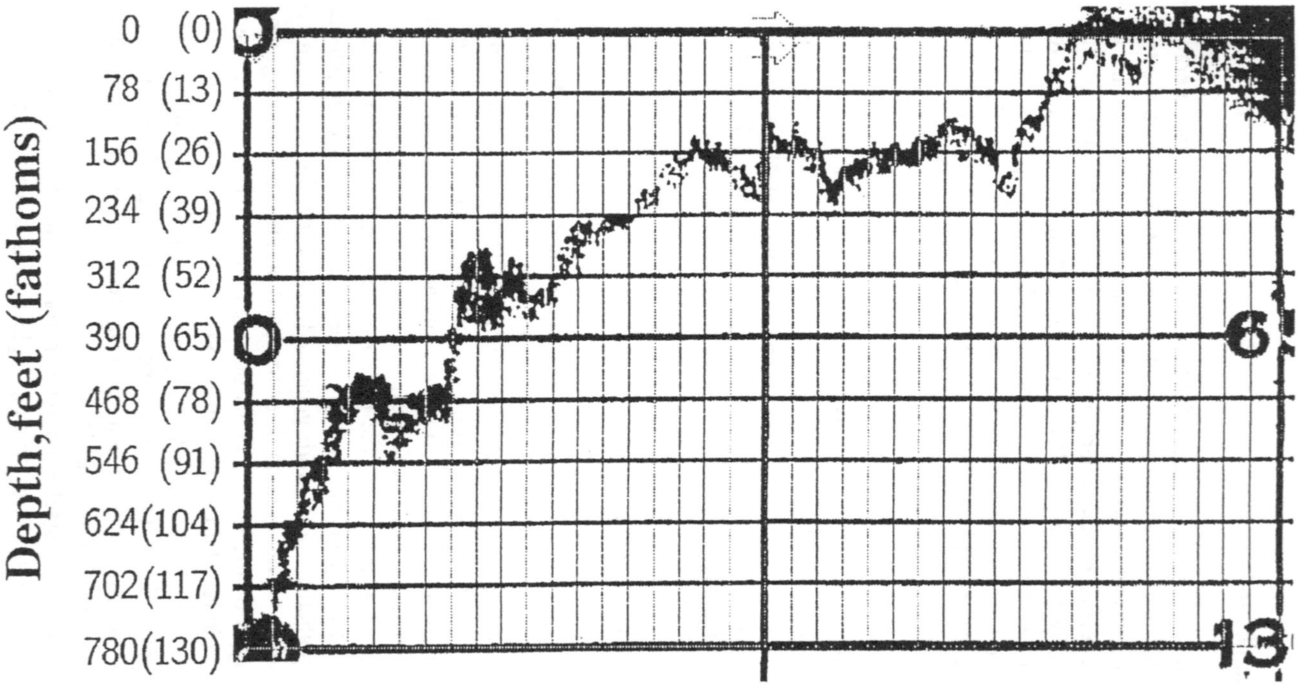

Fig. 17. Incident fathometer.

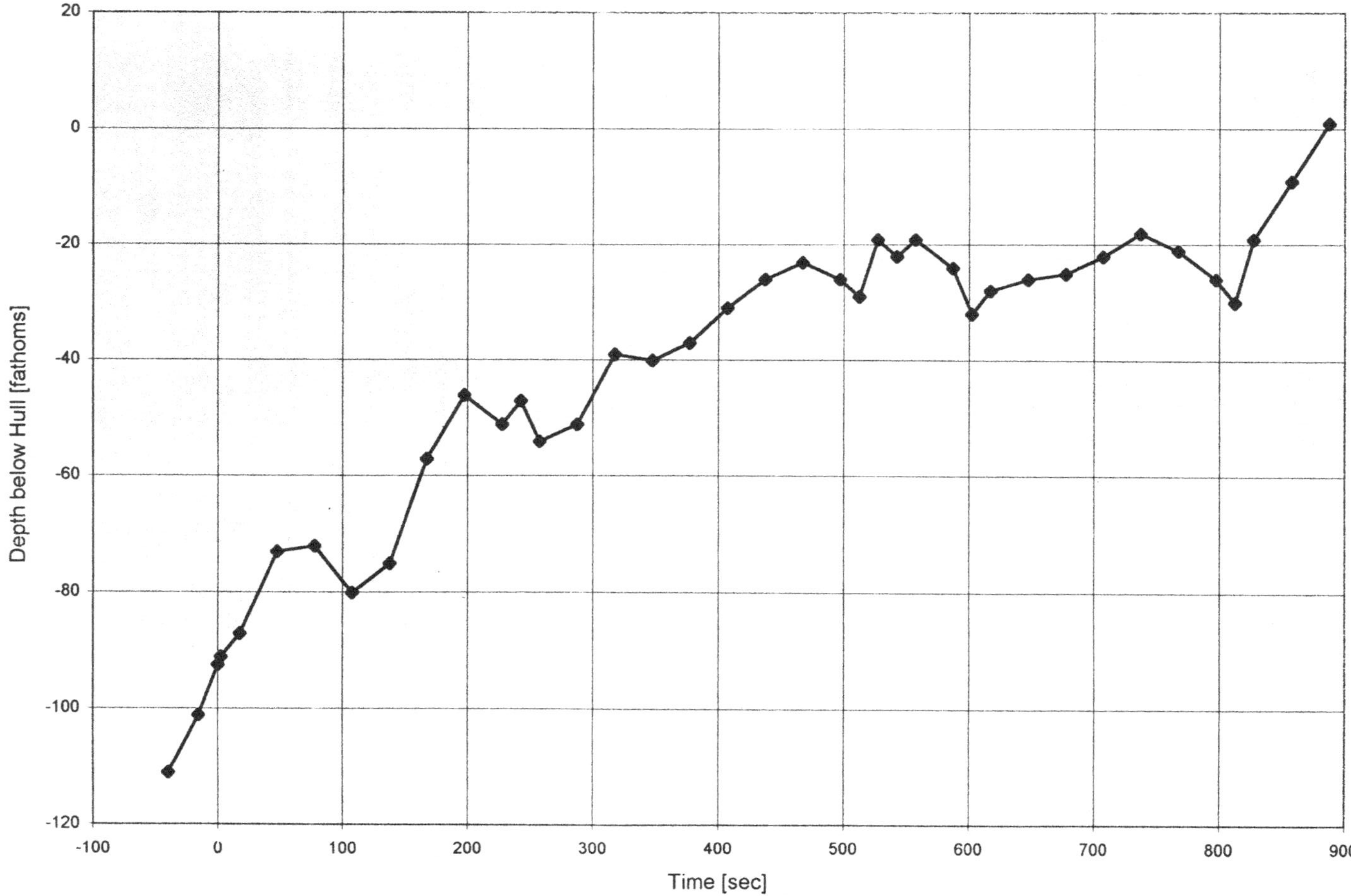

Fig. 18. Digitized incident fathometer trace.

the reef, when the vessel turns in response to a 10° rudder input. When the depth of water under the vessel for this trackline is generated by using the contour chart and then compared with the digitized vessel fathometer chart from the night of the grounding as shown in Figure 20, the best fit is achieved for a trackline with a constant 10° rudder deflection.

10 Trackline results

The above analyses were synthesized to reconstruct the vessel's trackline from the time of approximately 2339 to final grounding, as depicted in Figure 21. Figure 22 shows a table listing data for the important events marked on Figure 21.

2339 fix
A fix was entered on the *Exxon Valdez* bridge chart at 2339. Review of the bridge chart and a calculation of time, speed and distance, allowed us to establish the spatial coordinates of the ship at this time. The heading of the vessel at 2339 (199°) was obtained from the *Exxon Valdez* course recorder while the ship speed was obtained from an interpretation of sea trials data.

Turn to 180° heading
The vessel heading was changed from 199° at 2340 to 180° at 2342, as determined from a review of the *Exxon Valdez* course recorder.

2352: LPU initiation
According to the bridge log (bell book), load program up (LPU) was initiated at 2352. Using data from the sea trials, vessel speed was determined as shown Figure 22. The ship heading remained constant at 180° and the vessel position was established by a velocity, distance and time correlation from 2352 to 2355.

2355 fix
The crew entered a second fix on the vessel's bridge chart at 2355. This fix corresponds to a measurement made from the ship's bridge port wing, while the ship was abeam Busby light. Using a copy of the bridge chart, we computed the fix location to be approximately 1.06 nautical miles west of the Busby light. We performed a separate calculation which established that the ship was approximately 1.05 nautical miles west of the Busby light. This computation was based on a match between the ship fathometer chart and a hypothetical fathometer chart which was generated by passing a 180° ship track over the navigational chart.

The vessel speed and engine rpm were established from analysis of sea trials data and the time elapsed since LPU initiation.

2355 to beginning of turn
From the point abeam of Busby light, the ship continued along a constant 180° heading with speed and engine rpm increasing as a function of elapsed time from LPU initiation. Based on a review of the course recorder prior to the beginning of

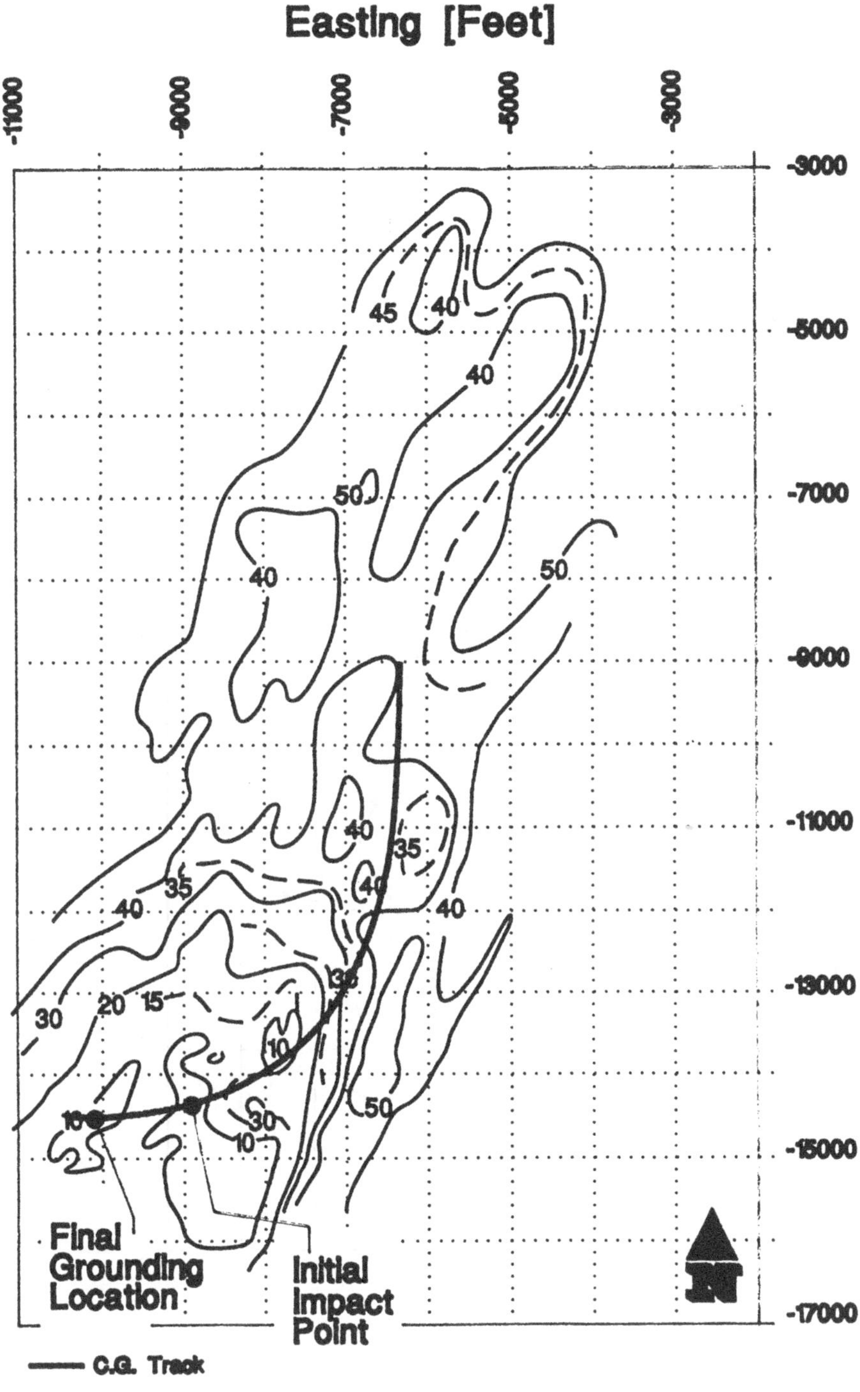

Fig. 19. Model-generated track: 10° rudder.

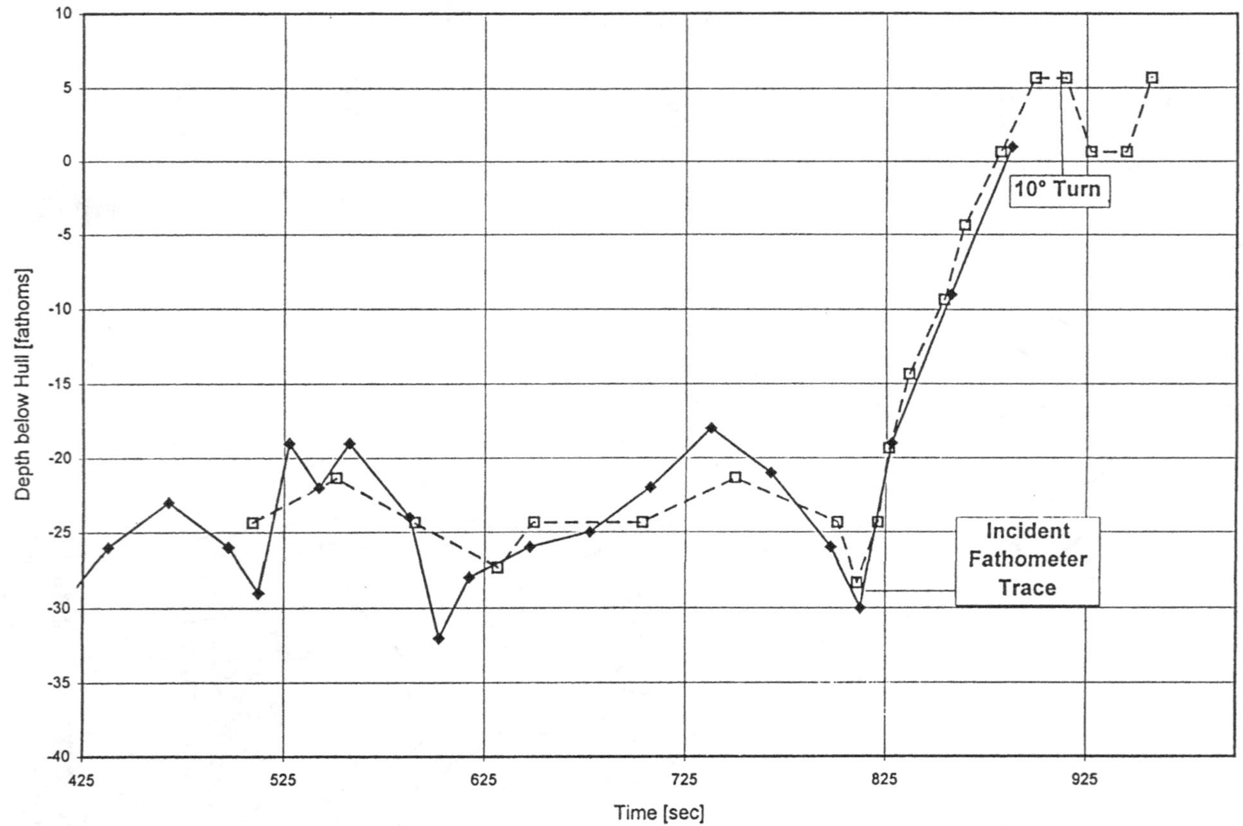

Fig. 20. Incident fathometer trace vs. model 10° turn.

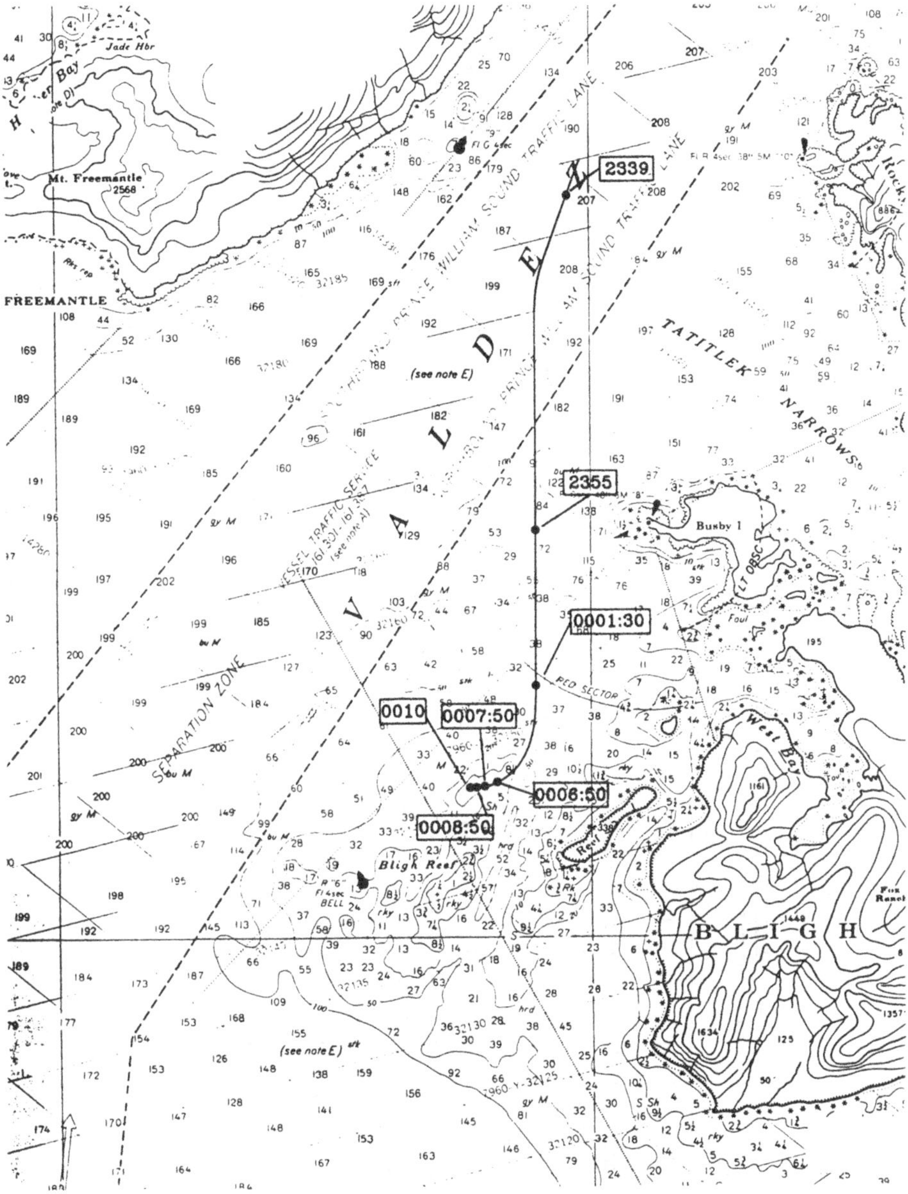

Fig. 21. Vessel track.

EVENT	ALASKA TIME	HEADING (degrees)	SPEED (knots)	SHIP'S COORDINATES X (feet)	SHIP'S COORDINATES Y (feet)	DISTANCE FROM PREVIOUS POSITION (feet)	TIME FROM PREVIOUS POSITION (min:sec)	RUDDER ANGLE (degrees)
POSITION FIX	2339	200	11.1	4420	−18170	–	–	
POSITION FIX	2355	180	11.1	6380	370	18880	16:00	
VESSEL BEGINS TURN	0001:30	180	11.3	6380	8940	8570	6:30	10 RIGHT
FIRST IMPACT AT BOW	0006:50	245	9.3	8580	14200	6030	5:20	20 RIGHT FOLLOWED BY HARD OVER
VESSEL OVER 1st GROUNDING RIDGE	0007:50	251	4.5	9230	14510	720	1:00	↓
FINAL GROUNDING INITIATION	0008:50	264	4.5	9680	14570	450	1:00	↓
FORWARD MOTION STOPS	0010	297	0	9980	14680	320	1:10	↓

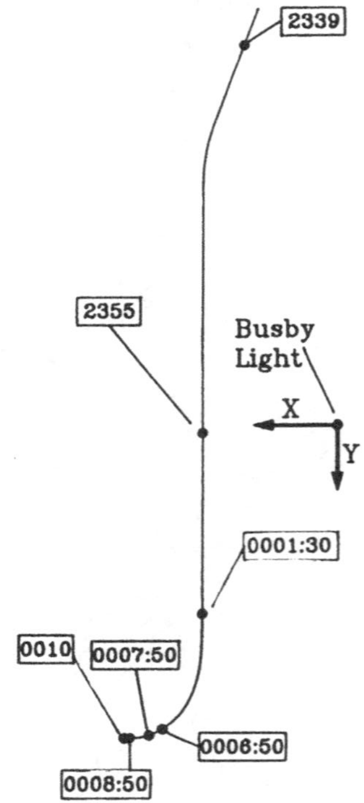

Fig. 22. Track data.

the turn, and a comparison of this chart with sea trials results, we concluded that more likely than not, the ship was operating on autopilot prior to the beginning of the turn.

Beginning of turn

Our analysis of the course recorder shows that a right turn was initiated at approximately 0001:20. This result was also confirmed by an analysis of the ship's predicted track and ship bow position at initial grounding. The turn began at approximately 1.05 nautical miles west of and approximately 1.50 nautical miles south of Busby light. The vessel speed and engine rpm at that point were established through the analysis of sea trials results and the time elapsed since LPU initiation.

The turn

The track of the vessel during the turn was predicted with numerical simulation using the hydrodynamic model. The effects of depth of water under the hull were integrated into this model. The proximity of underwater terrain to the port and starboard sides of the hull was also considered. The model was then used to predict the position, time of position, speed, and heading of the vessel for various hypothetical rudder inputs and engine shaft speeds established by LPU. Results from this analysis showed that the ship was turning in response to a 10 degree right rudder input which was held constant from initiation of the turn until first grounding. Figure 23 shows a heading versus time comparison of both 5° and 10° hypothetical hydrodynamic analytical model turns with the actual vessel digitized heading versus time. This comparison clearly shows that the vessel turned under a 10° rudder input.

Using the navigational charts, reef survey charts, and the track line predicted for the 10° turn, a water depth under the hull versus time chart was generated and compared to the *Exxon Valdez* fathometer chart. Agreement between the two charts provided independent confirmation of the steady 10° right rudder turn. Finally, the ship track for a constant 10° right rudder turn predicts a first grounding location which corresponds to the first grounding location identified by the reef survey and a heading at initial grounding which conforms with the long gouge damage to the vessel.

Review of the course recorder shows that at approximately 0005:30 the rate of change in vessel heading decreases from the value predicted by the hydrodynamic model for the vessel holding a steady 10° rudder in deep water. Based on review of the fathometer chart and reef survey, the vessel had moved into a region where the water depth was not more than 3.5 times the vessel draft, or not more than approximately 195 feet or 33 fathoms. In these conditions the hydrodynamic performance of the vessel differs from the performance of the vessel in deep water. The difference in performance effects was accounted for in the vessel hydrodynamic model and was based on published data obtained from literature of systematic model tests. When these effects are simulated using the hydrodynamic model, the resulting ship heading, with a 10° constant rudder input, is consistent with the change in course recorder heading at approximately 0005:30.

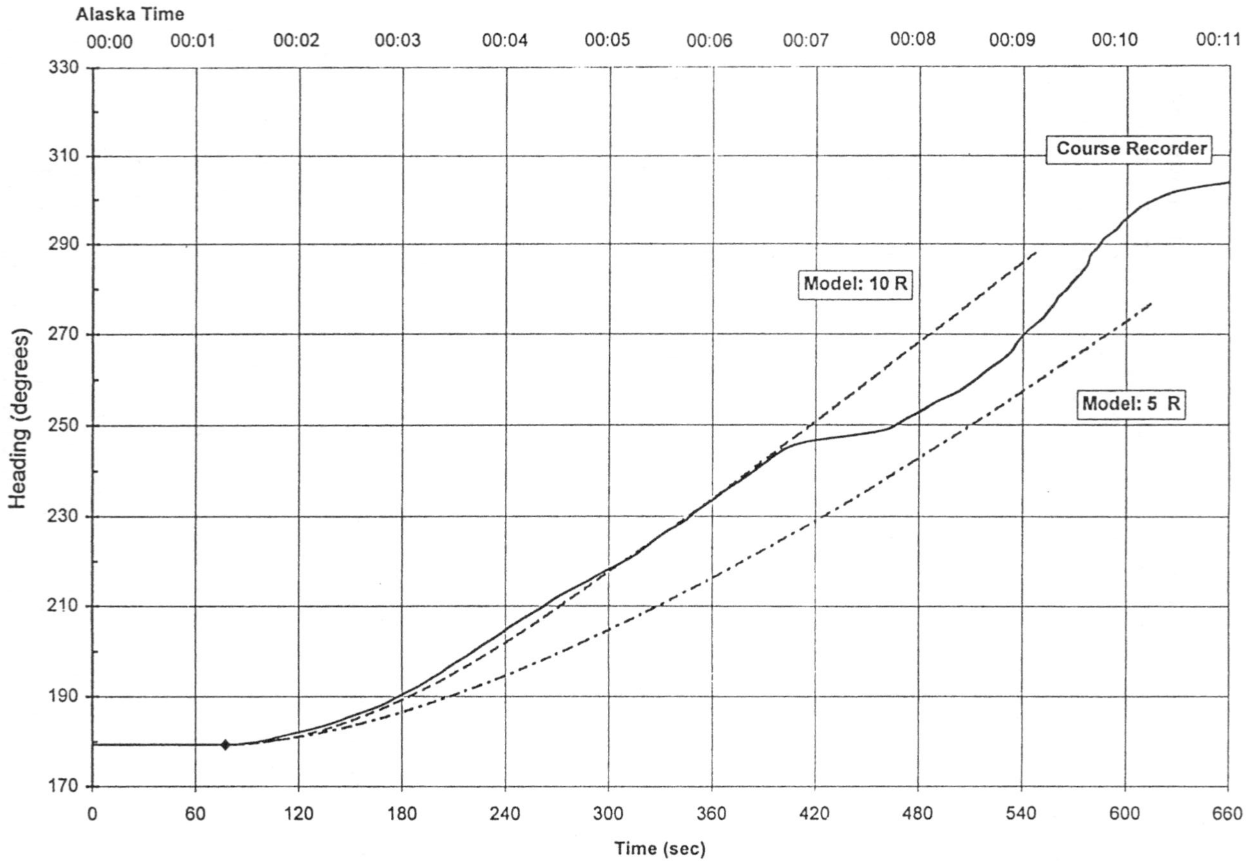

Fig. 23. Vessel heading comparison.

First grounding

Our diver survey of Bligh Reef identified the location of the initial vessel impact with the reef.

The predicted ship track, superimposed onto a hydrographic map of the area, showed that the initial impact of the bow occurred at approximately 0006:50. The ship speed and heading at this time were established by computation. The predicted heading matches the *Exxon Valdez* course recorder heading at the corresponding time. Further, this heading at first impact correlates with the observed hull damage. This first grounding event caused the substantial decreased rate of heading change observed on the course recorder at approximately 0006:50.

Between first and final grounding

Following the first grounding, the ship returned to a free floating condition and during this time, the ship continued to increase its heading, consistent with a right turn, at a rate of approximately 20° per second. Hydrodynamic effects, where the depth of water to draft ratio is less than 3.5, continue to be present in this portion of the track.

Final grounding

Forward motion of the ship stops at approximately 0010 based on analysis of the course recorder and results from the reef survey which identified the vessel's point of rest position.

11 References

1. Kadlec, R., Bayan, F., Beres, E. (1995) Hydrodynamic modeling of the *Exxon Valdez* using full-scale sea trails. *Structural Failure, Product Liability and Technical Insurance, Proceedings of the Fifth International Conference.*
2. Lewis, E. V. (1989) *The Principles of Naval Architecture*, The Society of Naval Architects and Marine Engineers.
3. Tuck, E.O. (1978) Hydrodynamic problems of ships in restricted waters. *Ann. Rev. Fluid Mech.* Vol. 10, pp. 33-46.
4. Abkowitz, M.A. (1980) Measurement of hydrodynamic characteristics from ship maneuvering trials by system identification. *SNAME Transactions,* Vol. 88, pp. 283-318.
5. Abkowitz, M.A., Liu, G. (1988) Measurement of ship resistance, powering, and maneuvering coefficients from simple trials during a regular voyage. Society of Naval Architects and Marine Engineers.
6. Abkowitz, M.A. (1984) *Measurement of Ship Hydrodynamic Coefficients in Maneuvering from Simple Trials during Regular Operations.* U.S. Department of Transportation.

7. Abkowitz, M.A. (1980) *Measurement of Hydrodynamic Characteristics from Ship Maneuvering Trials by System Identification.* The Society of Naval Architects and Marine Engineers.
8. Inoue, S., Hirano, M., Kijima, K. *Hydrodynamic Derivatives on Ship Maneuvering,* pp. 112 - 125.
9. Jack, R.L. (1973) *Sea Trials: Some Recommended Practices, Marine Technology,* The Society of Naval Architects and Marine Engineers, 380 - 391.
10. Kadlec, R., Schoof, C., Beres, E. (1994) Crude Oil Supertanker (VLCC) Structural Response To High-Energy Groundings. *Fourth International Conference on Lessons From Structural Failures.*

12 Appendix

Table A1. Unit conversion table

1 inch =	0.0254 meters
1 foot =	0.3048 meters
1 nautical mile =	1.852 kilometers
1 square foot =	0.093 square meters
1 gallon =	3.785 liters
1 short ton =	907 kilograms
temperature (°C) =	[temperature (°F) - 32]/1.8

19 HYDRODYNAMIC MODELING OF THE EXXON VALDEZ USING FULL-SCALE SEA TRIALS

R. KADLEC, F. BAYAN and E. BERES
Failure Analysis Associates, Los Angeles, California, USA

Abstract
This paper presents an analytical hydrodynamic modeling tool developed to predict the performance of a very large crude carrier (VLCC), the *Exxon Valdez*, which ran aground on Bligh Reef in Prince William Sound.

The model consists of simplified hydrodynamic equations of motion with coefficients calibrated from sea trials. The model also incorporates hydrographic data of Prince William Sound to include shallow-water effects, data from a survey of Bligh Reef, current effects, the vessel geometry, and the tanker's draft and trim on the night of the grounding.

The final form of the hydrodynamic equations derived for the model are presented, along with the derived values of the six coefficients that characterize the dynamic properties of the vessel in water.

Full-scale sea trials of the vessel were conducted over a range of rudder and speed inputs at various vessel drafts. A matrix optimization was employed to determine the derivatives that best characterize the *Exxon Valdez* based on a systematic comparison of model-predicted positions with data collected during sea trials.

For inputs of rudder position and engine shaft speed, the model predicts vessel position and heading as functions of time for various trackline reconstructions, including the one followed by the vessel prior to grounding.
Keywords: current calibrations, *Exxon Valdez*, hydrodynamic derivatives, hydrodynamic model, load program up, model optimization, sea trials, shallow water

Structural Failure, Product Liability and Technical Insurance, Proceedings of Fifth International Conference. Edited by H.P. Rossmanith. Published in 1996 by Taylor & Francis, 2 Park Square, Milton Park, Abingdon, Oxon, OX14 4RN.
ISBN 0 419 22080 1

1 Introduction

The tanker *Exxon Valdez* ran aground on Bligh Reef in Prince William Sound on March 24, 1989. As part of our reconstruction of this event, we developed an analytical hydrodynamic modeling tool to predict the performance of this very large crude carrier (VLCC). The model was then calibrated against performance data derived from full-scale vessel sea trials. Results of the sea trials, combined with the predictions from the hydrodynamic model, were used to carefully reconstruct the tanker path in its last turn before grounding, and to predict the rudder configuration during this turn. This paper presents the development of the hydrodynamic model, and its calibration with full-scale sea trials.

Modeling of the vessel turn prior to grounding was one of the building blocks in the comprehensive vessel grounding reconstruction described in Reference [1].

2 Vessel Description

The *Exxon Valdez*, pictured in Figure 1, is a modern VLCC of all-welded structural steel construction. The vessel measures 987 feet (301 meters) long overall, 166 feet (51 meters) wide, and 88 feet (27 meters) deep from the main deck to the flat keel. The tanker has a maximum draft (fully laden) of 64.5 feet (20 meters), with a corresponding dead weight of approximately 210,000 long tons (213,360 metric tons). At maximum draft, the vessel can transport approximately 1.48 million barrels (235 million liters) of crude oil per voyage.

On the night of the grounding, the vessel contained approximately 1.31 million barrels (208 million liters) of crude oil resulting in a draft of 56 feet (17 meters).

Fig. 1. *Exxon Valdez.*

3 Methodology

Unlike commercial aircraft transports, tankers are not equipped with sophisticated data recorders which can be used to precisely reconstruct position, path, and control inputs. In the case of the *Exxon Valdez*, on-board data-recording instrumentation included the following:

- A course recorder: produces a strip-chart recording of vessel heading versus time. The course recorder trace from the night of the grounding is shown in Figure 2. This heading trace shows that the beginning of the turn occurred 77 seconds after midnight (9 a.m. Greenwich mean time). A jog in the trace at approximately 7 minutes after midnight indicates the first impact of the vessel with the reef.
- A fathometer: records water depth below the bottom of the hull versus time. The incident fathometer trace is shown in Figure 3.
- A bell logger: records engine revolutions per minute versus time. This information, not presented here, was used to identify the beginning of the ship's computerized acceleration process (LPU) in relation to the beginning of the turn.

Control inputs are not recorded or stored, although in this case, it was known that a right turn had been initiated prior to grounding. With the limited information described above, developing a model which could predict the ship path for a given rudder input became necessary. The model could then be used to predict the vessel's path and control input on the day of the incident, as well as predict other paths for different rudder positions.

The hydrodynamic model computes the lateral forces and yaw moments for a given velocity profile and rudder input to determine the trajectory of the ship. The model's analytical foundation consists of hydrodynamic equations of motion, the final form of which contain six coefficients unique to the vessel being studied. For this analysis, the numeric values of these derivatives were derived through a systematic comparison of predicted positions with data collected during sea trials. The sea trials data were obtained through full-scale testing of the *Exxon Valdez* in deep water, and the collected data were corrected to account for current effects present during the sea trials.

Once the model was calibrated for deep water conditions, reconstruction of the tanker path on the night of the accident was performed, taking into account case-specific information obtained from the on-board data recording equipment, the draft of the vessel, and the ocean floor topography in the accident area. It was known from the bell logger that the ship's computerized acceleration process referred to as "load program up" (LPU) was initiated prior to the beginning of the final vessel turn. Sea trials provided a speed rise profile for LPU conditions and a speed decay profile for a given turn at a given rudder deflection. These profiles were combined appropriately and incorporated in the model. Finally, as the ship approached the grounding site and entered shallow water, the vessel hydrodynamic characteristics were altered. The hydrodynamic derivatives which characterize the vessel

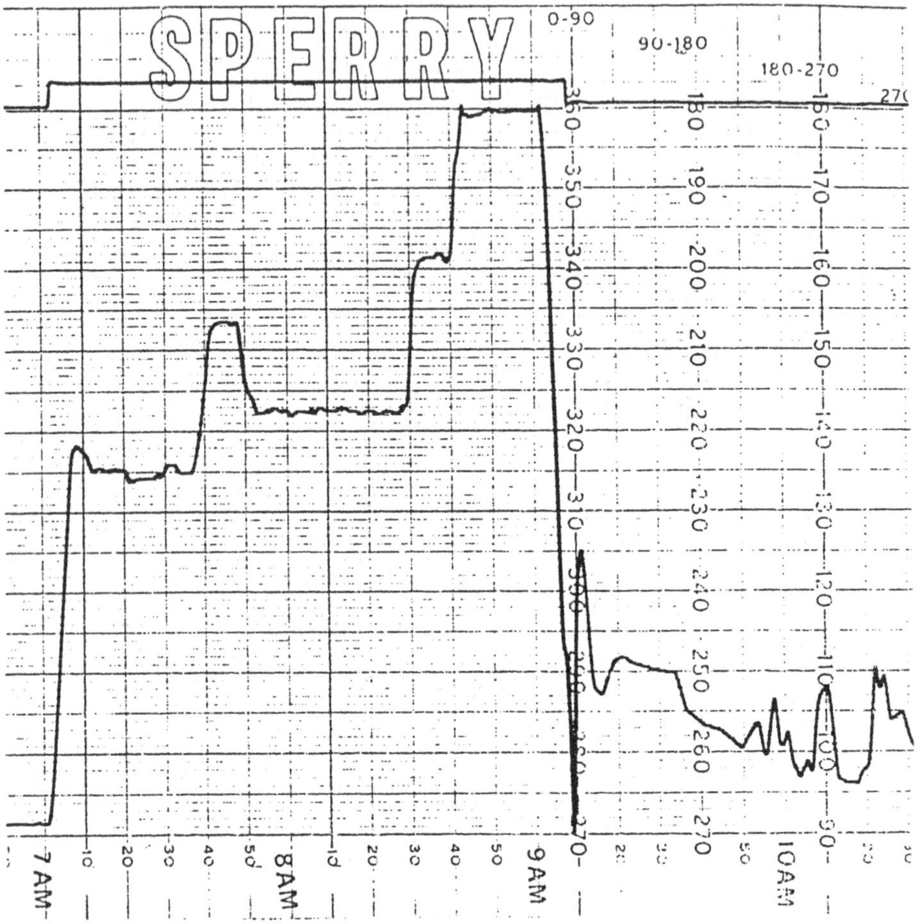

Fig. 2. Incident course recorder trace.

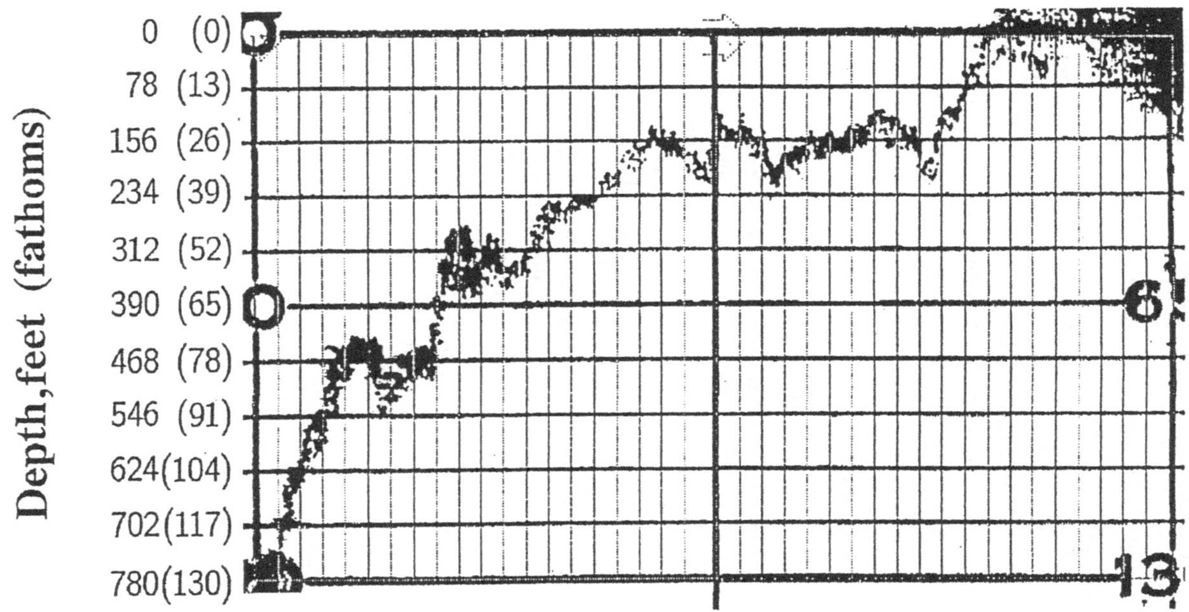

Fig. 3. Incident fathometer.

hydrodynamic properties were modified accordingly for this portion of the ship track and incorporated into the model.

The calibrated model, modified to include all incident-specific information, was then used to predict position and heading for various rudder inputs. A match between predicted heading and the incident course recorder heading trace provided the basis for identifying the rudder position commanded on the night of the incident. Additionally, the match between the calculated depth under the hull for the predicted track as compared to the incident fathometer trace provided further basis for establishing the actual control input on the night of the accident.

4 Mathematical Basis of the Model

An analytical model described in Reference [2] was adapted to this analysis. Figure 4 depicts a hypothetical vessel track which has been labeled with the nomenclature and coordinate system used in the analysis. Axes x_o, y_o, and z_o define an earth-fixed left-handed coordinate system, and axes x, y, and z define another left-handed coordinate system, fixed at the vessel's center of gravity (CG). The angle between the x_o and x axes is denoted as Ψ, and is also known as the heading angle.

The motion of a vessel at sea is dictated by the forces and moments which act upon it. When studying the track of a surface vessel on calm water, as was the case in this incident, we are interested only in the two-dimensional motion which describes its path. Thus, the only forces and moment which enter into the equations that define the track are the resultant surge and sway forces in the x_o and y_o directions, the magnitude of which are defined as X_o and Y_o, respectively, and the total yaw moment, of magnitude N, about the z axis. Identifying the position of the ship's CG in the earth-fixed coordinate system as x_{oG} and y_{oG}, and designating the mass of the ship as m and its mass moment of inertia about the z_o axis as I_z, the forces and moment can be expressed as

Surge $$X_o = m\ddot{x}_{oG} \tag{1}$$

Sway $$Y_o = m\ddot{y}_{oG} \tag{2}$$

Yaw $$N = I_z\ddot{\Psi} \tag{3}$$

where the two dots above the symbols represent the second derivatives of those variables with respect to time.

Transforming the forces from earth coordinates to ship coordinates yields

$$X = X_o\cos\Psi + Y_o\sin\Psi \tag{4}$$

$$Y = Y_o\cos\Psi - X_o\sin\Psi \tag{5}$$

Using u and v to represent the magnitude in the x and y directions of the ship's velocity at its CG, we can write

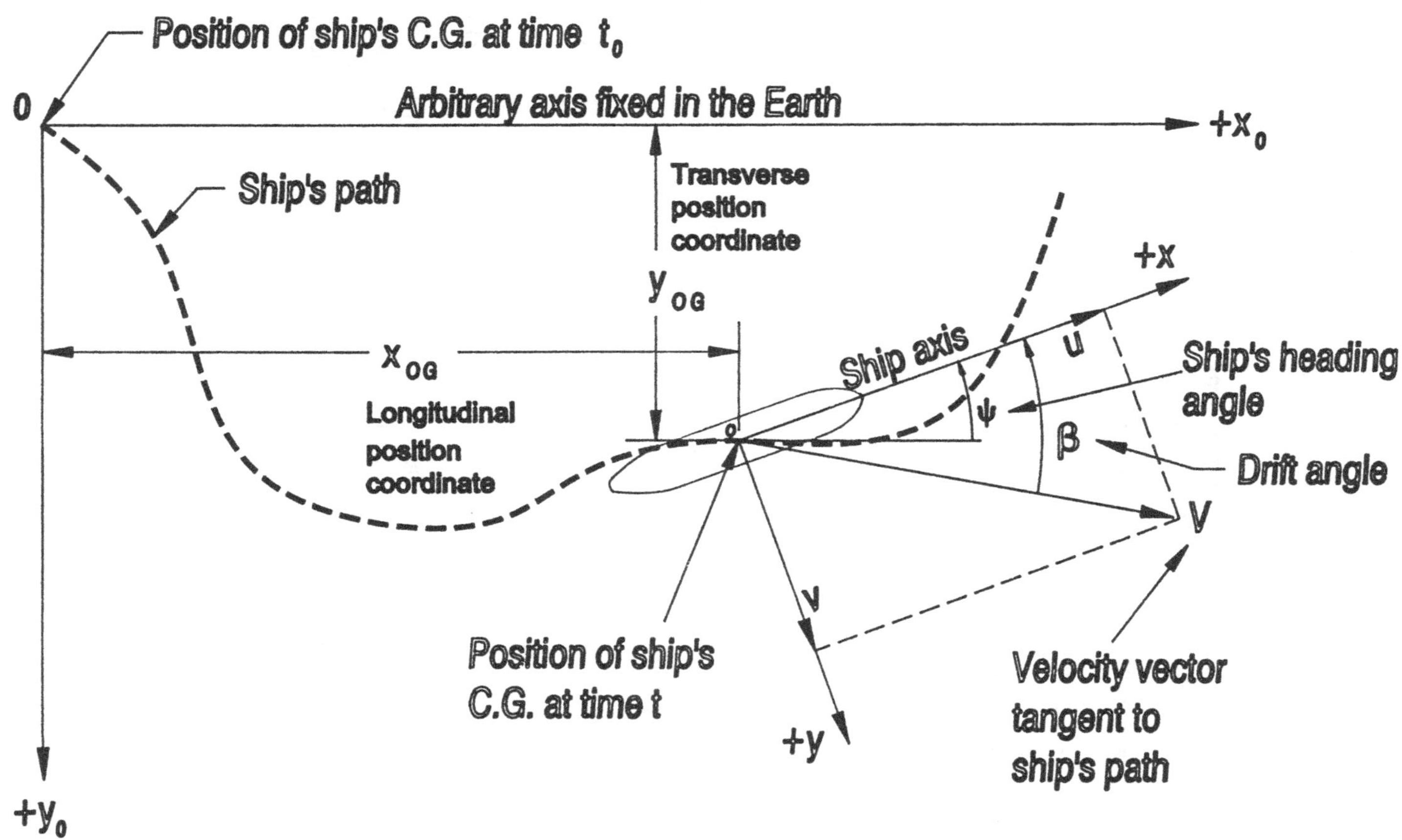

Fig. 4. Nomenclature and coordinate system.

$$\dot{x}_{oG} = u\cos\Psi - v\sin\Psi \tag{6}$$

$$\dot{y}_{oG} = u\sin\Psi + v\cos\Psi \tag{7}$$

where the dot above the symbols denotes the first derivative of those variables with respect to time.

Differentiating these two velocity equations with respect to time yields the following expressions for acceleration in ship coordinates:

$$\ddot{x}_{oG} = \dot{u}\cos\Psi - \dot{v}\sin\Psi - (u\sin\Psi + v\cos\Psi)\dot{\Psi} \tag{8}$$

$$\ddot{y}_{oG} = \dot{u}\sin\Psi + \dot{v}\cos\Psi + (u\cos\Psi - v\sin\Psi)\dot{\Psi} \tag{9}$$

Substituting Equations (8) and (9) into Equations (1) and (2), provides expressions for X_o and Y_o which can then be used in Equations (4) and (5). The resultant equations in ship coordinates then become

$$X = m(\dot{u} - v\dot{\Psi}) \tag{10}$$

$$Y = m(\dot{v} + u\dot{\Psi}) \tag{11}$$

$$N = I_z\ddot{\Psi} \tag{12}$$

The forces X and Y, and the moment N are functions of the ship's velocities and accelerations, and can be expressed as

$$X = F_x(u, v, \dot{u}, \dot{v}, \dot{\Psi}, \ddot{\Psi}) \tag{13}$$

$$Y = F_y(u, v, \dot{u}, \dot{v}, \dot{\Psi}, \ddot{\Psi}) \tag{14}$$

$$N = F_\Psi(u, v, \dot{u}, \dot{v}, \dot{\Psi}, \ddot{\Psi}) \tag{15}$$

These functional expressions can be converted to a useful mathematical form through Taylor expansion. Assuming an initial equilibrium condition of straight-line motion at constant speed, with zero angle of attack, $\dot{u}_1 = \dot{v}_1 = \dot{\Psi}_1 = \ddot{\Psi}_1 = v_1 = 0$, where the subscript 1 references the initial equilibrium condition. Further, noting that cross-coupled derivatives such as $\partial Y/\partial u$ are zero because of the ship's symmetry, the equations become

$$X = \frac{\partial X}{\partial u}(u - u_1) + \frac{\partial X}{\partial \dot{u}}\dot{u} \tag{16}$$

$$Y = \frac{\partial Y}{\partial v} v + \frac{\partial Y}{\partial \dot{v}} \dot{v} + \frac{\partial Y}{\partial \dot{\Psi}} \dot{\Psi} + \frac{\partial Y}{\partial \ddot{\Psi}} \ddot{\Psi} \tag{17}$$

$$N = \frac{\partial N}{\partial v} v + \frac{\partial N}{\partial \dot{v}} \dot{v} + \frac{\partial N}{\partial \dot{\Psi}} \dot{\Psi} + \frac{\partial N}{\partial \ddot{\Psi}} \ddot{\Psi} \tag{18}$$

In order to simplify the notation, shorthand symbols for the derivatives are used such that $r \equiv \dot{\Psi}$, $\dot{r} \equiv \ddot{\Psi}$, $X_u \equiv \frac{\partial X}{\partial u}$, $Y_v \equiv \frac{\partial Y}{\partial v}$, etc. Employing this notation, and substituting Equations (16), (17), and (18) into Equations (10), (11), and (12), respectively, the equations of motion become

$$-X_u(u - u_1) + (m - X_{\dot{u}})\dot{u} - mvr = 0 \qquad \text{(Surge)} \tag{19}$$

$$-Y_v v + (m - Y_{\dot{v}})\dot{v} - (Y_r - mu_1)r - Y_{\dot{r}}\dot{r} = 0 \qquad \text{(Sway)} \tag{20}$$

$$-N_v v - N_{\dot{v}}\dot{v} - N_r r + (I_z - N_{\dot{r}})\dot{r} = 0 \qquad \text{(Yaw)} \tag{21}$$

Nondimensional forms of the equations can be formed by dividing the force equations by $\frac{\rho}{2} L^2 V^2$ and the moment equations by $\frac{\rho}{2} L^3 V^2$, where ρ represents the density of sea water, L represents the overall ship length, and V represents the ship's velocity.

Making use of non-dimensional terms, the sway and yaw equations become

$$-Y'_v v' + (m' - Y'_{\dot{v}})\dot{v}' - (Y'_r - m')r' - Y'_{\dot{r}}\dot{r}' = 0 \tag{22}$$

$$-N'_v v' - N'_{\dot{v}}\dot{v}' - N'_r r' + (I'_z - N'_{\dot{r}})\dot{r}' = 0 \tag{23}$$

The equation for surge force can be used to calculate forward acceleration, and integrated to calculate ship speed. In the case of the *Exxon Valdez* reconstruction analysis, this is not necessary because speed is obtained directly from sea trials.

Finally, the force and moment created by rudder deflection are considered, and the following terms are introduced:

δ_R = Rudder deflection angle, measured from x-z plane of ship to the rudder plane; positive deflection corresponds to a turn to port.

Y_δ, N_δ = Derivatives of Y and N with respect to rudder deflection angle δ_R.

When applying the equations of motion in a model, we are considering small perturbations from the initial state, and thus small rudder deflections. This means that derivatives including Y'_v, N'_v, Y'_r, and N'_r are evaluated at $\delta_R = 0$, and can be assumed not to change at other admissible values of δ_R. Thus the equations of motion can be written in final form as

Moment (Yaw): $n'_z\dot{r}' - N'_v v' - N'_r r' = N'_\delta \delta_R$ (24)

Force (Sway): $m'_y \dot{v}' - Y'_v v' - (Y'_r - m')r' = Y'_\delta \delta_R$ (25)

where

$$n'_z = I'_z - N'_{\dot{r}} \cong 2I'_z \quad (26)$$

$$m'_y = m' - Y'_{\dot{v}} \cong 2m' \quad (27)$$

The approximations in Equations (26) and (27) can be made because the derivatives $Y'_{\dot{v}}$ and $N'_{\dot{r}}$, also known, respectively, as virtual mass and virtual moment of inertia, are associated with the added forces required to accelerate a body through a fluid.

An examination of Equations (24) and (25) indicates that the ship track and attitude are defined by six coefficients that characterize the dynamic properties of the vessel in water: three derivatives for the lateral forces $(Y'_v\ ,\ Y'_r\ ,\ Y'_\delta)$ and three derivatives for the yaw moments (N'_v, N'_r, N'_δ).

Having established the relationships of Equations (24) and (25), a computer program algorithm can now be developed. Equations (24) and (25) provide relationships for the translational and rotational vessel accelerations which can be integrated once with respect to time to obtain velocities, and integrated again to yield position and heading.

5 Model Calibration from Sea Trials Data

To apply this mathematical model to the *Exxon Valdez*, appropriate values of the above six coefficients had to be determined. The first step in deriving these values is the execution of full-scale vessel sea trials. The known track for a specific rudder input and velocity profile as recorded during sea trials can then be compared to the track as predicted by the model for the same conditions. Based on this comparison, the coefficients are then modified through an iterative approach until good agreement is obtained between the predicted and sea trials tracks. This is the procedure that was followed in this analysis.

The vessel's performance data were obtained by connecting electronic data acquisition equipment to the ship's instruments. Parameters recorded in this manner included heading, speed, rate of turn, and rudder angle. In addition, the ship's position was recorded by installing a global positioning system (GPS) on the ship for these tests. This satellite tracking system provided position data accurate to within approximately 1 meter. The GPS antenna was mounted to the port wing of the bridge. Position data provided by this system were then transformed to coincide with the vessel's center of gravity.

It was known that the ship performed a right turn immediately prior to the grounding event. The full-scale vessel sea trials performed in connection with this analysis included a complete set of maneuvers executed at the same draft and speed conditions as existed at the time of the incident. Right turns were performed at rudder inputs ranging from 5° through hard over, in increments of 5°. A heading change of 180° was achieved for each test, and the engine speed was held constant throughout the turn. Additionally, it was known that prior to the turn, the ship was slowly ramping up to sea speed under the command of its engine computer control system (LPU). To obtain vessel performance data for this condition, sea trials were also performed with the LPU system activated.

5.1 Current Calibration

Before the data gathered from the sea trials could be used in comparison with the model predictions, the data were calibrated to account for current effects. Prior to each sea trials turn, the vessel was maintained at a constant 90° (east) or 270° (west) heading. Data from this pre-turn period of each test were analyzed to isolate the average north component of the vessel velocity, which corresponds to the north component of the current velocity.

The east component of current velocity was determined by comparing the speed data acquired at constant headings of 90° to that at constant headings of 270° for tests performed in the same geographical location. The east component of current velocity was then calculated as the difference between the average east velocity and average west velocity. Current speed components were integrated to calculate the ship drift, which was then subtracted from recorded sea trials vessel position. Figure 5 compares the track of each starboard sea trials turn before and after current calibration.

5.2 Hydrodynamic Model Coefficient Optimization

Once the sea trials data were calibrated for current effects, a matrix optimization was performed to determine the hydrodynamic model coefficients that best characterize the performance of the vessel. The optimization was performed for all six sea trials right turns. The coefficients were refined until the hydrodynamic model's predicted heading change and track converged with those from the calibrated sea trials data. Figures 6 and 7 demonstrate the final match obtained between model-predicted and sea trials track and heading change. Optimum agreement was obtained for rudder angles of 5°, 10°, and 15°. The optimized nondimensional derivative values are presented in Table 1.

Table 1. Final derivative values

Derivative	Value
Y_v	-0.00564
Y_r	-0.00241
Y_δ	0.00408
N_v	-0.00195
N_r	-0.00148
N_δ	-0.00194

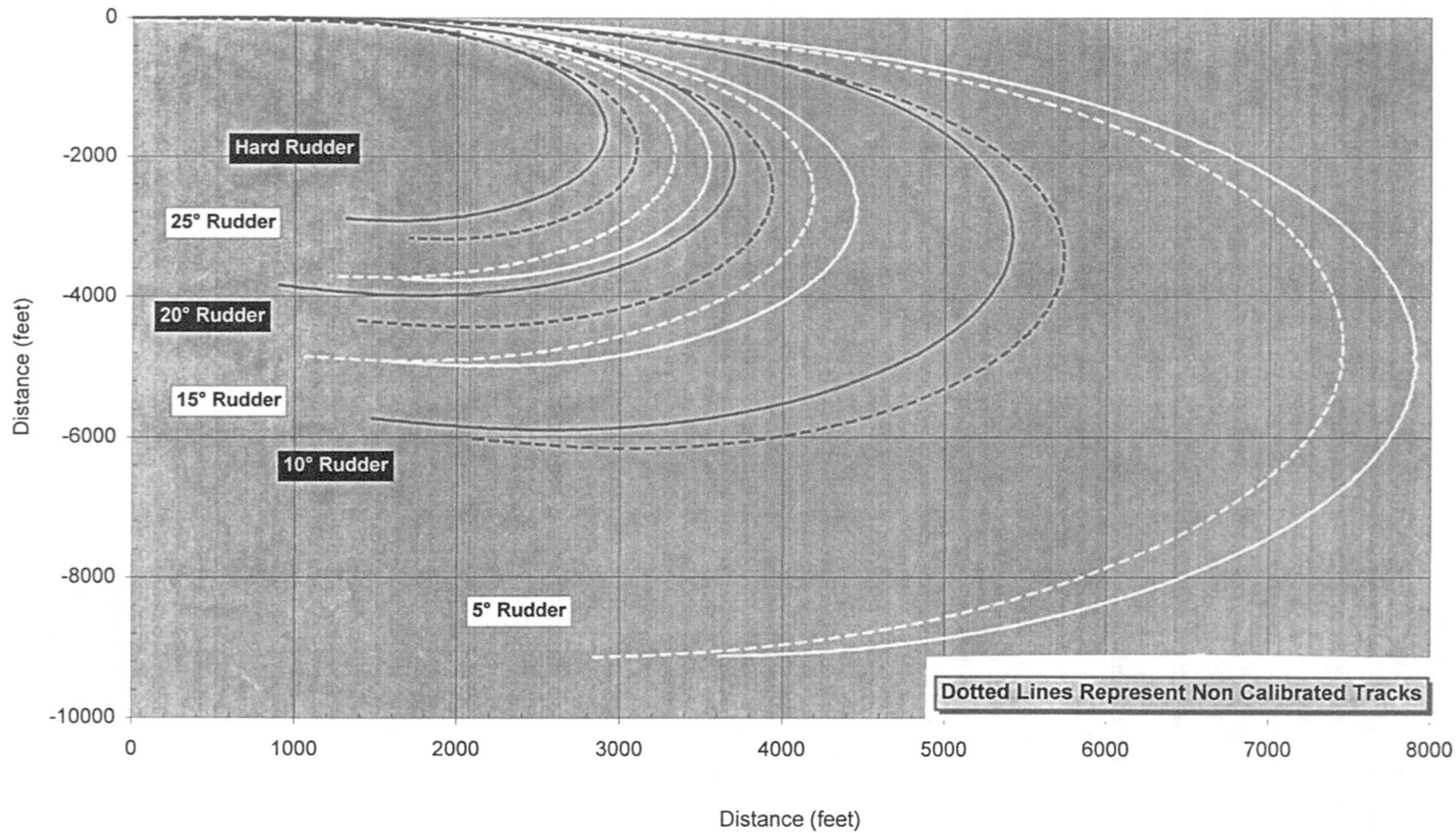

Fig. 5. Effects of current calibration to sea trials track data.

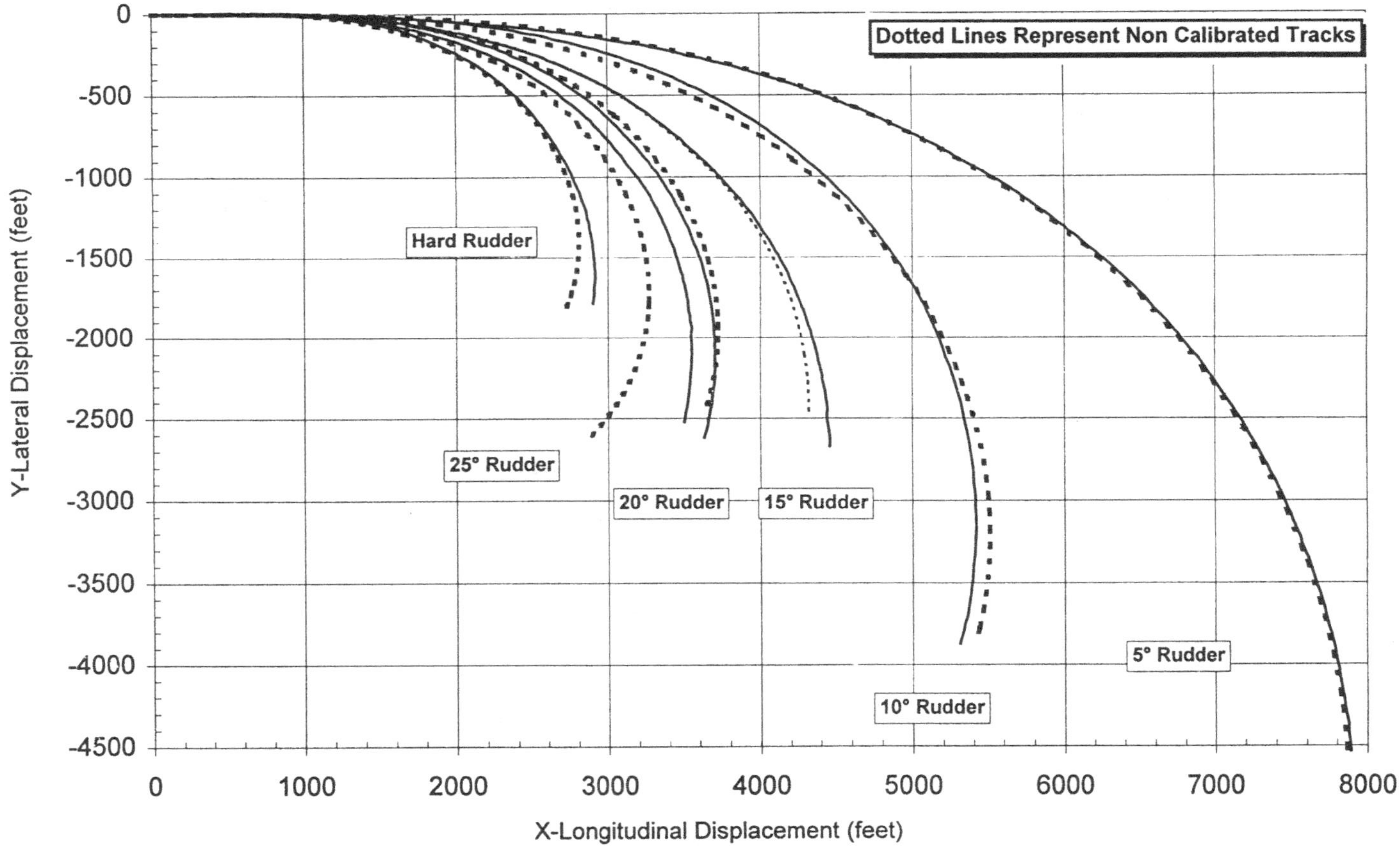

Fig. 6. Hydrodynamic model deep water track prediction vs. calibrated sea trials tracks.

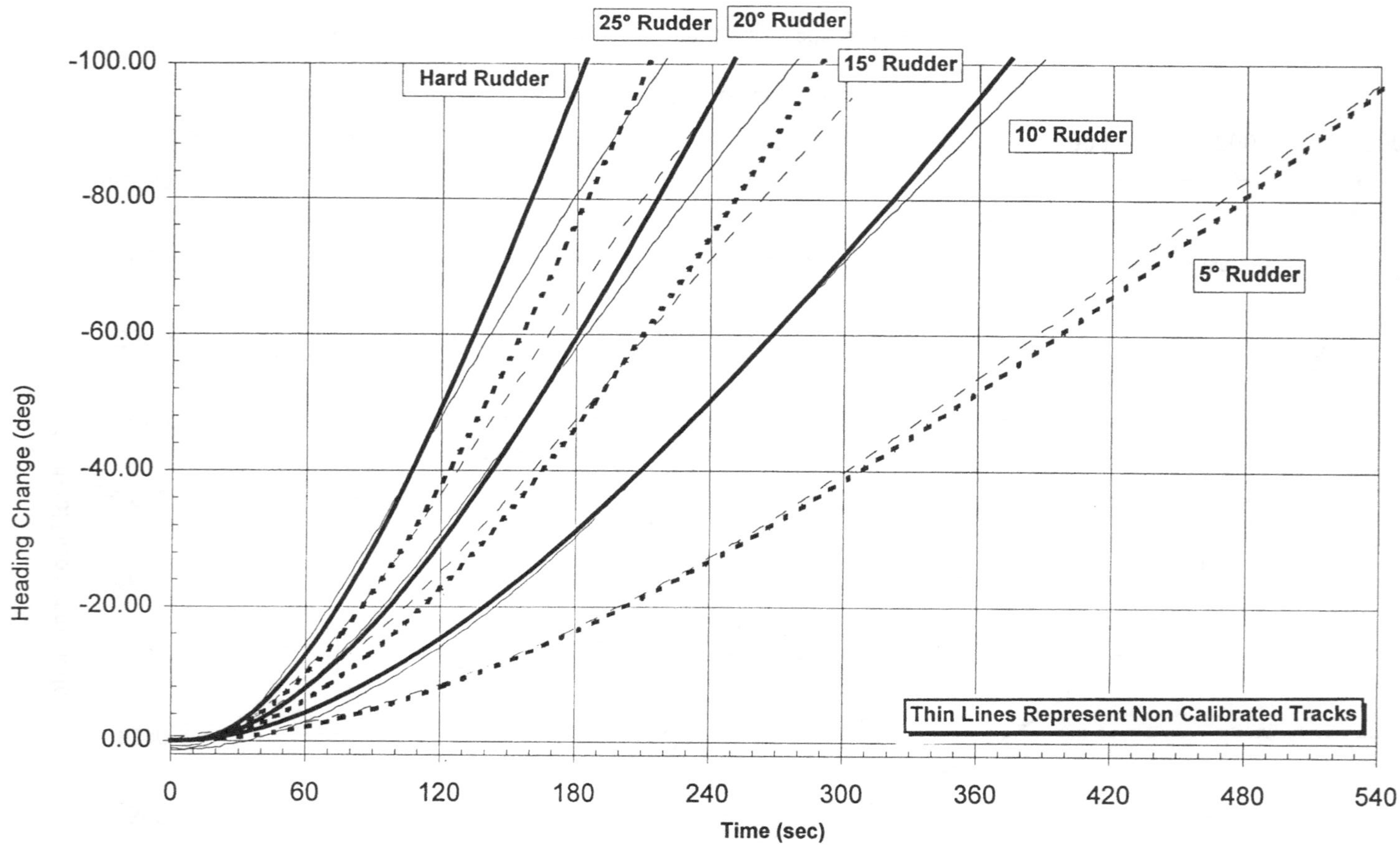

Fig. 7. Hydrodynamic model deep water heading change prediction vs. sea trial heading change.

5.3 Shallow Water Calibration

The ship coefficients discussed above model a ship traveling in deep waters. The forces and moments that are created as a ship passes through shallow water are substantially different from those that arise when it travels in deep waters. Further, when a ship enters a shallow water region, the ship rudder efficiency is also affected. These changes must be duplicated in the hydrodynamic model if it is to accurately characterize this change in dynamic behavior.

Based on a review and analysis of the literature on this subject (References [2] [3] [4]), as well as analysis of the course recorder, fathometer, sea trials data, and ocean floor topography, the shallow water effects were incorporated into the hydrodynamic model by adjusting the ship coefficients related to rudder input (Y_δ, N_δ) by a factor of 0.9, and adjusting the remaining coefficients (Y_v, Y_r, N_v, and N_r) by a factor of 1.5. These changes were effected in the model at the point when the ship entered a region where the water depth decreased to less than 3.5 times the vessel draft.

Figures 8 and 9 depict the difference between the track and heading change predicted by the model with and without shallow water calibrations.

5.4 Load Program Up Calibration

As discussed earlier in this paper, it was known that the ship's computer-controlled engine acceleration program (LPU) had been activated prior to the grounding. The elapsed time between LPU initiation and the start of the turn was calculated from examination of the bridge log and analysis of incident course recorder data. Knowing this elapsed time, the ship's condition at the start of the turn was then determined from the sea trials LPU test data.

The LPU sea trials tests were performed along a straight-line course. To account for velocity decay during a turn, a turn decay factor was derived from the 10° right rudder sea trial velocity profile. The final velocity profile was produced by superimposing the LPU and turn decay profiles:

$$V_{final} = V_{LPU} - (\text{turn decay factor}) \times (\text{time}) \tag{28}$$

The start of the turn was used as the matching point for the two velocity functions. Figures 8 and 9 demonstrate the effect of this calibration on the track and heading change predicted by the model.

6 Model Implementation and Results

In its final calibrated form, the hydrodynamic model predicts vessel position, translational velocity, rotational velocity, and heading as functions of time, based on input values of rudder position and engine shaft speed.

By systematically comparing the model's output data for different rudder inputs to actual vessel performance data recorded at the time of the incident, we were able to identify the rudder commands which resulted in the recorded vessel responses.

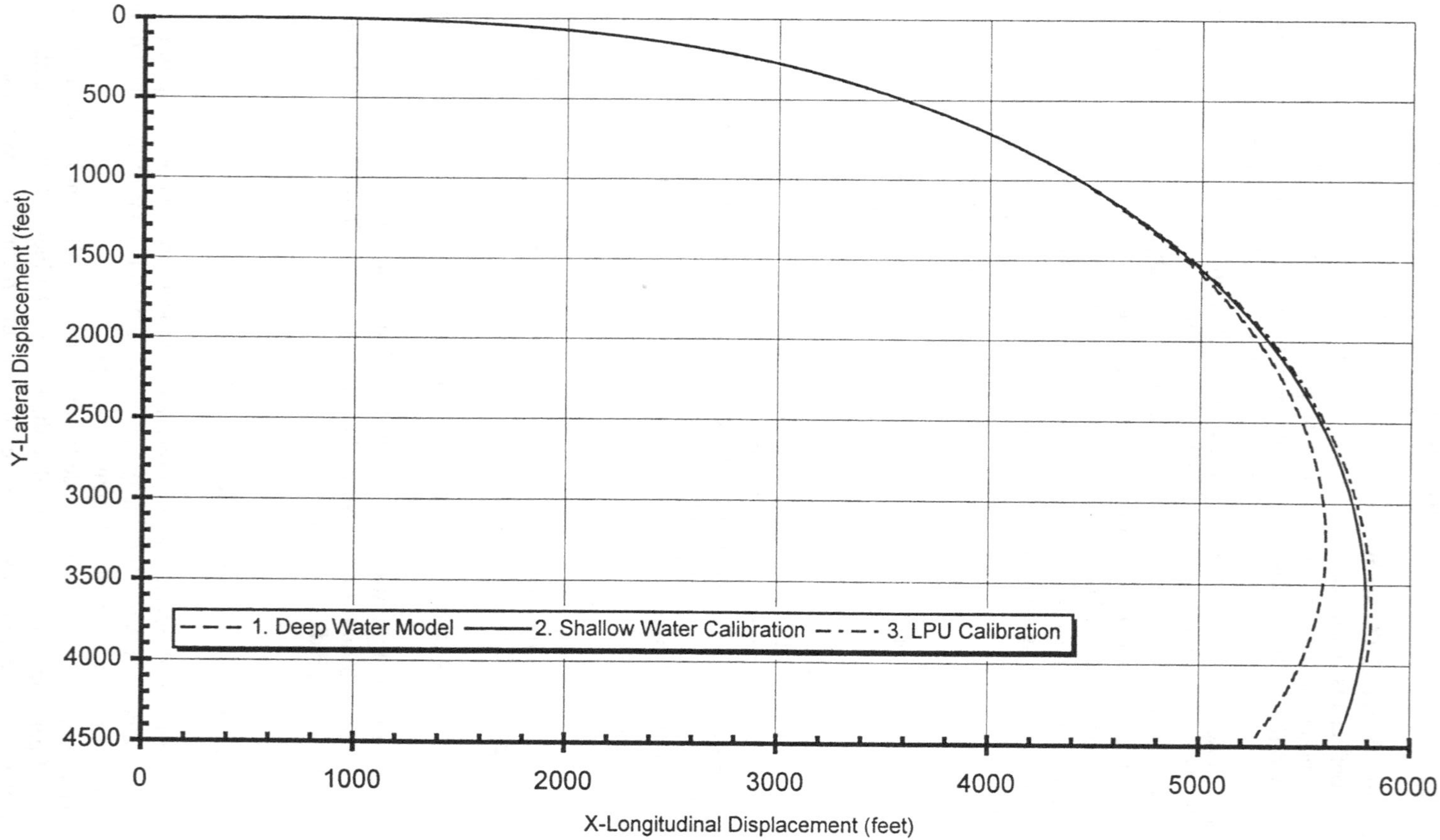

Fig. 8. Hydrodynamic model prediction evolution: track.

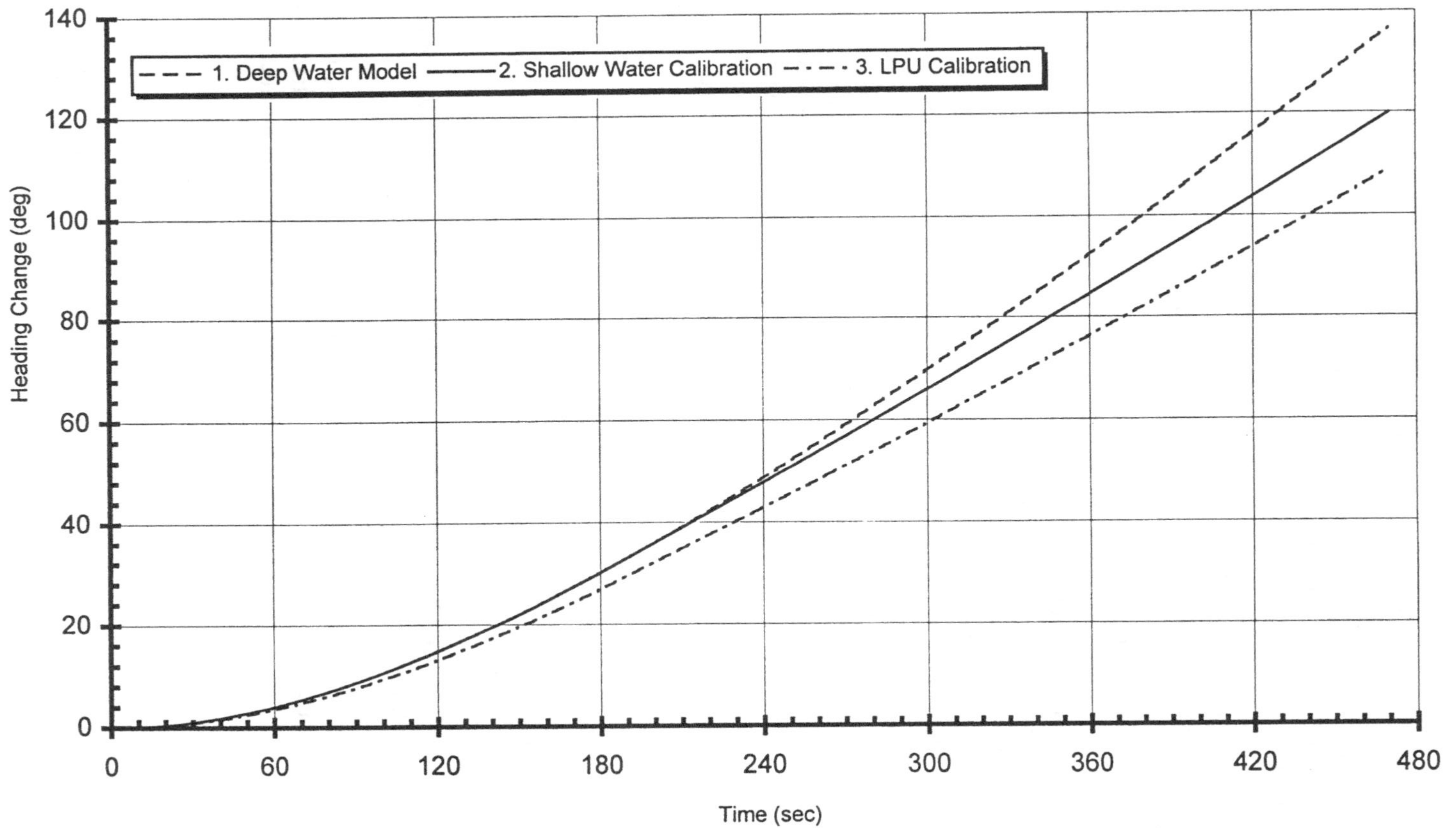

Fig. 9. Hydrodynamic model prediction evolution: heading change.

Shallow water effects were incorporated into the model. Data from a survey of Bligh Reef were used to determine the point where the ship enters shallow water.

The locations where the vessel contacted the reef were identified during the reef survey. By analyzing the incident course recorder to determine the length of time between turn initiation and reef impact, and by superimposing the tracks predicted by the model onto the reef contour map, the turn was established to be the result of a 10° rudder input. Figure 10 compares the vessel heading over the grounding site as predicted by the model for 5° and 10° turns to the incident heading data from the vessel's course recorder. As this plot demonstrates, the course recorder data very closely match the heading predicted by the model for a 10° turn. The change in slope in the course recorder plot at approximately 400 seconds occurs as a result of the vessel's first impact with the reef. Figure 11 shows that the track for a 10° turn passes through the reef impact locations.

Additionally, the best agreement between the incident fathometer trace and the water depth below the hull calculated for a model-predicted track, occurs for a rudder input of 10°. Water depths below the hull based on track predictions for other rudder inputs, and those based on track predictions which exclude LPU and shallow water effects, did not match the incident fathometer data as well as did the predicted track for 10° rudder input with LPU and shallow water effects. A comparison of the incident fathometer data to water depth for a 10° model-predicted right turn is shown in Figure 12. A least squares method, as shown in Figure 13, was used to determine the best match between data traces. The vessel's draft and the tide conditions at the time of the incident were accounted for in the analysis of fathometer data.

7 Conclusions

This paper presents the elements of a hydrodynamic model which can be used to predict the track of VLCC vessels such as the *Exxon Valdez*.

This model was calibrated by using data obtained from full-scale sea trials, which allowed the required hydrodynamic coefficients to be fully derived.

The calibrated model was then refined to include accident parameters specific to the *Exxon Valdez* grounding. Use of the model accurately predicted the ship track on the night of the incident as confirmed by correlation with the course recorder and fathometer traces.

8 References

1. Kadlec, R., Beres, E., and Hayes, D. (1995) Engineering analysis of the *Exxon Valdez* trackline path and grounding. *Structural Failure, Product Liability and Technical Insurance.* Proceedings of the Fifth International Conference.

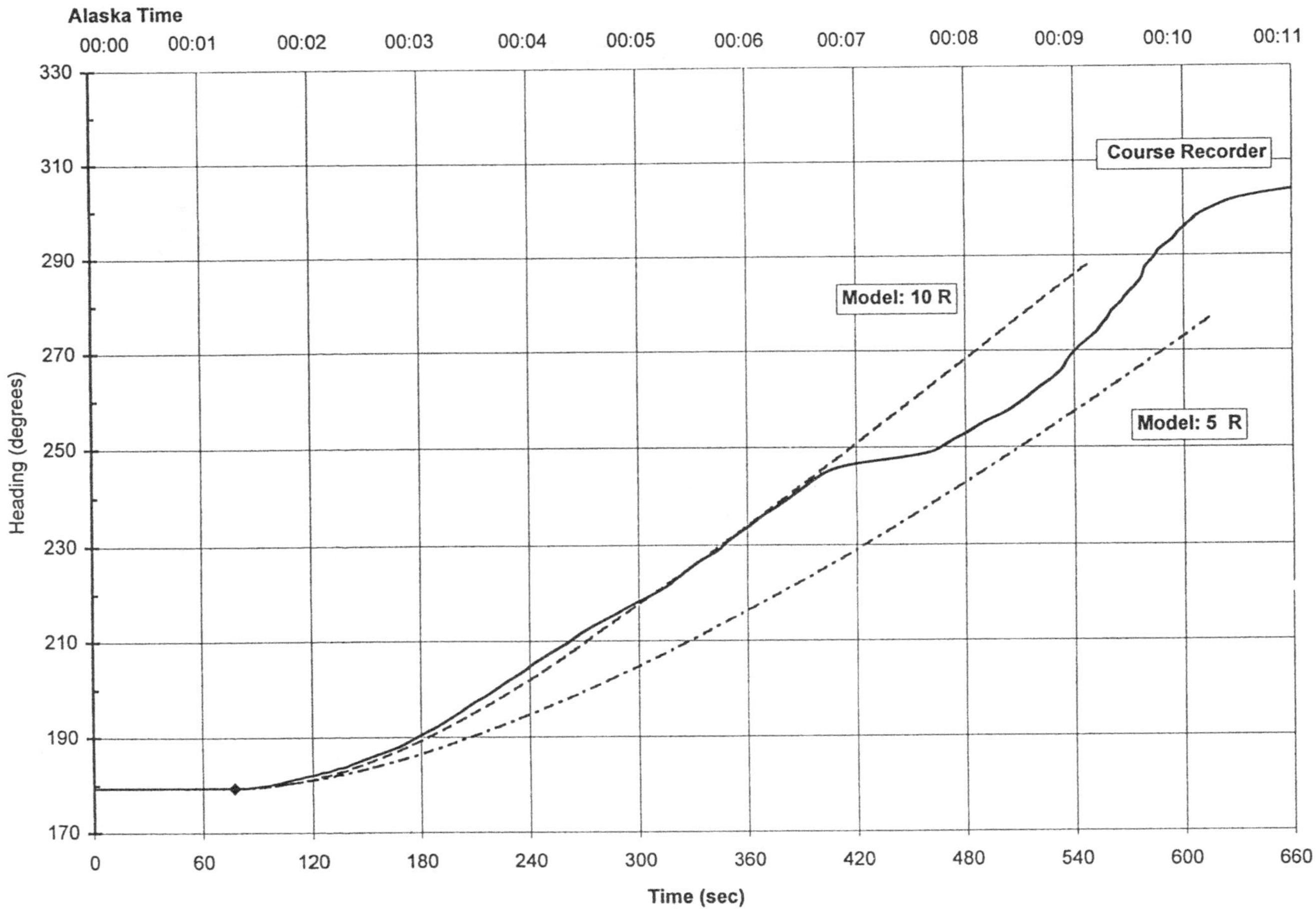

Fig. 10. Vessel heading comparison.

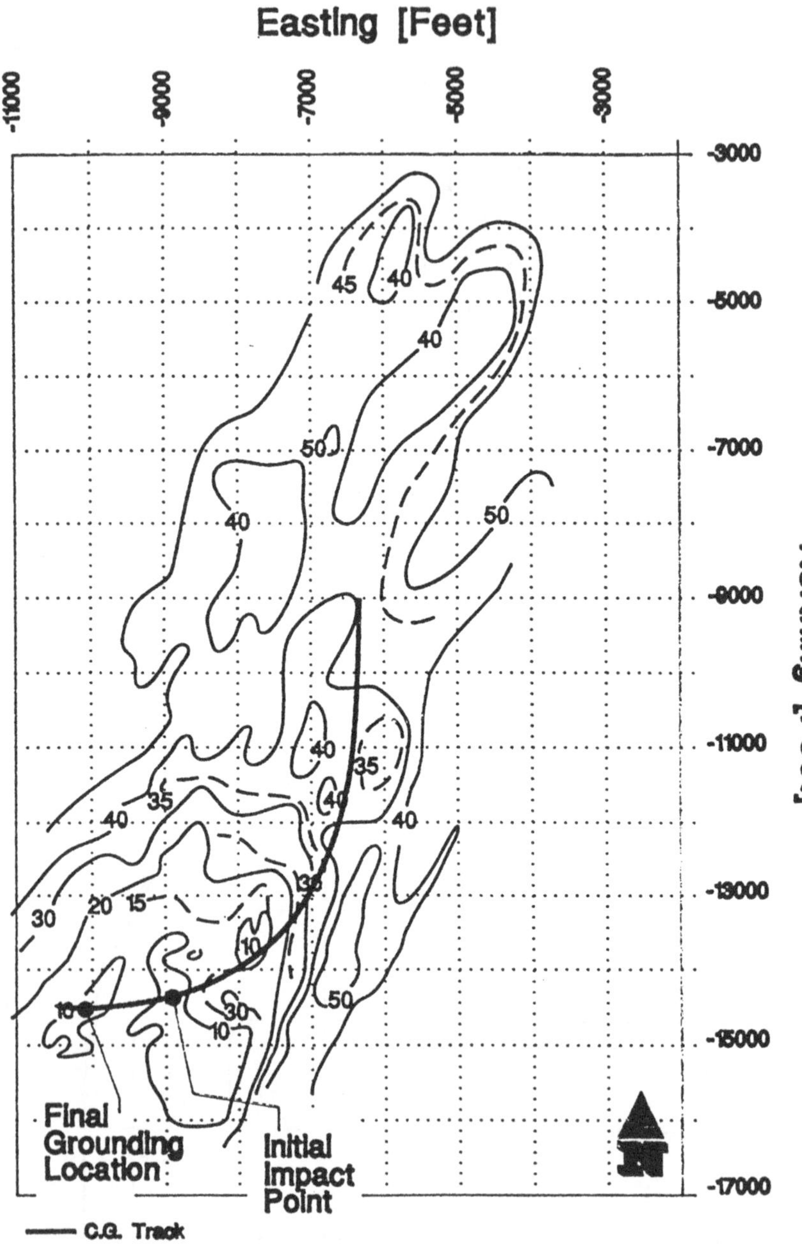

Fig. 11. Model-generated track: 10° rudder.

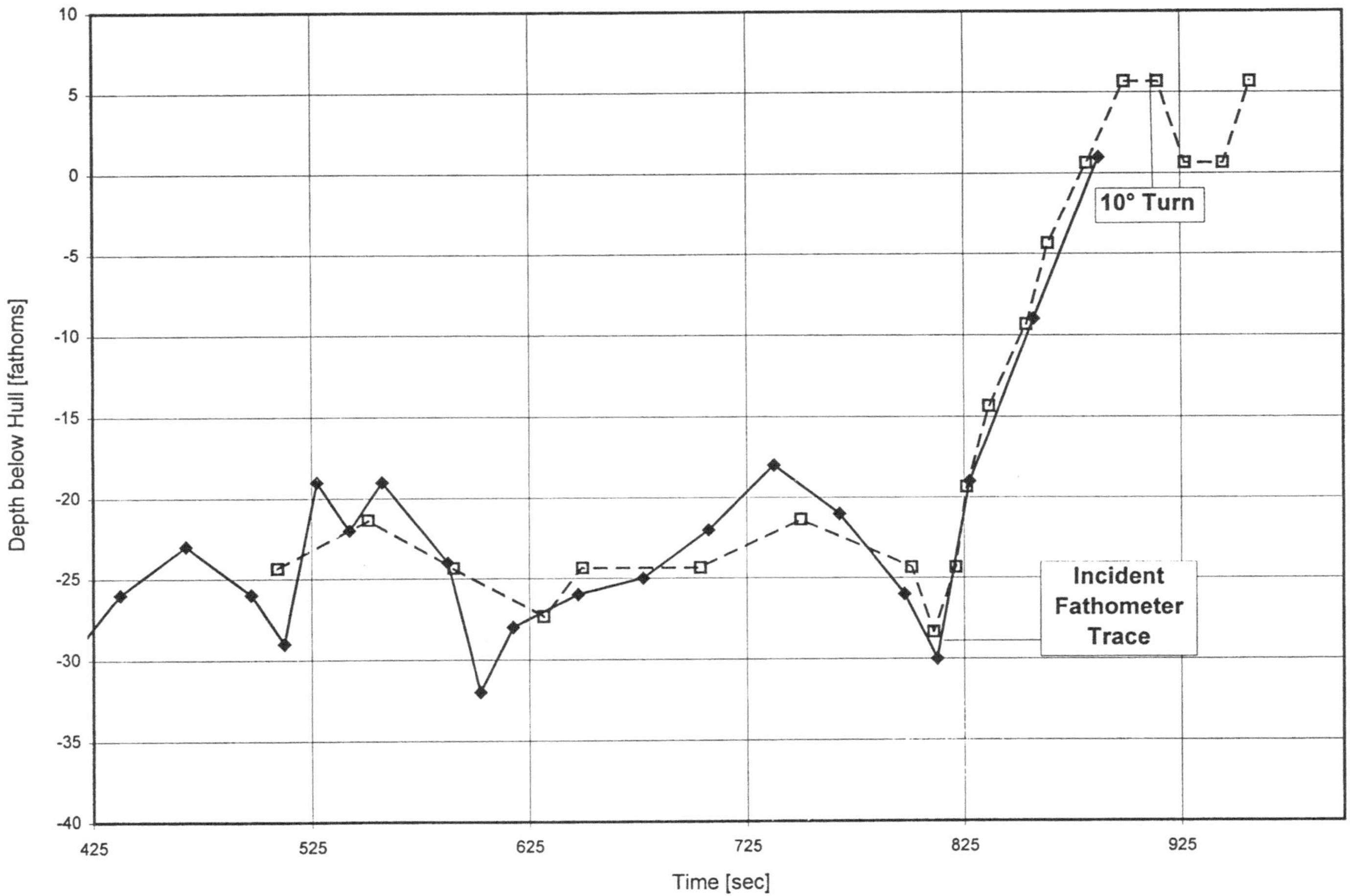

Fig. 12. Incident fathometer trace vs. model 10° turn.

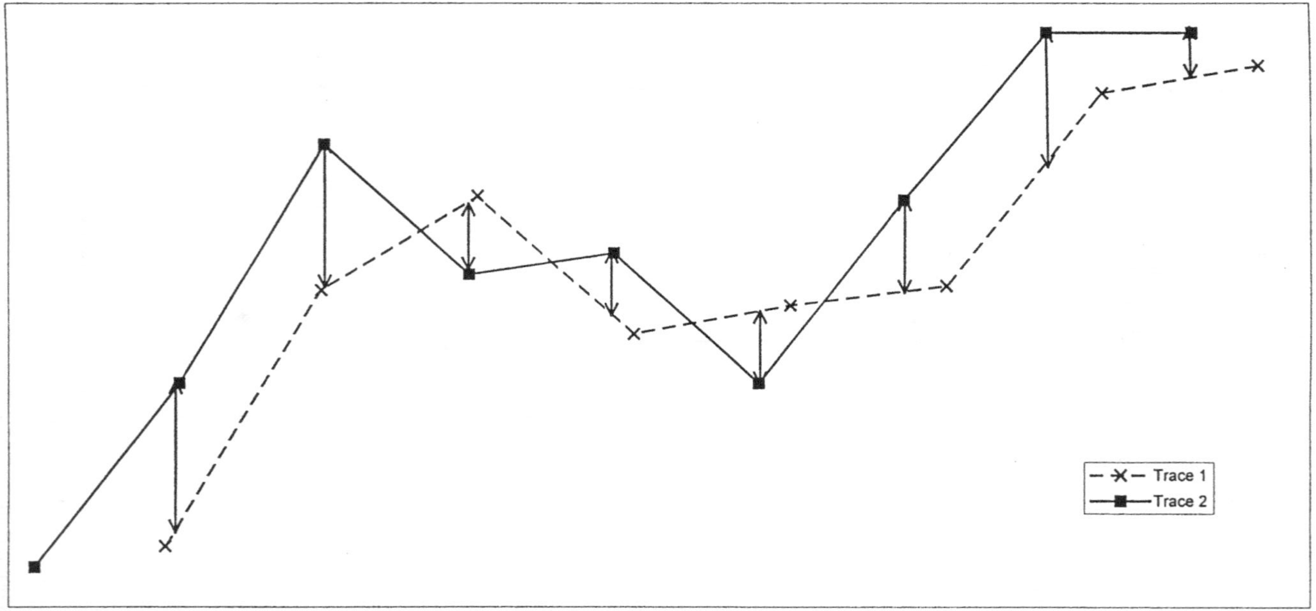

$$SSE = \sum_{i=1}^{n_2} (y_{2,i} - y_1)^2$$

n_2 = Number of points in data set 2

$$\overline{SE} = \frac{SSE}{n}$$

Fig. 13. Fathometer data matching.

2. Lewis, E.V. (1989) *The Principles of Naval Architecture*, The Society of Naval Architects and Marine Engineers.
3. Tuck, E.O. (1978) Hydrodynamic problems of ships in restricted waters. *Ann. Rev. Fluid Mech.* Vol. 10, pp. 33-46.
4. Abkowitz, M.A. (1980) Measurement of hydrodynamic characteristics from ship maneuvering trials by system identification. *SNAME Transactions,* Vol. 88, pp. 283-318.
5. Abkowitz, M.A., Liu, G. (1988) Measurement of ship resistance, powering, and maneuvering coefficients from simple trials during a regular voyage. Society of Naval Architects and Marine Engineers.
6. Abkowitz, M.A. (1984) *Measurement of Ship Hydrodynamic Coefficients in Maneuvering from Simple Trials during Regular Operations.* U.S. Department of Transportation.
7. Abkowitz, M.A. (1980) *Measurement of Hydrodynamic Characteristics from Ship Maneuvering Trials by System Identification.* The Society of Naval Architects and Marine Engineers.
8. Inoue, S., Hirano, M., Kijima, K. *Hydrodynamic Derivatives on Ship Maneuvering,* pp. 112 - 125.
9. Jack, R.L. (1973) *Sea Trials: Some Recommended Practices, Marine Technology,* The Society of Naval Architects and Marine Engineers, 380 - 391.

20 LIMITATION OF OPERATIONAL FORCES IN POWER-OPERATED DOORS AND GATES

L.D. ANDERSON and H.O. TORSTENSSON
Swedish National Testing and Research Institute, Borås, Sweden

Abstract

In the European directives minimum requirements of safety but no test method is described, and no admissible values. Test methods and admissible values are supposed to be developed in harmonised standards.

When developing harmonised standards for power-operated doors and gates it became necessary to measure crushing forces between closing and counterclosing edges and state maximum admissible forces.

Various methods and equipment for measuring of crushing forces have previously been developed, but there were significant differences between the results. This resulted in different levels of crushing forces.

Methods were compared and sensoric tests were performed, which resulted in a proposal for a European method for measuring of crushing forces, where effects on the human body are taken into account.

Keywords: Force measurement, personal protection, power-operated doors, machinery.

1 Introduction

In the European Community Directives of the "New Approach" type, minimum or basic requirements of health and safety are given. Those minimum requirements apply to most products used by consumers and enterprises. The basic, or essential, health and safety requirements are set with the aim that products must not be harmful or dangerous to the human body or to the environment.

Together with the "New Approach" the "Global Approach" is a means to specify and certify that the products fulfil the basic health and safety requirements. In harmonised standards for the different types of products more specified requirements can be described and used for the EC-type examination and the following EC-marking.

Structural Failure, Product Liability and Technical Insurance, Proceedings of Fifth International Conference. Edited by H.P. Rossmanith. Published in 1996 by Taylor & Francis, 2 Park Square, Milton Park, Abingdon, Oxon, OX14 4RN. ISBN 0 419 22080 1

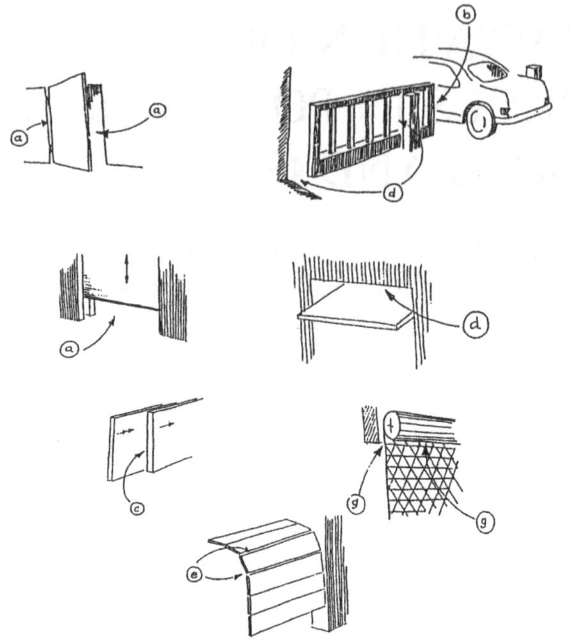

Fig 1 Hazardous locations

2 Safety standards

When developing harmonised standards for power-operated doors and gates it became necessary to reduce hazards caused by impact, crushing, shearing, drawing in and entanglement of people. The first step to be taken is to avoid the dangerous points by shaping of leaf surfaces and other parts or projections that may cause injuries. In Fig. 1 typical hazardous locations for power operated doors and gates are shown.

If it is impossible to avoid the hazardous locations, then *safety distances, safeguarding of danger points* and *limitation of operating forces* are other ways to reduce the risks.

Safety distances with specified minima are given in the European standard EN 349 [1]. Crushing, shearing and drawing-in points are avoided if safety distances are provided and sufficiently dimensioned concerning the endangered parts of the human body.

Safeguarding of danger points generated by the door leaf can be achieved by:

- Hold-to-run control (dead man control)
- Guards, such as enclosures, covers, fixed protection leaves
- Pressure sensitive protective devices (PSPD) which have a sensing function being actuated when mechanical pressure is applied to them, such as sensitive edges, mats and floors. PSPD are so called contact detection devices.
- Electro-sensitive protective devices (ESPD) which have a sensor function being actuated by interruption or reflection of beams, such as photo electric devices, light

grids, light detectors, or ultrasound devices. ESPD are so called presence detection devices.

Crushing points can also be avoided by *Limitation of operating forces*, caused by a moving part of the door or gate that impacts the human body or parts of it.

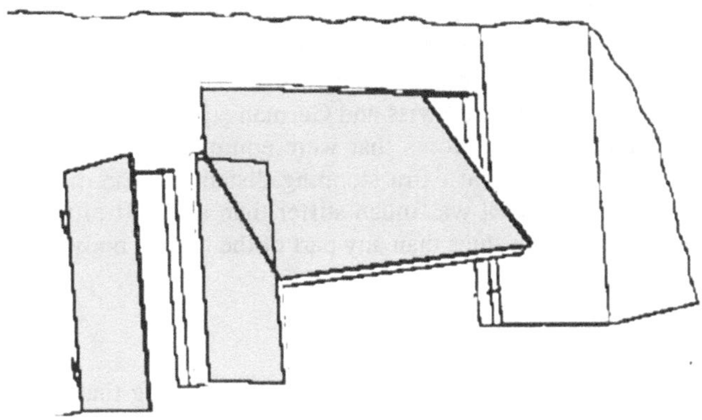

Fig 2 Example of safeguarding measures.

3 How to measure crushing forces

To establish the admissible values of crushing forces it became necessary to carry out research. Methods and equipment for measuring of crushing forces in doors and gates had been developed in Germany, Switzerland and Sweden since many years but there were significant differences between those methods and equipment, which gave differences in test results and therefore different levels of admissible crushing forces.

The German test equipment was a device equipped with a spring, with a spring ratio of 25 N/mm, placed on top of a electric load cell, connected to an amplifier and a printer.

The Swiss test equipment was a device consisting of a spring with the same spring ratio as the German equipment. The spring was equipped with a scale indicating the force as a function of the spring compression.

The Swedish test equipment was of the same type as the German one but there was no spring mounted on the load cell. Instead the amplifier was equipped with a low pass filter LFC 1000 to avoid the ringing phenomena which can occur when parts are impacting each other.

The Swiss equipment had the advantage that it was very easy to carry and use "on site" because of low weight, few parts and no need of electricity. Both the German and the Swedish equipment were much more expensive and needed more or less well trained personal and of course electrical power.

As mentioned earlier it was soon realised that those three equipment gave very different values when measurements were performed on different doors and gates. On some doors the results were almost the same, while on others the Swedish method gave much higher values than the other two.

If the door was equipped with a deformation zone on the closing edge, normally a rubber moulding, and the door came to a stop before the closing edge was totally compressed the three methods gave reasonable values compared to each other. But if the door was not equipped with a rubber edge or the rubber edge was totally compressed before the door had come to a total rest the methods gave different results with the Swedish method giving much higher values.

Different parts of the human body have different stiffness, for example the head compared to the body or an arm. The stiffer the parts are, that are being crushed, the higher the forces will be if the crushing edge is not equipped with a soft deformation zone. The disadvantage with the Swiss and German equipment was that the spring was so weak that it favoured some doors that were equipped with too small deformation zones. They gave the door leaf extra stopping distance. The disadvantage with the Swedish equipment was that it was much stiffer than any part of the human body and therefore gave higher force values than any part of the human body could experience.

4 Admissible forces

4.1 Tests performed

Wayne State University in the USA had results showing that the stiffness (spring ratio) of a human head can vary a lot but on elderly people the cranium may crack at a load of 4 kN and a deformation of 4 mm. It was decided that the head, being the most critical part of the human body, must be protected and this made it obvious that if the test equipment should be equipped with a spring it must be much stiffer than the German and Swiss spring.

A joint European measuring device equipped with a spring with a ratio of 500 N/mm, was developed in Sweden and Germany.

With a harmonised test equipment giving comparable results the level of admissible crushing forces should be established.

The base for the Swedish requirements on admissible crushing forces was the results from tests performed on young pigs, 12 to 30 kg, in 1978 [2]. Those tests were performed by a medical team from Östra Sjukhuset in Gothenburg and staff from the Swedish National Testing and Research Institute, SP. The pigs were crushed between the floor and a crushing steel edge with a width of 50 mm. The results from the tests indicated that even with very low crushing forces applied on the chest of the pigs they began to suffer.

The medical team decided that crushing forces of *450 N*, measured with the old "Swedish" method, was the maximum allowed value. This value was established under the presumption that it was only allowed during a period of maximum 0.5 seconds. After this time no force larger than *150 N* was allowed and this force must be reduced to zero within 4.5 seconds.

Germany and Switzerland had values corresponding to those, but the differences in measuring methods made it uncertain if they were comparable. It was therefore decided that sensoric tests on human beings should be conducted by BIA in Germany and SP in Sweden in order to establish admissible values.

Test groups consisting of 30 - 50 people, 65 % males in the age between 18 and 60 years were set up in both Germany and Sweden. Their task was to evaluate admissible forces individually according to the criteria "Well bearable", "Bearable" and "Not bearable". The test persons were crushed over the hip and over the fingers. The hip was chosen in order to have a large area of a softer part of the body, while the fingers were chosen in order to simulate a stiffer part such as the head. Tests were also conducted over chest and stomach between the folding panel and a rigid wall. The test group could withstand very high forces because of the large compression of the body, but the tests were stopped when it became evident that there was a great risk for the head to be crushed when the body was compressed maximally.

Two types of doors were used. One was a sliding door with a 8 mm steel edge, door panel mass 43.4 kg, and the other a folding door, door panel mass 148 kg. The sliding door was used for crushing the fingers and the folding door for crushing of the hip. The doors were using a constant static force (a weight pulling the door panels) instead of machinery and the crushing forces (impacts) were increased by letting the panels accelerate over longer and longer distances. The forces were measured with the new developed equipment. The spring used had a ratio of 460 N/mm.

When crushing the fingers the test person held his fingers in contact with the door frame and the sliding panel was set to accelerate against the fingers. The force (impact) was increased until the test person refused to continue and judged the force as not bearable.

When crushing the hip the folding door was mounted so that a clearance of 250 mm between the door panel, in open position, and a rigid wall was obtained. The width of the panels were 1200 mm. and the crushing forces were measured 1000 mm from the hinge between the rigid wall and the panel. The test persons were crushed in the same position and were in contact with the rigid wall.

Performing sensoric testing in this way is very difficult, mostly because the result depends to a great degree of the way the test person is treated before the test. It is essential that he or she gets a neutral treatment. The result of the test can differ a lot if on one hand the test group makes a competition to see who can withstand the highest forces or on the other hand the test person is afraid because of the big door and all equipment.

It was therefore surprising to see that the results from BIA and SP were going in the same direction and indicating the same level of admissible forces [3, 4].

4.2 Result

The result, from crushing the fingers, showed that *43 %* of the test population found a crushing force of *380 N* well bearable, *37 %* found it bearable and *20 %* not bearable when being crushed in the sliding door. When the crushing force was increased to *480 N, 17%* found it well bearable, *28%* bearable and *54%* not bearable. This indicated that the limitation of crushing forces should be somewhere between 380 and 480 N.

The results from crushing the hips in the folding door gave much higher forces, *25 %* of the test population in Germany judged *1600 N* as well bearable, *48 %* as bearable and *27 %* as not bearable. The Swedish test population gave values lying 300 N lower.

Those results highlighted the need of different values for different distances. It was therefore proposed that the admissible forces in gaps up to 500 mm should be *400 N* and in gaps larger than 500 mm *1400 N.*

Those values should be regarded as maximum values and are only allowed for a maximum period of 0.5 seconds (experience gained in the early Swedish tests on pigs). After this time no static force larger than *150 N* is allowed and this static force must come down to near *0 N* after 4.5 seconds (both the old Swedish test on pigs and sensoric tests at BIA establish this level). Remaining forces of not more than *25 N* will not be considered.

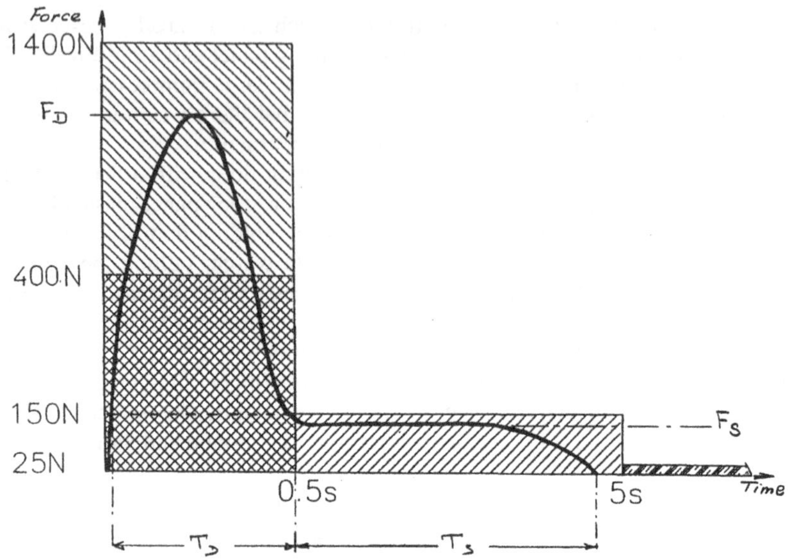

Diagram showing operating force versus time

Fd : maximum dynamic force
Fs : maximum static force
Td : maximum time for dynamic force application
Ts : maximum time for static force application

Fig 3 Diagram showing maximum crushing forces versus time

5 Variable radii of crushing edges.

5.1 Sensoric tests

Questions were raised whether different radii of the crushing edges would give different admissible (bearable) forces.

It was therefore decided that further sensoric tests should be performed with variable radius on crushing edges. Those tests were carried out on a test group of 42 persons, 39 males and 3 females, in May 1995.

The door used during the tests was a sliding door equipped with closing edges of different radii, see figure 4.

The tests were performed so that the fingers of the test persons were crushed between the rounded closing steel edges and the crushing force was slowly increased until the test persons found the pain not acceptable and stopped the test.

The maximum static crushing force was recorded for four different radii of the closing edges, 1, 3, 6 and 10 mm.

After each crushing test the test persons became more sensitive which affected the next test. In order to avoid this, different parts of the fingers were crushed. This method affected the result too much because of the fact that the different parts of the fingers did not have as much muscles as other parts. It was therefore decided that the same parts of the fingers should be crushed but half of the test group started with the largest radius and the other half with the smallest.

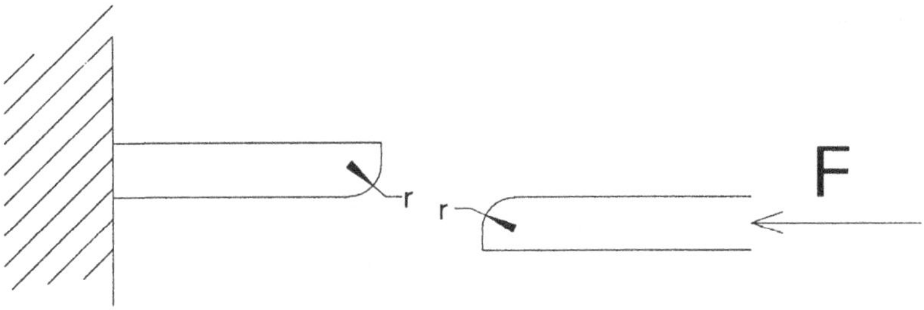

Fig 4 Closing edges of different radii

5.2 Result

As can be seen in the diagram no significant differences in force or pain as experienced by the test group, for the four used radii, has been recorded. The span between test persons is much larger than the variation between different radii.

The result is shown in the following diagram where the median level and the 25th and 75th percentile levels are indicated:

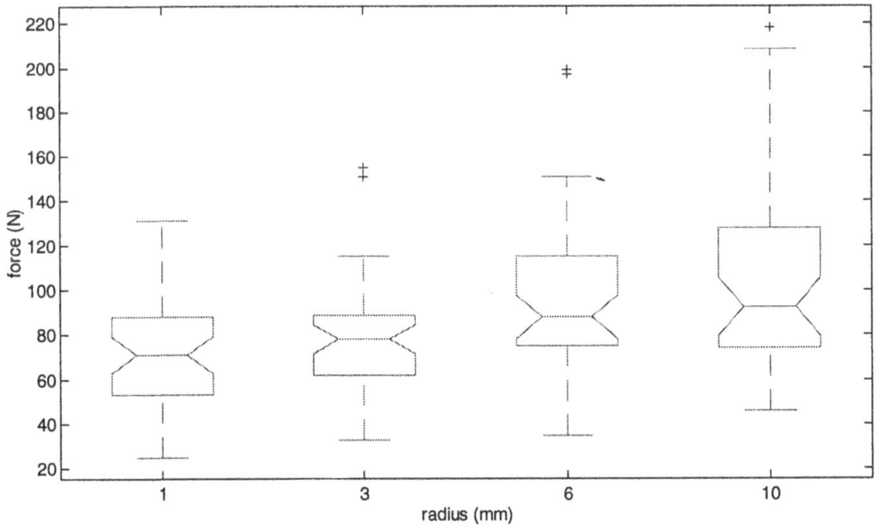

Fig 5 Diagram showing tolerable forces for different radii of crushing edges

6 Summary

As can be concluded from this discussion, the method of measuring the critical forces, determines to a very large extent the admissible values for impacting or crushing the human body.

It is therefore extremely important that the measurement method is compatible with the dynamic behaviour of the critical parts of the human body.

Radii between 1 and 10 mm do not give significantly differences in admissible forces.

7 References

1. European Standard EN 349: Safety of machinery - Minimum distances to avoid crushing of parts of the human body, CEN
2. Ehrning, U., B. Romanus and K. Ringsberg: Investigation concerning a proposal for rules for the prevention of crushing injuries at power-operated gates. ILAB Report I and II (in Swedish), Gothenburg 1978
3. Andersson, L.: Limitation of forces. SP Project report, Borås 1992
4. Mewes, D.: Limitation of forces. BIA Report No. 9300029, St. Augustin, 1992

21 STRUCTURAL DYNAMICS OF STEP LADDERS: FAILURE OF DEFECTIVE WOOD STEP

M.A. EISENBERG
Department of Aerospace Engineering, University of Florida, Gainesville, Florida, USA

Abstract
Dynamic loads on step ladders are the result of interaction between climber and ladder. The complex dynamical behavior of the climber must be experimentally determined. An analysis of the dynamical response is presented to allow the extrapolation of experimental results to ladders of grossly different structural dynamical parameters from those on which the data are obtained. It is demonstrated that the climber characteristics dominate the dynamical response. Reliable quantitative measures of data applicability are obtained, not withstanding significant uncertainties regarding the details of the mechanics of motion. The results of the dynamical analysis are combined with published experimental observations of dynamical loads to predict the range of anticipated loads to the bottom step of a wood step ladder which was the subject of a products liability suit. The particular step was known to be severely cross-grained near the center of the span. Described herein are the results of an analysis of the effect of the material defect on the strength of the step, a stress analysis of the rod-reinforced step, and an analysis of the post-fracture deformation of the structure.
Keywords: Cross-grain, defect, dynamic load, failure, ladder, products liability, structural dynamics, wood.

1 Introduction

The development of dynamic loads on step ladders is the result of interaction between the climber and ladder. Characterization of the structural dynamical behavior of the ladder is relatively straight forward, but the predictability of the properties of the climber is subject to much greater uncertainty. Experimental data on the manner in which ladder users actually climb ladders and quantitative measurements of the resulting dynamical

Structural Failure, Product Liability and Technical Insurance, Proceedings of Fifth International Conference. Edited by H.P. Rossmanith. Published in 1996 by Taylor & Francis, 2 Park Square, Milton Park, Abingdon, Oxon, OX14 4RN. ISBN 0 419 22080 1

loads are a scarce commodity. One of the few studies reported is that of Fox [1], who reports results of approximately 400 experiments in which climbers ascended and descended an instrumented six-foot aluminum step ladder. Among the results reported were the maximum vertical forces transmitted to the feet of the rails to which the steps are affixed.

Not surprisingly, maximum dynamical forces were measured during descent as the climbers transferred their weight from the second to first step. Loading the bottom step increases the proportion of the static load carried by the front rails. Loading the bottom step minimizes any contribution of rail flexibility to the shock absorption capability of the structure. The flexibility of the bottom step is the primary indicator of the contribution of the ladder to the dynamic loads created by the man-structure interaction. Since commercially marketed step ladders typically have flared side rails and since all steps typically have the same cross-sectional geometries, the bottom step functions as a beam with the longest unsupported span, thereby rendering it the most vulnerable to structural failure. Note that the lower ladder steps receive the most wear and tear since they are always loaded even during partial ascents. It should also be noted that the psychology of climbing plays a role. One's confidence increases as one nears the perceived safety of terra firma. Thus, the climber typically strides more confidently and forcefully as the ground is approached. All these factors contribute to the preferential failure of the lower steps of ladders.

The specific motivation for the study reported herein was an attempt to assess the causative factors relating to an accident involving the failure of a twelve-foot wooden step ladder in which the bottom step was unfortunately severely cross-grained at or near the center of the span. Given the fact that the only extensive data set for dynamical loads was gathered for an aluminum ladder of half the height of the subject wooden ladder, how could one apply these data to the inherently softer ladder of much larger step span? Essentially, the problem boils down to one of scaling -- to estimation of biomechanical parameters of dubious accuracy -- to the application of fundamental principles of structural dynamics -- and to the assessment of the sensitivity of the dynamical response to these poorly understood parameters.

2 Problem statement

As indicated above, the research reported herein was motivated by engineering analyses related to an accident involving use of a step ladder. Since the specifics of the data relevant to the particular case are of little general interest, we shall concentrate on issues and procedures that are broadly applicable.

2.1 Accident scenario

A climber was using a 12-foot Type I (industrial grade - 250 pound capacity) wood step ladder to perform routine repairs. The ladder was unexceptional but for a severe cross-grain defect near the center of the bottom step. The climber allegedly descended the ladder in a normal manner, and, in the course of transferring his weight from the second step to the bottom step, fell as a result of sudden structural failure of the step. In the course of attempting to arrest his fall to the ground he suffered serious injuries, the details of which are irrelevant to this study. Subsequent to the accident, the bottom step was found to be fractured along the grain at the defect and the reinforcing rod intact but severely deformed. There was ancillary damage to supporting braces and neighboring structure.

2.2 Structural configuration

The typical step ladder consists of a set of relatively massive front rails pin-connected via a top cap to a set of more flexible rear rails. The front rails, in turn, support the steps, which, to a first approximation, may be idealized as simply supported beams. The front rails behave like beams pinned at the point of contact with the floor and supported by an elastic restraint or spring at the top.

Ladders manufactured for the U. S. market are almost exclusively built in accord with standards promulgated by the American National Standards Institute (ANSI). They specify that the front rail slope be a minimum of 3.5" per foot measured along the rail. The minimum has become the norm, so most step ladders are sloped at an angle, $\theta = \cos^{-1}(3.5/12)$.

The bottom step of the subject ladder is an essentially rectangular slab of wood fitted into shallow grooves in the side rails and restrained against fore-aft motion. Below the step is a steel rod that serves two functions. First, it serves as a tensile member, inhibiting the inappropriate spreading of the rails and the potential of the step to fall out of the shallow grooves in which it is set. Second, the rod provides reinforcement of the step via truss blocks that are slipped between the rod and the lower face of the step. There are two truss blocks placed approximately at the one-third span marks. The step is further reinforced by diagonal braces riveted to the under side of the step and the insides of the left and right rails.

2.3 Elements of the failure analysis

The subject ladder was manifestly defective relative to generally accepted standards for ladder construction. The alleged usage was manifestly appropriate to the purposes for which the ladder was marketed. The fundamental questions to be resolved were: Was the defect the proximate cause of the accident? Was the post-accident condition of the ladder consistent with the alleged circumstances of usage? To answer these questions one needs to assess the quantitative influence of the defect on the strength of the step, the effect of step strength on the load carrying capacity of the structure, and the nature of the anticipated dynamical loads to be supported by the step under the testified usage. These issues are addressed in the following sections.

3 Effect of defect on the wood step

Since the tensile and bending strengths of wood are significantly higher along the grain than transverse to the grain, wood beams and steps of wood ladders in particular are nominally designed to be manufactured with the grain running horizontally from left to right. Given the natural variability in wood the ideal is not achievable and there have ben standards established restricting the permissible slope of grain [2]. The permitted slope of grain is 1-in-12 for this class of ladders.

Visual inspection of the subject step revealed a severe cross grain defect (1-in-2.27) near the center of the bottom step. Bodig and Jayne [3] report the influence of slope of grain on bending strength for slopes ranging from 1-in-6 to 1-in-20, which is taken as the reference state. Stalnaker and Harris [4] report data on strength perpendicular to the grain. The effect of cross grain on strength is summarized in Figure 1. The grain slope permitted by ANSI [2] results in a step with approximately 70% of the ideal strength. The step with the observed defect has a strength approximately 18% of the ideal. Thus, the step strength is significantly affected by the defect.

4 Structural analysis of the rod-reinforced step

The step configuration is described in Section 2.2. It is reasonable to model the step as a beam simply supported at its ends, but subject to additional intermediate supports. The reinforcing rod provides vertical point supports at approximately the one-third span locations. The angle braces provide point supports approximately three to four inches inboard of the ends. The climber's foot may reasonably be approximated as a point load. Since there are six support loads the beam is statically indeterminate.

An upper bound on the influence of the diagonal braces may reasonably be approximated. Assume the braces to be infinitely stiff and infinitely strong. Let the free ends of the front rails be modeled as beams cantilevered from the bottom step. Alternatively, consider the rails to be beams pinned at the second and first steps and overhanging below the first step. The second assumption results in a slightly softer support. By this process, the diagonal braces may be approximated as spring supports whose stiffnesses can be reasonably estimated.

The reinforcing rod contribution to step stiffness is problematic. Its effectiveness depends upon the pretension applied to the rods during manufacture and to the precise placement of the truss blocks that are held in place only by friction. The rods are typically made of low carbon steel with high ductility and low hardening characteristics. Under

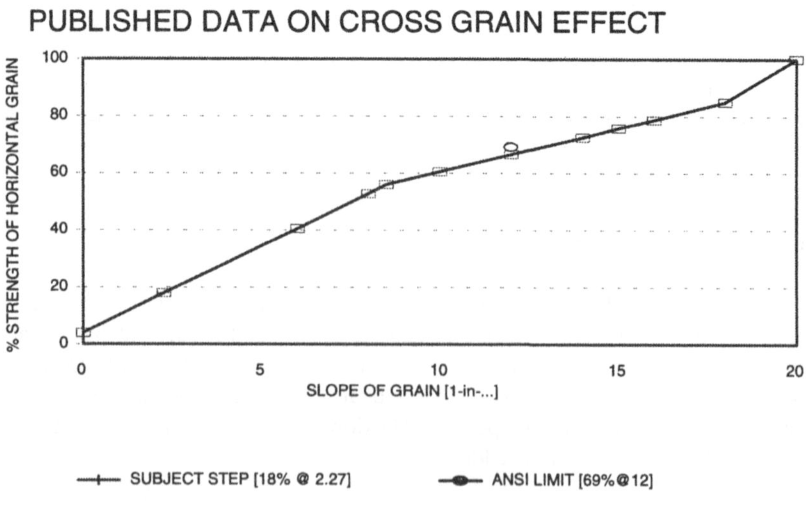

Fig. 1. Effect of cross-grain on strength of wood step.

normal usage, the rods are readily loaded into the plastic range, thereby rendering them unreliable as reinforcement under continued small deformation conditions. They do play a major role in the event of fracture of the wood step. Under these conditions, the rod support is a nonlinear function of deformation. With increasing rod angle, the load carrying capacity increases dramatically.

Figure 2 shows the load-deflection characteristics of the bottom step analyzed in accord with the above assumptions. Concentrate first on the lowermost curve. It represents the load-deflection characteristics of the reinforcing rod alone. The wood step is assumed to be present only to serve as a spacer to maintain the distance between the left and right rails to which the rod is attached. Careful examination of the figure reveals a cusp just below the 100 pound level. Below this cusp the rod is assumed to behave elastically. It's load-deflection behavior is dominated by the linear increase in stress associated with rod strain. Above this point the rod is assumed to be perfectly plastic and the hardening is that associated with nonlinear large deformation geometric effects.

The line with the lower horizontal transition represents the behavior of the defective wood step in conjunction with the reinforcing rod. At somewhat greater than 200 pounds the step begins to fracture. As the crack propagates rapidly upward along the sloped grain from the lower surface of the beam toward the top surface the effective beam stiffness rapidly goes to zero as the load is transferred from the wood step to the steel reinforcing rod. The sudden extension under constant load is consistent with the experience to which the climber testified and to which he reacted with resulting injury.

The upper line indicates the theoretical load carrying response of a non-defective wood step. All calculations presume that the rod is plastically deforming but not broken, that its connections with the rails remain intact, and that the rails also remain intact. Thus, these represent upper bound estimates of the load carrying capacities of the defective and non-defective structures. Note, however, that central loading of the step is assumed. The actual foot placement is unknown. To the extent that the climber chooses to place his feet toward the rails, the load carrying capacities will be larger. Alteration of other modeling hypotheses will also result in minor quantitative changes with no probative consequence

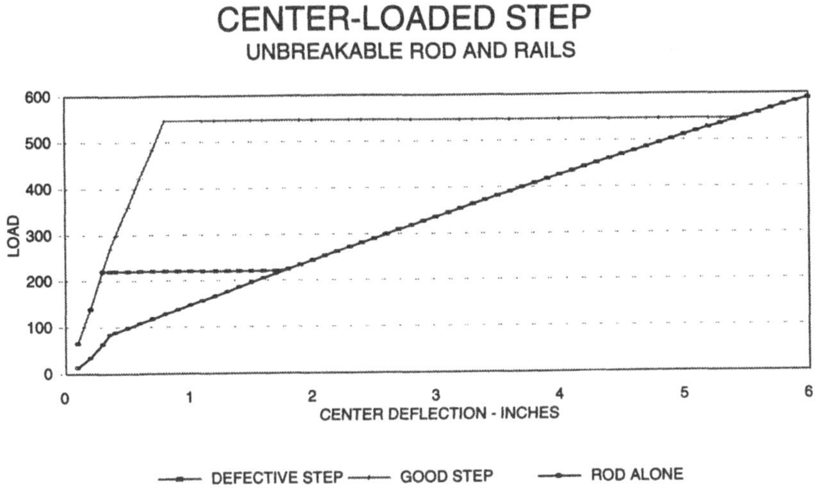

Fig. 2. Load-deflection characteristics of rod-reinforced step.

regarding the predicted outcomes. It will be shown that inherent variability in dynamic response renders further refinement in structural analysis unnecessary.

5 Dynamic load assessment

It remains now to determine the nature of the dynamic loads applied by the climber. In [5] Eisenberg demonstrates in detail the validity of the assumption that dynamical load characteristics of a ladder subjected to a climber descending to the first step may adequately be characterized by the properties of the step alone. In the following sections, the dynamical loads are demonstrated to be consistent with the loads necessary to cause failure of the defective ladder but not to cause failure of a ladder manufactured to specification.

5.1 Dynamical response model for man-ladder interaction

The essence of the dynamical response may be characterized by a simply supported beam of length, L, centrally loaded by a point mass, m, falling from a height, h. The mass is not in free fall, but subject to a time varying upward restraining force, $F(t)$. Here, L represents the effective span of the bottom ladder step, m, the mass of the climber, k and k_C, the effective stiffnesses of the ladder and climber's musculoskeletal suspension system, respectively, and $F(t)$, the restraining force represented by the bearing of the other foot on the second ladder step plus any vertical support provided by the hands which are grasping the side rails. This model is too crude and the data are too difficult to estimate to permit confident prediction of the dynamical forces. But, we are not attempting to make a prediction of the outcome. We have in our possession experimental data that fully characterize the response for one set of parameters (even though we are not sure of their actual values). We are attempting to extrapolate the measured response to significantly different conditions.

Let δ and δ_C represent the respective deflections of the ladder and musculoskeletal springs. The net work done by the mean restraining force, $\bar{F}$, is then equal to the change in potential energy of the system,

$$-\bar{F}(h + \delta + \delta_C) = \frac{1}{2}k\delta^2 + \frac{1}{2}k_C\delta_C^2 - W(h + \delta + \delta_C) \tag{1}$$

where $W = mg$. Since the springs are arranged in series, $k\delta = k_C\delta_C$, so

$$\frac{1}{2}k\delta^2(1 + k/k_C) - (W - \bar{F})\,[\delta(1 + k/k_C) + h] = 0 \tag{2}$$

Let $\delta^* = W^*/k$, where $W^* = W - \bar{F}$. Then

$$\frac{\delta}{\delta^*} \equiv K = 1 + \sqrt{1 + \frac{2h}{\delta^*(1 + k/k_C)}} \tag{3}$$

or

$$(K - 1)^2 - 1 = 2h/[\delta^*(1 + k/k_C)] \tag{4}$$

But $K = \delta/\delta^* = (\delta/\delta_{ST})(\delta_{ST}/\delta^*) = D\beta$, where $\beta \equiv W/W^*$, δ_{ST} is the static deflection of the step, and D is the dynamic load factor, δ/δ_{ST}. Let Eq 4 be applied to two different ladders: one wooden, and the other made of aluminum. Then,

$$\frac{(\beta D_W - 1)^2 - 1}{(\beta D_A - 1)^2 - 1} = \frac{k_W}{k_A}\frac{(1 + k_A/k_C)}{(1 + k_W/k_C)} \tag{5}$$

where the subscripts, A and W, refer to the aluminum and wood ladders, respectively. Since the restraining force, $F(t)$, varies from W to zero as the climber's weight is transferred from the second to the first step, to a first approximation the force may be assumed to be linearly varying with the distance traversed. Thus, $\bar{F} = W/2$ and $\beta = 2$.

5.2 Dynamical loads on structures of grossly disparate stiffnesses

The experimental data on dynamic load factors were gathered for an aluminum ladder. We require an estimate of the dynamic load factors for a wood ladder. The following data are reasonable approximations to the relative pertinent structural parameters: E_A / E_W = 5.88, $L_A / L_W = 0.5$, I_A / I_W = 0.431. Since stiffnesses are proportional to $48EI/L^3$, $k_A / k_W = 20.28$. To assess the stiffness of the climbers' musculoskeletal systems appears to be an order of magnitude more difficult. Surely the stiffness is strongly dependent on the voluntary state of muscular tension and we have no reliable data on such. But is it truly that difficult? Imagine a 200 lb man hopping barely into the air (or off a very short curb). By how much will he permit his knees to flex in order to absorb the anticipated shock? Two inches? Three? Four? This suggests an effective musculoskeletal stiffness of 50 (for 4" knee flex) to 100 lb/in (for 2" knee flex) as a reasonable estimate for k_C. At 4" knee flex, $k_A / k_C = 117$ and $k_W / k_C = 5.78$. With this information, Eq 5 can be used

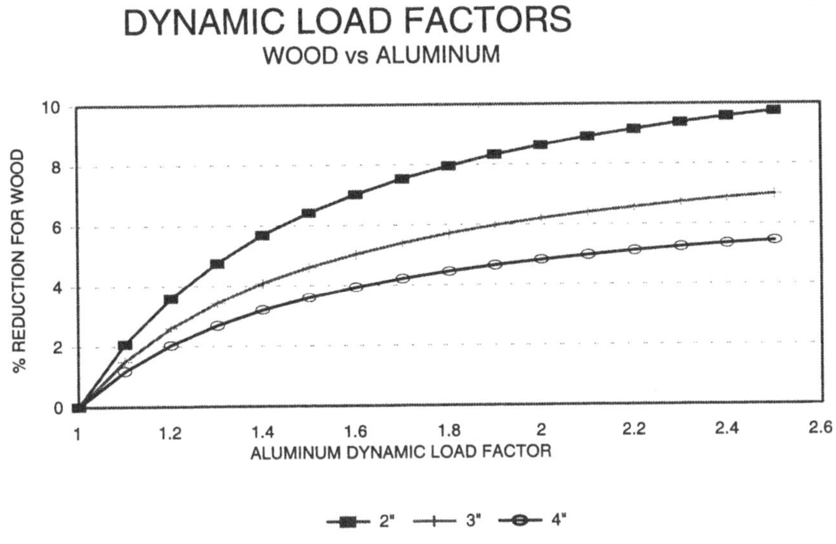

Fig. 3. Dynamic Load Factor Reduction for Softer Structure

to predict the effect of substitution of the 20-fold softer wood ladder for the aluminum ladder.

Figure 3 shows the per cent reduction in dynamic load factors for the 12-foot wood ladder relative to that of the 6-foot aluminum ladder. Aluminum load factors are plotted for a range of values typical of those determined experimentally. The three curves correspond to the degree of knee flex as a gross measure of musculoskeletal spring stiffness.

Our knowledge of k_C is surely speculative, but Figure 3 shows that the reduction in dynamic load factor is insensitive to the actual value of k_C. It is sufficient to recognize that the differences in stiffnesses of the wood and aluminum ladders, however large they may be, have a second order effect on the dynamical response of the man-ladder system. The cushioning effect of the musculoskeletal structure (however imprecise may our knowledge of it be) dominates the dynamical response. We may confidently apply the results of the aluminum ladder tests to the wood ladders and assert that the measured dynamic load factors are applicable to within 10% error.

5.3 Experimental data on dynamical loads

Figure 4 shows the results of the previously cited tests by Fox on behalf of the Consumer Products Safety Commission. Each of the stacked bar components represents the results for a single subject of the weight indicated by the legend below.

The "Dynamic Load Factor" represents the ratio of the maximum load sensed under the front ladder feet to the load that would have been predicted under static loading of the first step. Note the great subject-to-subject variability, as well as the lack of consistency with which each individual subject descended in the course of their 100 attempts. Yet, for all the variability, a clear picture emerges of the magnitude of the expected dynamical loads. Figure 3 demonstrated that even allowing for the great uncertainties of our comparative dynamical analysis, the CPSC data derived from an aluminum step ladder can be applied to a wood step ladder of grossly different stiffness with errors of less than 10%.

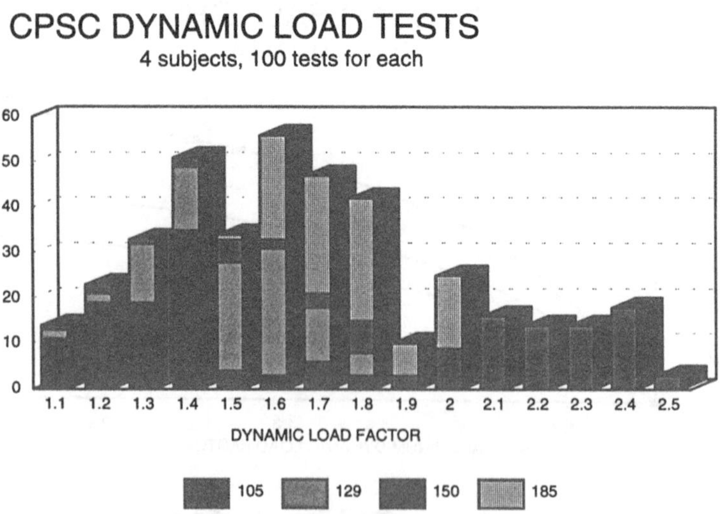

Fig. 4. Measured loads relative to static loads on descending to bottom step of a six-foot aluminum stepladder.

6 Conclusion

Unstated above was the weight of the climber in the subject products liability action. The climber actually weighed approximately 165 pounds. Given the range of observed dynamic load factors and their demonstrated applicability to the subject wood ladder, it is apparent that the observed failure is consistent with the testified usage, and that but for the observed defect the accident would not have occurred.

The primary thrust of mechanics research is the application of highly sophisticated mathematical methods to highly idealized "clean problems" with well-determined parameters and amenable to the development of well-posed initial boundary value problems. The foregoing analysis illustrates the continuing value of simple but fundamental methods of mechanics to "dirty problems" with fuzzily understood system parameters, initial and boundary values. It illustrates how creative application of mechanical principles can allow one to "look under the lamp post" by redirecting its light to dark corners in which answers are sorely needed. In the particular problem studied an existing body of experimental work was given new life by demonstrating its applicability to a greatly expanded set of applications. Mechanicians, if they are to be successful in applying their discipline to solving "real world" problems, must cultivate a broad spectrum of analytical, numerical, and experimental tools.

7 References

1. Fox, R.L. (1977) *A Report on the Dynamic Testing of Ladders*, Report CPSC-C-76-0052, Consumer Product Safety Commission, Washington DC.
2. *American National Standard A14.1-1982 for Ladders - Portable Wood - Safety Requirements*, American National Standards Institute, Inc., New York, NY.
3. Bodig, J. and Jayne, B.A. (1982) *Mechanics of Wood and Wood Composites*, Van Nostrand Reinhold.
4. Stalnaker, J.J. and Harris, E.C. (1989) *Structural Design in Wood*, Van Nostrand Reinhold.
5. Eisenberg, M.A. (1995) Structural Dynamical Modelling of Step Ladders, Invited contribution to *Applied Mechanics Reviews*, Special Issue - November 1995.

22 CASE STUDIES IN LIFTING EQUIPMENT

C.E. NICHOLSON and P.F. HEYES
Health and Safety Laboratory, Sheffield, UK

Abstract
In the UK the Health and Safety Executive is committed to carrying out research to improve safety in the workplace. The Health and Safety Laboratory (an Agency of HSE) investigates industrial accidents and carries out research on behalf of HSE. This paper presents a review of recent incidents involving lifting equipment and describes some of the research which has been required subsequently. The subjects covered are dynamic loading of wire ropes, non-metallic slings and self-locking lifting hooks.
Keywords: Dynamic loading, lifting, non-metallics, self-locking hooks, slings, wire ropes.

1 Introduction

Lifting equipment is used daily in all industries and in most places of work. Devices range from large cranes, designed to lift thousands of tonnes, to lightweight slings which support only a few kilogrammes. Although many items of lifting equipment are of traditional design (eg chains and wire ropes), there are many modern applications where, for a variety of reasons, new designs and new materials are desirable.

In the UK, the Health and Safety Executive is committed to carrying out research to improve safety in the workplace and, as would be expected, this research effort is linked to the pattern of industrial accidents. The Health and Safety Laboratory (HSL), now an agency of the Health & Safety Executive, investigates many industrial accidents each year and approximately 25% of these involve some form of lifting equipment. Consequently a significant proportion of HSL's research effort is devoted to projects which aim to improve the safe use of lifting equipment and thereby reduce the number of workplace injuries .

Structural Failure, Product Liability and Technical Insurance, Proceedings of Fifth International Conference. Edited by H.P. Rossmanith. Published in 1996 by Taylor & Francis, 2 Park Square, Milton Park, Abingdon, Oxon, OX14 4RN. ISBN 0 419 22080 1

In this paper the authors review relevant incidents, investigated over the last three years, and proceed to discuss research work arising from some of these investigations.

2 Review of incidents

Incident investigations, carried out by engineers and scientists at HSL, are requested by HSE Inspectors working in many different industries (for example, construction, mining, offshore, etc). The following table gives an overview of lifting equipment incidents investigated during the period 1992 - 1995:

	Number of investigations	Percentage
Cranes and hoists	15	22
Lifts (passenger & goods)	6	8
Slings	5	7
Fork lift trucks	6	8
Hooks	2	3
Chains and chain blocks	6	8
Wire ropes	10	14
Shackles and eyebolts	2	3
Access & work platforms	7	10
Others	11	17
TOTAL	70	100

The majority of crane failures are due to lack of a safe system of work and/or a failure to maintain and, although this forms the largest category, it is not dealt with in this paper as avoidance of such failures is a matter of education and training rather than research. The reader is referred to HSE Publication "Management's responsibilities in the safe operation of mobile cranes", HMSO, Third Impression, 1989.

3 Case studies

3.1 Dynamic loading of wire ropes

3.1.1 Incidents

HSL investigates about ten serious wire rope failures each year, most of which have resulted in loss of life or extensive damage to property. A significant proportion of failures occur under circumstances where dynamic, or impulsive loading, is a factor.

For example, two men were killed during construction of an oil rig when a lifting operation went wrong. A tubular brace, weighing approximately 80 tonnes, was being manoeuvred into position using a system of five wire rope slings suspended from a crane jib when several slings failed causing the brace to fall onto the men working below. A similar lift is shown in Figure 1.

All the slings had been made from 52mm diameter steel wire rope and the breaking load of each sling was approximately 165 tonnes, hence any one sling should have been capable of supporting twice the weight of the tubular brace!

Figure 1 Tubular brace lift

Why then did a catastrophic failure of the lifting equipment occur at such a relatively low load ? The explanation was provided by a detailed examination of the failed items and a careful analysis of the lifting operation. During the initial part of the operation, the tube was being lifted at an angle of 10^0 to the horizontal when the upper end caught on the underside of some scaffolding. As the lift continued the tube pivoted about the trapped end causing one leg of the sling to become slack. Eventually the trapped end of the tube slid free resulting an impulsive load being applied to the assembly. This sequence of events was demonstrated in the laboratory using a scale model and calculations showed that the sling system had been called upon to absorb the rapid application of 1.9MJ of energy. As this was well in excess of the energy absorption capacity of the sling arrangement, the sudden failure was explained.

3.1.2 Offshore loading trials

Although not directly connected with this incident, the Offshore Safety Division of HSE expressed concern about the effects of dynamic loading on the performance of wire rope slings used offshore in routine lifting operations. A team of HSL engineers spent several days on an oil rig monitoring the forces experienced by typical sling sets using a specially developed radio transmitting load cell incorporated in the load line. Figure 2 shows a typical lift where an offshore container is being uplifted from a supply vessel. The load cell can be seen just to the right of the vessel's funnel.

The loads experienced by the fifth leg of a wire rope sling set, when lifting payloads up to 5500 kg, were recorded for two types of lift, (a) various payloads lifted from supply ships and (b) loads lifted on the deck of a rig. The maximum force measured in any lift was 129.90 kN and this lasted for approximately 100 milliseconds. This force is in excess of the rating of the wire rope sling set used (78.5 kN). On only four occasions did a maximum shock force exceed double the container maximum gross weight, twice during payload lifting and twice during deck lifting, however, the proof load of the fifth leg of the sling set was never exceeded.

The average factor by which loads lifted from supply ship decks increased, by virtue of ship movement and crane snatch, was approximately 1.7 (that is, the force was 1.7 times greater than the container weight). This factor increased to approximately 3.5 when lighter loads were lifted, although more data are required to establish a firm relationship between container weight and 'shock load factor'. Lifts on the deck of the rig showed the same relationship as supply ship lifts, light containers having higher shock load factors. The average value of the dynamic factor for rig lifts was 1.5, that is slightly less than for supply ship lifts.

The performance of chain and wire rope slings was compared for this application and there were no significant differences when lifting from the rig deck. Wire rope slings attenuated the forces more and hence showed slightly lower maximum forces and shock load factors than chain slings. Rapid lifting of the containers increased the maximum forces by a factor of 1.5 when compared with steady lifting for both evenly and unevenly loaded containers. An uneven load distribution had no significant effect on the maximum force seen by the fifth leg of the sling set, although the cycling period of the load as it oscillated may vary from that observed for an even load distribution. Empty containers handled during rig deck lifts appeared not to have a high shock load factor unless they were hoisted at a rapid winding rate when on occasions the shock load factor increased by a factor of 1.8. Stiffer lifting systems (eg use of the main block or a stiffer crane) increased the maximum force observed, although this was offset by the reduced lifting speed. Analysis techniques are being developed to quantify the effects of high loads during a lifting operation, with emphasis being placed on the number of load peaks of a certain magnitude. This may have additional relevance when considering fatigue or incremental damage.

Figure 2 Lifting an offshore container

3.1.3 Large scale dynamic tests

HSL has a large scale purpose-built impact test track on its Buxton site, shown in Figure 3. This test facility is comprised of 265m of four-rail track laid in such a way that a specimen truck (on the inner rails) and a hammer truck (on the outer rails) can run freely down the opposite slopes of a natural valley. The trucks are released automatically at selected positions so that they collide on the lowest section of the track, thus subjecting the specimen to a large impact force. Scientific equipment to measure the physical effects of the impact is contained in a control building in the valley. The truck masses and speeds are such that samples can be subjected to impact forces in excess of 300MN at speeds upto $27ms^{-1}$. The maximum usable impact energy is 1.8MJ.

Figure 3 Impact track at the Buxton site

In recent years this facility has been used to carry out large scale impact tests on various items of lifting equipment, examples include chains and wire ropes. Currently locked-coil wire ropes used in the mining industry for cage suspension purposes are being tested and the rig has been modified recently to allow tests on pre-tensioned ropes. Preliminary results have identified a considerable reduction in the energy absorption capacity of these ropes compared with quasi-static tests.

3.2 Non-metallic slings

3.2.1 Incidents

Non-metallic slings are finding increasing use in industry as they are light, easy to store and relatively inexpensive and, as a result, they have replaced traditional steel chain and wire rope slings in a number of applications. However they are vulnerable to heat, mechanical damage and, occasionally, chemical attack. There have been a number of fatal accidents in recent years arising from misuse and/or the presence of pre-existing damage in non-metallic slings. In the example shown in Figure 4, a 12mm three strand polypropylene rope sling was being used to install a heavy folding door in a warehouse. The door had been positioned horizontally on the floor, the rope had been wrapped several times around it, looped through the door handle and then tied onto a lifting hook. However, as the door was lifted the sling slid along the edge of the door producing a cutting action under load which severed the rope causing the door to fall, fatally injuring one of the workmen. Examination of the failed rope ends in the laboratory, one of which is shown in Figure 5, revealed that all three strands had broken at the same position where they had been severed by a relatively sharp cutting action. There was no evidence of pre-existing damage at the failure location.

Examination of the individual filament ends revealed straight cut edges with no indication of any tensile failure, suggesting that failure was due to dragging along a sharp edge rather than stretching across the edge. Although the rope had become worn during previous use, causing its tensile strength to be reduced somewhat, failure had occurred at well below the safe working load due to a sliding and cutting action whilst the rope was under tensile loading.

3.2.2 Research

As we have seen, the effects of load sharpness on the performance of non-metallic slings can be very significant. Comparative factors of safety for these slings have not been established experimentally for anything other than axial tensile loading conditions. In particular the effects of loading in choke hitch configuration are not well known.

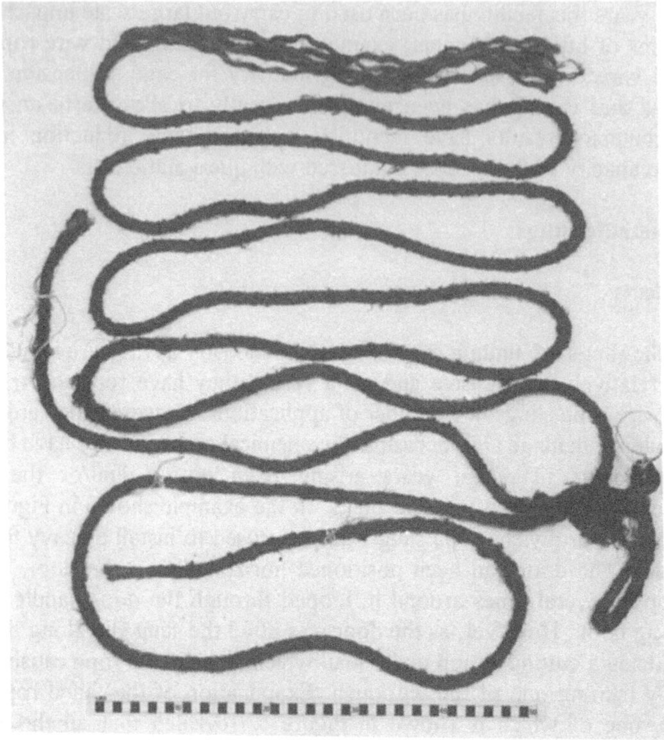

Figure 4 Failed rope sling

Figure 5 Close-up of cut end

For this reason the Health and Safety Executive initiated a research project at HSL with the aim of measuring the breaking load and energy absorption characteristics of three types of non-metallic sling (polypropylene fibre rope; polyester webbing; polyester continuous filament). The performance of these slings was to be compared with that of traditional chain and wire rope sling in three test configurations: axial tension; choke hitch around a sharp load; choke hitch around a smooth load. Slings approximately 2m in length, with a lifting capacity of around 2 tonnes were tested. Tests were carried out on a purpose built long-stroke test frame mounted on the universal test bed at HSL's Sheffield laboratory. Slings were tested horizontally, the load was applied using an hydraulic actuator with a maximum force capability of 400kN and a maximum displacement of 2m. The latter was needed to accommodate the large extensions of the non-metallic slings.

The results showed that in axial tension the actual breaking loads of slings, both metallic and non-metallic, are generally in excess of the minimum values quoted by the manufacturers. When loaded in choke hitch configuration around a smooth object, the breaking load of non-metallic slings could be less than expected from the relevant British Standard (ie the derating factor is sometimes optimistic), however the factor of safety was, in all cases, acceptable to the Standard, typically 5:1. In choke hitch configuration around an I-beam without edge protection the non-metallic slings failed prematurely due to the damage created at the sharp edges. In general, packing in accordance with manufacturers' recommendations prevented this, although in one or two cases the packing chosen was not adequate. It was shown, for example, that soft wood was unsuitable for stranded fibre ropes because gouging and splitting could occur thus exposing the sling material to the sharp edge. With reference to the above incident, the use of a suitable packing material may well have prevented the failure.

The metallic slings all broke at the choke position irrespective of the shape of the load used. However these slings were also damaged by the sharp edges of the I-beam and hence their performance on subsequent lifts could be affected adversely if packing is not used in practice.

Tests on non-metallic slings also yielded useful information on the fracture appearance resulting from different loading configurations. This will assist HSE Inspectors is diagnosing the cause of failures in service.

Perhaps the most striking result was the reduction in the energy absorption capacity of all sling types when tested in choke hitch configuration irrespective of the geometry of the load. This was a consequence of the large reductions in extension which occurred during choke tests. In the worst case, that of polypropylene rope slings, the energy absorbed was only 20% of that absorbed during the axial tensile tests. The dynamic performance of these slings is to be investigated in a future project using the impact test track facility described previously in section 3.1.3.

3.3 Self-locking lifting hooks

3.3.1 Incidents

HSL has recently investigated two incidents in which self-locking lifting hooks opened inadvertently during load transfers between oil platforms and supply vessels. In both cases the hooks opened as the load was put down, possibly as a result of the impact of another part of the lifting assembly on the hook. Self-locking hooks are designed to close automatically when the hook is loaded and have a locking mechanism which is activated when the hook is closed. The user must then unlock the hook manually before it can be opened. Both failed hooks were of the eye type with a pivoted lock, as shown typically in Figure 6, and both had been made allegedly to the requirements of a draft CEN standard.

Detailed examination in the laboratory, mechanical testing and study of high-speed video of the hook mechanism revealed that these hooks could open on impact and the risk of opening increased dramatically when the bearing surface of the lock is worn. A static analysis of the forces involved indicated that an opening rotational moment was transferred to the lock through its bearing surface onto the lug of the latch or eye, hence wear of the bearing surface increases the likelihood of unintended opening. Thus it was established that hooks made to the draft CEN standard could fail by this mechanism and, furthermore, as the standard does not include tests on worn hooks, this problem is unlikely to be detected.

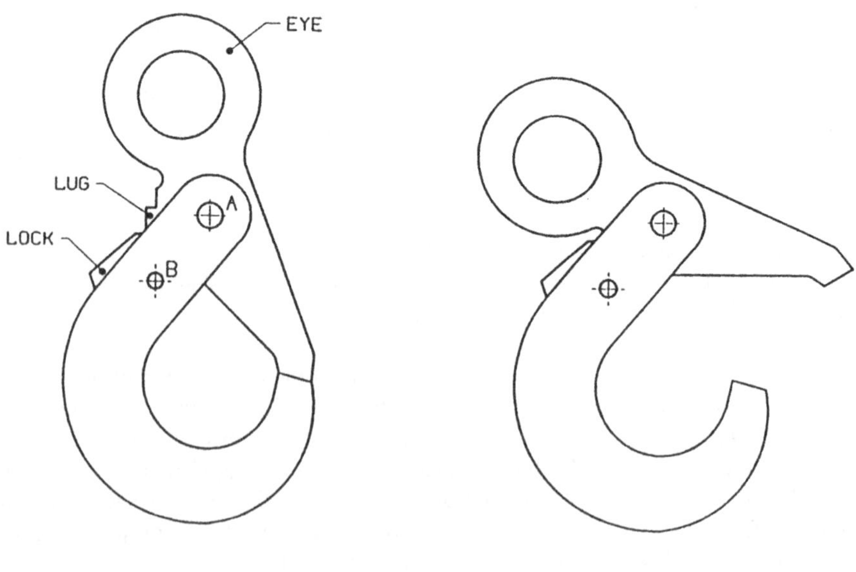

Figure 6 Self-locking lifting hook

3.3.2 Research

In order remedy these deficiencies, HSL undertook a research programme with the aims of examining the design of hooks of this type and assessing appropriate test methods. Four hook manufacturers co-operated in this study, allowing their products to be analysed and tested. A number of possible design improvements were recommended by HSL:

1. Move the position of the lock pivot so that a closing moment instead of an opening moment is produced when the eye of the hook is struck
2. Reduce the size of the lug on the eye to move the point of action towards or perhaps beyond the lock pivot
3. Reduce wear by increasing the hardness of the lock material
4. Change the angle of bearing surfaces of the lock and eye

Figure 7 Pendulum impact rig

The present CEN standard includes a quasi-static tensile test with the load being applied to the hook transversely across its widest part. The HSL tests clearly demonstrated that this test is not capable of detecting those hooks which, because of wear in the mechanism or poor design, might be prone to opening when subjected to an impact load. Consequently, a pendulum impact test, based on the Charpy impact test for metals, was proposed by HSL as an alternative. The equipment, designed and manufactured at HSL, was required to be sufficiently versatile to test a range of hook sizes (ie hooks used with chains made from bar in the diameter range 5mm to 28mm) and a range of designs. Variable impact energy and a flexible hook gripping arrangement was therefore needed.

Following design and manufacture of a suitable test device, a total of 560 tests were carried out on twenty eight hooks. It was concluded that the arrangement shown in Figure 7, was most suitable for identifying suspect hooks without the risk of damage to non-defective hooks. As expected, the worn hooks from the incidents failed this test but, in addition, a number of new hooks were shown to be prone to opening. The manufacturers of these hooks are currently considering design modifications and the test procedure has been submitted to the appropriate CEN committee for inclusion in the standard.

4 Summary and further work

The need to analyse the underlying cause of lifting accidents must be emphasised. Only in this way can we learn from mistakes and omissions and, sometimes following more detailed research, utilise the results to produce safer products and practices. Thus the repetition of accidents may be prevented.

It is clear that the degree of derating required for certain lifting operations, to allow for impulsive or dynamic loads, bad lifting practices, different methods of hitching or attachment, etc., is not fully understood or quantified. Such data that are now available should be incorporated into both national and international standards and be readily available in the form of guidance for users.

New materials and new designs, which initially are thought to offer significant improvements, sometimes introduce unexpected disadvantages, for example the vulnerability webbing slings to mechanical damage. The use of risk assessment techniques may enable these problems to be identified and overcome before accidents arise.

An additional outcome of HSL's research is the production of training videos (eg on fork lift truck operation) and specialist training courses, for example on wire rope technology. It is anticipated that future areas of study will include the dynamic performance of non-metallic slings; further assessment wire rope terminations; fracture mechanics applied to lifting equipment, and defect assessment.

23 FRETTING CRACKING AND ITS PREVENTION: APPLICATION IN TRANSMISSION LINE FATIGUE

Z.R. ZHOU
Tribology Laboratory, Southwest Jiaotong University, Chengdu, PR China
A. CARDOU and S. GOUDREAU
Departement de Genie Mecanique, Universite Laval, Quebec, Canada
M. FISET
Department de Mines et Metallurgie, Universite Laval, Quebec, Canada
L. VINCENT
Department MMP, Ecole Centrale de Lyon, France

Abstract
Fretting occurs in quasi-static joints or at surfaces in contact under low relative motion. It has been proven as one of the main sources of engineering structure failure. A great amount of fundamental fretting test investigations on several metals have shown that rapid crack nucleation and its early propagation is located in mixed fretting regime which is the most critical to service failure. The remedies for fretting cracking are proposed from the classical fretting test point of view. Occurrence of fretting in transmission lines is a typical fretting fatigue example in the electrical industry. Individual aluminium wire breakages due to inter-fretting frequently occur in the clamp vicinity. Therefore, the effective palliative against fretting cracking is very important to be searched. In this paper, two factors, clamping conditions and lubricant, are discussed. The results show that fretting cannot be entirely prevented, but that fretting crack growth can be delayed which increases service lifetime.
Keywords: Fretting, Fretting cracking, Prevention, Conductor

1 Introduction

It is well known that fretting[1] occurs in quasi-static assemblies such as wire rope, splines, sliding shafts and in nuclear power plants. Fretting can promote the initiation and growth of cracks. These cracks may propagate under external cyclic loading and lead to catastrophic service failure.

Fretting is very complex. Many parameters concerning mechanics, material and environment can affect fretting behaviour. In this paper, fundamental fretting behaviour obtained from classical fretting test is presented. Several remedies are then suggested for fretting cracking. For a typical fretting fatigue example - fretting in a conductor-two palliative methods are suggested to minimize the risk of fretting cracking.

Structural Failure, Product Liability and Technical Insurance, Proceedings of Fifth International Conference. Edited by H.P. Rossmanith. Published in 1996 by Taylor & Francis, 2 Park Square, Milton Park, Abingdon, Oxon, OX14 4RN. ISBN 0 419 22080 1

2 Fretting behaviour

Many fretting tests have been carried out on classical test rigs(2) in laboratory controlled conditions. Several metals, especially, different types of aluminium alloys, have been tested. All information on the load - displacement variations have been recorded during testing. Subsequent metallographic examinations have been performed on each specimen after test.

2.1 Contact conditions

For most metals, when the imposed amplitude is varied, three classical tangential load (F)-imposed displacement (D) curves have been obtained, indicated in Fig.1.a.

(1) Closed F-D cycle: this case generally corresponds to a very low imposed displacement. No relative movement occurs and both specimens are stuck. Displacement is essentially accommodated by elastic behaviour. (2) Quasi-rectangular F-D cycle: this is a typical gross slip case. In general, the imposed displacement is high and more relative slip is obtained. (3) Elliptical F-D cycle: this kind of cycle generally develops after a certain number of fretting cycles. The stage before the formation of elliptical cycles is not stable and F-D cycles may be closed and rectangular cycles. This case (mixed fretting regime) is often observed for an intermediate imposed amplitude between the sticking and gross sliding regime. It is noted also that the tangential load or the ratio of the tangential load to the normal load is the highest compared to the other two regimes.

The above evolution is obtained as a function of the imposed amplitude. A similar evolution may be observed when one of other parameters is varied such as normal load, frequency, specimen rigidity. To the sake of brevity, only normal, one of important parameters in combination with displacement amplitude are considered. Fig.1.b shows an important fretting map, called Running Condition Fretting Map(RCMP)(2). This map clearly separates three fretting regimes as a function of amplitude and normal load.

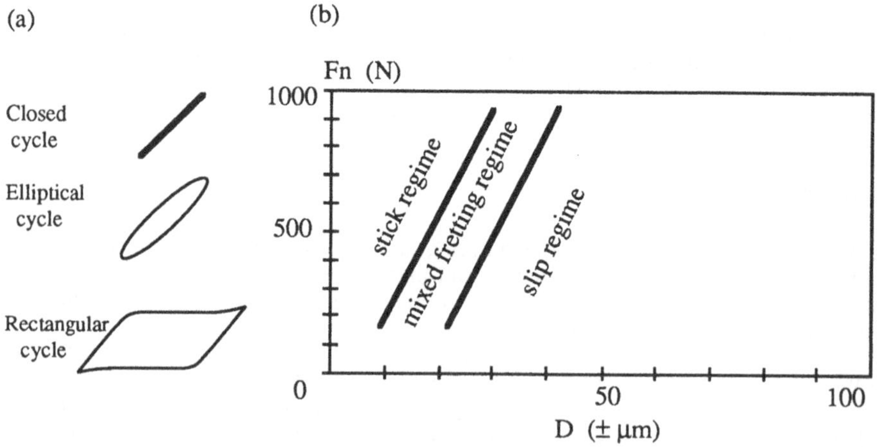

Fig.1. Contact conditions: (a) three types of cycle; (b) Running condition fretting map (RCFM).

2.2 Fretting damage behaviour

There are different degradation mechanisms from a microstructure point of view during a fretting test. However, two essential, macro-degradation modes can be observed: detachment of particles or surface wear damage, fretting cracking growth or volume damage. It is clear that the latter mode is the most dangerous for service failure.

Referring to the above fretting map, metallographic examinations show three zones as indicated in the Material Response Fretting Map(MRFM)(2): severe detachment of particles, fretting cracking with more or less detachement of particles and slight or even no surface degradation. Fretting cracking zone between lines ab and ac initiate within the mixed fretting regime (Fig.2) and increase (between line ed and fg) towards the sticking and gross sliding regimes as the number of cycles increases. Crack depth in the mixed fretting regime increases as a function of the number of cycles and the normal load.

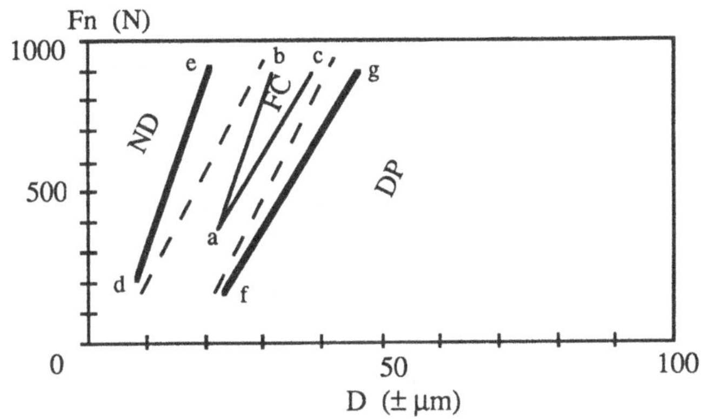

Fig.2. Material response fretting map (MRFM).

Fretting cracking mode at the higher number of cycles is quite different for the three regimes: in the sticking regime, crack initiation direction is quasi perpendicular to the interface; in the slip regime, oblique cracks can be observed from the worn contact surface: more complex cracking occurs in the mixed fretting regime although early crack growth is always at angle with the interface.

3 Prevention of fretting cracking

It has been recognized that early crack formation is controlled by local fretting and it is a main cause to reduce component fatigue lifetime reduction. Therefore, how to prevent or alleviate early cracking is the most important step to increase service fretting fatigue strength.

According to both kinds of fretting maps, the mixed fretting regime is the most critical with respect to fretting cracking. Thus, several measures against fretting cracking are proposed as follows:

(1) Elimination or reduction of the mixed fretting regime

The following remedies are very effective to eliminate or reduce mixed fretting regime, in other words, a sticking (or quasi-sticking) or a gross sliding case replace the above regime.:
-- The ideal remedy for fretting cracking is to eliminate or reduce external vibration, which is the cause of slip between the fretting surfaces. This will minimize not only fretting cracking but also detachment of particle from the interface, because relative cyclic motion between the surfaces is the the essential ingredient in the fretting.
-- Increasing the friction or the normal load (or modifying other mechanical parameters) are effective to eliminate or diminish slip amplitude. For instance, in the light of the RCFM, for a given imposed amplitude, the mixed fretting regime can be moved to the sticking regime as more normal load is applied.

(2) Increasing contact material strength against fretting cracking
-- Increasing material fatigue strength. The fretting fatigue endurance can greatly increase by applying a contact material with higher fatigue strength. Fretting cracking nucleation will be much delayed or even disappears. Here, it is noted that this is not effective against detachment of particles.
-- Contact surface treatment. Some methods such as contact surface treatment by laser, induction of residual stresses also can prevent or delay fretting cracking nucleation. As the above, these methods are effective not much by reducing surface fretting degradation but rather by improving fretting fatigue strength.

(3) Applying third body protection
-- Lubrication. Higher cyclic tangential load is predominant to fretting cracking. Application of the lubricant can greatly reduce or delay fretting damage. How to penetrate as fully as possible to the areas of real contact and how to relubricate the fretted surface is the key in selecting the type of the lubricant.
-- Surface separation. Complete separation of surface by inserting a material of high elastic strain limit, such as rubber or a polymeric material(3), will prevent fretting, provided the inserted material can elastically absorb the vibrational amplitude

As indicated in the above, there are many effective remedies for fretting cracking. However, in practice, these methods are not useful due to many limitations from its service structure, material physical characteristics, external nature etc. Remedies for different fretting cases will be very different.

4 Application to transmission lines

Overhead electrical conductors play an important role in electric utility industry because of their power transmission, good flexibility and ability to transfer load. However, it is well known that fatigue lives of the conductor are much shorter than those of rods of equivalent cross-sectional area produced from the same material(4). This is because of fretting within the conductor under aeolian vibrations (wind excitation)(5). Further investigations have shown that fatigue failure frequently occur at fret locations in the vicinity of clamps. Fretting has been regarded as an important factor in determining the service life of the conductors, being responsible for shortening the period prior to the initiation of fatigue cracks in the conductor.

4.1 Fretting behaviour of conductor bending fatigue

A great number of bending fatigue tests have been performed on overhead electrical conductors. The test bench has already been described(6). For the sake of clarity, a typical suspension clamp and the Drake ACSR (Aluminium Conductor Steel Reinforced) conductor with simple geometry structure are used to present fundamental fretting behaviour. The suspension clamp-conductor system is shown schematically in Fig.3. The clamp is made of aluminium alloy. The Drake is composed of two aluminium layers and a steel core .

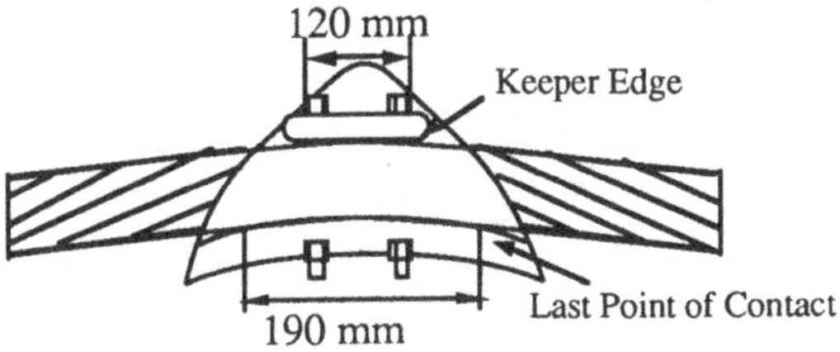

Fig.3. The suspension clamp.

4.1.1 Contact modes and fretting marks

In a clamping region, there are several distinct contact zones, each having its own characteristics:

-- Outer layer contact with clamp. In the suspension clamp, metal-metal contact occurs, either with the clamp groove, or with the keeper.

-- Contact between two aluminium wires. Three types of wire-wire contact can be found. They are: outer contact, inner contact, lateral contact(6).

-- Contact between aluminium wire and steel wire.

After a specimen has been cycled in bending, contact marks have undergone metallographical examination by optical and scanning electron microscope. Such examinations have shown that aluminium wire-clamp contact zones have become elongated, strip-like fretting marks. Lateral contact gives very narrow helical linear marks. Other contact types, which are nominally pointwise crossed-cylinder contacts, become elliptical. Some of them are purely plastic marks, others result entirely from wear, while a third category is intermediate between these two extremes.

4.1.2 Fretting charts

The aforementioned contact types are found in surface patches which can be described qualitatively through so-called "fretting charts". Aluminium wires belonging to the same layer can be shown in a plane and fretting marks can be plotted as in Fig.4. Since clamps and test benches are practically symmetric, and because fretting occurs in the clamp vicinity, a fretting chart will show a 200 mm conductor length from clamp plane of symmetry. Lateral contact damage may also be plotted, provided wires are shown rotated by a 90° angle, with the fretting mark facing upward.

Such charts can be used: (1) to show more clearly global surface degradation and type as well as critical zones, thus allowing comparison between clamp types; (2)

to compare more easily fretting and wear behaviour under different test parameter values; (3) to infer aluminium wire fretting cracking possibility from fretting mark types.

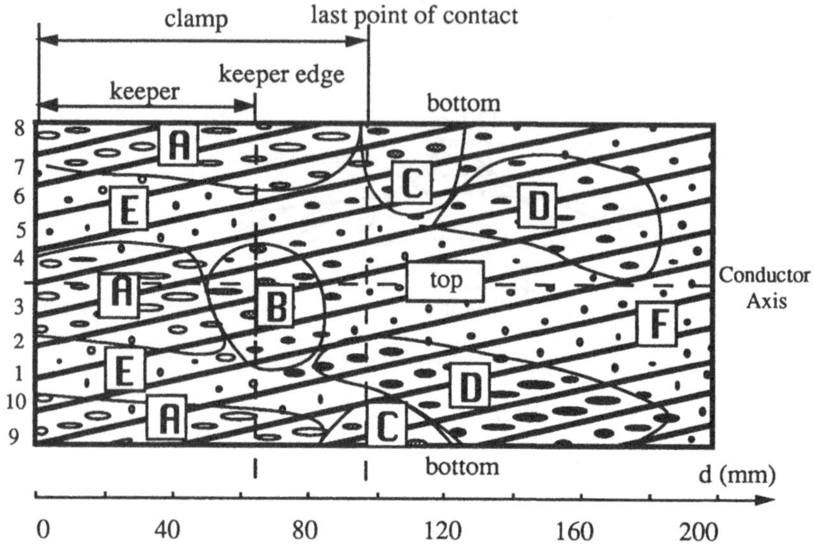

Fig.4. Fretting chart for the inner layer of the Drake.

4.1.3 Fretting regime and fretting cracking

In the light of metallographical observations, the preceding marks show three characteristic fretting regimes:

-- Stick regime (Zone A, E in Fig.4). In general, all contact marks between outer layer and clamp, as well as wire-wire contact points within the suspension clamp belong to this regime. It corresponds to high radial load on the conductor normal load at contact point. Micro-slip occurs only at the contact mark edge and contact wear (corresponding to some detachment of particles) is very slight. It takes a large number of cycles (of the order 1 to 10 millions cycles) to initiate a crack.

-- Slip regime (Zone D, F in Fig.4). As opposed to the previous case, global cyclic sliding occurs in such a regime. It corresponds to lower radial load and higher relative slip at contact points. Severe particle detachment can be observed. Fretting mark size increases rapidly with number of bending cycles. Corresponding zones are found outside a clamping zone. In terms of crack initiation and fatigue, this regime is of little consequence, in spite of the severely worn out mark appearance (from about 10μm to 200μm).

-- Mixed fretting regime (Zone B, C in Fig.4). It is observed in the last point of contact zone. The number of contact marks related to this regime is relatively less than in the stick or slip regimes. As it can be expected, contact mark degradation is intermediate, but closer to the stick regime case. Mark cross-section examination show that the mixed fretting regime is the most critical with respect to crack initiation. In fact, short cracks (about 100 µm) have been observed after a low number of bending cycles (about 10^4 cycles). Besides, surface plastic cycling is much more pronounced, as evidenced by the so-called "dark zones" which are not observed in the other two regimes.

4.1.4 Wire break behaviour

A fatigue test has been terminated after a certain number of wire breaks had been recorded. Subsequently, individual broken wires have undergone the following examination:

-- Wire break facies. Two widespread fracture patterns have been observed: the first is a quasi-planar one, following closely the wire normal cross-section; the second is mostly a plane inclined at 45° to the wire axis, with a small plateau perpendicular to that axis. The former pattern is generally found emanating from stick regime marks, while the latter is mostly found in mixed fretting regime.

-- Wire break location. Each fractured section is transversely positioned. Most of wire breaks are located between keeper edge and last point of contact as shown in Fig.5.

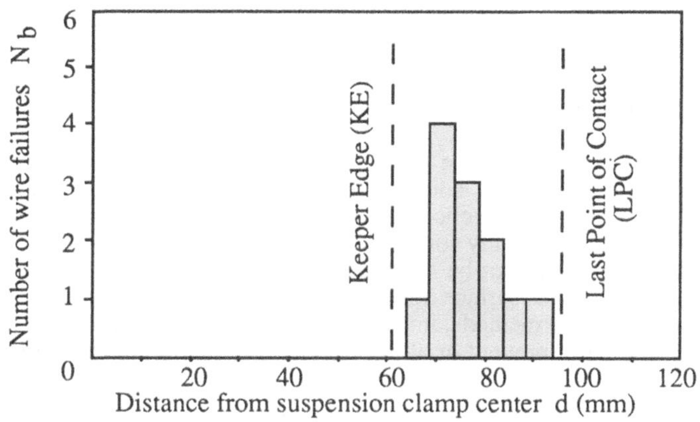

Fig. 5. Number of wire failure as a function of distance from the clamp center.

-- Layer sensitivity. Fretting cracking is related to a specific type of fretting regime: the mixed fretting regime. In conductor bending, fretting regime behaviour on a given layer can be drastically modified by a slight change in bending amplitude. For instance, at low amplitude, inner aluminium layer and steel core are found to behave as a solid with practically no fretting marks. Wire fracture probability on that layer is then almost nil. With the increase of the bending amplitude, mixed fretting regime instead of sticking condition is developed at the inner layer. As seen in Fig.6, wire break percentage on the inner layer increases with increasing amplitude, and decreases on the outer layer.

The above analysis have shown that fretting behaviour within the conductor is very complexe. However, three fretting can be identified and the mixed fretting regime has been proven to be most critical for individual aluminium wire break.

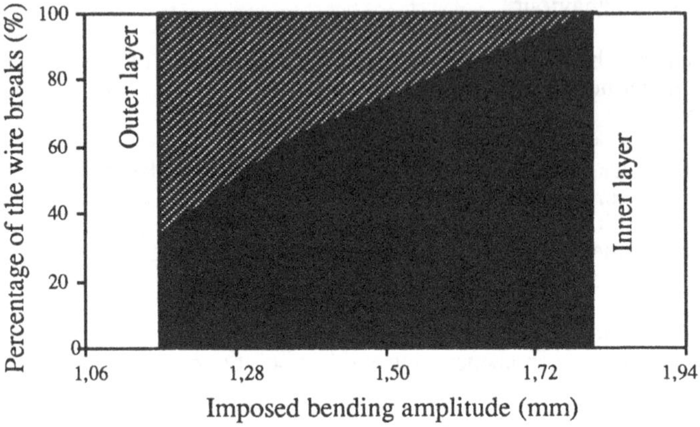

Fig.6. Wire break occurence as a function of bending amplitude (measured at 89 mm from last point of contact)

4.2 Palliative methods

According to the above study on the simple fretting test, many remedies for fretting cracking may be taken. However, some measures are not completely adapted to this practical case. For example, the cyclic bending amplitude of external vibration, closely related to fretting behaviour is only controlled by natural aeolian vibration; the material such as aluminium wire can not be changed due to its proper electrical properties. Surface separation, as well as surface treatement is impossible to be carried out.

Here, two palliative methods, improving clamping conditions and applying the lubricant are attempted to prevent or alleviate fretting cracking in transimission line fatigue.

4.2.1 Clamping mode

In order to compare the influence of clamp system, fatigue tests have been performed on the Zebra ACSR with two types of spacer clamp. Two spacer clamps are used. The first spacer clamp (clamp A) is shown in Fig.7a. It is a nutcracker type of clamp consisting of two pivoting half cylinders, locked in place with a force fitted pin. Elastomeric cushions prevent metal-metal contact. The second spacer clamp (clamp B) (Fig.7b) is really not a clamp but, rather, a half-cylinder, held in place by a set of four pre-twisted rods, of 8 mm diameter, which are about 1.1 m long. They fit tightly on the conductor. It is also fitted with an elastomeric cushion.

Metallographic examination has shown that not only fretting mark behaviour at the fretting chart is different but also their endurance limit are very different for each clamping system. Fig.8 is a fretting fatigue diagram that shows the first four wire breaks with the corresponding number of cycles to fracture during fatigue test for a given bending amplitude. This diagram is similar to the S/N or Wöhler diagrams. From this figure, it may be noticed that clamp B shows a much higher endurance limit, in spite of more fretting wear.

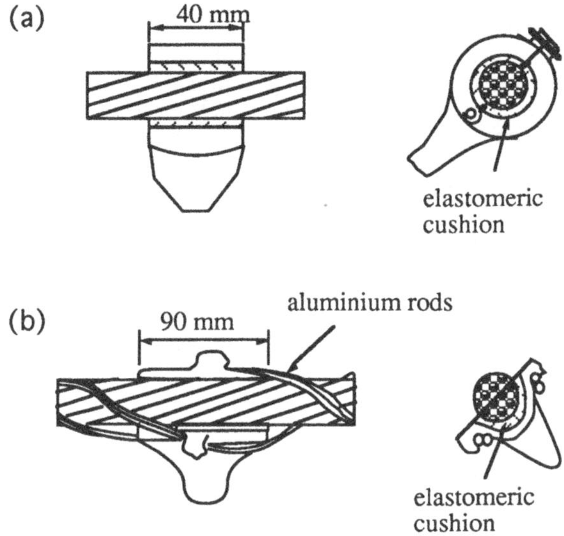

Fig.7. Spacer clamps A and B.

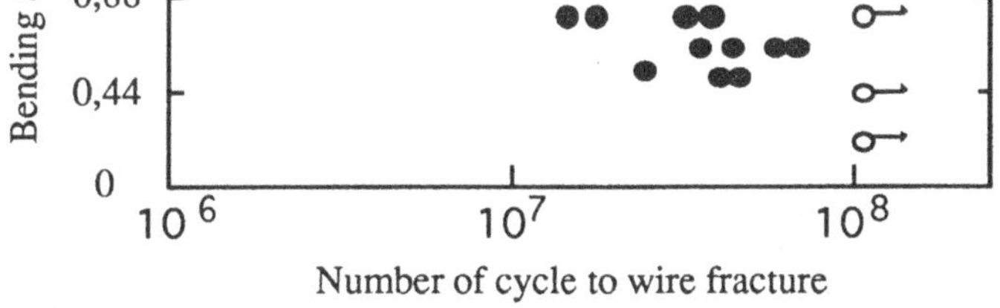

Fig.8. Fretting fatigue diagram for clamps A and B (Zebra ACSR).

In fact, in the clamp vicinity, clamp modes have a strong influence on wire-wire and wire-clamp contact conditions. The above study shows that the occurrence of a specific fretting regime is closely related with normal contact load value. The fretting regime for each wire-wire contact within the conductor can be modified by different clamping systems and the so-called mixed fretting regime, the most critical to fretting cracking may be reduced drastically for clamp B.

4.2.2 Lubricant

The Zebra ACSR conductor is a lubricated conductor. Lubricant within the conductor is the Castrol Rustlo 431 commercial brand. It is soft and semi-solid. Recommended film thickness is 0.5mm. To interpret the palliative effect, lubricated and dry specimens of that conductor have been tested.

Our experimental data show that number of cycles to first wire break should be about five times larger in the lubricated case than in the dry case. Moreover, wire break and fracture surface appearance correspond much more to the plain fatigue case, showing that fretting influence is much reduced in the lubricated conductor.

In fact, the coefficient of friction being smaller in the lubricated case, tangential forces at contact points are much smaller due to the accommodation properties of the lubricant-particles mixture. This has been clearly shown in single wire testing in case where the normal load and relative displacement is not high(7). From the metallographical examination, in the lubricated case, little or no dark area is observed in the fretting mark. Since it is known that this dark area is the result of high cyclic tangential load, which is associated with the critical fretting regime where cracking occurs, it appears that the lower tangential force will clearly retard crack nucleation.

5 Conclusion

According to the fundamental fretting behaviour of classic fretting test, three global remedies for fretting cracking are suggested: - modifying frettting regime behaviour; - increasing material or contact surface strength; - applying third body protection. However, the above measures are not completely adapted to one practical case such as fretting fatigue of transimission lines.

Two factors, clamping mode and lubricant have been studied. One fitting clamp mode and application of the lubricant can greatly increase the conductor service lifetime althrough their palliative mechanisms are very different. It is noted here that fretting can not be entirely prevented, but that fretting crack nucleation can be delayed.

Acknowledgements

Support from the Natural Sciences and Engineering Research Council of Canada and from National Grid Technology and Science Laboratories, U.K. is gratefully acknowledged

References

(1) Waterhouse, R.B. (1981) *Fretting fatigue*, Elsevier Applied Science, London.
(2) Zhou Z.R., Fayeulle, S. and Vincent, L. (1992) Cracking behaviour of several aluminium alloys during fretting wear *Wear*, Vol.155, pp 317-330
(3) Metals Handbook (19) Failure Analysis and Prevention, ASM Vol. 10.
(4) Harris, S.J., Waterhouse, R.B. and McColl, I.R (1993) Fretting damage in locked coil steel ropes *Wear*, Vol. 170 pp 63-70
(5) Rawlins, C. B. (1979) Fatigue of overhead conductors *Transmission Line Reference Book, Elec. Pow. Res. Ins.*, Palo Alto, 51-80.
(6) Zhou, Z. R., Cardou, A., Fiset, M. and Goudreau, S. (1994) Fretting fatigue in electrical transmission lines *Wear*, Vol. 173, pp179-188
(7) Zhou, Z.R., Goudreau, Fiset, M and Cardou, A (1995) Single wire fretting fatigue tests for electrical conductor bending fatigue evaluation" *Wear*, Vol. 181-183, pp. 537-543

24 ANALYSIS OF A POLYETHYLENE PLANT EXPLOSION

S.W. WINDER and D.R. HAYES
Failure Analysis Associates, Houston, Texas, USA

Abstract

On October 23, 1989, a massive release of polyethylene and combustible gases resulted in an explosion and fire at a large polyethylene production facility in Houston, Texas. The fire and explosion caused the deaths of 23 workers and resulted in damage and lost revenue of over two billion dollars (Figure 1). The resulting investigation ultimately focused on maintenance operations as the cause of the release. Because two of six workers who were directly involved in the maintenance operations were killed, and the others could not recall significant details, the exact facts and circumstances of how the release occurred will never be completely known. However, the physical facts and accident investigation point to human error as the most probable cause of the release of explosive gases. This paper discusses the engineering investigation and findings of the facts and circumstances pertaining to the incident. Also, the concept of Crew Resource Management (CRM) is introduced as a means of eliminating or reducing the frequency of these incidents. CRM also provides an analysis tool for investigating accidents that may involve human error.
Keywords: Accident investigation, blocking valves, command authority, Crew Resource Management, explosion, human error, polyethylene, reactor loop, release of explosive gases.

1 Background

The polyethylene plant made high density polyethylene (HDPE) from ethylene gas dissolved in isobutane which is reacted in two long continuous vertical loops at elevated pressures (4.14MPa/600 psi) and temperatures (110°C/230°F). Figure 1 shows the vertical reaction loops. Different chemicals are added to the process stream to cause the formation of polyethylene particles which have the appearance of a fine white powder. The particles settle out at the bottom of one of the loops and are removed from the loop through six settling legs (Figure 2). At the top of these legs are blocking valves which

Structural Failure, Product Liability and Technical Insurance, Proceedings of Fifth International Conference. Edited by H.P. Rossmanith. Published in 1996 by Taylor & Francis, 2 Park Square, Milton Park, Abingdon, Oxon, OX14 4RN. ISBN 0 419 22080 1

are air actuated ball valves that are open during normal operation (Figure 3). Below the blocking valves are product takeoff valves which continuously cycle between the on and off position to effect a controlled continuous flow of polyethylene particles from the vertical reactor loops (Figure 3). On occasion these settling legs become clogged with polyethylene "logs" and have to be cleaned. Standard procedure would be to close the blocking valve and secure it with a lockout device so it could not be accidentally opened (Figure 4), separate the downstream piping and remove the polyethylene plug. After the incident, the blocking valve was found open and the piping removed providing a direct release of explosive gases from the reaction loop to the atmosphere (Figure 5). This allowed the release of approximately 38,500 kg (85,000 lbs) of highly flammable products. The time to release these flammable products was estimated to be two to three minutes. Other explosions occurred ten to fifteen minutes later when two isobutane tanks exploded. Each explosion damaged other units which in turn exploded (Figure 6). It was reported that ten explosions were noted over a two hour period after the initial explosion.

The aftermath of the explosion revealed that all the polyethylene plant manufacturing plant was destroyed, and all the superstructure for the reactors had to be torn down and rebuilt. The rebuilding process took over a year to complete causing a significant expense in construction costs and loss of revenue.

Fig. 1 Aerial view of polyethylene plant damage.

2 Investigation

2.1 Gas release site observations

Because of the total destruction of the plant and the numerous investigations underway, a month passed before we were able to focus on the site of the initial release. Once the release site was determined and inspected the following observations were noted (Figure 7):

1. The blocking valve was open;
2. There was no lockout device on the blocking valve;
3. The blocking valve is an air driven valve and the flexible air hoses were connected to the valve driving ports; and
4. Although the flexible air hoses were severed in the incident, remnants were found attached to the instrument air lines and the 1/4 turn valves were found in the on position indicating the flexible hoses were capable of receiving air and actuating the blocking valve (Figure 8).

According to standard operating procedures, the following conditions should have existed on this valve and settling leg:

1. The blocking valves should have been closed and locked out to prevent an inadvertent opening;
2. The air hoses should have been removed from the blocking valve; and
3. The instrument air line 1/4 turn valves should have been closed.

Fig. 2 Reactor loop and settling legs.

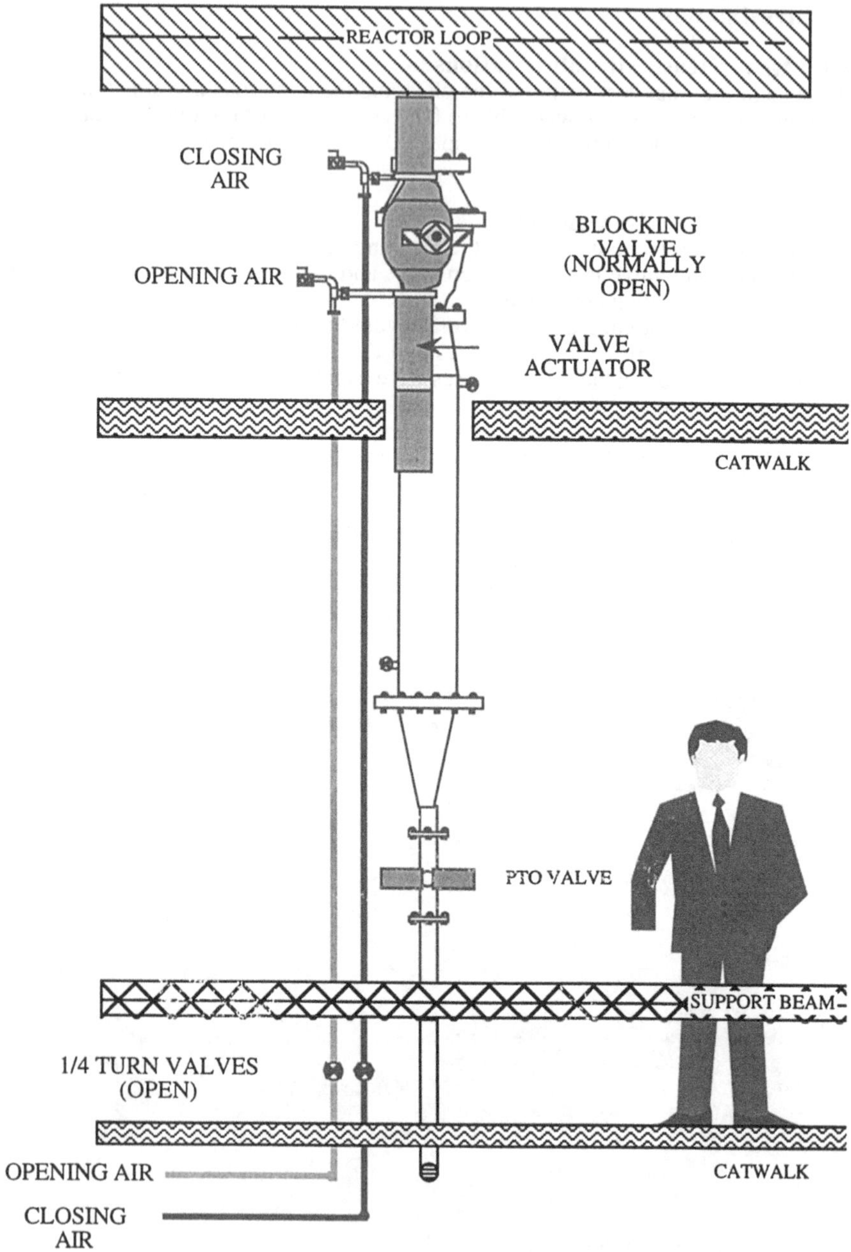

Fig. 3 Settling leg configuration during normal operation.

Fig. 4 Example of actuator operation.

Fig. 5 Settling leg with piping removed.

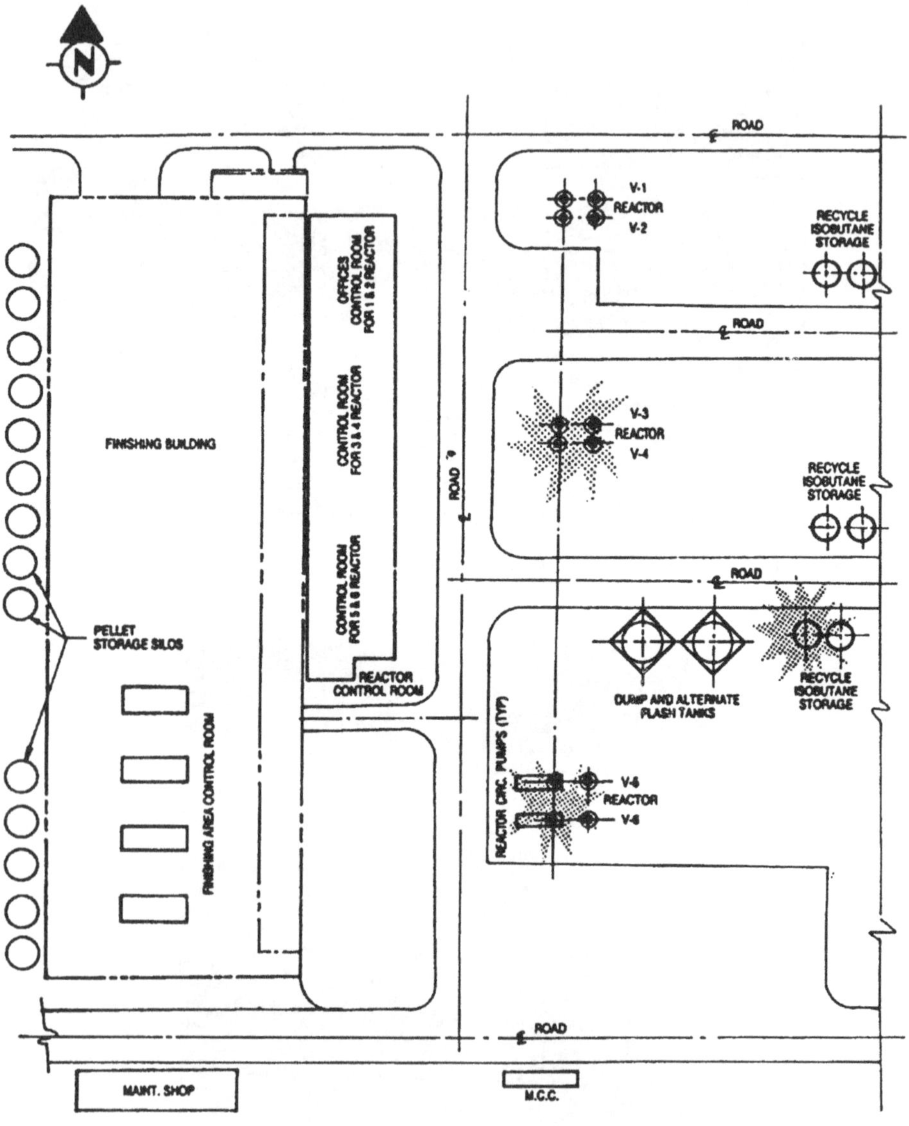

Fig. 6 Polyethylene plant explosion sites.

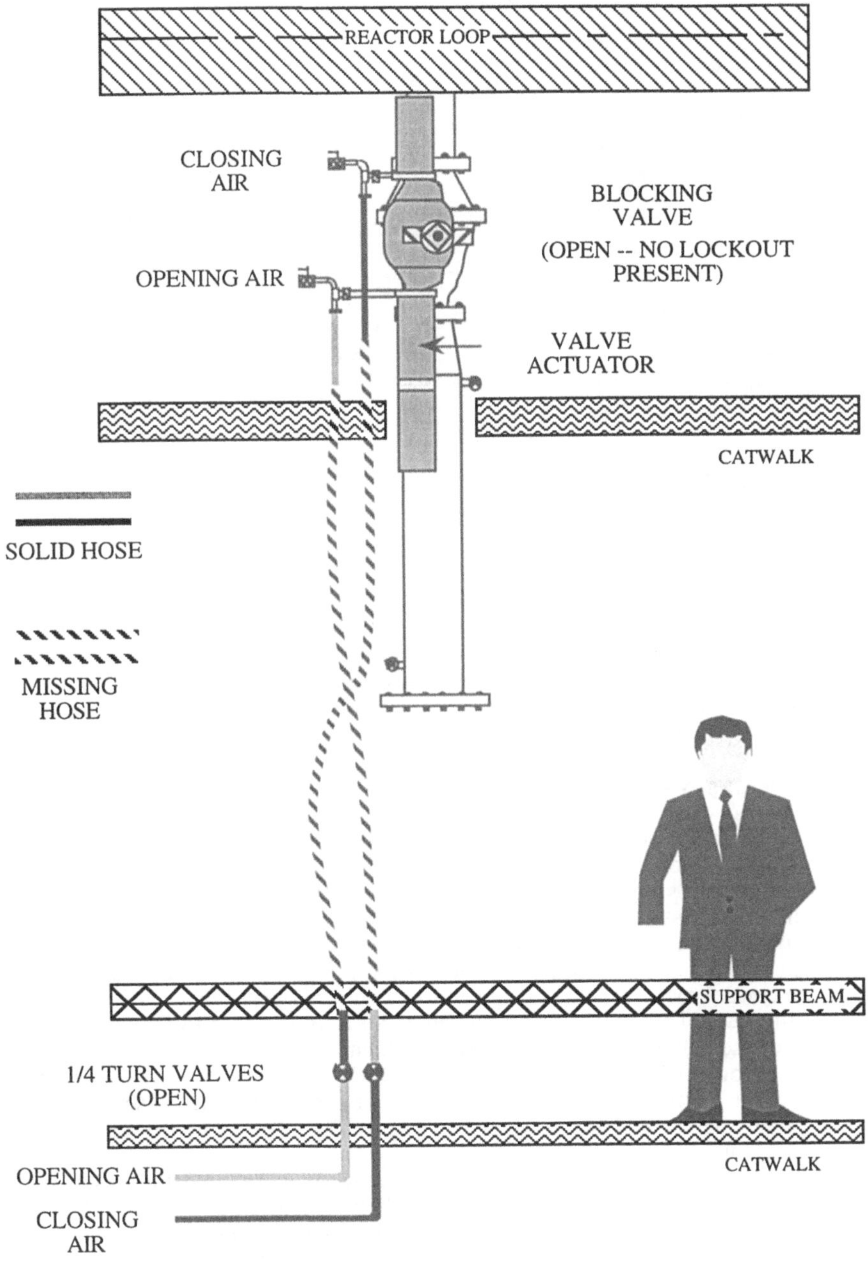

Fig. 7 Valve and hose positions after gas release.

Fig. 8 Quarter turn valves in "on" position.

The conditions after the accident were exactly the opposite from what they should have been. In a subsequent analysis of the air hose remnants, the hose connected to the actuator "closing" air port was attached to the "opening" instrument air line and the hose connected to the actuator "opening" air port was attached to the "closing" instrument air line. Thus, the controls for closing would actually open the blocking valve and the controls for opening would close the blocking valve.

2.2 Blocking valve control

The blocking valve is a pneumatically actuated ball valve. The valve operates by supplying air to a piston which in turn converts linear motion into rotational motion and changes the position of the valve by 90 degrees. There are two pistons opposing one another so by applying air to one piston drives the ball in one direction while applying air to the other piston reverses the 90 degree rotational motion (Figure 9). The ball valve will either be opened or closed: no intermediate positions are permitted.

The control logic is such that the blocking valve can be pneumatically closed only from the control room. The blocking valve can only be opened by a lever stationed near the settling leg. This logic was perceived as a safety measure to prevent inadvertent opening of the valve from the control room which has no direct visual contact with the settling leg or blocking valve. With the air lines reversed, the valve could be only opened in the control room and only be shut by an operator near the settling leg. The switch near the settling leg is shown in Figure 10. The switch is in the down or activated position.

Subsequent statements by other workers who maintained these settling legs indicated that at times the plugs in the line, if they were near the valve, would be blown out by the pressure (4.14 MPa/600 psi) in the reactor loop. However, on these occasions, the piping would all be connected so there would be no leak of gases to the atmosphere. Also, other shift workers working on the incident settling leg stated that the night before the incident the settling leg had been properly safeguarded with a lockout in place and the air hoses removed.

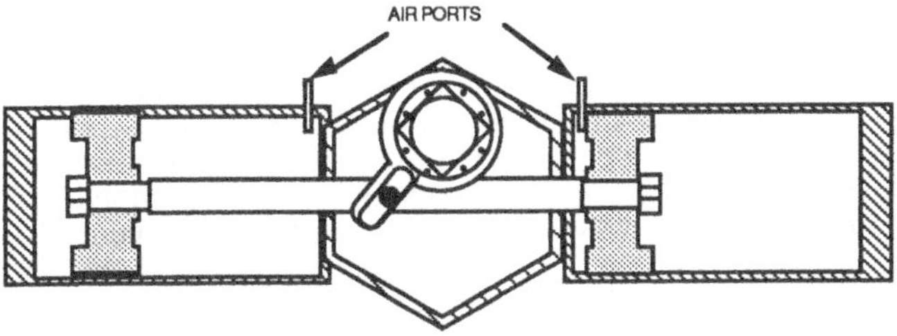

Fig. 9 Actuator for blocking valve.

Fig. 10 Actuator switch near settling leg.

The incident blocking valve and actuator were removed, functionally checked, and disassembled for inspection. The valve and actuator functioned properly and no evidence was found that would produce a valve malfunction.

2.3 Most probable accident scenario

From the physical evidence, the most probable cause of the gas release was an attempt to remove a polyethylene plug from near the blocking valve in the settling leg by opening the valve from the control room and intending to shut the valve from the switch near the settling leg. Once the plug was dislodged it either severed the hoses or the rushing gases severed the flexible hoses preventing the valve from being closed. With the valve open, the release could not be prevented.

The ignition source was never determined. However, there were several possible sources in the vicinity of the settling leg:

1. Numerous vehicles;
2. Welding operations;
3. Open flames from waste gas flares; and
4. The control room air intake.

2.4 Possible design changes

As with any accident of this magnitude, there have to be lessons learned as to how accidents like this can be prevented. Additional standard operating procedures will not help. Had standard safeguards been followed, the incident would never have happened. What additional things could have or could be done to reduce or eliminate such problems? Some mechanical and technological innovations such as different connectors for opening and closing air ports on the valve would have prevented switching the air hoses. However, the valve still could have been opened at the settling leg. Flexible air hoses could have been replaced by steel piping which might have withstood the force of the escaping gases and allowed the valve to close. However, this is not guaranteed. The actuator controlling the valve could have been constructed to close if instrument air is lost. Such actuators are made by the same supplier of actuators found on the incident valve. However, the effect of such a design change on the overall safety of the reactor loop has not been evaluated. The key causal element in this incident is the human interaction with the blocking valve control system.

3 Crew Resource Management

3.1 CRM concept

CRM is a concept to improve the information exchange between workers to identify possible problem areas and eliminate them before they lead to accidents. The concept concentrates on awareness by the employees as to what is happening around them and communicating effectively with supervisors and fellow workers their concerns that something may be wrong. Equally important is the role of the person in charge. He must be receptive to the possible problems identified by his workers. Failure Analysis Associates (FaAA) has recently launched a program in conjunction with Flight Safety International to introduce the concept of CRM to the petrochemical industry as a means of reducing incidents caused by the human interface with chemical process equipment. These concepts apply equally well to maintenance as to the actual running of the plant processes.

The work done by Flight Safety International centered on increasing the safety of commercial airline flights. Data collected (Figure 11) showed the airline accident rate had reduced significantly between 1960 to 1982 but had reached a constant value from 1982 to 1993. Careful examination of accident data showed these accidents were largely due to human errors. Also, the majority of aircraft accidents were in the landing and takeoff phases of flight. A careful study of chemical process indicated most accidents happen during the shutdown to startup phases of operation (Figure 12) when the workload is most demanding because of pressures to bring the plant or unit to a fully operational status. In the case of the polyethylene settling leg release, there was pressure to get the legs cleaned so the reactor unit could again reach full production capacity.

3.2 Situational awareness

An examination of past aircraft and chemical process accidents showed there were several clues or situations which were evident before incidents took place which, if acted upon, would have prevented the accident. A summary of these situational awareness clues is produced in the Appendix. Not all of these clues or situations are present in each accident, or if present may not be obvious, but an analysis of aircraft accidents showed that on average seven of these clues were present prior to the accident with a minimum of four. Analysis of refinery accidents with this technique gives similar results.

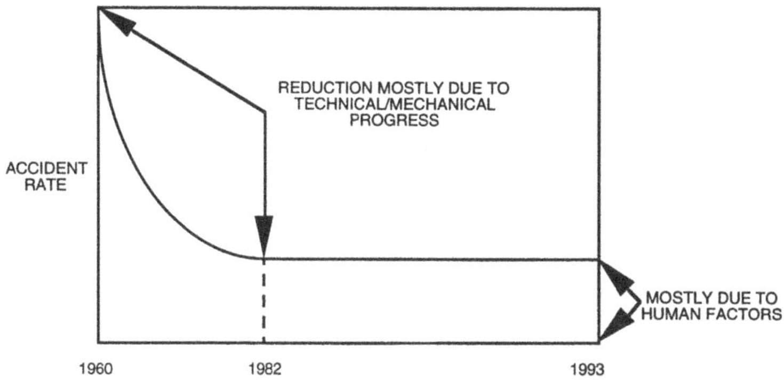

Fig. 11 Aircraft accident history.

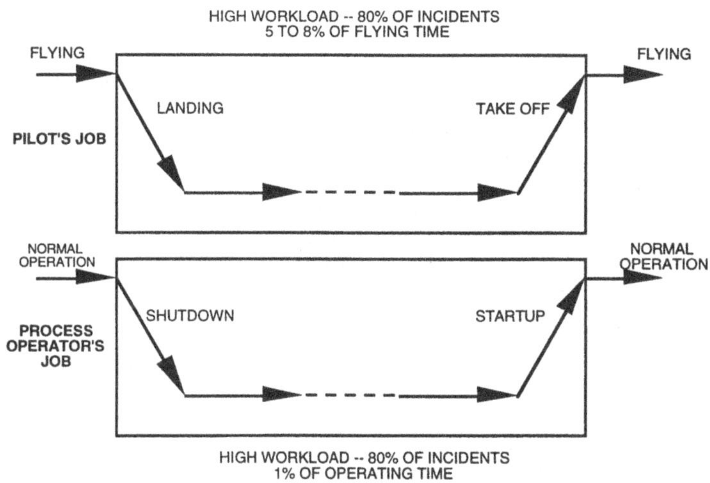

Fig. 12 Comparison of flying and process operation.

In the polyethylene plant accident we can identify at least four of these conditions and can speculate on others. These clues are as follows:

1. Undocumented procedure
   The procedure used to blow the plug from the settling leg was not part of any published procedure.
2. Departure of standard operating procedures
   It was clear from the post accident investigation that there was a violation of the published standard operating procedure for securing the blocking valve.
3. Violating maximums or other limitations
   There was no reason to have a release of gases to the atmosphere even if it was thought the release would be minimal or only last for a very short duration.
4. Preoccupation and distraction
   There was pressure to complete the repair and to bring the unit back to full production. Distractions such as this can result in poor decisions.

There are other clues we can speculate on as being contributory as a result of looking back with "20/20" hindsight.

5. <u>Communications breakdown</u>
   Several of the contractor crew working on the repair were Hispanic and the degree of language skills they possessed is not known. Even if they sensed something was wrong, they might not have been able to express their concerns. Also, the work crew consisted of five contractor employees and one plant employed operator/supervisor. The plant employees were all union members while the contractors were not. The union workers felt all maintenance workers should be in the union and be regular plant employees. It was obvious there was some friction between the two groups of workers.
6. <u>Confusion or apprehensive feeling</u>
   We feel that at least one of the six men in the maintenance crew must have had some misgivings about what was going to happen if they were knowledgeable about what was going to take place and the relative dangers involved. Why didn't they speak up or demand the process be halted?

3.3 Command - Authority

In any group performing work tasks, whether it be in the chemical process industry or flying an aircraft, an authority figure must be in charge. In the case of an aircraft, the authority is the captain. In the case discussed in this paper, the authority is the operator employed by the plant. From cockpit voice recordings of several aircraft accidents it was evident that others in the flight crew recognized the clues but were reluctant to present theirs views to the captain in a manner that alerted him to the problem. The captain or person in charge had to take the information, process the contents, and then make the final decision. If the captain is preoccupied at the time, he may hear the information but not digest its importance until it is too late to avoid an accident. Also, it is important to note that authority figures actions and demeanor are important. If the authority figure is very autocratic, communications will be curtailed between him and his crew. If, on the other hand, the authority figure listens to everyone but never makes a decision, the communications are excellent but the information is useless. Clearly, somewhere in-between is where the command authority should be (Figure 13).

There are important lessons to be learned from these human interactions and how they cause accidents. If CRM principles are followed, there is no reason they should not be effective. However, just training people in these concepts won't be any better than introducing more procedures or technology . Everyone in the plant from top management down to the lowest contract worker must accept the concepts, be trained, practice the concepts, and have periodic review to insure the concepts are being followed.

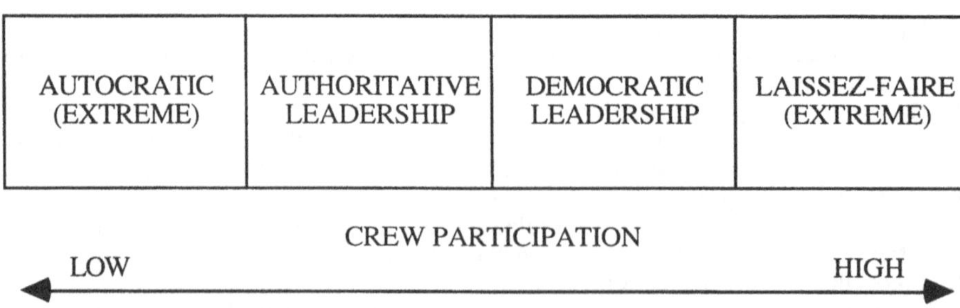

Fig. 13 Command-Authority leadership styles.

4 Conclusions

The gas release and subsequent investigation showed the most probable cause to be due to human error. No mechanical or engineering defects could be found which were significant contributors to the uncontrollable releases of gases from the reactor. The one design exception would be to have an actuator which would automatically close if instrument air is lost. However, such a modification would require a complete revamping of the control logic and a thorough system analysis to insure that such a modification would not introduce other possible problems that could be equally as dangerous.

The Crew Resource Management concepts discussed provide a means of reducing or preventing human error induced accidents. Had any one of the clues identified been acted upon by a good exchange of information between workers, the gas release could have been prevented.

CRM concepts also provide an effective analysis tool to investigate accidents. Engineers involved in investigating accidents tend to focus mainly on the technological aspects of the accident without being as probative as they should as to what human interactions may have contributed to the accident. Examination of the situational awareness clues provides insight as to what may have been happening prior to the accident. Careful interrogation as to interaction between the workers and the supervisor could also provide important insight.

APPENDIX
SITUATIONAL AWARENESS CLUES

Failure to meet targets

A process is a sequence of operations. Any process, be it taking off in an airplane or starting up a hydrotreater, has certain milestones along the way. These milestones may be temperatures, pressures, rates, velocities, etc. coincident with various steps in the process. For example, in a Fluid Catalytic Cracking Unit Regenerator, the temperature must be at least 800°F before introduction of torch oil or it will not ignite. Certain other combinations of target variables are typical of successful startup operations. If all variables but one are in the typical range, then it may be as simple as a bad instrument or, more importantly, we may have a "Failure to Meet targets" that is indicative of a more serious problem.

Undocumented procedure

This is a procedure that is not part of a facility's collection of well thought out written Standard Operating Procedures. It may be a procedure conceived on the spur of the moment. It may be that series of actions about which we later remark, "It seemed like a good idea at the time". The Air Florida flight that struck the 14th Street bridge and fell into the Potomac River tried to melt ice from the aircraft by getting close behind another aircraft in the hopes the heat from its engines would remove some of the ice from their own plane. This is an undocumented procedure that did not work.

Departure from standard operating procedure

This is similar to the "Undocumented Procedure", but in this instance a deviation occurs from an existing, documented procedure in use for a process usually in progress. For example, a procedure for firing of an air heater required that the air bypass valve be closed, thus forcing all of the air through the heater. When the bypass valve stuck in the open position, a decision was made to start the heater firing anyway. Insufficient flow through the heater resulted in a large hole burned in the heater outlet.

Violating maximums or other limitations

All operating processes have an envelope of acceptable operating parameters that were determined by the process and mechanical designers. When we deviate from these acceptable parameters, we enter a region not considered by the designers. We may not be in an area of safe operations. We don't know the consequences and our Situational Awareness is very low,

No one in authority actively supervising

Several aircraft accidents have resulted when there was no one at the controls. Absence may be due to misunderstanding, distraction, boredom, failure of the auto pilot, or other factors. In a process plant, board operators may be absent from their control duties for similar or other reasons. For example, a lack of understanding of potential process hazards can be a reason also. A control board team on a Fluid Catalytic Cracking Unit left their post for a champagne toast upon getting feed into the unit, unaware of the extreme criticality of operations at that moment. Quick action by an engineer brought them back to their posts before a serious incident occurred.

No one looking out the window

Despite the sophisticated instrumentation available, there is still no substitute for the human observer in a complex process. Someone in the airplane must watch where they are going. Distractions sometimes preclude this. A chemical process operation is not moving through space, but it is moving internally and through time. The outside operators may see, hear, feel and smell subtle changes in appearances, sounds, vibrations and odors. These can be invaluable clues to problems developing in the process. The outside operator may be distracted by a heavy workload in an isolated area, may be in the control room drinking coffee, or just recovering from exposure to inclement weather. Whatever the source of the operator's distraction, Situational Awareness is low and incidents are more likely,

Communications breakdowns

Breakdown in communications can be as simple as language problems or misunderstandings, personality conflicts, failure to communicate, or even malfunctioning communication equipment. During a maintenance procedure the experienced head operator was called away to another job due to a malfunctioning radio. During his absence, the inexperienced operator opened the wrong valve and was badly burned.

Ambiguity

On an aircraft as well as in a process plant, we can be the recipients of ambiguous information. This is information from various sources that does not make sense. The co-pilot of an airplane that crashed just after takeoff noted to his captain that the throttle position was wrong for the indicated airspeed. In a process plant, instrument malfunctions are common, and sorting out ambiguous readings from various instruments is a common problem during startups and upsets.

Unresolved discrepancies

In some instances, suspect information or conditions can be ignored. A contributory cause to an ethylene fire was a boiler feed water leak that had gone unrepaired for a week. Our psychological nature is such that we tend to disbelieve information that is very unusual. Ignoring alarms seriously out of limits because, "I know that can't be right" is a common thread in many incidents.

Preoccupation or distraction

People can be distracted by an intense workload in a specific area of responsibility. They also can be preoccupied or distracted by personal problems, family and personal health problems, too many people in the operating area, noise, fatigue, trivial conversation, urgent but unimportant duties, etc. Distraction due to excessive pressure to meet schedule, production, or economic targets can result in poor decisions.

Confusion or apprehensive feeling

This can be a feeling of confusion or not knowing what is happening. It also can be a "gut feeling" that something is wrong. Psychologists tell us that we should respond to these feelings. Our subconscious picks up on subtle signals and sets off our internal alarms. Either of these feelings are signals that our conscious Situational Awareness is low. The converse is not true. Feeling good about a situation is not necessarily indicative of high Situational Awareness. In fact, some have suggested that "Unbounded Optimism" is another clue to low Situational Awareness.

25 FAILURE ANALYSIS OF A VALVE AND COURT ROOM CONSEQUENCES

C.P. HOOGENSEN
Hoogensen Metallurgical Engineering Ltd, Calgary, Canada

Abstract
This paper describes a failure analysis of a 914 mm gate valve. The failure resulted in an explosion and subsequent fire which destroyed a large gas compressor station. The cause of the failure had been traced back to a grossly under-designed 914 mm loose-type flange. The conclusions of the failure analyses resulted in litigation.

The trial judge had to digest over 10,000 pages of transcript of evidence plus several reports. The judgement was in favour of a large dynamic event. The judgement was vacated in appeal, but one of the judges decided to evaluate the transcripts and make a ruling.

This paper presents an account of the failure analysis and courtroom consequences. The paper supports the view that the judgement of technical methods should be conducted by professionals trained in technical matters. Rulings in court should be changed to reflect the present era, rather than an era where a persons' knowledge could encompass all the art and sciences.

Keywords: Brittle failure, compressor station, experimental stress analysis, flange design, fracture flange connection, frost heave, gate valve failure, trials of technical matters.

1 Introduction

On February 26th, 1980 at approximately 6:00 p.m., an explosion rocketed through, what was in 1980, the largest compressor station in the province of Alberta,Canada. The station had construction in progress, but at the time of the explosion, only four people were at the plant.

Structural Failure, Product Liability and Technical Insurance, Proceedings of Fifth International Conference. Edited by H.P. Rossmanith. Published in 1996 by Taylor & Francis, 2 Park Square, Milton Park, Abingdon, Oxon, OX14 4RN. ISBN 0 419 22080 1

A gas compressor station is protected by a number of safety valves. None of the automatic failsafe systems actually worked to shut off the gas flow to the burning compressor station.

Manual closure of "block valves" 25 to 30 km away from the station subdued the fire only after a couple of hours.

It had been immediately evident that a 914 mm (36") discharge line near a gate valve had ruptured.

When an industrial accident occurs which involves the loss of insured equipment, an adjuster is called upon to assess the cost of the damage. The adjuster needs to know the cause of the accident. Most insurance policies contain some limitation of coverage. As an example, corrosion can be an exclusion and in order to establish the cause, the adjuster retains an engineering firm.

Figure 1 shows an aerial view of the plant taken on March 4th, 1980 after the explosion.

Figure 1: Aerial view of the compressor station after the incident; viewed from the north. Photo taken by B. Boyce, March 4, 1980.

The destruction of the plant was extensive. The arrow shows the location where the failure occurred. The ongoing construction is shown at the lower left hand corner. A ditch which was dug just south of the valve can also be seen.

2 Layout of the Compression Plant

The valve[1] was clearly at the initiation of the explosion. The valve body itself was still intact. On top of the valve was the valve operator. The explosion disabled the operator from closing the gate. Had the gate closed immediately, it would have prevented the damage which resulted from the enormous torch created, thus destroying the plant.

A compressor station is a facility where the natural gas transported from the supplier to the user is regulated.

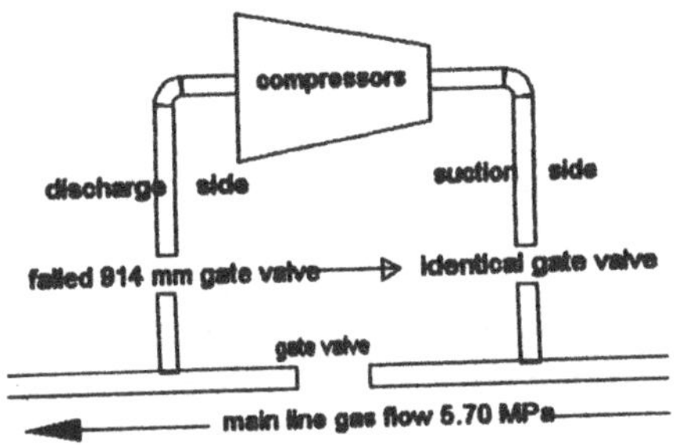

Figure 2: Schematic of the compressor station.

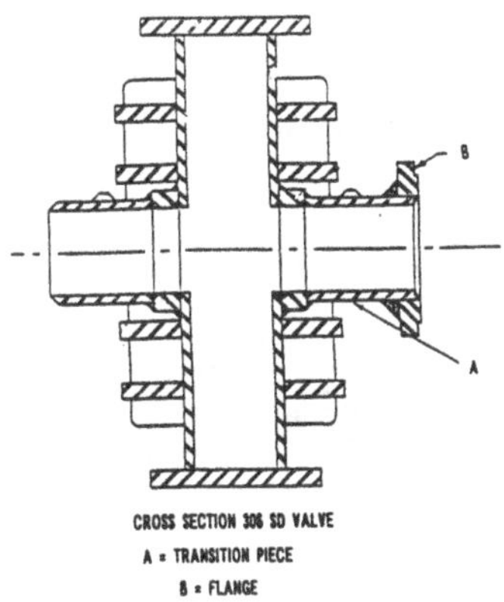

Figure 3: Construction of the valve.

When the pressure in the main line is too low, a compressor will increase the pressure to the required main line pressure. The suction valve and discharge, as shown in Figure 2, is then opened and the valve on the main line closed. The gas is thus directed through the compressors. The schematic of the gate valve is shown in Figure 3. The average temperature in the main line is 3-4°C. The gas, when compressed, has a temperature of about 27°C measured at the compressor station.

3 Physical Properties of Transition Piece

It was not the valve body which had failed, but rather a transition piece which connects the valve body with a flange. The valve had been in service for 10 years.

Figure 4 shows a schematic cross-section of the valve as it was observed at the site before and after failure.

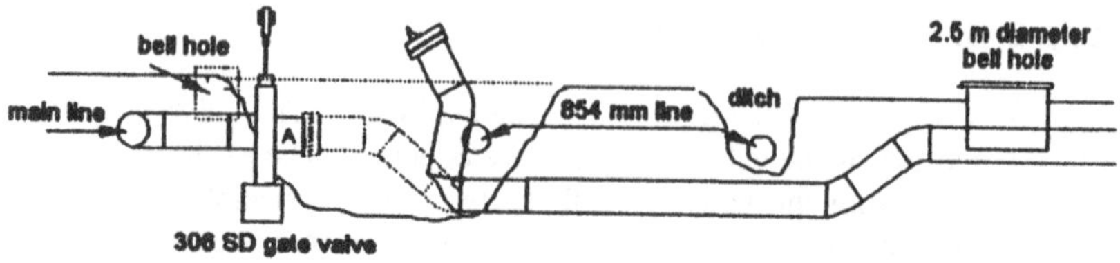

Figure 4: The transition piece is marked by the letter "A".

Only the chemical and mechanical properties of the transition piece are reported here. All values are averaged from 3 samples and are pertaining to both discharge and suction valves. The properties are shown in Table 1 below. The material corresponds with an ASTM 572 Grade 70.

Table 1.

Chemistry:	C=0.21%; Mn=0.91%; P=0.012%; S=0.025%; Si=0.015%; V=0.08%; Nb=0.025%; Mo=0.045%
Tensile:	Tensile Strength=675 MPa Yield Strength=456 MPa Elongation=36%
Charpy Impact:	0°C = 7.5 J; 20°C = 11 J; 65°C = 40 J; 100°C = 70 J
CTOD:	@ 0°C = 0.024 - 0.028 mm

4 Flange Design Calculations (ASME)

When an equipment failure is analyzed, it is imperative to verify the design criteria. Although design calculations do not establish the failure stress, they are conducted to ascertain whether the flange was reliable for service according to the governing design rules. When the basic design requirements are not adhered to, the integrity of the structure is compromised. In order to investigate this, several well accepted analysis techniques were employed. One can use either straight code calculations or a experimental stress analysis technique.

Two analytical methods were used to ascertain the stress levels to which the fillet weld of the flange was subjected. The stress is usually calculated using the conventional code formulas for bolted flange connections. This method of calculating stress was developed in the late 1920's by E.W. Waters and J.H. Taylor (1, 2, 3), and was subsequently accepted and incorporated by the ASME code. This code is in use worldwide with many national codes accepting the ASME standards.

The second method establishing the stresses at the fillet weld using the finite element method (FEM). The FEM can be considered an alternative and also as a superior method to the conventional stress calculations. The stresses using the finite element method can be presented on a global scale, as well as in detail, such as the edge of the fillet weld toe. It has to be emphasized that the results must be carefully interpreted. The analysis used was an elastic analysis. This means that when the material exceeds the yield stress and the material deforms plastically, the FEM still computes the results as elastic. Thus, values above the proportionality limit are linearly extrapolated stresses, rather than actual stress based on the materials' stress-strain relationship.

Beside the two (2) theoretical stress calculations, a physical stress analysis was also conducted. The best known method measures the strain with electrical resistance strain gauges. The experimental stress analysis was conducted on a dimensionally identical valve.

1. ASME Code Section VIII
2. Modern Flange Design, Bulletin 502, Edition VII
3. Waters, E.O. and Taylor, J.H., The Strength of Pipe Flanges, Mech. Eng. 49 (1927), Pages 531-542

4.1 Calculations

The rules for bolted flange connections are found within the ASME Boiler & Pressure Vessel Code, Section VIII, 1968, Division 1. These calculations would normally be carried out by any flange designer in order to obtain a flange which satisfies the code.

ASME Section VIII, Division 1 (paragraph 2.4 of appendix (3)) states that for optional type flanges, loose type flange calculations can be used, provided the following four (4) criteria are to be met.

1. The wall thickness of the transition piece not to exceed 16 mm (5/8 inches).
2. The inside pipe diameter over wall thickness B/g_0 not to exceed 300.
3. The gas pressure not to exceed 2,070 kPa (300 psi).
4. Operating temperature not to exceed 370°C (700°F).

Criteria 1 was not met since the wall thickness for this transition piece was 19 mm (3/4 inch). Criteria 2 was not met because the limiting design pressure of 2.07 MPa (300 psi) was exceeded.

Since only two (2) of the four (4) criteria are met, the maximum stresses for this flange must be calculated using the method described for an integral flange. This implies that the attachment of the flange ring to the pipe wall is considered to act as a single unit.

The valve model and class were on the purchasing order. They were confirmed also on tags attached by the manufacturer of the valve. The tag on the 306SD valve had disappeared, but the valve on the suction side, 306SS, bore a tag which besides the manufacturer's I.D., read:

ASA Class 600
W.P. = 1,440 psi (9.93 MPa)

Based on these data, the ASME flange calculations were performed for the pressure 5,700 kPa and 9,900 kPa. The test with strain gauges were performed at 3,900 kPa, stress calculations were also performed at this pressure.

4.2.1 FEM Analysis

The longitudinal stresses, as they occur, are graphically shown in Figure 5 for the flanges which were joined (i.e.: the failed flange and the mating flange). A flange connection identical to the flange connection which failed with a flange connection as the mating flange which was an integral flange.

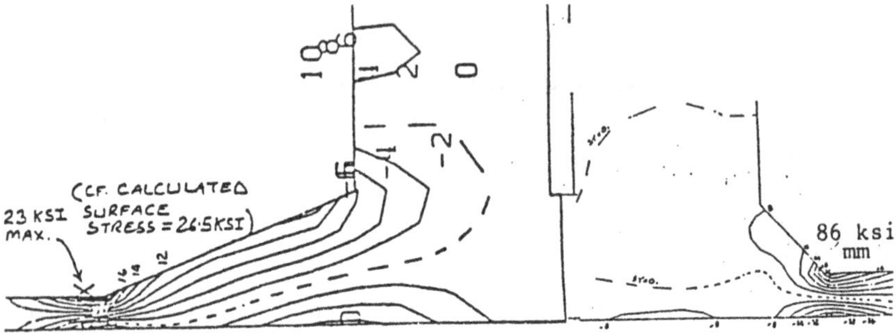

Figure 5: Illustrating the large difference in stress at the toe of the fillet weld and the toe of the weld neck.

Although the stress values found with this method are larger than calculated by using the conventional analysis, it can be concluded that the code calculations are prudent for the design of flange configurations considered in the ASME Code, 1969, Section VIII, Appendix III.

4.3 Strain Gauge Analysis

After the failure of the flange from the discharge valve, and the confirmation of the under-designed flange, the gas company investigated whether there were more such valves in use.

One (1) valve of the same make was found in another section of the pipeline system. Not surprisingly, it was decided to remove this valve. The flange from the suction side (shown in Figure 2) had immediately been cross-sectioned after it had been excavated. As a result, no actual physical test on the flange as a whole had been performed. With this new located valve, special care was taken to obtain as much data as possible, before excavating the valve. It was considered prudent not to put the gauges on at the operating pressure of 5,700 kPa. It was lowered to 3,900 kPa.

Strain gauges were mounted on the transition piece after the soil was removed sufficiently to permit installation. The strain gauges were mounted on the transition piece near the fillet weld, where the fracture had occurred on the failed discharge valve. By doing so, it was possible to measure the strain relaxing during depressurizing. These values were analyzed and the actual surface stresses obtained. For safety reasons, the pressure on the line had been reduced to 3.9 MPa. The location of the strain gauges are shown in Figure 6.

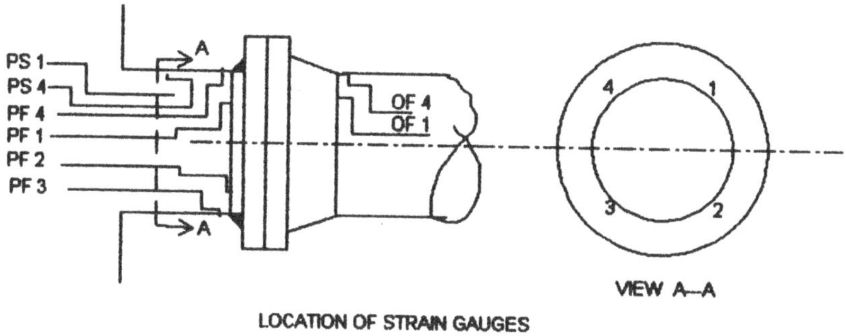

Figure 6: PF1, PF2, PF3 & PF4 are directly adjacent to the fillet weld. PS1 & PS4 are on the transition piece, 80 mm away from the weld.

At the side of the mating flange which was an integral type flange, strain gauges were also attached, but only at the top (OF1 and OF4). Attaching the gauges at the bottom was difficult due to close quarters and soil conditions.

The measured strains were converted to stress in order to compare the results with the standard and FEM methods.

The result of the standard calculations, as well as finite element method and strain gauges are shown in Table 2. It shows that with any method used, the flange used for this valve was unsuitable. The stress analysis using strain gauges, not only confirmed this, but showed that even at pressures as low as 3,900 kPa, the stress at the toe of the fillet weld was too high. The bolting-up had caused the material at the toe to stretch. This is consistent with the small cracks found in the 306SS flange. The smooth contour at the weld toe of this flange had deformed beyond the elastic limit, but no cracks were found. The difference between the weld contour from the tested flange and the failed flange is illustrated in Figures 7 and 8 .

5 Fracture Mechanical Analysis

The analysis and especially the experimental and FEM determined that the stresses at the toe of the fillet weld were very high. Small cracks had been found at the toe of the sectioned suction and discharge valve flanges, as shown in Figures 9 and 10.

The material toughness, as has been shown, was very low. The stress intensity factor determined from the CTOD values varied from 50 MPa$\sqrt{m}$ to 70 MPa$\sqrt{m}$. The critical defect size was determined using the program from FractuREsearch[2]. The geometric factor, β, as determined by this program, was modified to reflect the stress concentration at the toe of the weld, as determined by the experimental stress analysis. The result is graphically shown in Figure 11. A small defect of 1 to 2 mm would be critical. Compared with the stresses found by any of the methods used, it is surprising that the weld connection did not fail earlier.

Table 2. Stresses at the Edge of the Fillet Weld (600# Flange)

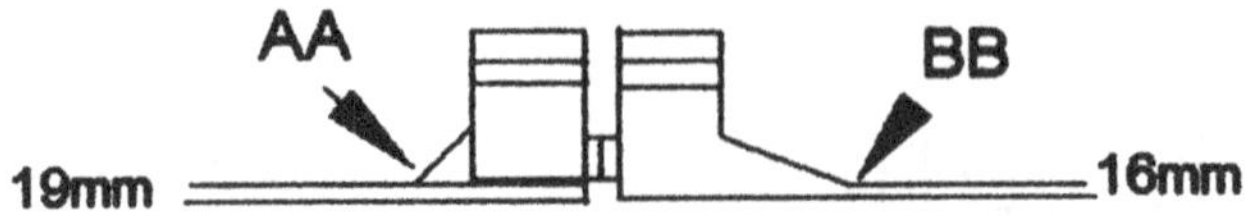

	ASME	ASME Calculated Stress			FEM
	Maximum Allowable Stress	Design Pressure 9.95 MPa	Operating Pressure 5.70 MPa	Test Pressure 3.9 MPa	Hydrostatic Pressure 5.70 MPa
Longitudinal Stress MPa	207	648	372	259	595

	Strain Gauges			
	Stress Release from 3.9 MPa to 0 MPa at AA	Stress Release Unbolting Flange at AA	Total Stress Release at AA	Total Stress at BB
Longitudinal Stress MPa	62	723	785	241

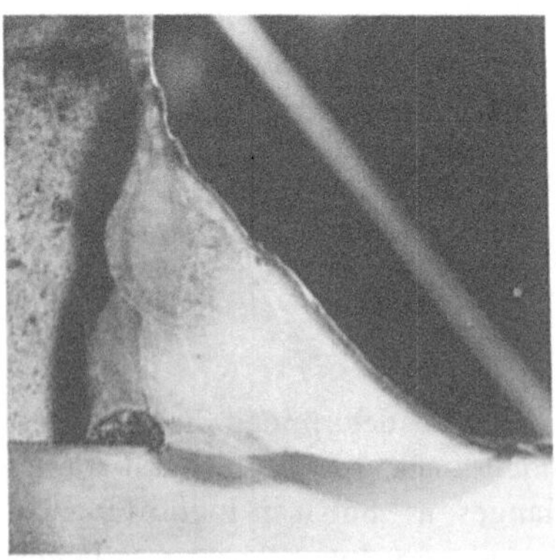

Figure 7: Cross-section of a fillet weld from the same flange of an identical valve, which had been removed after design stresses were known.

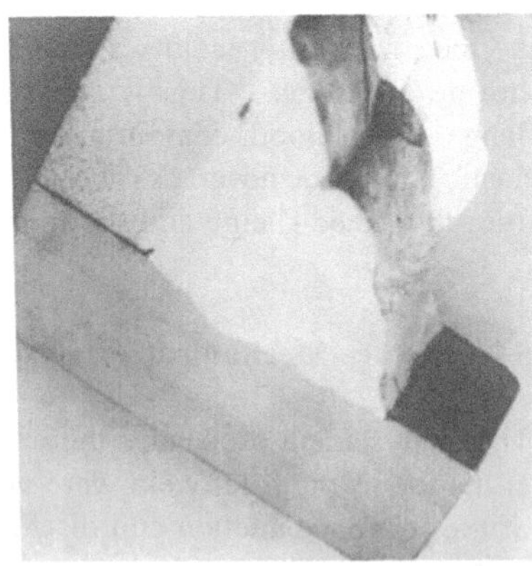

Figure 8: Toe of the fillet from the failed flange.

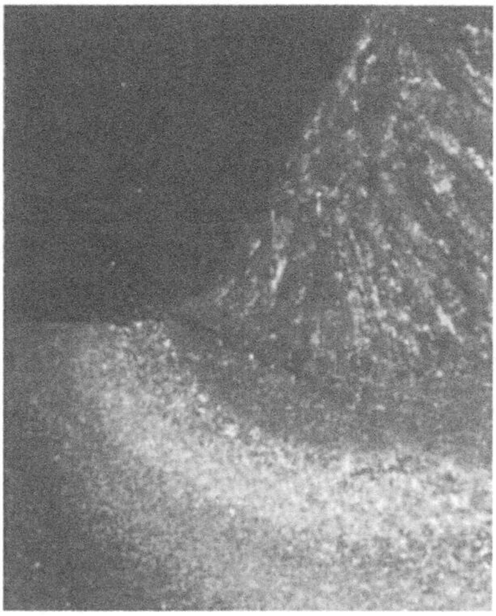

Figure 9: Fillet weld from the 306 suction valve flange.

Figure 10: Fillet weld from the 306 failed discharge valve flange.

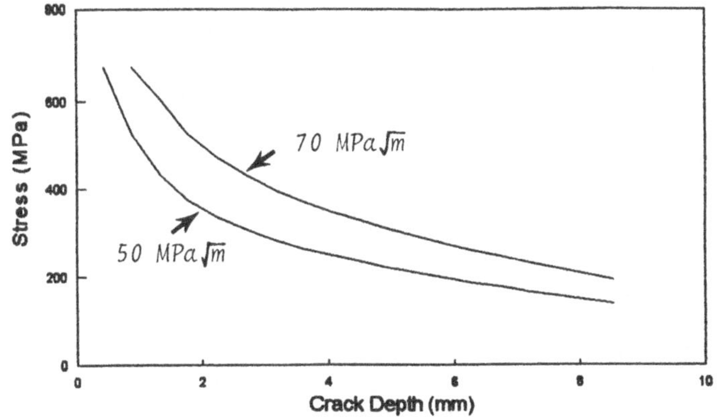

Figure 11: Critical crack size versus stress.

6 Stress Source

In order to fail, a force is required regardless whether a piece of equipment is designed properly or not.

When a pipe line is buried, it is subjected to the internal pressure. An additional stress such as an externally applied load, had to be present to cause the pipe to fail, or alternatively, a change in the material properties, material deterioration had to occur.

As had been established, the fracture mechanical properties of the material were very poor and the welding was of a low quality for pressure piping.

When the fracture mechanical properties and the inadequate design are considered, it was merely waiting until a condition arose which triggered the fracture.

At the time the failure occurred, construction was on-going at the site. It was considered that the transport of the vessel across the line during the days before the failure could have affected the stresses on the flange connection, but it was shown by analysis, that such a loading would have had a minimal effect.

Lowering of the temperature alone could have caused the flange connection to fail, or the on-going construction could have introduced an additional bending on this part of the pipe line. A gas leak was also considered, but because of all the bell holes, piping and ditching, it was considered not realistic that any amount of pressure could build up to lift the heavy valve. A comprehensive study was conducted by a geotechnical engineering firm on the soil conditions.

The average monthly temperatures in 1979 were:

October	+12.9°C
November	-2.25°C
December	-4.64°C
January	-13.39°C
February	-8.94°C

The pipeline was covered with about 1.5 metres of soil. The soil temperature will be a gradient through the depth of the frozen soil. At a depth of about 20 cm. below the frost line, the temperature will be 1-4°C. When the compressor is operating, the gas temperature at the supply side is also approximately 4°C. After compression, the gas temperature is 27°C. This temperature is measured at the compressor building. It is unknown how much drop in temperature would occur over a distance of about 100 metres, especially when it passes cold zones such as an 2.5 metre diameter bell hole and the recently dug ditch. The compressor had not been working for 84 hours. The mainline was at an estimated distance of 8 to 10 metres from the flange and flows at a temperature of 4°C. The valve with its' heavy steel body did extend through the frozen soil, reaching the surface of the ground. About 1 metre behind the valve was another bell hole of 90 cm. in diameter. With all these variables, it was unrealistic to establish an exact temperature of the pipeline at the location of the flange. The frozen soil would have cooled the body of the valve and thus, it must be expected that the temperature of the flange was close to 0°C after 84 hours.

It is well known that in northern climates such as Canada, frost heave may occur, provided that the soil conditions are right. These conditions were investigated by Nixon[3], who established that the soil did have the capacity to cause heaving by frost action. Nixon determined that frost heave action was cumulative. This means that whenever, at a given year, frost heave occurs, the object which experiences the action will not return to its' original position, but some of its' lift action is permanent. This phenomenon had been confirmed by other studies. The study by Nixon revealed that frost heave may have caused a lift of 40 mm. The frost depth in the direct vicinity of the valve is shown in Figure 12. This is the basis for reasoning that frost heave was the most probable source of additional stresses acting on the flange connection.

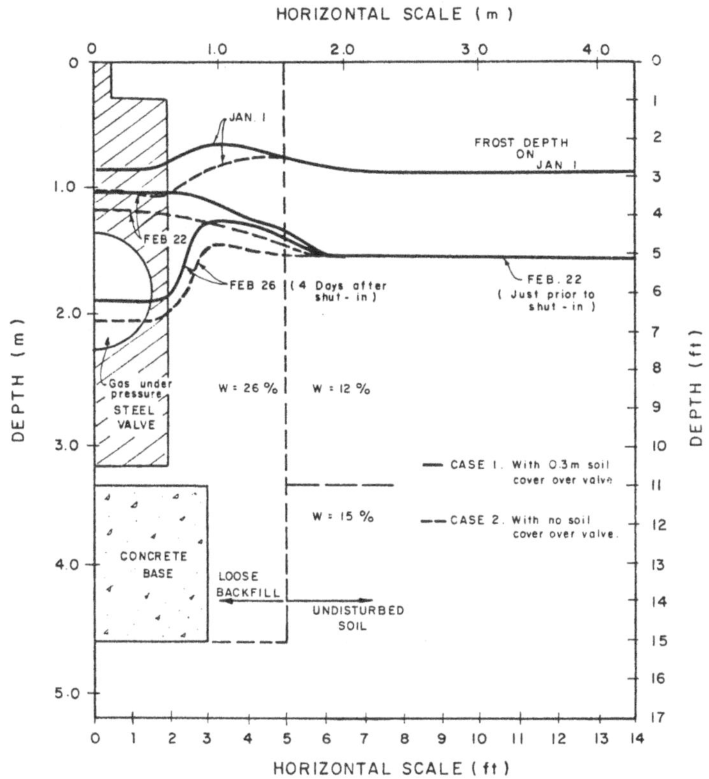

Figure 12: Frost depth profile determined by D. Nixon. This model suggests that the temperature was below 0°C.

7 Analysis of Fragments

Two (2) ruptures in the gas lines were observed. This is shown in Figure 13. One (1) of the failed lines was, however, easily recognized as a secondary rupture, resulting from the torch-like flame which came out of the valve and was aimed at the 862 mm gas line.

The rupture and explosion had strewn the fragments over a large area. The fragments were identified by letters, as shown in Figure 14. One fragment C was found in the crater which had been formed. This fragment had been exposed to high temperatures from the gas flame streaming out of the valve. Fragment M, which had been situated adjacent to the fragment found in the crater, was found at a distance of approximately 305 meters, and was in fact recovered only several months after the explosion by a farmer who had found it in his field. Fragments C and M are roughly of the same size and both were located at the bottom of the transition piece.

Figure 13: The valve rupture and the second ruptured 862 mm gas line.

In order to understand what happened, it is required to visualize the situation during the rupture at different stages. Naturally, it is not possible to picture each detail at every moment in time. But just as it is possible to determine sequential locations on a low ductility fracture surface, it is possible to reason what happened during a fracturing process.

The location of the fragments reveal important information about the manner in which the fragments had separated. Although the material, as evident from the properties, was of low ductility, it did not spontaneously fracture. Had the fracture occurred spontaneously, the fragments would have followed a path and direction which would have been severely limited by the proximity of the flange and valve. Fragment E is a good example.

Fragment E, for example, was located at the upper part of the transition piece, but it was propelled only a short distance away from the valve. Fragment M on the other hand, located at the lower half of the transition piece, was thrown more than 300 metres.

The initial velocity of fragment M could be calculated from the distance it travelled. An approximation of the required energy to travel the distance could be found, since the mass of this fragment was also known. It would not be possible for a large fragment located at the bottom to be propelled out of the ground, because the flange of the pipe line would be in the flight path. Fragment M thus, came out of the ground simultaneously with the flange. The pipe with flange bend upwards, as shown in Figure 13, by the gas pressure. When the pipe was bent and slammed into the 864 mm pipe line, the flange acted as a sling shot, catapulting Fragment M all across the compressor station into the farmers field.

The distance and the direction of all fragments were related to the way the transition piece came apart.

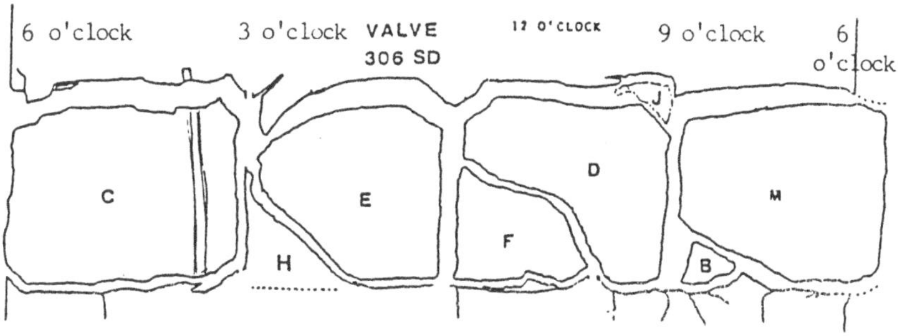

Figure 14: Projection is at the inside of the transition piece. The approximate clock positions are shown.

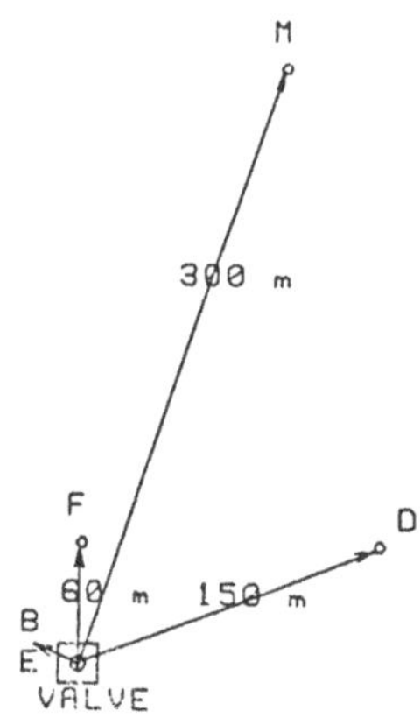

Figure 15: Approximate locations of fragments from 306SD.

8 Fracture Origin

All the fragments showed chevron markings on the fracture surface, except at the 12 o'clock position of the pipe (i.e.: the top). Although the transition piece was fragmented, there was no question that the fracture initiated at the flange and not at the valve, it was unequivocal that the fracture initiated at the toe of the fillet weld, where the transition piece was welded onto the flange. Figure 16 shows the refitted transition piece and Figure 17 shows the fracture at the 12 o'clock position. The rubbing or slide markings on the fracture surface at this location showed that there was initially, no dynamic action, but stable crack growth as shown in Figures 18 and 19, and this can only be when it is the origin, because the further fracturing is dynamic.

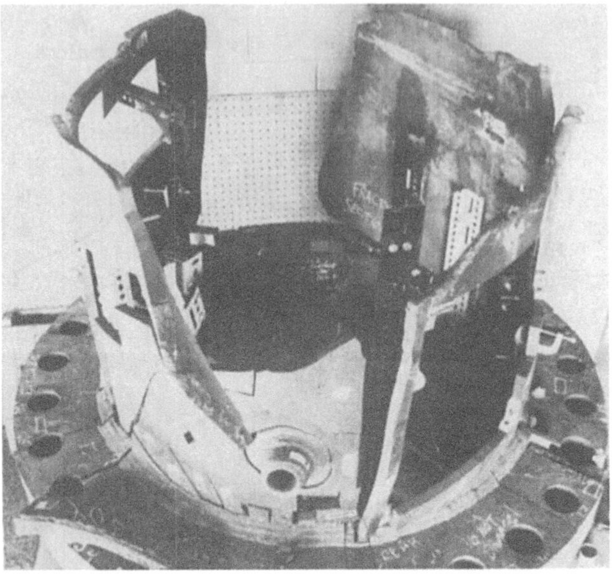

Figure 16: The fragments re-fitted after return from the court. Only the outline of some fragments remain after the interior had been cut out for testing.

Figure 17: Location C shows where the fragment hit the stud when it broke out. When fit together, the direction of break-out can be followed (see arrows).

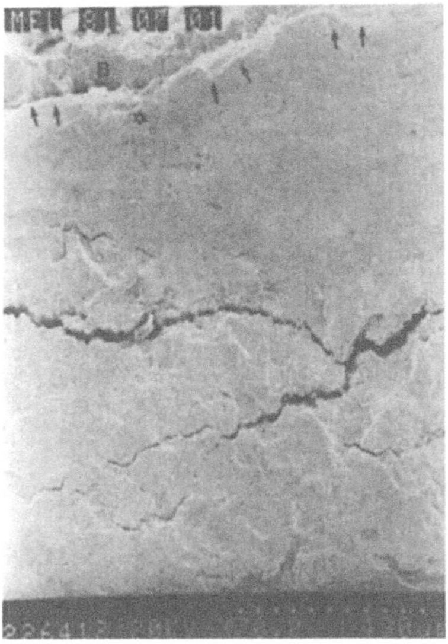

Figure 18: Edge of the crack is indicated by arrows B correspond to B in Figure 19. (Mag X 70)

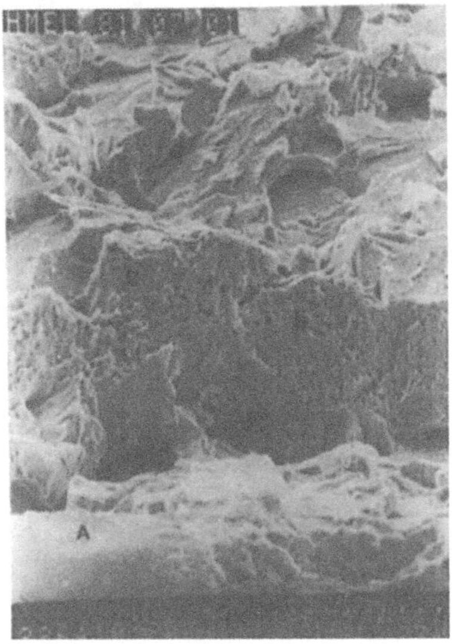

Figure 19: Note the dimple structure at B. "A" is the edge of the fracture surface. (Mag X 350)

9 Fracture Sequence

In the previous sections, it has been shown that a stable fracture initiated at approximately the 12 o'clock position, and the fragmentation sequence has been illustrated. There is one more highly significant aspect which completes the analysis of the transition piece fragmentation.

When the transition piece fractured, the fragments were propelled away, or as in the case of Fragment C, it remained at the same location in the crater. The fragments showed damaged. This damage could have occurred when a fragment had collided with a vessel or pump, etc., but not by falling in an open field. Thus, damage had to be incurred during the explosion and fragmentation. One example of the fragments colliding with the valve is shown in Figure 20.

Fragment E collided heavily with a rib of the flange. Figure 20 shows how this fragment hit the rib. The lower part was still attached to the flange. All fragments were analyzed in a similar fashion.

With this information, the fracturing initiation and sequence could be summarized. This is illustrated in Figure 21 by using the lay-out of the fragments described throughout this study. The initiation occurred at the location indicated by the numbers 000. The fracture then progressed along path 1 which removed fragment E. The gas caused a large back pressure, resulting in a rupture at approximately the 4:30 position shown as fracture path #2. Fragment C and M were separated at this instant. The bolt-up load caused a high stress in the 3 and 9 o'clock position, causing fracture #3. Fragment D fractured through path #4 which isolated fragment F. At that instant, there was complete separation between the valve and flange.

Figure 20: Fragment E colliding with the reinforcement rib.

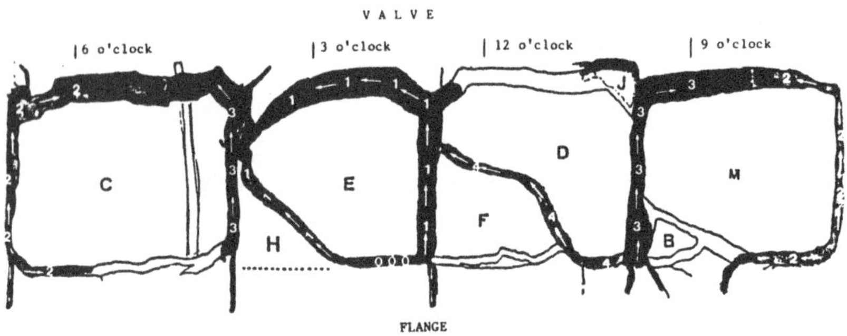

Figure 21: The sequence of fracturing is shown by number and colour, as developed from all the material and analyses.

Fragment C remained partly attached to the flange and D attached to the valve, they collided when the pipe moved upwards, causing deformation on the fragments consistent with the collision. Fragments F and M were catapulted away when the pipe bent upwards and collided with the 862 mm pipe.

10 The Court Ruling

The decision was in favour of the defense.

When the case was appealed (No. 19382 Alberta Canada), the judgement was vacated unanimously. One appeal judge stated on page 13 of the appeal:

"A judge, or a juror, must have reasons. It is manifestly an error in law to make a decision without reasoning, by which I mean a step-by-step movement through issues with appropriate analysis and decisions. A decision-maker in our system cannot decide a case by the toss of a coin. Judges, and jurors, take an oath to try issues fairly and this means to think through the case."

"I cannot comprehend why he decided that it was not practical for him to do so. He thereby not only deprived the parties of the appearance of justice, but also of the assurance that he had, in this complex case, adequate reasons. His approach demands a huge, and in my view, unreasonable, act of faith. He asks for more than we can give. Accordingly, the judgement must be vacated."

Another appeal judge remarked that a major reason for differences in testimony was that the witnesses disagreed whether any of the flaws exceeded the size of 0.5 mm in depth. If any had been much over that size, it could have been sufficient to cause the fracturing. This judge recommended a new trial.

The judge chose to make an assessment of the evidence which was before him. This would be a monumental task for anyone to undertake. It is logically a far more difficult task for a given person, who albeit learned, has no technical training. The judge made a grievous error when he stated on page 18 of the appeal:

"Nevertheless, I have decided to make a fresh assessment of the evidence in this case which questionably involves a choice among experts."

This judge states that he makes a choice among experts. This is, as an open admission, that he is unable to judge technical matters as can possibly be made.

On page 30, the appeal judge states:

"Cracks and fractures result from a combination of factors internal to the steel and from the external factor, of course, of building pressures. (Engineers do not speak, I have discovered, of building pressures. They speak of loading stresses. For fear it might be important, I adopt the terms.) I will discuss internal factors in detail in a moment, but first shall discuss stress. Stress may try to push the steel down or up, stretch it, or pull it from side to side, or any combination of these. Engineers calibrate and calculate stress and stress factors."

Clearly, this judge was not qualified to evaluate and render judgement on an engineering report of any depth.

Once a position is taken, it is naive to entertain the thought that the position is likely to change. This is illustrated in the testimony of a professor who acted as a expert witness for the defense[4].

He produced a theory whereby high pressure gas would accumulate underneath the valve and cause an explosion.

Some of his testimony is noted below (from page 5104 of the transcript):

Q. "Now sir, would you explain to the court your opinion as to how the failure of the 306SD valve occurred at Princess, Alberta in February of 1980."

A. *"I believe that there was a source of energy distributed near the bottom of the valve from the north side of the valve to the elbow on the south side of the sag bend. This is -- this would mean that distributed energy filled a zone roughly 5 to 10 feet wide, 25 to 30 feet long and that this energy was what caused the motion of the soil from around the valve and the 19 pipe to generate the crater that I see in the pictures that I have been shown. The source of this energy, I think, was probably high pressure gas. Secondly, I think, that when this soil was ejected from around the pipe and the valve to cause the symmetric crater I see on the north of the valve and south of the lower elbow of the sag bend that a large force was applied to the bottom of the valve and the pipe and that this in turn lifted the pipe and stretched and bent the transition piece. I think that the force of the gas in the pipe."*

Clearly, the professor did not present a shred of evidence. His beliefs and thoughts were strongly based on a picture, as shown in Figure 13, which has a somewhat circular appearance.

There were many dry wells in the direct vicinity of the valve which failed. A dry well is an access directly to the pipe for servicing purposes. As is evident in Figure 22, there were many dry holes varying from 3 to 8 feet.

Figure 22:

It is complete folly to assume that the gas would not escape through the large openings even if the freshly dug large ditch was ignored. Yet, the response was in court (page 5235, line 16.....):

Q. "And are you aware of how many gas wells there are - I'm sorry, dry wells there are in that pipeline trench that, for instance, runs in a north-southerly direction on the discharge line?"
A. "I'm really only aware of the gas well -- of the dry well that was labelled or was just north of the valve 306SD."
Q. "I would ask you to assume that there were other dry wells. Would that effect the ability of this gas to reach that point at 306SS and have that pressure of three to 400 pounds per square inch?"
A. "I believe it would have."
Q. "Might even eliminate the likelihood of it reaching SS?"
A. "I think that they would be likely sealed up on the bottom much like the valve was -- the dry well was on the north of 306SD."
Q. "All right. So you are assuming those dry wells would all be sealed up; is that right?"
A. "Obviously the one north of the valve 306SD was."

Occurrences as described, can be prevented when matters which belong to a particular profession are evaluated by those practicing the profession.

References

[1] 914 mm thick conduit gate, M & J, Model M303 ASA 600 valve.
[2] Fracture Mechanics & Damage Tolerance Analysis, FractuREsearch.
[3] Nixon, Field Sampling & Specialized Frost Heave Testing of Soils, Nova Princess Gas Compressor Station, near Princess, Alberta.
[4] Page 5104, transcript Nova versus Daniel, Court of Alberta.

26 EXPLOSION OF AN OIL TANK AT A THERMOELECTRIC POWER PLANT

O.F. ALY
Eletropaulo - Electricidade de Sao Paulo S.A., Sao Paulo, Brazil

Abstract

This paper describes how the explosion of an oil tank of 1800 m^3 capacity constructed in 1954 took place at a thermoelectric power level of 470 MW.

The explosion has caused serious damage but, fortunately, no casualties were registered. Three men working above on the service steel frame platform could have been seriously injured, however, the platform's steel frame protected them from the deformation of the tank.

The ambient temperature was approximately 26°C and vicinity was filled with 80% fuel oil. The purpose in discussing this accident is to provide the readers with some recommendations derived from the experience gained in this case.

Keywords: Explosion, fuel oil tank, thermoelectric power plant.

1 Introduction

In February 1991 - summertime in Brasil - an oil service tank of 1800 m^3 capacity, 12.2 m high and 14.6 m in diameter which was constructed in 1954 exploded at the Piratininga Power Plant. This power plant generating 470 MW consists of 4 units and is located in Sao Paulo in Brasil. The ambient temperature was about 26°C and the oil level indicator was showing the tank to be 80% full with fuel oil.

Due to the circumstances no personal injuries were inflicted to three workers as these men were working above a steel frame platform located above the tank.

Structural Failure, Product Liability and Technical Insurance, Proceedings of Fifth International Conference. Edited by H.P. Rossmanith. Published in 1996 by Taylor & Francis, 2 Park Square, Milton Park, Abingdon, Oxon, OX14 4RN.
ISBN 0 419 22080 1

In addition, the fire was rapidly brought under control by the Plant Fire Brigade which arrived at the site immediately after the call. The fire was extinguished minutes after the arrival of the fire fighters. It is worth mentioning that next to the site of the accident there is no other service oil tank or any other flammable source. However, the tank is located at about a distance of 50 m from the Water Treatment Building of the power plant and the plant's Restaurant Building.

Onlookers, bystanders and eyewitnesses that were gathering around the site at the time of the explosion reported that, when the explosion took place, a strong air displacement occurred concurrently with the burst into a blaze from the tank (both phenomena to be expected in this case) but this happened without any major consequences.

As the tank does not possess a heater the warmed-up oil is stored with the oil heated at the Transference Station as well as in the oil recirculation from the heaters of the pumping station of the unit (within a distance of about 100 m from the tank). This pumping station receives oil from the tank which passes in sequence through filters, low pressure pumps and heaters to reach the low pressure head (data: 10^6 N/m^2; temperature between 110°C and 121°C). After having passed this sequence of processes high pressure transference pumps push the oil to the burners with the excess oil being returned to the service oil tank.

2 Tank Description

The tank has a cylindrically shaped main body with a conical section and its stationary top casing is inclined at 6.25%. The tank is steel plated with bottom plates of 6.25mm thickness. The cylindrical section up to a height of 2.4 m is 8 mm thick whereas the top plates are 4.8 mm thick. The central column welded plate is of 4.8 mm thickness and measures 457.2 mm in diameter.

The level measurement must be performed through the floating system with a steam tracer. The tank possesses two stairs of which one is helically wounded whereas the other one is a vertical service staircase. On the day of the accident the fuel oil weighed a total of 1612 tons at a temperature of 79°C. The flammability point is 108°C and the flash point was 82°C when closed and 98°C when open.

3 Accident Description

On February 28, 1991, at about 8:30 a.m. three workmen were occupied with repairing a leakage through the level set steam tracer. The ambient temperature on such a summer day was approximately 26°C.

When the welder approached the welding rod, an arc spark ignited followed immediately by an explosion. The top and side casing junction of the tank were blasted open at several places as shown in Figure 1 and set the oil on fire.

The fire fighters employed three water jets to cool the tank whereas the fire was extinguished with a chemical foam spray system. As a result of the explosion boiler

No.1 as well as the fuel oil pumps fell to the ground surface.

On the same day, at 10:45 a.m., the tank was then set up again for provisional operation by means of the oil recirculation inlet. The boiler was started-up again at 2:45 p.m.

Figure 1 General view of the tank after the explosion. On the right hand side on can see a water jet from the fire brigade directed at the tank in order to cool the casing. Note the deformation of the top casing and the inclination of the steel platform. The photograph was taken from the restaurant building at a distance of 50 m.

4 Diagnosis of the Accident

Considering the high potential risk of the accident and the several causes and reasons that led to the accident, the power plant administration appointed an audit crew consisting of three (3) engineers of which one of them - the author of this paper - works at the administration office and the other two members of the audit crew belong to the operation and maintenance offices, respectively. This group of experts was in charge of carrying out the investigation and of preparing recommendations.

The crew, mentioned above, at present is also in charge of inquiring, interrogating and cross-examining several persons involved in the accident, as well as workers of the service staff and maintenance supervisors. Moreover, they are incumbent upon writing reports to be handed in by each one of the persons involved in the accident. The main factor which contributed to the accident were identified by the audit crew as:

- Two leaks were detected on the day before the accident occurred. The Maintenance and Supervision of Equipment Sector submitted a service order to the Operations Sector requesting that the equipment be repaired the following morning.

- The supervisor in charge of the area mentioned above requested the welding super-visor to organize a repair crew of three men: a welder, a mechanic and an assistant.

- The mechanic reported that he had been at the site on a previous occasion to find out what the service requirements were.

- Repair had not been made in accordance with the plant's safety regulations. Also, nothing had been written on the service order form regarding the service method (welding, etc.) to be applied in order to eliminate the existing leaks.

Finally, the audit crew came to the following conclusions:

- The most probable cause of the accident is the arc-opening of the welding rod as well as the spark which set to fire the gas exhaust pipe through the clearance exis-ting between the steam trace inlet pipe and the top casing plate of the tank. This is schematically illustrated in Figure 2.

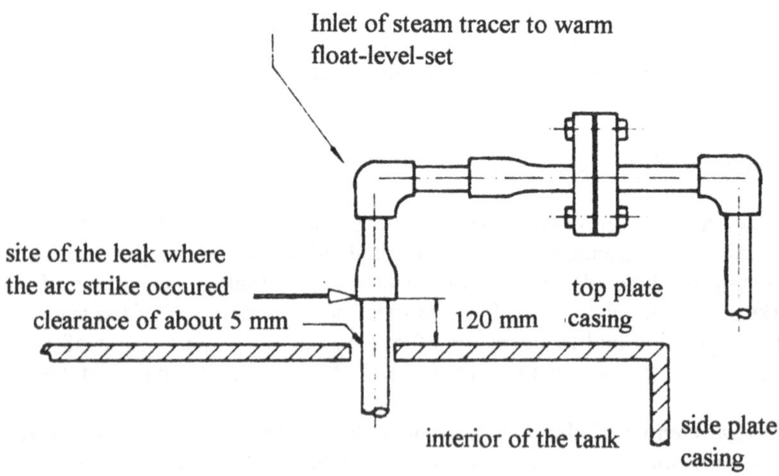

Figure 2 Schematic illustration of the leakage of the tank

- Obviously, the risk had been underestimated by all supervisors. Moreover, there had been also - on the part of the supervisors - excessive confidence regarding their own experience as well as a lack of information concerning the series of risks involved in this specific job.

- The majority of persons working on the project was unaware of the safety regulations required to be followed in the performance of their work. It should be kept in mind that all employees having newly arrived at the plant should be handed out and possess a copy of the safety regulations which contains general recom-mendations for each specific kind of service.

- Up until then safety regulations regarding this kind of service were totally unknown.

- In addition, safety procedure instructions for welding operations of this type of service were not available.

- The repair crew consisted of trainees having very limited and inadequate training and knowledge about safety regulations in this field.

- The welder had not had adequate supervision by the two maintenance supervisors involved, namely, the Field Equipment and Welding Supervisors.

- The service order did not contain nor provide proper information for the operation crew with regard to the type of repair (in this case: welding) to be performed.

- Similar cases of risk had occurred at other welding maintenance sectors at that particular time.

- When the service was being performed (at the time of the occurrence of the accident) there was no supervisor at the work area.

5 Recommendations

The audit crew proposed the following recommendations at the time the final report about the accident was being drafted:

- Reconstruction, on an emergency basis, of the tank's upper casing in order to prevent the risk of additional sparks (thunderbolts, etc.) and the emanation of flammable gas around the field including a strong smell invading the restaurant and operations area, etc. Expenditures amounted to approximat-ely US$ 250,000.--.

- Provisional emergency piping from tank No.2 was made to feed units Nos.1 and 2. The following recommendations may be useful to other companies with similar equipment and facilities.

- A safety training scheme should be drawn up especially for the crew working at the power plant including the technical, operational and maintenance staff. This scheme or plan should be kept active on a full-time basis. Moreover, continuing educational courses should be provided to the workforce active on such a plan. Certificates should be awarded to the personnel who have taken courses on safety regulations.

- A complete list of safety procedures for welding or other hot jobs whose service is potentially hazardous as they involve materials such as fuel oil should be provided. We have learned through our own experience how essential it is for us to develop our own procedures notwithstanding the existence of different procedures in other power plants.

- Defining and assigning well defined responsibilities to the personnel in charge of the task as well as to the engineering and technical staff.

- Moreover, meetings should be held periodically with regard to new and more difficult tasks as well as to whom such tasks should be assigned to.Tasks should be thoroughly discussed and guidelines be established for the procedures to be adopted for each kind of service.

- The firing system of tanks and other hazardous systems should be periodically tested.

- The plant's fire brigade should be adequately maintained and trained.

- Finally, as a general recommendation, we propose to tank builders a system which should be designed in order to minimize risks and to provide the proper procedures for the operation and maintenance of this kind of equipment (e.g. another type of set level measurement, etc.).

6 References

This paper is based on the "Final Report on the Accident which Occurred on February 28, 1991 - Explosion of Oil Tank No. 1 at Piratininga Power Plant" by O.F.Aly, P.A.Jiurgiu and H.A.Nogueira, ELETROPAULO, 1991.

27 STATISTICS OF TRANSMISSION PIPELINE FRACTURES

V.V. POLYAKOV and V.V. KHARIONOVSKY
VNIIGAS Scientific Technical Center, Moscow, Russia

Abstract
A statistics of fractures of transmission pipelines is presented where the fracture of a large diameter pipeline is regarded as a catastrophe whereas fracture accidents involving smaller diameter pipes are considered to be less dangerous. It is found that hydrogen causes brittleness in the outer layers of the pipes. Therefore, critical crack lengths for pipes have been calculated by means of linear fracture mechanics techniques. It was found that a crack of any length may become critical. The contrary opinion claiming reliable operation of large diameter pipes (larger than 1420 mm in diameter) will be discussed.

1 Introduction

This contribution reports about statistics of transmission pipeline fractures which occurred during the period 1980-1994. Fractures of pipes with diameter of 1420 mm have been thoroughly studied. Hydrogen field distributions have been measured by our scientific group using vacuum extraction techniques and anodic dissolution. A laser method has been used by the scientific group in Sverdlovsk.

Large diameter pipes made of low-alloyed steels are more sensitive to hydrogen embrittlement than those made of carbon steels. The reasons for high hydrogen concentrations in some surface layers of the fractured pipes [1,2] will be discussed. The authors checked hydrogen fields after the fractures of 1420 mm-pipelines in Kurgań (South of Sverdlovsk; 1990) and in the Krasnaya Tura district (North-east of Russia; 1993). Primarily, hydrogen is trapped under the surface layers. The high cathodic potentials result in hydrogenation and in practice it is difficult or even impossible to control hydrogen embrittlement because in Russia and in the other republics of CIS there are several hundred thousand kilometers of buried pipelines.

Structural Failure, Product Liability and Technical Insurance, Proceedings of Fifth International Conference. Edited by H.P. Rossmanith. Published in 1996 by Taylor & Francis, 2 Park Square, Milton Park, Abingdon, Oxon, OX14 4RN. ISBN 0 419 22080 1

Even low hydrogen concentrations caused fracture of pipelines in North America [3]. The total length of pipe ruptured in West Alberta was approximately 420 m where the main line had been in service since 1962 and was constructed of 36" (914 mm) diameter and 0.406" (10,31 mm) thick API 5LX Grade X-52 pipe material. A hot tap was installed in 1981 by NOVA procedure. After brittle fracture on January 8, 1992, gas was ignited and continued to burn for approximately 12 hours. The fracture surface associated with the origin of the rupture revealed vertical Chevron markings which originated from the pre-existing defect located on the outside surface.

2 Statistics of Fractures

In America the no-fracture life time (t) of buried transmission pipelines with a diameter of 1067 mm is estimated with 12-17 years [4]. In the former USSR the situation is practically the same for pipes with diameter of 1000 mm.

The 1993- statistics for transmission pipeline fractures in Russia contains the following information where the life-time is given in years and the rupture length in meters:

Pipes of 1420 mm diameter: 8,7,7,11,7,7 years and 142,150,64,33,125 meters
Pipes of 1220 mm diameter: 18,4,15,17,17,17,17,19,16 and 50,25,47,70,45, 52,130,90,40 meters
Pipes of 1020 mm diameter: 21,28,19 years and 37,-,48 meters
Pipes of 820 and 720 mm diameter: 14,11,11,28,18,10 years and 61,-,weld, 1, 23,13 meters
Pipes of 530 mm and less diameter: 10,1,13,13,21,13,22,9 years and 14, 1,16,500, 500,450,0.5,24 meters.

Some data in the official statistics of fractures of the former USSR and the data concerned are presented in Table 1 and in Figure 1.

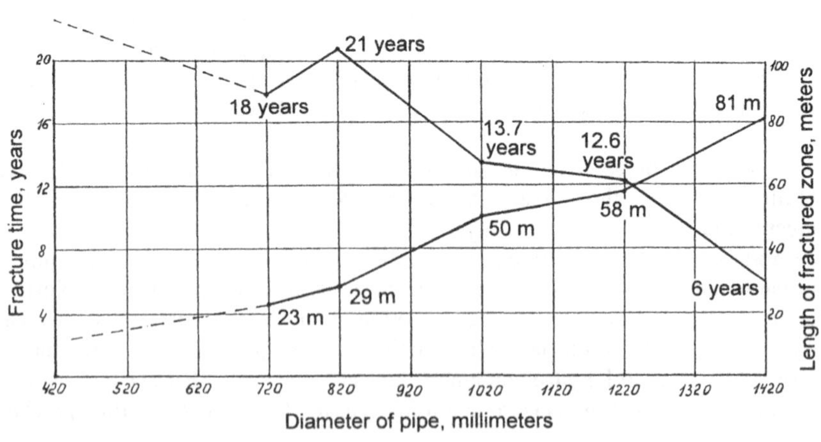

Figure 1 Statistics of life time and rupture lengths of fracture accidents of transmission pipelines in Russia (1993)

It is necessary to note that not all cases of fracture were fixed: cracks and fracture zones during preliminary hydrostatic tests, cracks from cathodic protection [5] and small cracks. 50% of the fracture accidents for 1420 mm diameter pipes occurred during the first three years of operation. The fracture times of all transmission pipelines is unknown. The rupture length of approximately 50% of all 1420mm-pipes fractured ranged between 30 and 80 meters. However, some cracks (fractured pipe zones) were 100,200,400, 500 or even 600 meters in length. The average fracture time increased from 6 to 12.6 and to 13.7 years when the pipe diameter decreased from 1420 mm to 1220 mm and to 1020 mm, respectively. The mean lengths of zones fractured decreased from 81 to 57.8 and to 50.4 meters, respectively. For 13 years Soviet officials have fixed only 70 large fractures for 1420 mm diameter pipes (Table 1), about 200 large fractures for 1020 mm diameter pipes and 140-150 fractures for 1220 mm diameter pipes [1,2,5]. The effects of fractures in large and small diameter pipes differ considerably.

Table 1: Data on some fractures of transmission pipelines

Diam, mm	Mean value of t, years (variation of data)	Mean value of L, m (variation)	Total L, (number.)
1420	0.8 (0.1-2)	63 (8-124)	317(6)
	2.8 (0.5-9)	106(2-342)	849(8)
	6 (1-11)	78 (34-132)	392(6)
	5.3 (1-9)	104(57-183	313(3)
	6.8 (0.5-13)	41 (35-47)	100(3)
	8.3 (0.5-14)	37 (1-66)	150(3)
	7.1 (0.5-15)	22 (1-86)	110(6)
	4.5 (2-9)	45 (11-70)	181(4)
	12 (5-22)	67 (35-92)	300(5)
	5.9 (0.5-12)	170(1-600)	171(7)
	7 (4-9)	155(33-400)	618(4)
The mean value for t..is 6 years for l is 81 meters.			

Why is fracture of a large diameter pipeline considered to be a catastrophe ? Several reasons will be worked out in the following:

Scale effect: The increase of the diameter of a pipeline results in a decrease of fracture time and in an increase of the length of the fractured zone. In fact, the main cracks are very large as can be seen from Figure 1.

Loss of gas: The loss of gas after catastrophic failure of a pipe of 1420 mm diameter amounts to 10 million cubic meters, for a fractured pipe of 1020 mm diameter the loss amounts to 1 million cubic meters. During the period 1980-1992 the catastrophic failures of large diameter pipelines lead to a loss of gas in the order of 4-10 million cubic meters. Official data for pipeline accidents for 1993 in Russia only the loss was about 100 million cubic meters. The use of pipes with diameters of 720 mm or 820 mm can help reduce the loss by two orders of magnitude.

Ecology: The ecology of the Earth atmosphere suffers greatly after a pipeline rupture catastrophe, because gases of light fractions with density less than that of the air go up to the atmosphere and gases of heavy fractions settle on the Earth's surface.

Transportation safety: Large diameter pipelines do not provide a safe long distance transportation of large quantities of gas. The occurrence of a fracture at a single location within the system may stop gas transfer completely. The reliability of a single large diameter pipeline system is lower than that of several parallel pipes of smaller diameters.

Economic loss: The economic losses due to fracture of large diameter pipes amounts to millions of dollars every year.

Casualties: Explosions of gas pipelines cause numerous deaths each year. Therefore, the operation of such projects near cities and towns is dangerous.

Geomechanics: The geomechanical interaction between the buried pipe and the surrounding soil has to be taken into account in the analysis and design of the pipeline.

Dynamics: The dynamic character of a gas pipeline explosion must be recognized and be properly accounted for in the analysis.

Operation and Maintenance: The operational staff must be attentive of possible problems in connection with possible failures of large diameter transmission pipelines. The management of gas industry demands and request that designer, manufacturer and scientists must be in a position to guarantee and assure reliable operation of gas pipelines.

Hydrogen concentration: The high level concentration (10...20...30...50 ppm) of hydrogen in the outer layers of the pipe is a serious factor and, for a large diameter pipe system, it may become catastrophic.

Gas explosions: In soils gas explosions produce craters and holes with sizes of 10-100 meters in diameter and 5-10 meters in depth.

Fires: Gas pipeline explosions do cause forest fires and conflagrations of buildings nearby.

3 Case Study: Pipe Fracture Accident (5 April 1994)

The Kakva river transmission pipeline was built during the period April 1982 - June 1983. In June 1983 the pipeline was tested with respect to strength according to Soviet standards (SNIP III-42-80) at a water pressure of 8,25 MPa and with natural gas at a pressure of 7.5 MPa in order to check the seals. The operating pressure during service was maintained at 7.5 MPa. No repair was undertaken in the zone where the accident was to occur.

The scheme of the pipeline between the compression station at 1248 km and the valves near the Kakva river at 1273 km (the zone where the fracture accident occurred) is presented in Figure 2. The ambient temperature just before the accident occurred was 4°C. The pipe capacity was 90 million cubic meters of gas per day. The design characteristics were such that the pipe was operated very close to burst conditions for the stress. The pipe material was X70 steel and the pipeline was insulated with NIITO-53 and NIITO-56 materials.

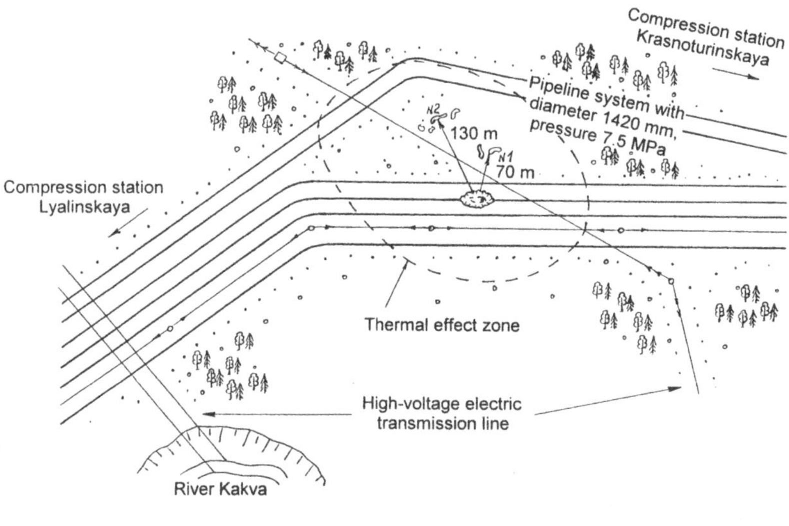

Figure 2 Scheme of the site of Krasnoturinskaya pipeline accident which occurred on April 5, 1994 at a position 1250.2 km from Urengoy-Uzhgorod

Upon explosive fracture of the gas pipeline the rupture zone of 22 meters in lengths was investigated by official experts. The crater created measured 32x13 m^2 and its depth was 5 m. A section of 47 m of the pipe was replaced. The thermal and heat affected zone resulting from burning gas and forest fire extended 250 meters to the North-East and 150 meters to the South measured from the site of the accident. The supports and insulators of the electric transmission line were damaged. The fractured section of the pipe was located in the crater generated. There is no data available about geodynamic effects (according to the design and accident expertise !). One of the Moscow based Institutes of Mining and Geology has investigated 1500 fracture accidents involving gas and oil pipelines and came to the conclusion that in more than 1100 cases the accidents occurred due to geological and geomechanical factors. The fracture accidents occurred due to cracks in the rock and soil layers (joints and faults). During the period April 1988 - April 1994 there occurred 8 fracture accidents in this region and at the same place (1251 km) did an accident occur on June 21, 1992 where a X70-steel pipe of 1420 mm diameter and 15,7 mm thickness failed. The length of the rupture near the very same location was 47 meters. The surfaces of the pipe fragments were free of corrosive products and there were many cracks on the surface of the pipe in the hydrogenated zones.

The authors [1,2,5,6] emphasize that hydrogenation of pipe steel is not a sufficient condition for stress corrosion cracking. There are technologies to obtain galvanic and chemical coatings where hydrogenation occurs. On the contrary, corrosion can take place without hydrogenation of the material. Cathodic reactions without hydrogenation of steel include:

Generation of hydrogen molecules at the surface of the pipe:

$$H + H \Rightarrow H_2$$

Cu-ion reduction:

$$Cu^{++} + e^- \Rightarrow Cu^+,$$

Reduction of oxygen:

$$O_2 + 4\,e^- + 2\,H_2O \Rightarrow 4\,OH^-$$

Fessler (USA), Gutmann-Gareev (Israel-Russia) located a siderite type deposit at the surfaces of the fractured transmission pipelines. However, this fact does not suggest corrosion interaction of the steel pipe with the soil because a $FeCO_3$ deposit is a dense material and possesses protective properties as a coating.

4 Fracture Mechanics Application

The critical crack lengths in transmission pipelines can be estimated by means of linear fracture mechanics methods. The relation between critical stress and crack length is given by:

$$\sigma = \sqrt{\frac{2\gamma E}{\pi L}} \tag{1}$$

where γ is the specific surface energy density, L is the length of the crack and E is Young's modulus. The quantity " γ " can be presented in terms of the stress intensity factor K.

Let us estimate the orders of magnitude of the fracture length:

$$L = \frac{K^2}{\pi\sigma^2} \tag{2}$$

and with the data: $E = 2\cdot10^5$ MPa, the fracture toughness value K_c for pipeline steels ranging between from 50 to 150 MPa$\sqrt{m}$, σ = 170 MPa. Hence, $L \approx 0{,}3$ m . The pipes are operated under cyclic or dynamic loading conditions and critical threshold range of the stress intensity factor in fatigue, ΔK_{th}, decreased to a value of 2-10 MPa$\sqrt{m}$. Hence, L may even become smaller than 0,3 m. This is also justified by noting that K may become less than the level of 50-150 MPa$\sqrt{m}$. This is considered to be an acceptable result. However, hydrogenation will decrease the critical levels of the fracture length to 0.1....0,01....0,001....0,0001 m.

5 Independent Gas Pipeline Systems in Arctic Regions

One of the basic features of Russian gas fields is that they are situated in the Northern Territories. The majority of gas pipelines is situated in the Unified Gas Supply System with a length of about 230 000 km of transmission pipelines or about 65% of the total length of transmission pipelines. In Western Siberia there are only more than a hundred gas and gas condensate fields with total discovered resources of over 36,800 billion cubic meters with gas of Sakha (Yakutia) Republic, Northern Sakhalin, fields of the Barents and Kara Seas and fields of the Far North added.

For remote industrial areas autonomous gas pipeline systems are of great importance as they are the basic source of energy and heat supply. The region of Norilsk, Yakutia, and the Far East may be mentioned as examples. As the climatic conditions of these regions are extreme and the geological conditions are characterized by permanently frozen soils, the construction and operation of gas fields and gas pipelines should meet special requirements of reliability. The experience gained over the time interval since 1969 of Trans-Arctics with respect to analysis, construction and operation of the gas pipeline system supplying with heat and gas the largest Russian metallurgical group and Norilsk City with a population of more than 240 000 people is of practical interest.

The Norilsk gas pipeline system consists of 3 gas condensate fields which are about 35-40 km apart and connected with 720 mm diameter field collectors. The fields were put into operation in 1969, 1974 and 1982.

In order to make gas transport more effective a pressing outcompressor station was constructed at the Yuzhno-Soleninsk field in 1985. The transmission pipelines constitute a 3 line system of 720 mm diameter pipes with 325 mm diameter condensate pipeline.

The climate of the region is the most severe with lowest air temperature at -57°C during winter and +58°C during the summer months. Winter lasts for 8-9 months and the wind velocity is in the range of 45 m/sec. Permanently frozen soil is highly subsident with ice lenses 0.3-0.5 m thick, with temperature around minus 4-7°C and an ice content of 0.3-0.5.

The gas pipelines cross many springs, some small rivers and three large rivers (including the Western Yenisei). Under these conditions a system of above-ground pipelines on piles with slightly bent compensation sections was adopted as the basic construction element. Buried gas pipelines cross flooded areas of large rivers and on the plains there are elevated sections of several km's in length.

A serious accident was registered in February of 1979. It was preceded by durable frosts at a temperature of minus 45-55°C and wind velocities of 8-15 m/sec.
During the interval December 1987 to February of 1979 these conditions caused the defects in a "1-st" pipeline which led to fracture initiation in one of the pipes. A "1-st" pipe is a pipe which was constructed without any experience in gas pipeline construction in the Arctic in 1969 and which contained a large number of defects found during an investigation.

In this particularly interesting fracture accident the fracture initiation zone was located on the 1-st pipeline near the junction of two pipelines. The fracture zone of the first pipeline was very long (some 27 km to both sides from the 35-th km mark) while the fracture zone on the second pipeline because of its shorter lifetime of 5 years was only some 8 km long. The compensation sections have been ruptured in the event. The pipeline section have been thrown off their supports. Some parts of the pipelines have been found scattered around the site of the accident. A similar accident took place in November 1989 in two collector gas pipelines connecting two fields. An abrupt fall of temperature was the cause of the accident. The pipeline of the fractured zone became flat. The total length of the fractured zone was 20 km.

During the entire period of operation of the Norilsk gas system there occurred 95 accidents among which 4 accidents are associated with long distance fractures [7].

Table 2: Potentials (V, saturated $Cu/CuSO_4$ electrode) of two cathodic stations

Year	1983	1984	1985	1986	1987	1988	1989	1990	1991
Station 1	-1	-1.2	-2	-1.3	-1.3	-1.25	-2.3	-1.5	-1.4
Station 2	-1	-1.2	-	-2.75	0.9	-	-1.15	-0.9	-1.2

Year	1992	1993	1994 before fracture	1994 after fract.
Station 1	-2	-2	-1.5	-1
Station 2	-2.3	-2.1	- 2.2	-1.6

6 Conclusions

A crack-like defect of any length for large diameter transmission pipelines located in an ambience of hydrogenation at the outer surface of the pipe may become critical. Hydrogen induces accidents with fractured zones from 10 to 1000 meters, pipe explosions and soil craters of sizes 10-100 m wide and 5-10 m deep. During one such accident the loss of gas into the environment and being potentially explosive is in the order of 10 million cubic meters.

In order to prevent such catastrophes one must give up operating large diameter pipelines. Another recommendation is to disband the operation of cathodic stations. High cathodic potentials (Table 2) are dangerous due to hydrogenation.

7 References

1. Polyakov, V.N.: A new conception of estimating fracture time and crack lengths in transmission pipelines. In: progress in the Understanding of Prevention of Corrosion, Institute of Metals, London, pp. 1603-1609, 1993.
2. Polyakov, V.N.: Catastrophes of large diameter pipelines. Problemy Prochnosti (Strength Problems), No 1-2, pp. 137-146, 1995.
3. Chiovelli, S.C. et al.: Main line failure resulted from combination of minor factors. Oil&Gas, No 21, pp. 91-100, 1994.
4. Urednicek, M., Lambert S., Vosikovsky, O.: SCC-Monitoring and control. In: Pipeline Reliability, Calgary, Canada, pp. V11-2-1 - V11-2-19, 1992.
5. Polyakov, V.N.: in Gasovaya Promyshlennost, No 12, pp. 21-23, 1993.
6. Polyakov, V.N.; Criteria and models of SCC and CF. In: Transaction of the 9th Int. Congress Eurocorrosion, Espoo, Finland, Vol.1, pp. 573-579, 1992.
7. Kharionovsky, V.V. and V.M.Alexandridi: Independent gas pipeline systems in Arctic regions. In: Transactions of Gas International Congress, Rome, Italy, IGU/C1-94, pp. 1-16, 1994.

7 [illegible]

Materials and Failure Mechanisms

Structural Failure, Product Liability and Technical Insurance, Proceedings of Fifth International Conference. Edited by H.P. Rossmanith. Published in 1996 by Taylor & Francis, 2 Park Square, Milton Park, Abingdon, Oxon, OX14 4RN. ISBN 0 419 22080 1

28 A FAILURE OF A MATERIAL OR OF ADVANCED TECHNICAL EDUCATION - IMPLICATIONS OF AN UNEXPECTED MATERIAL FAILURE

R. FIEDLER
Department of Microelectronics, Technical University, Brno, Czech Republic

Abstract
This paper describes a case of unexpected material failure in an electronic component. The failure demonstrates the limitations of even the best quality-control programme. If European electronics industries are to remain competetive, total quality management within the narrow environment of the company will not be enough. The concepts of total quality management must be extended to the global environment surrounding production plants. This paper focuses on the implications of global quality management in the advanced technical education system.
Keywords: diagnostics, education, electronic component, iron-nickel alloy, material failure, quality control.

1 Introduction

Failure problems in modern electronic systems are usualy minimized thanks to the total quality management in the electronic industry. A very important stage in introducing the quality programme is the Failure Mode and Effect Analysis (FMEA) [1]. It helps the producer to find all the reasons for a given failure and take essential precautions against it. Due to the world wide implementation of Total Quality Management (TQM) the failure of any product is very harmful for producers who compete for ISO 9000 quality certification [2] and/or for a place on the Qualified Manufacturers List.

If by chance a repeated failure of an electronic component appears the manufacturer may choose from a broad spectrum of methods for material diagnostics. However, if the material failure is due to an error in a human controlled technology we do not speak only about a failure but also about the effect of human factor.

Structural Failure, Product Liability and Technical Insurance, Proceedings of Fifth International Conference. Edited by H.P. Rossmanith. Published in 1996 by Taylor & Francis, 2 Park Square, Milton Park, Abingdon, Oxon, OX14 4RN. ISBN 0 419 22080 1

The paper will briefly describe a case of material failure in an electronic chip component fabricated from imported parts. The results of material analysis, however, can be hardly understood in terms of accidental material failure alone.

2 The analysed problem

In the electronic industry bonding is generally the most expensive phase in the manufacturing process [3]. The quality of bonds is usually tested mechanically. In some electronic chip components used for Surface Mounted Technology, band leads are used instead of conducting wires for terminations. Recently, we have analysed reasons for the low strength of soldered bonds and for the apparently spontaneous fracturing of chip terminations. We found that the low quality of bonds was caused by a low adhesion of Pb-Sn solder due to unusually high amounts of gas present in plated layers.

Material engineers know that a high amount of gas is released during electrochemical plating. The technology of electroplating is so developed that the amount of gas in a plated layer can be controlled and easily kept at minimum.

The problem of poor adhesion was easily solved and the subcontractor promised to maintain the quality of plated layers. A couple of months later, however, another customer observed spontaneusly fractured termination on some chips. Standard methods of analytical Scanning Electron Microscopy (SEM) did not reveal any changes in the structure either in the composition of samples taken from the plated band of material used for manufacturing terminations.

N.B. The effect of poor adhesion and mysterious fracturing appeared when the fabrication of the electronic component was transfered to an other country with cheaper labour market. As the former subcontractors known for their high quality of production were still contracted the first conclusion of the top management was: "The cheaper the labour market the lower the quality of fabrication."

2.1 The structure and composition of analysed chip termination

The analysed band leads used for chip terminations are made from an iron-nickel matrix, plated on both sides with layers of nickel, silver and Pb-Sn solder (Fig. 1).

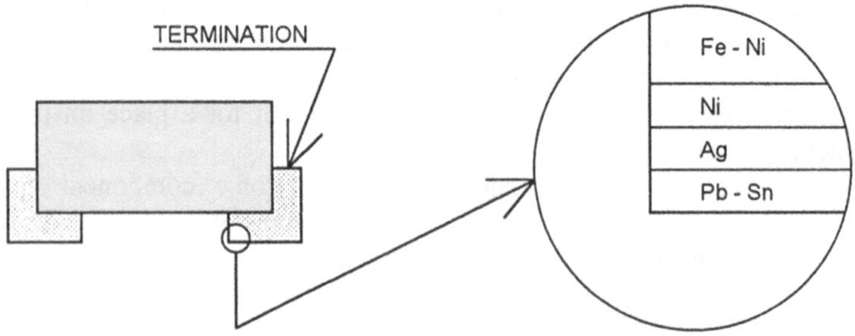

Fig. 1. A chip component and structure of termination schematic

2.2 Used experimental method and results

As the standard method of analytical SEM failed to find any natural reason for spontaneous fracturing it was necessary to take the unusual measure of preparing a sample for SEM observation. Trial and error was used to find a suitable etchant for etching the polished surface of a cross section through the sample. After proper chemical etching a very fine pitting of the surface was observed through SEM. The pits were 0,001 mm in size. The average distance between pits was 0,015 mm. X-ray spectral analysis showed high concentration of aluminium in pitted areas. The presence of aluminium, however, was not found in the sample prepared for SEM and X-ray spectral analysis by standard methods.

The only explanation found for this observation is based on the presumption that very fine alumina (Al_2O_3) particles must have been dispersed in the Fe-Ni matrix. The size of the alumina particles was estimated to be in order of 0.000 01 mm; just under the resolving power of a standard SEM. The presence of aluminium was later proved and documented through x-ray spectral analysis on the fractured surfaces. The fracture surface contained measurable concentrations of aluminium.

3 Discussion

Alumina particles are not ductile and their occurance in Fe-Ni matrix may have very harmful effect on the mechanical properties of terminations. During fabrication of the chip, the terminations are bend. The stress around non-ductile particle produces a microcavity and/or microcraks in the Fe-Ni matrix. This microcavity acts as a nucleus for later cracks and fracture.

Alumina particles might cause further damagewhen chip components are used in an electronic device. Due to the elastic strain around the alumina particles, the atoms of gas dissolved in the plated layers of the terminations would slowly migrate to the particle and form a void filled with gas. When the chip component is then heated up by Joule heat losses during operation, up-hill diffusion is enhanced, microcracks in a termination start to grow in size, and the electronic component after some time fails.

The alumina particles found in the Fe-Ni matrix of terminations are so very tiny that their presence cannot be detected by any standard method of material diagnostics. Their occurrance of this size and amount in Fe-Ni alloy is very unusual. They would have had to have been introduced to the material during casting and/or during rolling of the Fe-Ni plates. Having analysed the failure we have had to ask the old prosecutor's question: "Who benefited?"

3.1 The role and limitations of diagnostics methods

Physical and chemical methods of material diagnostics make it possible to determine both the structure and composition of a material. We have sufficient knowledge about relations between structure, chemical composition and properties of material to trace out reasons for material failure. We know how to manage structure and composition to produce a material with demanded properties.

Our diagnostics methods, however, are unable to analyse the human factor. We can only discover the probable motivations for erratic performance and/or malicious human behaviour using suitable psychological methods. At present, we can only speculate about the motivations for operator error. The results we get in this way are hardly unambiguous.

Our experience with "trouble shooting for industry" has shown that the main factors for the malicious behaviour are: insufficient respect for the value of finished work, insufficient appreciation of the limited avalability of raw materials, egoism with respect to nature, and very often the fetishization of money. These factors of human behaviour, however, can be altered by education.

4 Conclusion

The fact that the material failure was caused by the dispersion of very fine non-ductile particles, too small for detection by modern electronic analytical apparatus, together with the fact that the manufacturing of the component had been transfered to a cheaper labour market, prompts some speculation. This speculation may be based on knowledge of the character of Herr Schultze the chemistry professor known from Jules Verne's [4] "Stahlstadt".

Those of us concerned with shaping the future competetiveness of European industry should pay more attention to the human factor. This attention must extend beyond TQM at the business level, to include global quality management training [5] from the most basic levels of education. We should find, popularize and audit methods of education for schools to get new generation living in better harmony with nature and having strong sense of cooperation and tolerance. We should find a method of teaching the new generation to think and act globaly and from a global perspective.

5 References

1. Franke, W.D. (1989) *Fehlermoeglichkeits und einflussanalyse*, Moderne Industria, Landsberg/Lech.
2. Czech Standards Institution (Urad pro normalizaci a mereni). (1987) *Quality Management and Quality Assurance Standards. Guidelines for selection and use.* UNM, Praha. CSN ISO 9000.
3. Trigwell, S. (1993) Failure Mechanisms of Wire and Die Bonding. *Solid State Technology*, Vol.36, No. 5, pp.45-49.
4. Verne, J. (1900) *Les Cinq Cents Millions da la Bégum*, J. Hetzel , Paris. (In Czech: Ocelove mesto, Albatros, Praha 1989.)
5. Willborn, W., Cheng T.C.E. (1994) *Global Management of Quality Assurance Systems*, McGraw-Hill Inc, New York.

29 APPLICATIONS OF AUTOMATIC LEARNING OF FAILURE CASES TO OPTIMAL RELIABILITY

J. ZARKA, P. NAVIDI and M. TERRIEN
Laboratoire de Mecanique des Solides, Ecole Polytechnique,
Palaiseau, France

Abstract

We have proposed the principles of a new approach in the intelligent optimal design of materials and structures based on automatic learning techniques and special optimisation techniques. In this approach, we use softwares such as **L.E.S from LMS de l'X or NEUROSHELL from WARD Systems**, which automatically generate the rules and gives their degree of confidence; but any other software could be used instead. The input data of these softwares implies that an examples base has to be built at each time and the creation of this examples base requires a lot of expertise and time.

Here, we shall show how it is possible to prepare its particular application to various problems faced during the analysis of the reliability of a structure.

Keywords: Automatic learning, Optimisation, Structural analysis, Fatigue, Reliability.

1 Learning Expert Systems

We recall briefly the principles of any **AUTOMATIC *L*EARNING *E*XPERT *S*YSTEMS Generator** i.e. where the rules base is automatically extracted from the raw examples base given by the expert. They are important in technical domains where experts know they do not know the full **solution**. But these experts must be able to build an examples base, for which the solution is known experimentally or numerically sometimes with some fuzzy or missing informations.

The main problem is to provide a good description of such an examples base.

Basically such a learning system includes five main functions :

PREPARE : to transform the example files from user format (ASCII, dBase, Excel, ...) into the own format of the system and to handle the discretisation of the descriptors and the splitting of the initial data base into a training set and a test set.

Structural Failure, Product Liability and Technical Insurance, Proceedings of Fifth International Conference. Edited by H.P. Rossmanith. Published in 1996 by Taylor & Francis, 2 Park Square, Milton Park, Abingdon, Oxon, OX14 4RN. ISBN 0 419 22080 1

LEARN : to automatically extract a rules base from the training set according to the quality of available information (noise, sparseness of the training set,..).
TEST : to experimentally evaluate the quality of the extracted rule on the test set.
INCLEAR : to allow the expert to visualize IN CLEAR the extracted rules with the initial user format and to say what descriptors are kept.
CONCLUDE : from the description of a new case, to deliver a conclusion based on the extracted rules.
In all problems, it is necessary to consider one conclusion which may be a class or any continuous real number. Moreover, often, several conclusions may be considered together. The rules have to be automatically generated for each one of them.
Then a sixth function, **OPTIMISE**, based on genetic algorithms and other special optimisation techniques, may be used to solve the inverse problem i.e. when some conclusions and some descriptors have to belong to some given sets (or constraints), what are the possible solutions and in some particular cases what is the best solution if an objective function is given (cost, weight..) ?
This is the base of our new method for optimal intelligent formulations and processes of materials, the direct coupling with CAD systems in view for example of reducing or getting ride off the numerical simulations but by satisfying the given requirements for the structures (Optimal design, Control of Processes, Manufacturing...). The same methodology may be applied into other domains (biology, management, medecine, bank or insurance ...).

2 General Methodology

We are interested by problems when there is no solution and when the experts do not understand the problem in its whole. Moreover, the available data may be not statistically representative (i.e. are in limited number), fuzzy, qualitative and missing in part.
The approach that we proposed may help to improve the knowledge of the experts themselves by bringing them the rules explicitly and giving them the degree of confidence of these rules.
But chiefly, the solution of any new problem may be obtained and moreover an optimal solution may be reached too.
We need to follow the succesive steps :

2.1 Building the data base

The experts have underlined all variables or descriptors which may take a part starting from primitive descriptors (x_i) in a limited number to associated intelligent descriptors (XX_J) where there are putting all their actual knowledge. They have built a data base where they have prepared various examples with given conclusions.
The file that they have prepared may have the following format :
INPUT DESCRIPTORS :
Primitive description
x1 | x2 | . . . x5 |

Intelligent description

XX1 | XX2 | XX3 | . . . | XX29 | XX30 | . . . |

OUTPUT DESCRIPTORS or CONCLUSIONS :

C_1 | C_2 | C_3 | . C_I . .

Let us assume that these informations have been obtained for ≃ 50 examples of the same kind and that there are ≃ 20 to 500 intelligent input descriptors and ≃ 5 to 20 conclusions.

2.2 Generating the rules with LES in its numerical option

The data base, with its intelligent description, is presented to LES which :

says that only the three variables (intelligent descriptors) XX2, XX12 and XX29 are active to explain the actual value of the conclusion C_I and

generates the mathematical expression

$$C_I = 1.479132e{+}00 + 8.468159e{-}02\ XX12\ XX29 - 1.525532e{+}01\ XX29^{-2} - 1.270556e{-}04\ XX2\ XX29^2 + 1.178885e{-}03\ XX12^2\ XX2 - 1.892207e{-}02\ XX12^2\ XX29 - 7.679606e{-}06\ XX29^4$$

indicates that the error (similar to fiability) of this expression is for example of the order of 3 %.

2.3 Optimisation

With this expression, we may apply the OPTIMISE option according two hypothesis :

- we assume that the intelligent input descriptors are independant and we find that for the values XX12 = 2.13 , XX2 = 0. , XX29 = 13.7

the optimal predicted value of C_I is equal to 2.29 .

- we take into account the real nature of the intelligent input descriptors which are not independant but are linked to a much smaller number of the primitive descriptors. The optimal solution is obtained for the primitive descriptors x_2=34.5 for which the values of the intelligent descriptors are XX12 = 2.03 , XX2 = 0.4 , XX29 = 11.7 and the optimal predicted value of C_I is equal to 2.09 .

If with the first hypothesis, the optimal solution is very different from that relative to the second case, it is worth while looking after the discovery of new mechanisms able to give the special values of these intelligent input descriptors (although if we do not know from the beginning how to reach them from a technological point of view). If the optimal solution stays within the same order, it is better to keep only the independant primitive descriptors which garantee the technological existance.

But we can also be

2.4 Generating rules with LES in its symbolical option

The same data base may be treated with the Symbolic version of LES.

The predicted rules are then :

Rules driving to case 1 (C_I within [1.400000 , 1.500000]) ; Number of rules = 5

Rules of weight 52

If XX29 within [4.900000 , 6.100000] then Conclusion 1

Rules of weight 24

If XX12 = 1.000000 , and XX2 within [15.000000 , 20.000000] then Conclusion 1

....

Rules driving to case 5 (C_I within [2.250000 , 2. 30000]) ;

Rules of weight 16

If XX12 = 2.000000 , and XX2 = 0.000000 , and XX29 within [9.400000 , 10.500000] then Conclusion 5

....

The list of variables by increasing order of importance is :

XX29 (coefficient 8.069444)

XX12 (coefficient 7.774194)

XX2 (coefficient 7.338983)

based on an initial partition of the data base into a learning base and a test base, LES gives the fiability of 92 % for the rules set.

The function OPTIMISE can be used for the moment with the symbolic version.

Then, the use of L.E.S allows :

i) to understand what are the descriptors which drive to each global property (this may help the experts in the building of GOOD PHYSICAL, MICRO-MACRO MODELS too),

ii) and what is the most important for the future, to optimise the formulation and the process of the material which has to satisfy some given requirements and at a limited cost for example.

3 Application to some Problems in Reliability of Structures

Actually, two approaches are classically used in order to evaluate the reliabity of a structure :

The first one is the deterministic approach where we need to know the exact representation of the materials (for non-linear behaviour and with damage, the initial state of the structures and all the loading paths on the structures but also where we need to control the errors during the numerical simulations and where we need to define laws of cumulation of damages !!!!. In order to compensate our (still) bad knowledge, "safety factors" are used in the design rules.

The second one is the probabilistic approach where we produce a final (usually) very small number where we do not try to explain what was the cause of the incident when building the failure probability. Here, the number can not be verified since even if the probability is low everything may happens.

But based on the elements recalled in paragraph 2, we have proposed a general new approach for "Intelligent Optimal Design of Materials and Structures" where our simplified analysis of inelastic structures associated with simple rules of damage (they will be sufficient relative to our poor actual degree of knowledge on the materials, on the loadings.. .) are coupled with results from returns of tests on real structures.

Now, we underline how it is possible to look at the particular application to the problem of structural reliability.

We are concerned by the reliability of a general structure which is subjected to a general complex cyclic static or dynamical loading.

It is well known that even if there is no collapse or ratchetting or excessive deformation, after a while, there will be shakedown and then failure of the structure by fatigue. If there is elastic shakedown, the number of cycles is high and we have long term fatigue ; if there is plastic shakedown, on the contrary, the number of cycles is small and we have low cycle fatigue.

To obtain a deterministic solution to this problem, there are two steps to overcross :

i) the first one is the knowledge of the stationnary state of the structure after shakedown i.e. the limiting fields of strains and stresses in the structure,

ii) the second one is the knowledge of the rule to cumulate damage untill we reach failure.

Some years ago, we proposed a very simple and practical framework to solve the first step practically.

But the second one, until now, is an open problem. We propose the following treatment.

3.1 Recall of the framework for the kinematic hardening material for cyclic static loadings

All details may be found in the book **A New Approach in Inelastic Analysis of Structures,** by J. ZARKA et al, CADLM Editor (1990).

We consider a structure represented by a finite volume V with its boundary ∂V. It is subjected to the loading path :

Surface forces --- $\underline{F_i}^d(t)$ on $\partial_{F_i} V$,

Body forces --- $\underline{\vec{X}}^d(t)$ in V ,

Displacement --- $\underline{U_j}^d(t)$ on $\partial_{U_j} V$,

Initial strain --- $\underline{\mathbf{E}}^I(t)$ in V .

i) We start by analyzing the structure assumed to be made of a linear, purely elastic material with any Routine **ELAS**

$$\mathbf{ELAS}(V, \partial_{F_i} V, \partial_{U_j} V, \underline{\vec{X}}^d, \underline{F_i}^d, \underline{U_j}^d, \underline{\mathbf{E}}^I, \underline{\mathbf{M}})$$

to determine the fields $\underline{\vec{U}}^{el}(t)$, $\underline{\mathbf{E}}^{el}(t)$ and $\underline{\Sigma}^{el}(t)$.

In the real structure, we have :

$\underline{\vec{U}}(t)$, $\underline{\mathbf{E}}(t)$ real displacement and strain fields

$\underline{\Sigma}(t)$ real stress field $\equiv \underline{\mathbf{L}}\,(\underline{\mathbf{E}}(t) - \underline{\mathbf{E}}^I(t) - \underline{\mathbf{E}}^p(t))$

that we may decompose into :

$\underline{\vec{U}}(t) \equiv \underline{\vec{U}}^{el}(t) + \underline{\vec{U}}^{ine}(t)$

$\underline{\mathbf{E}}(t) \equiv \underline{\mathbf{E}}^{el}(t) + \underline{\mathbf{E}}^{ine}(t)$

$\underline{\Sigma}(t) \equiv \underline{\Sigma}^{el}(t) + \underline{\mathbf{R}}(t)$ with

$\underline{\vec{U}}^{ine}(t)$, $\underline{\mathbf{E}}^{ine}(t) = \underline{\mathbf{M}}\,\underline{\mathbf{R}}(t) + \underline{\mathbf{E}}^p(t)$ K.A. with 0 on $\partial_{U_j} V$. and

$\underline{R}$ S.A. with $\underline{\vec{0}}$ in V and $\underline{0}$ on ∂_{F_i} V .

The tool **ELAS** and the associated arguments :

$$\textbf{ELAS}(V, \partial_{F_i} V, \partial_{U_j} V, \underline{\vec{0}}, \underline{0}, \underline{0}, \underline{E}^p, \underline{\mathbf{M}})$$

gives $\underline{\vec{U}}^{ine}(t)$, $\underline{E}^{ine}(t)$, and then $\underline{R}(t)$.

ii) Since here, we have a kinematic hardening material associated with the Misès criterium ,

$f(\Sigma,y) \equiv \frac{1}{2} (S - y)^T(S - y) - \sigma_0^2 \leq 0$ we can define the internal parameters as the components of the global plastic strain tensor $\mathbf{x} \equiv \boldsymbol{\alpha} \equiv \mathbf{E^p}$ and the transformed parameters as $\mathbf{y} = C\ \mathbf{E^p}$.

Locally, the flow rules are written :

$$\dot{\mathbf{E}}^{\mathbf{p}} \in \partial\Psi_{C(\mathbf{y})}(\mathbf{S}) \text{ , or } \dot{\mathbf{E}}^{\mathbf{p}} \in - \partial\Psi_{C(\mathbf{S})}(\mathbf{y}).$$

Globally, we can introduce the **transformed parameters for the structure** by setting :

$$\underline{\sigma} \equiv \underline{\mathbf{S}}^{el} - \{\underline{y} - \text{dev } \underline{R}\} \equiv \underline{\mathbf{S}}^{el} - \underline{Y} .$$

The equations for the structure are :

$\dot{\underline{E}}^p \in - \partial\psi_{C(\underline{\Sigma}^{el})}(\underline{Y})$ with $\underline{Y} \in C(\underline{\Sigma}^{el}) \equiv C_0 + \underline{\mathbf{S}}^{el}$, a convex set which moves with the known translation $\mathbf{S}^{el}$, and which is locally constructed for each mechanism.

iii) Conversely, if the field $\underline{Y}$ is known, we come back to the classical fields by using only the operator **ELAS** :

The inelastic strain field

$$\underline{E}^{ine}(t) = \underline{\mathbf{M}}\ \underline{R}(t) + \underline{E}^p(t) \text{ is equal to}$$

$$\underline{E}^{ine}(t) = \{\underline{\mathbf{M}} + \underline{C}^{-1}\text{dev }\}\underline{R}(t) + \underline{C}^{-1}\ \underline{Y}$$

$$\equiv \underline{\mathbf{M}'}\ \underline{R}(t) + \underline{C}^{-1}\ \underline{Y} ,$$

where $\underline{\mathbf{M}'}$ is the new elastic matrix field.

$$\textbf{ELAS}\ (V, \partial_{F_i} V, \partial_{U_j} V, \underline{\vec{0}}^d, \underline{0}_i^d, \underline{0}_j^d, \underline{C}^{-1}\ \underline{Y}, \underline{\mathbf{M}'})$$

gives $\underline{\vec{U}}^{ine}$, $\underline{E}^{ine}$, and then, $\underline{R} = \underline{L}' (\underline{E}^{ine} - \underline{C}^{-1}\ \underline{Y})$, and, $\underline{\mathbf{x}} \equiv \underline{C}^{-1} \{\underline{Y} + \text{dev } \underline{R}\}$.

For an elastic isotropic material characterized by its Y_G, the Young modulus and its ν, the Poisson ratio, the new elastic matrix is isotropic too and is obtained by substituting :

$$\frac{\nu}{Y_G} \quad ==> \quad \frac{\nu'}{Y'_G} \equiv \frac{\nu}{Y_G} + \frac{1}{3C}$$

$$\frac{1+\nu}{Y_G} \quad ==> \frac{1+\nu'}{Y'_G} \equiv \frac{1+\nu}{Y_G} + \frac{1}{C}.$$

iv) During monotonic radial loading, an ultimate state is reached and can be constructed **exactly**. We take

$\dot{\underline{\mathbf{Y}}} = \dot{\underline{\mathbf{S}}}^{el}$ and, we return to $\dot{\underline{\mathbf{E}}}^{p}$ using the operator **ELAS**; for points where $\dot{\underline{\mathbf{E}}}^{p}$ is different from zero, the flow rule gives the particular position of **Y** on the sphere centered in $\mathbf{S}^{el}$

v) During cyclic radial loading, bounds are constructed with elementary rules. Only elastic or plastic shakedown can occur. There is elastic shakedown when the intersection C_ℓ of the two convex sets $C(\mathbf{S}^{el}_{min})$ and $C(\mathbf{S}^{el}_{max})$ is non-void. It is sufficient to take the projection of the initial state on C_ℓ.

There is plastic shakedown when the intersection is void. We may easily obtain the mean state and its range, or, the lower and the upper bounds.

The same simple treatments were also defined fo contact loadings (such rolling, shot-peening...) or dynamical loadings (such as sismic excitation, explosion...).

3.2 Obtention of rules to predict the number of cycles for fatigue-failure

We want to predict the time life of any structure.

For sake of simplification, here, we shall assume that the structure is made only of an homogeneous material and that moreover the loading on the structure is only a radial "static" cyclic loading. (The real loading, the initial state would never be known for sure !!). Since we have been working for several years in the field of fatigue, we shall consider that we are, ourself, the experts for this problem. Then our first step is to describe the problems with our best knowledge.

3.2.1. Fundamental data on the materials

We assume that we have for the material in the structure the following fundamental data which have been obtained after tests on **smooth tensile specimen** :

a) cyclic curve at $\simeq \pm 1$ % of total strain (in traction-compression or alternating torsion) that we represent by two straight lines and from which we measure the fundamental descriptors : the elastic constants : Young modulus, Y_G, Poisson ratio, ν ; the plastic constants : workhardening modulus, C and radius σ_0 (4 fundamental constants)

b) Woehler Curves

these curves have been obtained for various mean stresses (alternated cycles i.e. $\sigma_m = 0$ or $\sigma_{max} = - \sigma_{min}$, repeated cycles i.e. $\sigma_{min} = 0$, and some, may be, at least two other $\sigma_m = \frac{\sigma_{max} + \sigma_{min}}{2}$; these curves have been obtained with some probability factors (for example 75 % and 95 %) from which we deduce various curves that we represent at each time by two straight lines with the fundamental couples (i.e. we are making a discretization of each curve) σ_m , probability factor : $(\Lambda\sigma_1 , 10^2)$, $(\Lambda\sigma_2 , 10^4)$, $(\Lambda\sigma_3 , 10^7)$ ($\simeq 2 \times 4 \times 3 = 24$ new fundamental constants)

and we built for example, a simple endurance criterium based on the Dang Van or the Kakuno-Kawada expression :

$$C_a + a P_m + c P_a \leq b$$

where C_a is the amplitude of the loading (in the deviatoric stress space **S** i.e. based on $\sqrt{J_2}$)

and P_a , P_m are respectively the amplitude and the mean value of the mean pressure (based on $I_1/3$)

we introduce ($\simeq 3 \times 3 =$) 9 new fundamental constants. It is obvious that these new constants are directly linked to the previous ones ; but we shall consider them as new descriptors.

Note : It may be possible to apply the automatic learning techniques to these first data set in order to see if the data obtained from fatigue tests or Woehler curves might be predicted only from the four first fundamental constants during cyclic curves.

3.2.2 Application to a general structure

A structure is generally loaded in such a way that the loading is not homogeneous and that of course during cyclic loading there is a redistribution of the stresses.

Then for the radial cyclic loading, we use our simplified method to evaluate the two extremities elastic stresses fields σ^{el}_{max} **and** σ^{el}_{min} .

Now, it is necessary to define some elementary terms which could characterize globally these two fields. Indeed, it is needed to move from a primitive passive description of the structure, the material and the radial loading parameters to an intelligent active description in order to hope that the experimental results on one structure may be used into any other new structure.

It is necessary to adapt intelligent descriptions of geometries of structures, the materials behaviours, loadings ... which allow to use the automatic learning techniques. Many appropriated descriptions were described in J.M. Hablot (1990).

It is obvious that the descriptions could not never be absolute and simplifications or compromises have to be made :

i) limited but sufficient number of descriptors

ii) reasonnable cost of their construction from the primitive descriptors

iii) independance from non significative informations (such as coordinates systems..).

A geometrical domain Ω in $\mathbf{R}^n$ will be characterized with its size parameters, position parameters and shape parameters particularly with the help of morphological mathematics, images treatments, group technology, ...

The characterisation of a scalar field ϕ in Ω or a vectorial/tensorial field $\mathbf{T}_{m,n}$ order m on a domain of dimension n

$$t_{i_1,i_2,...,i_m} = \int_\Omega x_{i_1} x_{i_2} \dots x_{i_m} \, d\Omega \text{ with } i_1, i_2, \dots, i_m = 1,\dots,n$$

is more complicated.

It is necessary to describe the change of ϕ in the domain by looking at its extrema, the way that they are distributed for example their repartition into layers by dividing the interval $[\phi_{min}, \phi_{max}]$.

The tensor $\mathbf{T}_{2,2}$ on a plane domain Ω with the 4 terms

$$t_{11} = \int_\Omega x^2 \, d\Omega \; ; \; t_{12} = t_{21} = \int_\Omega xy \, d\Omega \; ; \; t_{22} = \int_\Omega y^2 \, d\Omega$$

has the same invariants than the inertia tensor $I_1 = I_{XX} + I_{YY}$ and $I_2 = I_{XX} I_{YY} - I_{XY}^2$.

These invariants have to be reduced to adimensionnal quantities by dividing them by a reference quantity such as

- radius of giration $r_g{}^2 = \dfrac{I_{XX} + I_{YY}}{S} = \dfrac{1}{S}\displaystyle\int_{\Omega} (x^2 + y^2)\, d\Omega$

- or radius of reference $r_r = \sqrt{\dfrac{S}{\pi}}$ where $S = \displaystyle\int_{\Omega} d\Omega$.

A scalar field ϕ varying with time (real time or parametric time) may be characterized by its minimum, maximum, mean value, moments distribution, spectral analysis, autoregression factors...

The loading a structure is the most difficult object to describe. A global synthesis is reached by taking the elastic response or the simplified response of the structure wich have a very small cost relative to the real incremental inelastic response.

It is necessary to look after zones with high stress gradients and their evolution with time

An equivalent stress field, s^{eq} may be selected such as the Mises one based on the deviatoric stress tensor $\mathbf{S}$: $s^{eq} = \sqrt{1/2\ \mathbf{S} : \mathbf{S}}$

from wich we deduce the following fields :

i) $\phi_1 = \mathrm{Sup}_t \left[\dfrac{s_{eq}}{\mathrm{Sup}_{\Omega_0}\ s_{eq}} \right]$

with layers 0.25 ; 0.5 ; 0.75 et 1.

ii) $\phi_2 = \mathrm{Sup}_t\ [\ s_{eq}\ ]$

with layers σ_0, 2 σ_0, 3 σ_0, 5 σ_0, 10 σ_0, 25 σ_0.

iii) $\phi_3 = r_r\ \mathrm{Sup}_t \left[\left\| \mathrm{grad} \left(\dfrac{s_{eq}}{\mathrm{Sup}_{\Omega_0}\ s_{eq}} \right) \right\| \right]$

iv) $\phi_4 = r_r\ \mathrm{Sup}_t\ [\ \|\ \mathrm{grad}\ (s_{eq}\ )\ \|\ ]$

v) $\phi_5 = \dfrac{\sum_i \mathrm{equiv}\ [\sigma_{i+1} - \sigma_i]}{\mathrm{Sup}_i\ [\mathrm{equiv}\ (\sigma_i)]}$

or $$\phi_5 = \frac{\int_0^t \text{equiv}(d\sigma)}{\text{Sup}_{\tau \leq t}\,[\text{equiv}(\sigma(\tau))]}$$

3.2.3 Application to notch specimen

Let us consider the case of simple tensile cylindrical specimen with a variable radius of the notch which are subjected to a radial cyclic axial load. We used the report of the CETIM where several tests are described and for which their experts have indicated that the gradient of the stresses at the tip of the notch is the main factor.

For this problem, we have the primitive descriptors of the tests:

the radius and the length of the cylinder,

the variable radius of the notch

the mean value and the range of the applied axial load

the properties of the material (elastic, plastic and fatigue properties).

and one conclusion

the fatigue life of th specimen.

Once again, it is needed to reach an intelligent description for wchich we could differentiate geometry, loading, material.

The knowledge of the expert is completed with informations coming from numerical simulations.

The numerical treatments were done within CASTEM 2000 system, the finite element code from DEMT/CEA since it is easy to introduce our simplified method and to compute any quantity based on the results of the analysis (we thank Mr Poupart for his help in this part of the work during his studies at the IUP of Evry).

We deduced new clever descriptors from the primitive descriptors :

- a first family of descriptors will be linked to the elastic stress gradient field
- a second family of descriptors will be linked to the elastic stress field itself to characterize particularly the volume of the specimen where there is a plastic shakedown regime (i.e. periodic plastic strains) and the volumes where the elastic stresses will be higher than 50 %, 80 % and 100 %

• a third family will characterize the volumes where for example the fatigue criterium is violated at 50 %, 80 % and 100 %

• a fourth family will charaterize le assymptotic shakedowdn state of the specimen.

As indicated, all these quantities are normalized internally to each family.

Let us recall that such a learning drives to the results that may be used to any new structure. Of course, it may be useful to improve them as it will be done in a following paper.

4 References

1. A. Brand ; Calcul des pièces à la fatigue, Guide du CETIM;, 1980.
2. J.M. Hablot : Construction de solutions exactes en élastoplasticité. Applications à l'estimation d'erreur par apprentissage; thèse Ecole des Ponts et Chaussées; juin 1990
3. P. Navidi and J. Zarka : Clever Optimal Design of Materials and Structures, in Proceedings of Second French-Korean Conference on Numerical Analysis of Structures, Seoul, september 1993.
4. Y. Papadopoulos, Fatigue Polycyclique des Métaux, une Nouvelle Approche; Thèse Ecole des Ponts et Chaussées, Paris 1987
5. M. Schoenauer and M. Sebag : Incremental Learning of Rules and Meta-rules, Proceedings of 7th International Conference on Machine Learning, 1990
6. M. Terrien : Systèmes experts par apprentissage en contrôle non destructif. Textes des conférences Les systèmes experts et leurs applications, Avignon 1991.
7. J. Zarka, J Frelat, G. Inglebert and P. Kasmai-Navidi, A New Approach in Inelastic Analysis of Structures; CADLM Press 1990.

30 INTELLIGENT SYSTEM OF FAILURE ANALYSIS

J. JAZWINSKI, K. WAZYNSKA-FIOK and J. ZUREK
Institute of Technological Research, Warsaw, Poland

Abstract
The paper describes the process of improving maintenance procedures, one that constitutes the intelligent portion of the system of collecting and processing information on aircraft failures. Phases of the process of improving maintenance have been pointed out, e.g. evaluation of tendencies of changes, evaluation of the need for improvements and selection of the technique(s) to improve maintenance, evaluation of the effectiveness of improvements, etc.
Keywords: effectiveness, maintenance, reliability.

1 Introduction

In course of operating the technical equipment individual devices usually show some structural and manufacturing defects. The wearing and ageing processes occur, as well. Events of both types influence quality characteristics. Such being the situation results in the need for a system to control the quality characteristics of the equipment. Many and various quality characteristics of technical devices can be defined. Considerations presented below have been limited to five factors only. They are as follows: non-vulnerability, safety, effectiveness, reliability, suitability. The system of collecting and processing information on events that occur in the course of

Structural Failure, Product Liability and Technical Insurance, Proceedings of Fifth International Conference. Edited by H.P. Rossmanith. Published in 1996 by Taylor & Francis, 2 Park Square, Milton Park, Abingdon, Oxon, OX14 4RN. ISBN 0 419 22080 1

operating the equipment, bound up with the equipment, determines specific quality coefficients. A system of statistical inference on tendencies of changes that take place while operating the equipment, based on analyses of the above-mentioned coefficients, is the first intelligent item of the information system. In case unfavourable tendencies are observed, e.g. equipment safety deterioration, the second region of the intelligent system is entered, i.e. "analysis of advisability of preventing unfavourable tendencies". If the preventive steps are not advisable, possibilities of either operating the equipment with the quality coefficient reduced, or taking it out of service are to be examined. If there is a chance to upgrade the equipment, the next step is to pass over to another intelligent sub-assembly, i.e. to develop a number of means and techniques to improve a given item of equipment or operating procedure. With the improvements introduced the forth intelligent sub-unit of the information system is used to check effectiveness of the efforts made. The system of controlling the operational characteristics of a device has been illustrated in Fig. 1.

2 Evaluation of tendencies to change(s) in quality coefficients

The algorithm of tendencies to change(s) in quality coefficients includes the following techniques of evaluating them:

- comparing the current quality coefficients with the required ones;
- comparing the quality coefficients with those featuring two or more time intervals;
- comparing the operational quality coefficients under the pre-set conditions with those under standard conditions.

Various techniques of evaluation are often used with reference to various quality coefficients. In outline, the algorithm of inferring on tendencies to changes in quality coefficients comprises the following stages:

1. determination of the basis for comparisons to give an interval measure of quality coefficient $[\underline{W}_x, \overline{W}_x; \beta]$, where: $\underline{W}_x, \overline{W}_x$ - lower and upper intervals of confidence of the coefficient; β - level of confidence;
2. determination of the comparative coefficient W_x;
3. if $W_x \subset [\underline{W}_x, \overline{W}_x; \beta]$ - there are no tendencies to any changes;
4. if $W_x \subset [\underline{W}_x, \overline{W}_x; \beta]$ - there is a tendency (either an unfavourable or a favourable one) to a change.

The basic coefficient refers to the previous period of operation, standard conditions of operation, current operation (while compared with the requirements). The comparative coefficient shows the current mean value of the quality coefficient, the mean value of the coefficient determined for the pre-set conditions of operation.

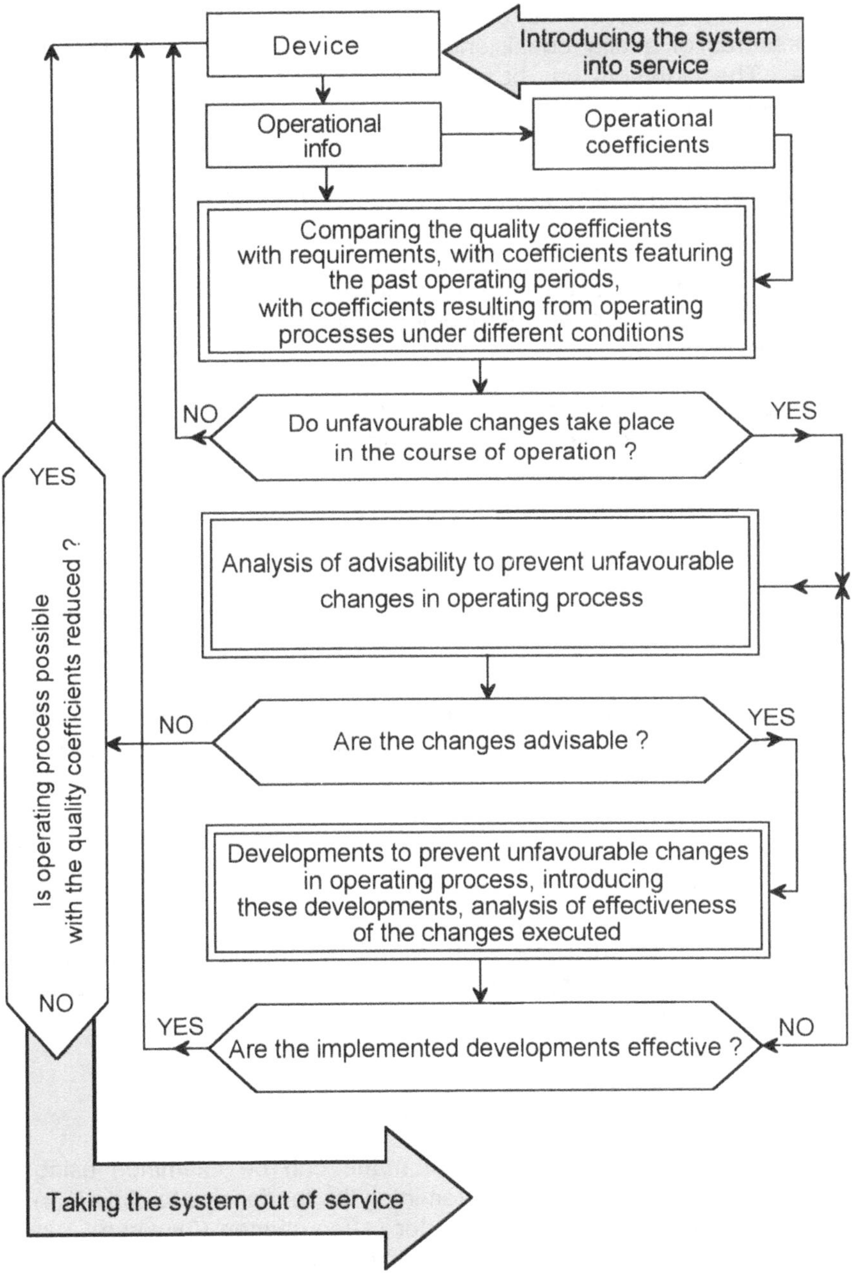

Fig. 1. Algorithm of controlling the operating process.

The algorithm includes a set of procedures to determine: the interval and point measures of quality coefficients of a device, and statistical inference methods. The algorithm has been included into the system of information collecting and processing.

3 Inferring on the need to improve (to select the technique of improving)

The following information is necessary to arrive at a decision on the need for improvement and to select an improvement technique from a set of many and various improvements possible:

W_x - unfavourable quality coefficient (frequency of events occurrence);
P_x - coefficient of effectiveness of the effort under analysis - $P_x \subset [0, 1]$;
S_x - losses due to deterioration in the quality coefficient x;
C_x - costs of having the tendencies in quality coefficient x grown better.

A decision is taken up on the basis of the following relation:

$$P_x \geq \frac{C_x}{W_x S_x}$$

The relation being satisfied means that the upgrading of the device under analysis or the process of operating it is advisable. The upgrading procedure should be selected such as to satisfy the following:

$$\max_i \left[P_{xi} - \frac{C_{xi}}{W_x S_x}\right]$$

The algorithm calls for additional information, i.e. knowing the coefficients C_{xi}, P_{xi} that describe the i-th improvement of the device. The coefficients are determined on the grounds of expert investigations.

4 Examination of the improvement effectiveness

The effectiveness of the improvement made can be examined using an algorithm parallel to that applied to examining the tendencies to change(s). In practice, the three-point measure: better (+)/no change (0)/worse (-) is not always sufficient. We often want to know how much better or worse the effect of the improvement is. That is why the measure of effectiveness of a given effort has been introduced from the interval [0,2]. Let us denote the measure of the effort effectiveness vs criterion X with Y_x (W_{x1}, W_{x2}), where W_{x1} - quality coefficient of the device prior to upgrading, W_{x2} - the respective coefficient after the upgrading. The presented measure makes the following sense: if $Y_x \subset [0 \div 1)$, the upgrading brings about negative effects; if $Y_x = 1$, the upgrading is to no effect; if $Y_x \subset (1 \div 2]$, the upgrading has proved effective. The coefficient Y_x higher, the level of effectiveness higher as well.

Performing the upgrading process according to the X_1 parameter can result in the change(s) of the parameters X_2, X_3, ... Improvement in the safety coefficient R_B can effect deterioration in the availability coefficient R_G. This is the reason to carry out the effectiveness evaluation by means of the complex coefficient.

5 References

1. JAŹWIŃSKI, J., BORGOŃ, J., (1989) *Niezawodność eksploatacyjna i bezpieczeństwo lotów,* Wydawnictwo Komunikacji i Łączności, Warszawa.
2. JAŹWIŃSKI, J., WAŻYŃSKA-FIOK, K., (1993) *BezpieczeNstwo systemów,* Państwowe Wydawnictwo Naukowe, Warszawa.

31 PRINCIPLES OF SYSTEM SYNTHESIS FROM THE POINT OF VIEW OF RELIABILITY AND SAFETY CRITERIA

J. BORGON, J. JAZWINSKI and K. WAZYNSKA-FIOK
Institute of Fundamental Research, Warsaw, Poland

Abstract
The paper gives considerations on the engineering system synthesis from the point of view of reliability and safety. The system reliability is shaped with various types of redundancy, e.g. structural, functional, parametric, informational, etc. redundancy. Increase in reliability reduces system vulnerability, which results from loosening correlation between system reliability and system safety. The safeguarding equipment to ward off the hazards that occur in the course of system operation is the primary factor to ensure the system safety.
Keywords: reliability, safety, system synthesis.

1 Introduction

Any engineering system, e.g. an aircraft, consists of some functional units. A functional unit performs a specific function within a system. At least five primary functional units can be distinguished. They are supposed to:

- perform basic functions of the system (a structural unit, a power plant);
- supply the system (with electric energy, fuel, gases, etc.);
- control the system (its attitude in space, landing, etc.);
- evaluate the system condition (speed indicators, fuel indicators, etc.);

Structural Failure, Product Liability and Technical Insurance, Proceedings of Fifth International Conference. Edited by H.P. Rossmanith. Published in 1996 by Taylor & Francis, 2 Park Square, Milton Park, Abingdon, Oxon, OX14 4RN. ISBN 0 419 22080 1

and to serve as

- safeguarding subsystems (fire protection, anti-icing subsystems, etc.).

Every unit and sub-assembly can be decomposed into components similar to those of a system. It is evident that individual subsets of functional units are not equivalent to each other in terms of reliability and safety. Different forms of redundancy predominate in various functional sets. In primary functional units structural, parametric and strength redundancies predominate, whereas in control ones - structural redundancy is uppermost. In the units intended to evaluate the system condition the informational and functional redundancies are predominant. Units to supply the system are featured with structural and parametric redundancies, whereas the safeguarding subsystems - with time redundancy. The pages to follow give analysis of only some of numerous aspects of system synthesis in terms of reliability and safety.

2 Applying correlation analysis to system synthesis

Let a functional unit be considered. Let the random variable X describe the functional unit's fitness for use X=1, and unfitness for use X=0. Let Y denote the random variable that features the functional unit's safety. Then, Y=1 - safety unreliability has not occurred, Y=0 - safety unreliability has occurred.

Let the following denotations be introduced:

$$P(X=1) = R \tag{1}$$
$$P(X=0) = Q = 1-R \tag{2}$$

The following relations between the random variables X and Y take place:

$$P(Y=1/X=1) = P_{1/1} \tag{3}$$
$$P(Y=0/X=1) = P_{0/1} \tag{4}$$
$$P(Y=1/X=0) = P_{1/0} \tag{5}$$
$$P(Y=0/X=0) = P_{0/0} \tag{6}$$

It can be shown [2] that:

$$P(Y=1) = R_B = RP_{1/1} + QP_{1/0} \tag{7}$$
$$P(Y=0) = Q_B = RP_{0/1} + QP_{0/0} \tag{8}$$

The R_B factor has been assumed to represent safety reliability, whereas the Q_B one - safety unreliability.

The factor of correlation between the random variables X and Y is given with the following formula:

$$\rho_{xy} = (1 - P_{0/1} - P_{1/0})\frac{RQ}{R_B Q_B} = \frac{RQ}{R_B Q_B}\frac{dR_B}{dR} \quad (9)$$

It can be shown that there is a relation between reliability R, safety R_B and the factor of correlation ρ_{xy}:

$$R_B = \frac{K}{1+K} \quad (10)$$

$$K = \left(\frac{R}{1-R}\right)^{\rho_{xy}} \quad (11)$$

Two approaches follow from (10):

- to increase unit reliability with high correlation between reliability and safety maintained, and
- to increase the number of safeguarding subsystems at the expense of reliability reduction.

3 Principles of synthesis of reliability and safety of sub-assemblies

Probability $P_{0/1}$ covers cases of safety unreliability due to reasons other than the unit failure. The reasons include cases of external factors influencing the unit, e.g. natural factors, climatic factors, influence of other units, etc. Synthetizing the system from the point of view of reliability and safety one usually strives for reducing the influence of reliability upon safety by means of reducing $P_{0/0}$. Reduction of $P_{0/0}$ induces increase in $P_{1/0}$.

The $P_{1/0}$ increase means occurrence of great redundancy in the unit in the aspect of safety. The term 'redundancy' is to be understood as both the programmable and unprogrammable redundancies. Structural redundancy is a good example of the programmable redundancy that consists in providing the unit with standby components to switch in and become operative in case the primary component(s) fail(s). In case of the programmable redundancy the term 'failure' means failure of the system that consists of the basic component, standby components and switch-over members. The structural redundancy increases the system redundancy. The unprogrammable redundancy, on the other hand, gathers all other kinds of redundancy, ones that result from the functional structure of the unit. The unprogrammable redundancy is usually determined by the probability $P_{1/0}$.

Let some examples of the unprogrammable redundancy be considered:

1. Functional redundancy - a subset of the unit components performs separate, self-assigned functiones; in case any of these components fails,

the others take over to perform this function within the predetermined range;

2. Parametric redundancy - exemplified by a powerplant (some aero-engines on one aircraft). With one engine out mission can be accomplished, less effectively, however;
3. Time redundancy - principles of operation of mechanisms originate some definite time relations (time redundancy) that enable an operator to control the unit in a stress-free way;
4. Informational redundancy - lots of various items of information on a specific subject are fed in course of operating the unit. The situation gives then specific 'infomational' comfort.

The above-mentioned and unmentioned kinds of redundancy cause that any failure of the unit member(s) is not followed by safety unreliability, i.e. probability $P_{1/0}$ increases.

The safeguarding systems, the so called 'safety systems', provide a specific form of redundancy. They serve the aim of warding off the hazards.

A simplified process of synthesizing the unit in terms of reliability and safety has been shown in Fig. 1. The algorithm clearly shows how complex the problem of the unit synthesis is.

5 References

1. BORGOŃ, J., JAŹWIŃSKI, J., SIKORSKI, M., WAŻYŃSKA-FIOK, K. (1992) *Niezawodność statków powietrznych*, Ośrodek Badania Jakości Wyrobów "ZETOM", Warszawa.
2. JAŹWIŃSKI, J., BORGOŃ, J., (1989) *Niezawodność eksploatacyjna i bezpieczeństwo lotów*, Wydawnictwo Komunikacji i Łączności, Warszawa.
3. JAŹWIŃSKI, J., WAŻYŃSKA-FIOK, K., (1993) *Bezpieczeństwo systemów*, Państwowe Wydawnictwo Naukowe, Warszawa.
4. WAŻYŃSKA-FIOK, K., JAŹWIŃSKI, J., (1990) *Niezawodność systemów technicznych*, Państwowe Wydawnictwo Naukowe, Warszawa.

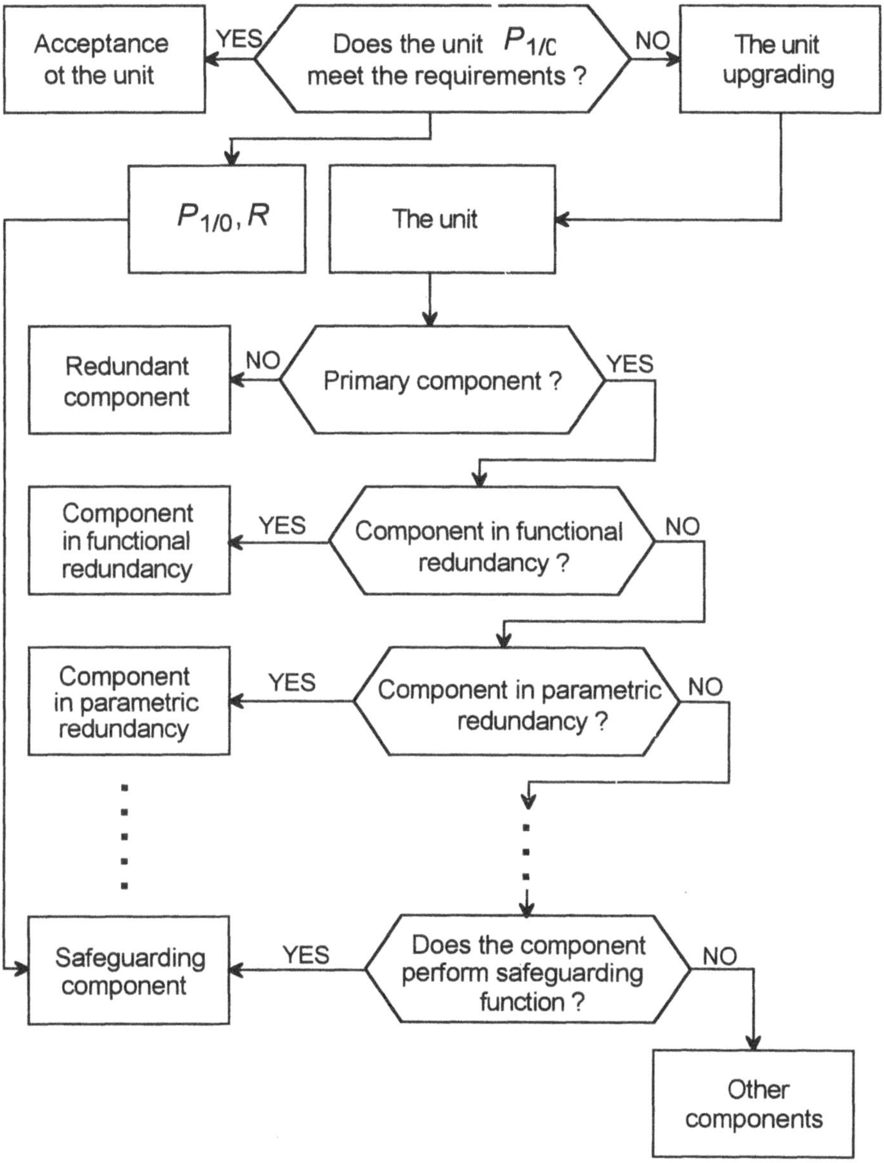

Fig. 1. Algorithm of the unit synthesis from the angle of reliability and safety.

32 ON PRESTRESSED CONCRETE ELEMENTS - REFLECTION ON CODE REQUIREMENTS AND PROPOSED SUPPLEMENTS

A. BUSSIBA and Y. KATZ
Nuclear Research Centre - Negev, Beer-Sheva, Israel
M. SZANTO
Rotem Industries Ltd and Ben-Gurion University, Beer-Sheva, Israel

Abstract
The current study is centered on prestressed concrete representing a high order composite material. In prestressed concrete elements, beside the constitutive properties of the dominant components, non-primary stresses are operating. In the case of interface debonding, crack initiation and growth or strands relaxation, the substructure boundary conditions may be violated due the redistribution of residual stresses.

Two levels of activities are demonstrated. First, in developing physical insight into the material behavior with emphasis on steel wires and strands. Second, activities along the corrosion-deformation interactive effects with emphasis on metal/hydrogen interactions. This was analyzed by continuum theory beside proposed models and simulations as related to environmental effects on the discrete lattice.

Finally, critical questions are raised supported by experimental findings and computational tools within continuum theory and fracture mechanics framework. These along testing methodologies vis-a-vis standards and codes while providing more of integrated views which might lead to the evaluation potential growth.
Keywords: Concrete, prestressing, mechanical properties of wires, stress relaxation, hydrogen/metal interactions, numerical models.

1 Introduction

Intentionally, the problem of a microstructural disordered material was selected which introduce some typical reliability aspects. Concrete as such, is a complex, brittle composite, while the superposition of residual stress field complicate even further the exploration of marginal assessments along the service life. The global fracture behavior of concrete is non-linear and the effect of the microstructural disorder on the mechanical behavior become progressively less important at larger scale. However, by following a localized approach to fracture (mainly on the meso-scale) additional aspects emerged.

Structural Failure, Product Liability and Technical Insurance, Proceedings of Fifth International Conference. Edited by H.P. Rossmanith. Published in 1996 by Taylor & Francis, 2 Park Square, Milton Park, Abingdon, Oxon, OX14 4RN. ISBN 0 419 22080 1

The linkage between the global response and the occurrence of local events on the microscopic or macroscopic scales require more of a lucite synthesis which becomes inevitable.

The main objective of the present study remains in the high strength steel wires and strands which are dominant components, facilitating the prestress process. In this context, aspects were developed as related to anelasticity, anisotrpic effects, time dependency, fracture resistance and environmental effects.

The present investigation was designed in establishing some data base from prestressed strands/wires tests following the existing codes and standards. This experimental activity was further stretched to simulation tests following thermo/mechanical treatments (all by induction heating under specified stress or strain, temperature and time), fracture mechanics of wires and environmental interaction. These mechanical findings were supplemented by of TEM and SEM observations, X-ray diffraction techniques and acoustic emission tracking (AE). These, stimulated incentives on experimental/theoretical interfaces in corrosion/deformation effects which were analyzed by continuum theory beside proposed models and simulations, concerning the environmental effects of a discrete lattice. As known, nearly all technologically important corroding materials in aqueous solutions evolve hydrogen. In fact, this motivated the study of external hydrogen and the eutectoid steel wires interactions.

By realizing the major features involved in prestressed concrete elements as related to the non-primary stresses, a numerical model was attempted to demonstrate the factors which are involved in marginal assessments against failure. A substantial amount of the aforementioned features are already reflected in the accepted standards. Even so, some arguments emerged particularly with respect to fatigue and stress corrosion tests as currently proposed in the standards literature. Here, the consistent approach still remains along the reliability aspects emerging from material properties and mechanical factors. Beside others, the key point in reliability growth seems to be in developing integrated views which are well explored, in order to refine the state of the art, without masking the inherent limitations.

2 Experimental and computational procedures

2.1 Experimental

As received, eutectoid high strength steel wires for prestressed concrete were selected mainly in the diameter of 4.2 mm to construct a 1x6 strands with appropriate thermo-mechanical treatment (TMT). The latter intends to match with necessary requirements of low stress relaxation values. The chemical composition consisted of (in wt.%): C-0.76, Mn-0.53, Si-0.16, Cu-0.08, Cr-0.08, Mo-0.001, O-0.01, N-0.007 and Fe-bal. Beside tests according to available standards (ASTM 416/87, BS 5896 or ISO 6934/91) the study was centered on simulation tests as illustrated in Fig. 1(a). The elaboration beyond the standard requirements allowed to deepen the general background in terms of experimental findings and trends. Along this framework standard mechanical tests in non prestressed (NPS) and prestressed (PS) wires were performed while considering the dominant TMT variables. In the context of comprehensive study, fracture and fatigue resistance was investigated utilizing fracture mechanics parameters. Briefly, additional activities were included along the time dependent processes. First,

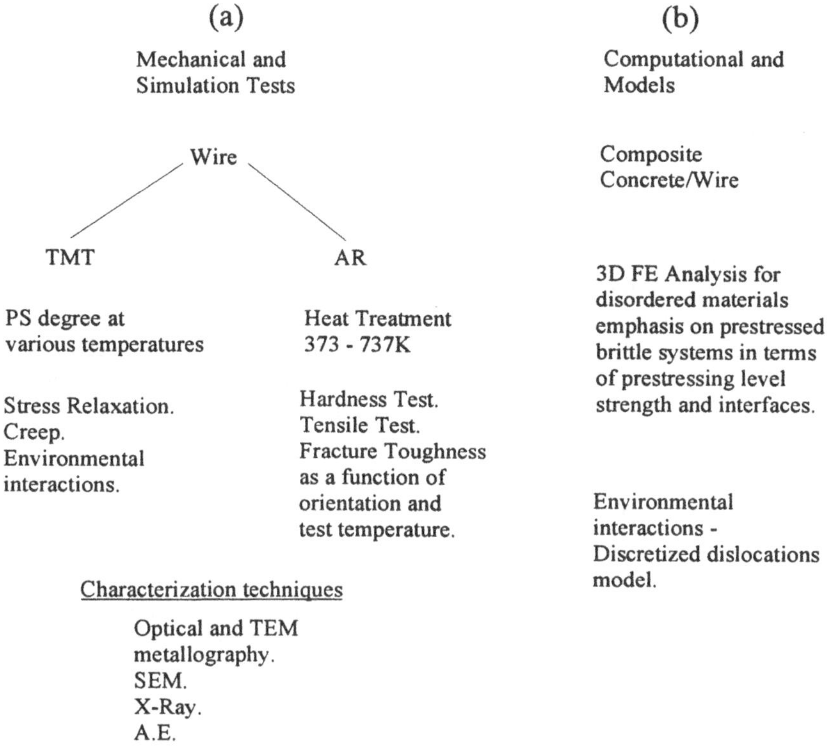

Fig. 1. Experimental (a) and computational program (b) - Schematic.

stress relaxation tests were conducted for various combination of PS levels and temperatures. stress relaxation tests were conducted for various combination of PS levels and temperatures. Additionally, creep tests were performed in order to establish an appropriate background for the role of micro-plasticity and activation processes. Second and still under the category of time dependent effects, hydrogen/metal interactions were investigated recognizing the fracture resistance of the wires as confirmed previously with no environmental interactions. Here, fatigue precracked wires with specified initial stress intensity factor were affected by external hydrogen of 1.2 atm. The subcritical slow crack growth regime was measured by calibrated COD gauge mounted near the crack mouth. With the add of AE the initiation and propagation stages were resolved to reveal the discontinuous behavior of the slow crack growth in the highly textured wires. Generally, along the mechanical characterization, microstructural observations were supplemented by optical, SEM and TEM observations. This beside fracture mode transitions and the relevant support to micromechanisms or/and kinetic considerations manifested by the environmental interactions.

2.2 Computational models [see Fig. 1(b)]

2.2.1 Hydrogen/metal interactions

Here, the theoretical background for the hydrogen embrittlement model is briefly described. The incentives to explore the micromechanism which control the microcrack onset and crack extension were two fold. First, to reveal the propensity of the high textured steel wires to environmental effects. Second, to provide experimental support to the hydrogen enhanced decohesion model in terms of fundamental consideration. Notice that the highly textured material is an intermediate case between single crystals and polycrystalline materials offering as such a particular interest. In this context, the model followed previous studies and developments regarding a cleavage model in semi-brittle materials [1-3]. Basically, it is a combined local approach in terms of the near crack-tip stress field and the local hydrogen concentration, affecting the cohesive strength of cleavage planes or interfaces.

Following this approach, the connection between the local tip stress intensity k_{tip} and the remote field stress intensity factor K_I under mode I loading was established. Thus, the fracture event is connected to the fracture toughness K_{ic} with the critical local Griffith value k_{IG}. As shown in Fig. 2, the modeling is a discretized dislocation simulation with organized dislocation structure representing the near-tip distribution and a super-dislocation which represents the far-field plasticity. As discussed elsewhere [4-5], a summation of forces of interactions between crack-tip, dislocations and applied stress allow a local stress field description. This by using the anisotropic elasticity theory of Atkinson and Clements [6] which gave the equilibrium position of all dislocations, the local value of k_{tip} and the stress distribution. The role of dislocation emission beside crack-tip shielding become the key issue in defining the fracture criterion along $K_I \rightarrow K_{Ic}$ when $k_I \rightarrow k_{IG}$.

In the case of hydrogen, the local K_{Ic} could be lowered to the local Griffith value for hydrogen embrittlement such that, $k_{IH} < k_I^{tip} < k_{IG}$ a condition which enabled to describe the subcritical growth even in a well defined kinetic. By realizing that k_{IH} is dependent on the local hydrogen concentration C_H, the threshold values for hydrogen interaction could be evaluated given by [7];

$$\frac{K_{IC}}{K_{Ith}} = \exp\left\{\frac{k_{IG}^2 - k_{IH}^2}{\alpha\sigma_{ys}}\right\} \qquad (1)$$

while α is a single adjustable parameter for a given mechanism.

2.2.2 Numerical model - prestressed concrete composite

In order to assist in exploring the dominant factors in prestressed concrete elements, a commercially Finite Element Code [8] was utilized. Basically, a three dimensional reinforced, pre-stressed concrete beam was considered with a failure envelope which allows

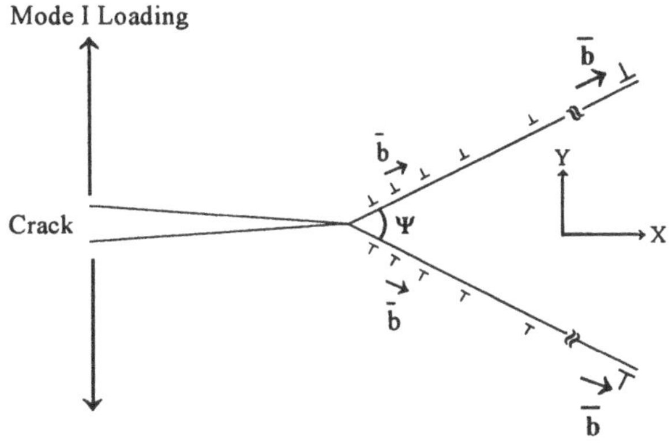

Fig. 2. Discretized dislocation model

to define the softening tail in tension (due to cracking and bridging) and the crushing conditions in compression. In tension, the failure is characterized by cracks onset in the plane perpendicular to the principle stress direction while obeying a critical stress criteria. Along this, closure and multiaxial effects were included. Under unloading condition, regardless the stress state (compression or tension) linear elasticity was assumed with a constant Young modulus E_0.

For the sake of simplicity, a supported concrete beam, (reinforced and prestressed by steel strand) while loaded gradually was analyzed (Fig. 3). Particular emphasis was given to the residual stresses and how the initial non-primary stress field prevail. The stress, strain and deflection was calculated from 3D, 20 node quadrilateral finite element mesh. Due to symmetry only a quarter section of the beam was discretized. The concrete layer was divided into 2880 elements and the steel strand into 1800 elements. In contrast to the non-linearity of the concrete, the steel was assumed to follow linear elastic behavior.

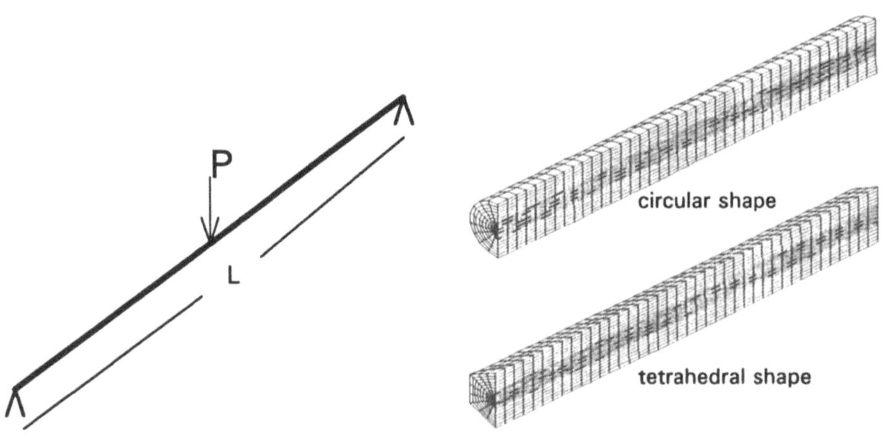

Fig. 3. Schematic - beam analysis by FEM.

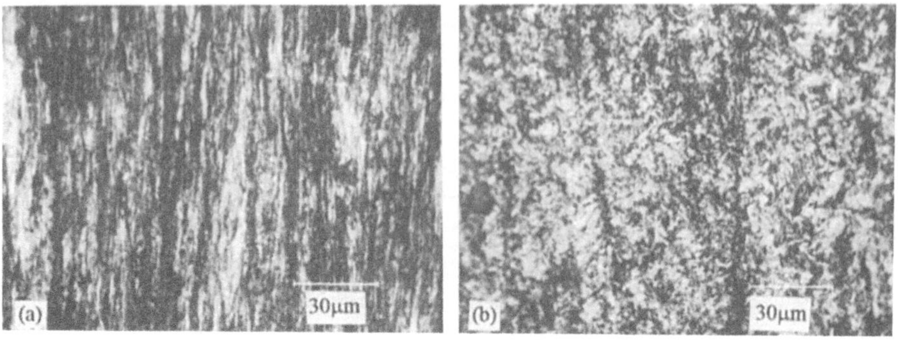

Fig. 4. Microstructures for the AR wire in the (a) L orientation and (b) T orientation

3 Experimental results

3.1 Microstructural and mechanical characterization

The microstructures of the as received (AR) condition for the Longitudinal (L) and the Transverse (T) orientations are shown in Fig. 4. The elongated fibrous morphology for the L is well demonstrated (Fig. 4a), in contrast to the T orientation (Fig. 4b). As the annealing temperature (AN) increases the grain boundary become more visible as well the pearlite and ferrite phases (Fig. 5a). The fibrous morphology of the pearlite obtained by TEM is illustrated in Fig. 5b.

Figures 6 and 7 demonstrate the effect of AN temperature on the hardness and tensile properties respectively. The hardness is almost unchanged up to AN temperature of about 500K, above this temperature drastic decrease of the hardness values occurred. This remained consistent for the yield and the ultimate stresses. Above 550K the stresses decreases sharply down to 700K.

Fracture toughness tests were performed for the L and T orientations. In contrast to regular Mode I loading for prefatigued single egde notched specimen in the L orientation, a wedge loading was utilized for the T orientation. Just for completeness, the effect of AN on fracture toughness was also examined (Figs. 8 and 9).

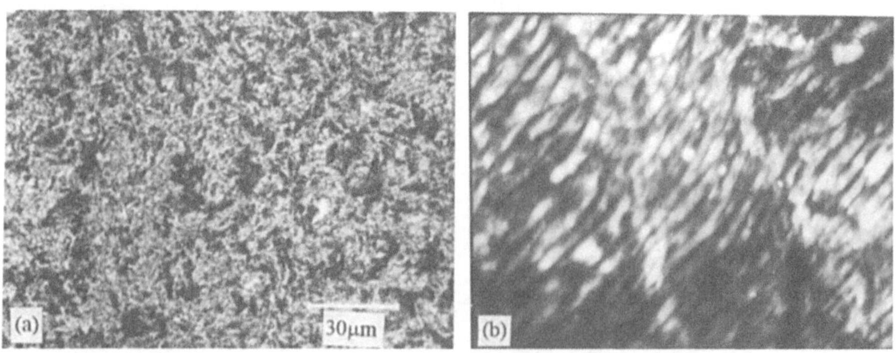

Fig. 5 (a) Microstructure after 773K/1/2h (b) pearlite morphology observed by TEM (x20000)

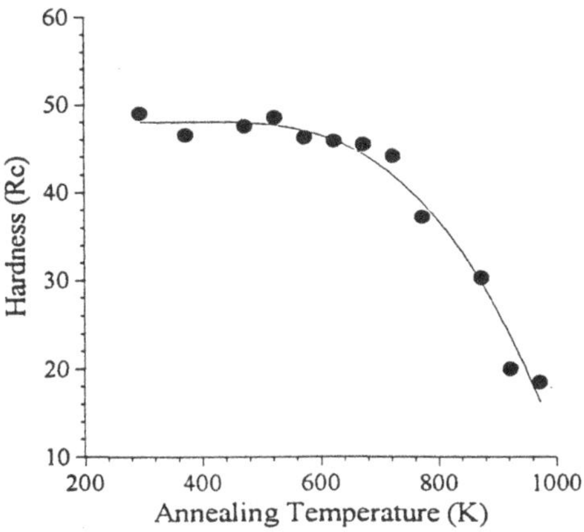

Fig. 6. Effect of AN temperature on the hardness values, determined at RT.

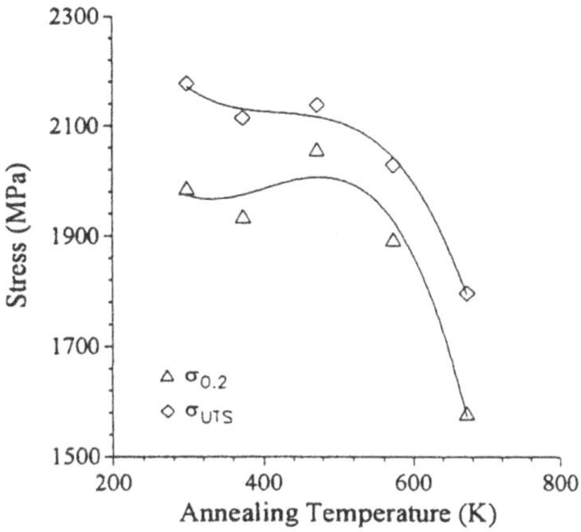

Fig. 7. Effect of AN temperature on the flow stress, determined at RT.

The anisotropic nature of the fracture resistance became apparent by the substantial differences between the fracture toughness values for the L and the T orientations. Here to mention that generally, the ductile-brittle transition (DBT) behavior remained soft caused by the available near-tip sources for dislocation emission, as addressed by Hirsch et al. [9] in terms of dislocation dynamic arguments. The sensitivity of K_{IC} to the AN is striking above 500K. This behavior is well illustrated in Figs. 10 and Figs. 11 while observing the fracture modes. In contrast to the AR wires in which the fatigue/slow crack growth interface was confined to discontinuous pop-in extension in quite limited sites the wires after AN treatment indicated typical subcritical growth, associated with relatively high energy fracture surface mode.

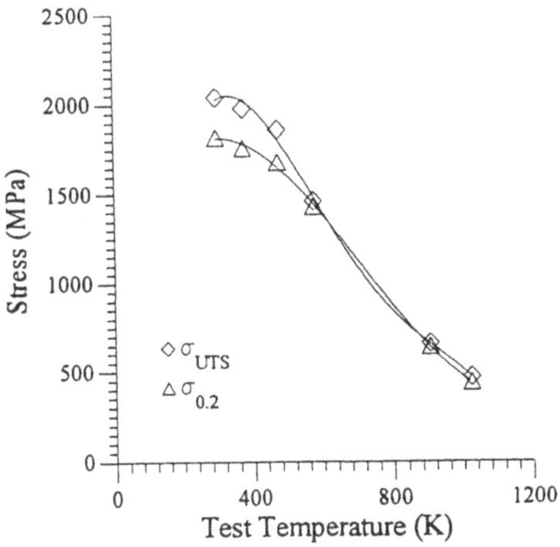

Fig. 8. Effect of test temperature on the flow stress for the AR wires.

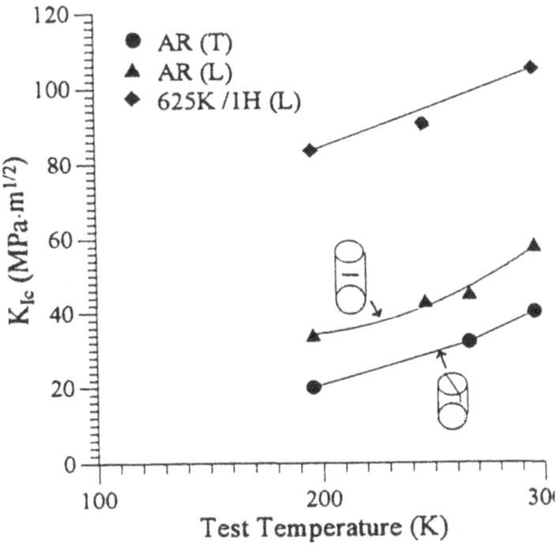

Fig. 9. Temperature dependence of K_{IC} for both orientations in AR and annealed wires.

Fig. 10. Discontinuous pop-in extension fracture mode in AR wire.

Fig. 11. Subcritical growth with relatively high energy fracture mode for AN wire (625K/1/2h).

Consistent features of stress relaxation by AN were indicated by X-Ray diffraction results. Table 1 summarizes some findings as related to the diffraction pattern in the transverse orientation determining the width, hights and the total areas of the selected peaks. As listed, increasing the AN temperatures decrease the peaks width and the total area values while the peaks hights increased monotonically.

Table 1. Analysis results for selected diffraction peaks.

Annealing Temperature	α_1 position	α_1 width	α_1 height	α_1 area	Total area
	°2θ	°2θ	counts, °2θ	counts, °2θ	counts, °2θ
100	82.369	0.550	43.97	29.40	40.99
	98.953	0.817	4.29	4.26	6.03
200	82.412	0.555	38.69	26.07	36.38
	99.107	0.729	5.52	4.89	6.93
300	82.401	0.481	39.09	22.84	32.07
	99.010	0.667	4.83	3.92	5.55
400	82.446	0.364	48.34	21.38	30.26
	99.037	0.457	7.98	4.43	6.29
500	82.435	0.267	58.17	18.90	26.86
	98.994	0.252	9.59	2.94	4.19

3.2. Prestressed wire findings

3.2.1 Tensile properties

As expected, the TMT (by prestress at specific temperature) mainly affected the microplasticity regime as long as other competitive relaxation mechanisms remain insignificant.

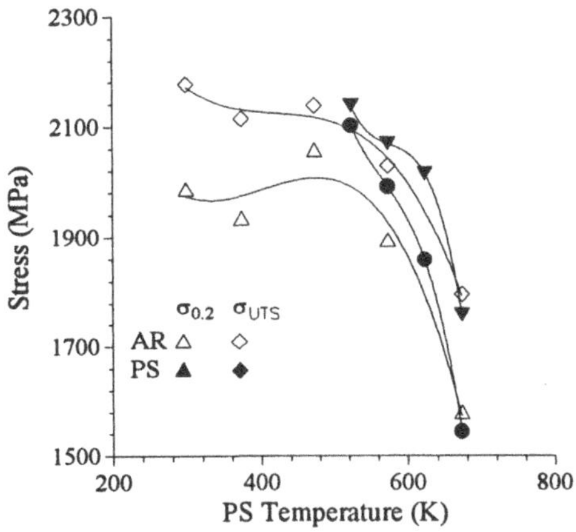

Fig. 12. Effect of PS by level of $0.6\sigma_y$ at various temperature. Compare to the NPS wire values.

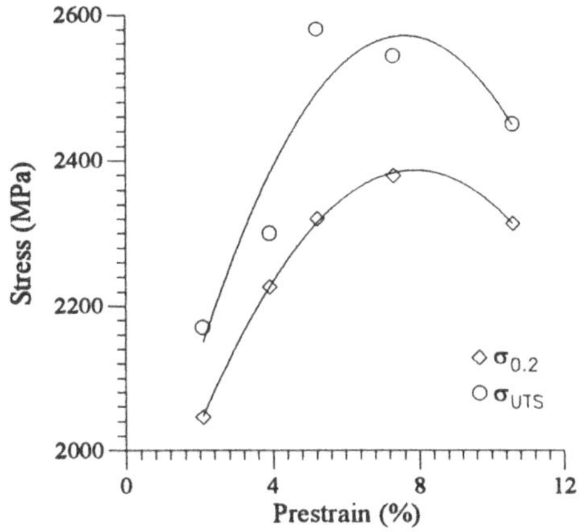

Fig. 13. Effect of plastic deformation degree at 473K on flow stresses.

The effects of PS of $0.6\sigma_y$ at specific temperature are demonstrated in Fig. 12 (the thermal dependency of σ_y was accounted in the determination of the PS stress value). As shown, the PS at the temperature range between 470K to 600K resulted in σ_y increase of 10%. With PS process almost ideal plasticity occrued in contrast to the parabolic plastic flow hardening behavior for the UPS wires. The role of the prestrain values on the yield and the ultimate stresses are illustrated in Fig. 13. In Figs. 14 and 15 the effects of PS on the yield stress and on the $\sigma_{0.2}/\sigma_{UTS}$ ratio are demonstrated respectively.

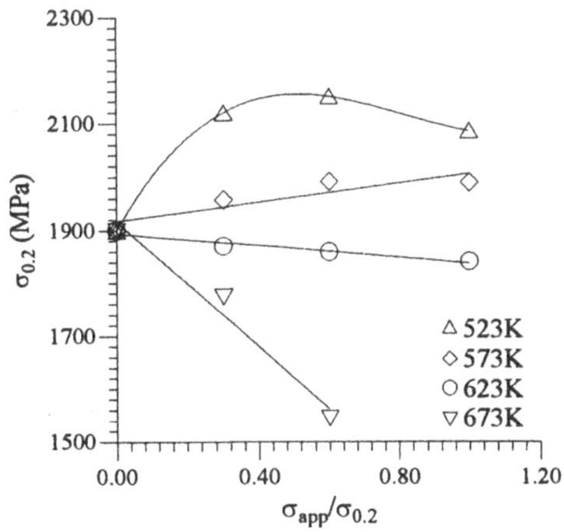

Fig. 14. Effect of PS level on yield stress for various temperatures.

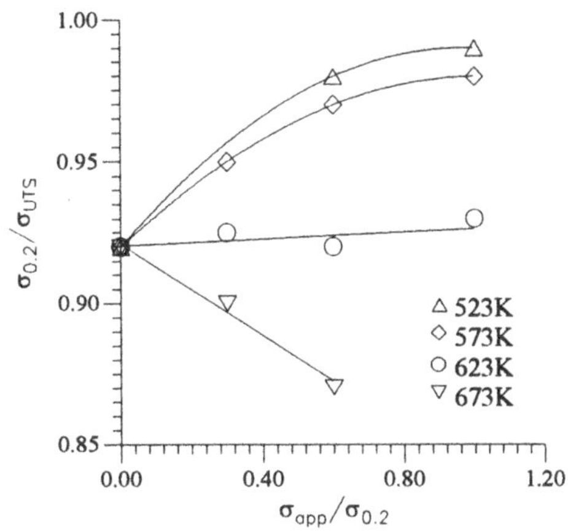

Fig. 15. Effect of PS level on the $\sigma_{0.1}/\sigma_{UTS}$ ratio for various temperatures.

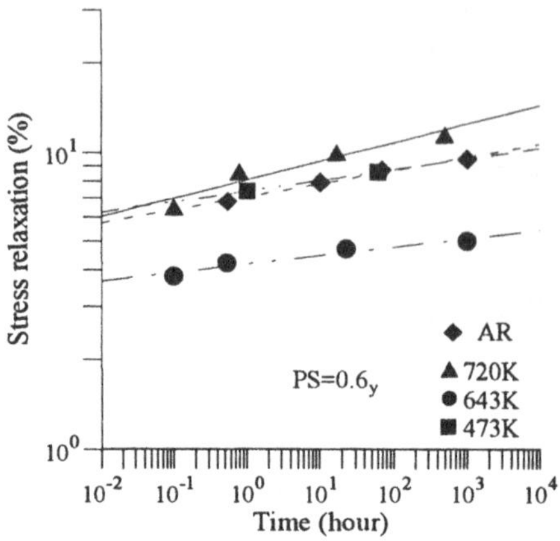

Fig. 16. Relaxation cyrve obtained at RT for AR and $0.6\sigma_{0.2}$ PS at various temperatures.

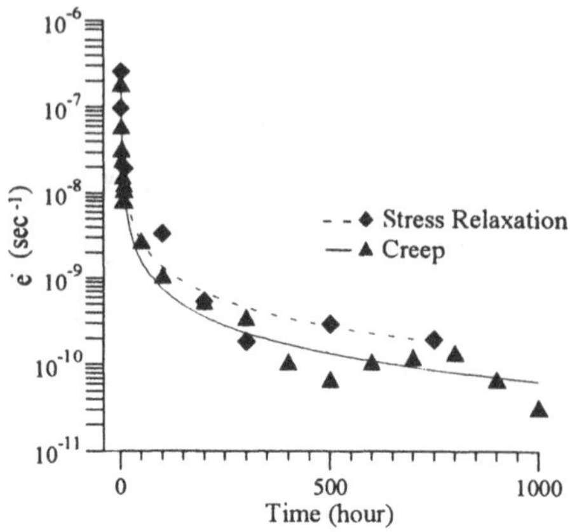

Fig. 17. Creep and stress relaxation rates in prestressed wire.

3.2.2. Stress relaxation and creep tests

The TMT process is mainly intended to achieve low stress relaxation (SR) by optimal dislocation/barriers - low energy structures. Some of these was confirmed experimentally as demonstrated in stress relaxation (fixed grips) and sustained load creep tests. As shown in Figs. 16, PS was clearly beneficial improving the resistance to microplasticity, but highly dependent on the TMT controlling variables, which shows an optimum at 643K. The TMT effects are well reflected on stabilizing the time dependent plasticity as compared to the AR specimens. Only for illustration, the strain rate values in creep are higher as compared to the SR results, as shown in Fig. 17.

3.3 Environmental interaction results

Even in low fugacity of external hydrogen interactions subcritical slow crack growth occurred at quite limited exposure times. As shown in Fig. 18 crack extension rates depends on the applied stress intensity factor indicating a near threshold values of $29\text{MPa}\cdot\text{m}^{1/2}$ (after 30h of exposure no crack extension detected). For high strength steel with σ_y of about 2000MPa the threshold value seems high. This issue will be addressed later on in terms of hydrogen embrittlement mechanisms. At this point it was found that the TMT resulted in similar effect a the AR wires with measured crack velocities up to 10^{-6} mm/sec. Particular interest was also in the AE tracking as performed for the L orientation.

Mainly it was found that the crack growth behaved in a discontinuous unzipping fashion as illustrated in Figs. 19 and 20. The typical features at an initial stress intensity factor of $66\text{MPa}\cdot\text{m}^{1/2}$ was, that the initiation-arrest reinitiation processes was associated with higher energy wave emission relatively to lower K at similar time exposures. The

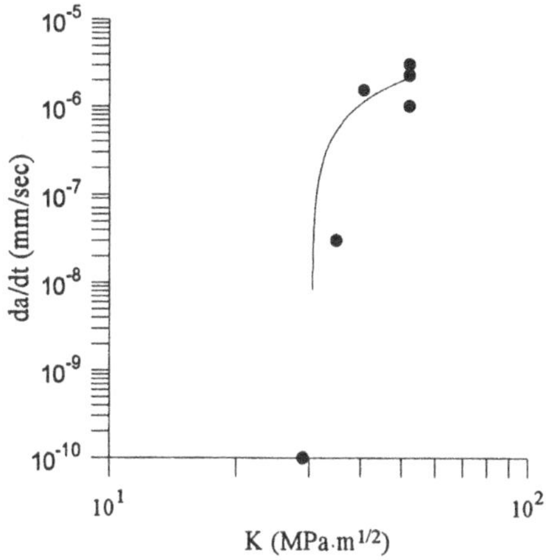

Fig. 18. Crack growth rate at H_2 for the AR wire.

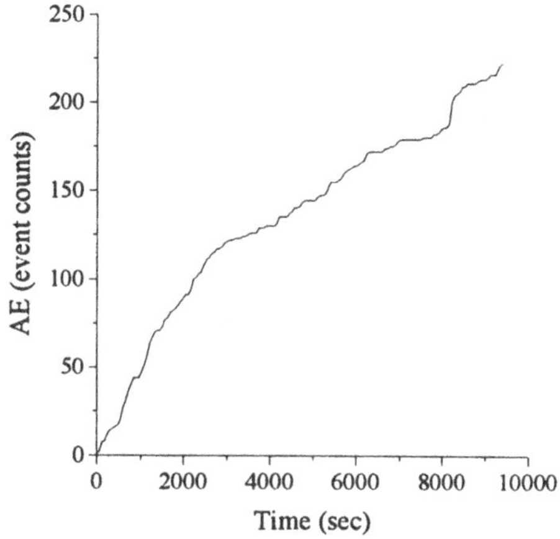

Fig. 19. Cumulative AE events in PS wire at $0.8\sigma_{0.2}$/650K for 3h H_2 exposure at K=66MPa·$m^{1/2}$.

key point was revealed by the fractographic findings in terms of fracture mode transition. In contrast to the consistent microvoid coalescence obtained in wires with no environmental interaction, transgranlular beside interfacial cleavage facets developed clearly enhanced by hydrogen (Fig. 21).

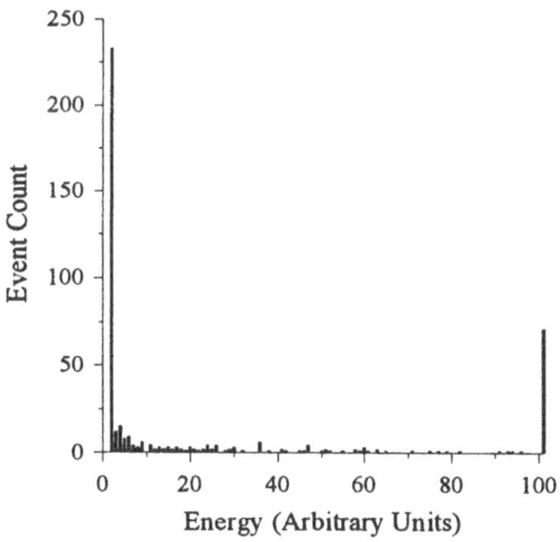

Fig. 20. Energy events historgam in PS wire at $0.8\sigma_{0.2}$/650K for 3h H_2 exposure at K=66MPa·$m^{1/2}$.

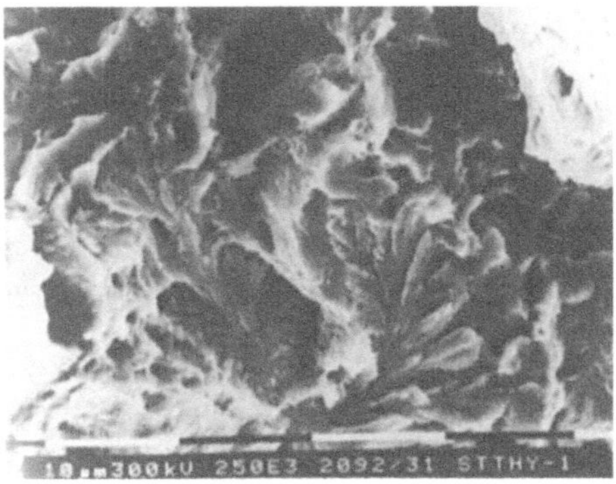

Fig. 21. Cleavage facets caused by hydrogen interaction in PS wire.

4 Numerical analysis and models

4.1 Environmental interactions

Considering the major phenomenological observations, it appears that the weight of evidence favors the hydrogen enhanced decohesion mechanism. Even in the high strength and textured wires, the crack propagation rates values are in the same order as observed in Fe-3%Si single crystals [10]. The combined influences of the intense driving force (from localized stress field computation by dislocation model), beside the fracture resistance degradation (caused by localized hydrogen concentration) allow the decohesion, as well as subcritical crack growth in defined kinetic and the threshold.

To summarize this section a local-two parameters failure curve is shown in Fig. 22. This presentation is quite different from the previous description assisted by the crack-tip stress intensity space illustration [7]. The advantage here is that basically the mechanism of hydrogen enhanced localized plasticity [11] is not excluded. However, it is the authors opinion that it still more consistent and easier to match the evidence with enhanced decohesion while consider plasticity as sequential event towards microcracking onset. The important point still remain that the strands integrity might be questioned under some circumstances. This clearly raises the problem of maintaining the non-primary stress field. Relaxation regardless the exact origins is a major issue as demonstrated in the coming analysis.

4.2 Discretization FEM analysis for a prestressed concrete element

By following the continnum beam theory [12] on the concrete/steel disordered composite as discribed in Fig. 23, the various initial strain copressive strains ε_{oc} were considered prior to central axial loading by force P.

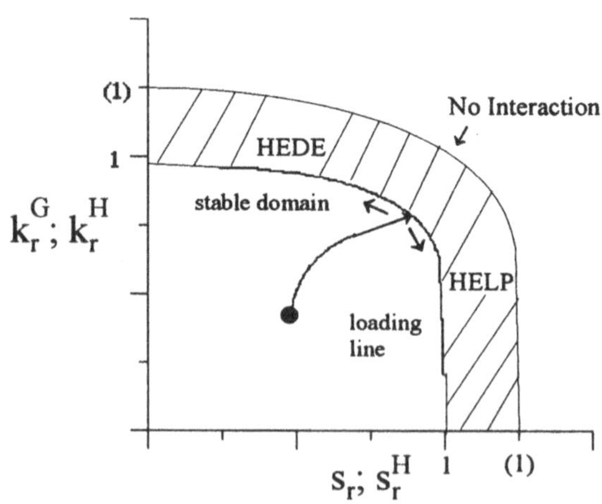

Fig. 22. Schematic- local criteria by two parameters failure curves. k_r - normalized crack-tip stress intensity, s_r - normalized crack-tip plastic instability.

Usually, the initial strain is in the linear range of the uniaxial stress-strain curve of the concrete material. Therefore the relation between ε_{oc} and the initial tensile strain in the steel ε_{ost} is given by:

$$\varepsilon_{oc} = \varepsilon_{ost} \frac{A_{st}E_{st}}{A_c E_c} \tag{2}$$

where: A_c - cross sectional area of the concrete, A_{st} - cross sectional area of the steel, E_c - Young modulus of the concrete (stress-strain relation in linear range) and E_{st} - Young modulus of the steel.

For a gradually increasing forces, the stress/strain and the deflection of the mid plane was evaluated.

Since the effect of the centrally located steel reinforcement on the deflection is minor, the equivalent values of E and I (momentum of inertia) can be replaced by the concrete values:

$$\frac{PL^2}{48E_C I_C} \approx \frac{\delta}{L} \tag{3}$$

Consider the normal stress redistribution, the maximum stress occurs at the points most remote from the neutral axis (tensile at one side and compression at the opposite side). The effect of the prestress, in the concrete, is to reduce the maximum tensile stress by adding compressive stresses. Since, particularly for concrete, the failure resistance to compression is much higher, prestressing become beneficial. Thus, increasing gradually the traction P, increase the damaged volume and induces non linear behavior on the load-deflection curve. A typical cracks evolution and distribution is plotted in Fig. 23.

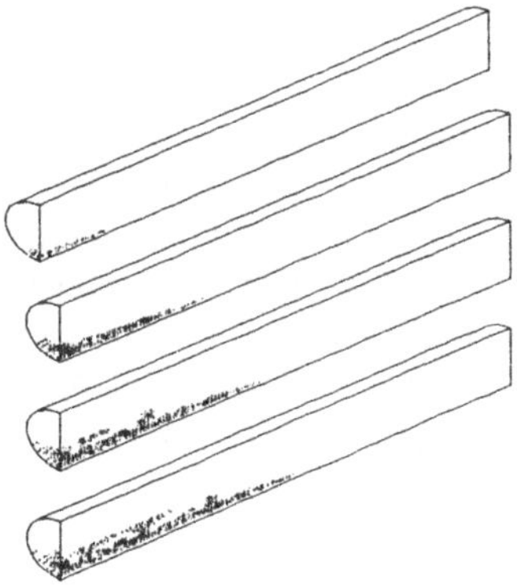

a - circular cross-section

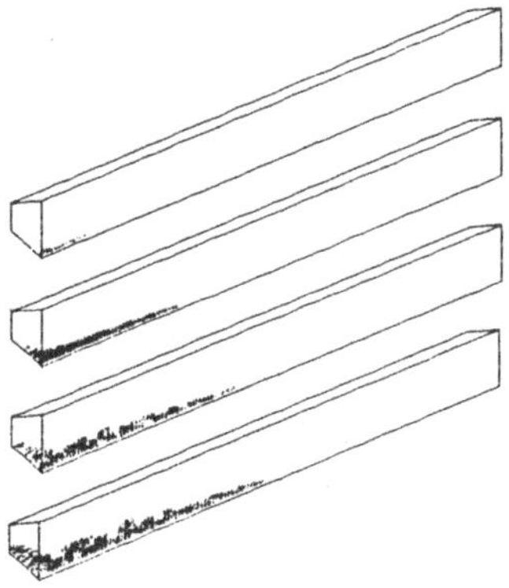

b - tetrahedral cross-section

Fig. 23. Crack evolution in prestressed concerte element.

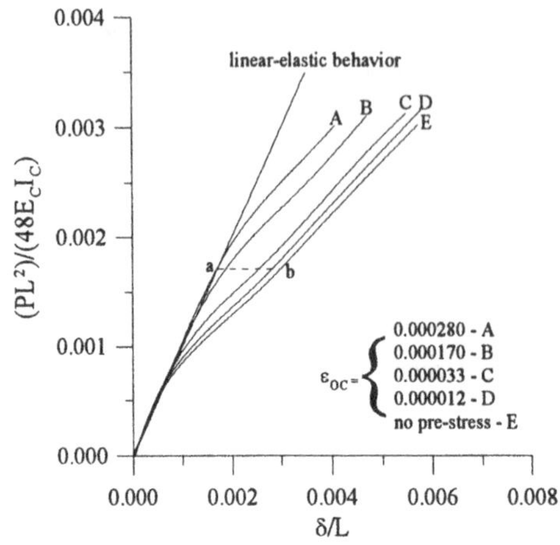

Fig. 24. Normalized load-deflection curves for various prestressed values.

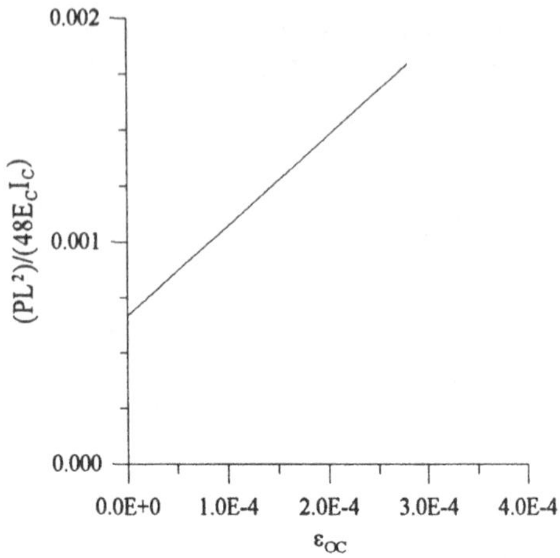

Fig. 25. The relative load at the transition versus the initial compressive strain.

The computational results of the normalized load versus the normalized deflection, for different values of the prestressing, is plotted in Fig. 24.

Accordingly, damage initiation is indicated as also the post cracking along the non-linear regime. The transition to non-linear behavior or the damage onset is a function of the initial compressive strain or the prestressing intensity (Fig. 25). As such, any lose of the non-primary stress field might follow the a-b line. Notice that initial condition in a non-damage regime in $\bar{a}$ might lead to highly damage regime in b. Moreover, this process might be self generative as the degree of damage can be accentuated as the loss of the initial strain is dynamically progressing. The origins for not maintaining the resedual field might be interfacial debonding, stress relaxation, wire or strands failures sometimes enhanced by environmental interactions.

5 Discussion

Damage analysis in concrete and rock has been addressed recently by Van Meir et al. [13]. As indicated by this overview, several approaches have been attempted with the emphasis to the lattice model development. By considering the additional complication of prestressing the cardinal issue of maintaining the residual stress state prevail, with origins in viscoelastic-plastic factor as environmental effects in monotonic and cyclic loading. While we understand the TMT processes or the internal microstructures of wires at high plastic deformation, better standards emerged including the low relaxation and the fatigue tests. For example, the relaxation test can be founded on the following first order analysis. Consider the long range internal stresses in constant grip test.

$$\dot{\varepsilon}_e + \dot{\varepsilon}_p = 0 \tag{4}$$

where $\dot{\varepsilon}_e, \dot{\varepsilon}_p$ - the elastic and plastic strains respectively.

Since, $\dot{\varepsilon}_e = \frac{\dot{\sigma}}{E}$ and $\dot{\varepsilon}_p = \lambda\rho\bar{b}v$ (5)

where; $\bar{b}$ - the Bugers vector, v - the dislocation group velocity, λ - constant and ρ- the avarage dislocation density.

While the dislocation velocity under isothermal condition is given;

$$v = A(\sigma - \sigma_0)^m \quad (6)$$

where σ_0 - is the long range stress; A - constant and m - the velocity stress dependancy degree, the following is formulated;

$$\frac{d\sigma}{(\sigma - \sigma_0)^m} = -F\bar{b}dt \quad (7)$$

where F is a constant (F=AEλ). Thus for a long time $t >> t_0$ and asymptotic solution for σ is obtained.

This relationship was experimentally confirmed which allows the required match between theory and expriments. In case of environment, the issue of deformation/corrosion is far from being mature with ample of contraversial issues. Consider the ASTM-A416 or the ISO-6934, the issue of environmental effects in other standards (for example BS-5896 Appendix B) are not under a general requirement. Even if requested, the evidence is described in general terms by a proposed immersion tests under 80% of σ_{UTS}. These aspects require further elaboration assisted that might be by fracture mechanics framework. Beside the principal variables which control hydrogen embrittlement, alloy chemistry as other microstructure features, might affect susciptability to embrittlement. While dealing with high yield stress values (about 2000 MPa) does not imply susceptibility increase. For example, one can both increase σ_y and control grain boundries chemistry and processing modifications. By this K_{Ith} may actually increase even though σ_y increases. These aspects have to be elaborated further particulary for the high texteured wires. (Consider the relatively high threshold value of about 30mpa·m$^{1/2}$ obtained in the present study). This may be attributed to the cleaner boundaries being the dominant feature. Thus, it is inappropriate to consider only one of those variables in isolation of the others. This can be explained also from eguation (1). For small degradation values of k_{IG} by environmental interaction might reduce K_{IC} by nearly an order of magnitude. In contrast, for an ultra-high strength steel with toughness already limited, the reduction might only be a factor of two or less. Notice that the high cycle fatigue test which is included, might also be affected by environmental considerations beside fatigue fretting effects. The dependancy between the scientific activities and the utilization of existing information, in the prestressed concrete application is quiet evident. Thus suplemented activity seems appropriate along the incorporation of fracture mechanics methods on environmental effects reflecting on the fatigue tests, surface conditions, interfacial strength and mixed mode features founded on more of a sound background. The role of non-primary stresses on damage analysis is complex, affecting the overall mechanical behavior [14]. In case of long term service the maintaining and monitoring prestressed levels become the key issue thus, the various relaxation origins require refirments and assessments.

6 Conclusions

1. Marginal assessment against failure in prestressed disordered materails involves additional complexities caused by the non-primary stress state.
1. In order to describe the residual compressive strss field emphasis on the eutectoid high strength strands properties is required.
3. The time dependent processes are not limited to the non-primary stress relaxation processes, but on environmental interactions as well along transgranular cleavage and fine scale subgrain interfaces.
4. Sub-critical crack growth was obtained in low fugacity hydrogen/wire intreactions with threshold stres intensity values of 30 $MPa{\cdot}m^{1/2}$
5. FEM analysis of a supported prestressed beam illustrates the possible transition transition from non-damage regime caused by residuals redistribution.
6. Environmental degradation which reflect also on high cyclic fatigue, fatigue fretting surface conditions or thermomechanical treatment is not expressed nowadays in the availabe standards.

Acknowledgments

The authors would like to acknowledge the experimental support by Mr. M. Kupiec, Mr. R. Shefi and Mr. A. Sobel from the NRCN, Beer-Sheva and Mr. D. Weiss from the Ben-Gurion University, Beer-Sheva.

The encouragement and assistance of Mr. R. Pessach and the technical staff from Mosserot in Netivot, Isreal is highly recognized and appreciated.

References

1. Rice, J.R. (1994) Philos. Mag,. Vol. A29, p.73.
2. Katz, Y., Chen, X., Lii, M.J., Lanxner, M. and Gerberich, W.W. (1991) Eng. Fract. Mech., Vol. 41, p. 541.
3. March, P.G. and Gerberich, W.W. (1994) Acta Metall., Vol. 42, p. 639.
4. Lii, M.J., Chen, X.F., Katz, Y. and Gerberich, W.W. (1990) Acta Metall., Vol. 38, p. 2435.
5. Gerberich, W.W. and Huang, H. (1994) Acta Metall., Vol. 42, p. 639.
6. Clements, D.L. and Atkinson, C. (1973) Acta Metall., Vol. 21, p. 55.
7. Gerberich, W.W., March, P., Hoehn, J., Venkatarama, S. and Huang, H. (1992) in Corrosion-Deformation Interactions, (eds. T. Magnin and J.M. Gras) Les editions de physique, Les Ulis Cedex A, France, p. 325.
8. Edina© (1994) Report ARD 92-8.

9. Hirsch, P.B., Roberts, S.G., Samuels, J. and Warren, P.D. (1998) in ICF7, Vol. 1 (eds. K. Salama, K. Ravi-Chandar, D.M.R. Taplin and P. Rama Rao) p. 139.
10. Chen, X., Foecke, T., Lii, M., Katz, K. And Gerberich, W.W. (1990) Eng. Frac. Mech., Vol. 35. P. 997.
11. Birnbaum, H.K. (1983) in Atomics of Fracture, (eds., R.M. Latanison and J.R. Pickens) PlenumPress, New York, USA.
13. Van Meir, J.G.M., Schlangeu, E., Vervuurt, A., and Van Vliet, M.R.A. (1995) in Mechanical behavior of materials (ed., A. Bakker) Delft University Press, Delft, The Netherlands, p. 101.

33 APPLICATION OF MONITORING TECHNOLOGY TO BRIDGE ASSESSMENT

J.T. DEWOLF
Civil and Environmental Engineering, University of Connecticut, Storrs, Connecticut, USA
R.G. LAUZON
Department of Transportation, Rocky Hill, Connecticut, USA

Abstract
After a decade of using different techniques to conduct field monitoring of bridges, researchers at the University of Connecticut and in the Connecticut Department of Transportation are undertaking a multi-phase effort to install continuous monitoring systems on different types of bridges in the State of Connecticut. Continuing advances in both hardware and software technology now allow for the development of permanent monitoring systems, both for safety and for the collection of related information needed to manage the highway infrastructure. This long-term study has begun with the development of a generic specification which serves as a guideline for using many different sensor types in an integrated monitoring system. Applications include existing bridges which are in need of continuous information and new, as well as rehabilitated, bridges. The work is beginning with the installation of 6 remote monitoring systems. The project is being carried out by a team effort, involving researchers at the University of Connecticut, engineers in the Department of Transportation who are involved in the operation of the highway infrastructure and manufactures of monitoring equipment. The goal is to demonstrate the feasibility of monitoring a bridge network using a variety of sensors and communication techniques.

1 Introduction

The U.S. National Science Foundation has defined four research areas which will assist in renewal and enhancement of the country's's infrastructure. These areas are deterioration science, assessment technologies, renewal engineering and

Structural Failure, Product Liability and Technical Insurance, Proceedings of Fifth International Conference. Edited by H.P. Rossmanith. Published in 1996 by Taylor & Francis, 2 Park Square, Milton Park, Abingdon, Oxon, OX14 4RN. ISBN 0 419 22080 1

institutional effectiveness and productivity. Bordogna [1] has noted that current methods for evaluation of the infrastructure are relatively primitive and unreliable, prompting conservative, often costly decisions.

Today, bridges are primarily monitored with visual techniques. It is not always possible to see all critical areas, with the result that problems go unnoticed until they become serious. Additionally, this approach does not provide information between inspections. Bridges with problems normally require more frequent inspections, requiring additional manpower.

Chase [2], in a report on research needs for bridge management, states that additional work is needed to integrate local and global quantitative nondestructive evaluation methods into bridge management systems. This work should provide bridge management with essential information and procedures to help identify and quantify short term and long term needs to identify cost effective maintenance, repair and replacement strategies. Research should also address the need for improved deterioration models and issues related to the collection, storage, retrieval, analysis and interpretation of large incongruous data sets which characterize bridge data.

This paper reports on use of recent advances in hardware and software monitoring technology to design and implement permanently installed monitoring systems on bridges. These will be used for safety determinations and collection of data needed to assist in the management of the bridge infrastructure system. This following topics are included: (1) a discussion of the monitoring approach, based on previous monitoring efforts; (2) design of a generic monitoring systems for installation of a variety of bridges, utilizing different sensors for application to a variety of bridges; (3) the goals of the work.

2 Monitoring Approach

2.1 General

Work during the past decade at the University of Connecticut has been directed toward field monitoring of bridges, involving determination of both vibrational and strain measurements. Included in this work has been the use of a fully operational vibrational monitoring system mounted on a Connecticut bridge and portable equipment for short term strain monitoring studies. This work has resulted in substantial savings to the State of Connecticut by providing information in the continued efforts to maintain and renew the state's bridge infrastructure. It has also provided design information needed for the development of expanded field monitoring systems. A history of this work is reviewed next, with examples demonstrating how monitoring has been used in the evaluation of bridge structures. This then provides the basis for development of the monitoring systems discussed in this paper.

2.2 Vibrational Monitoring

While many techniques are available for bridge monitoring, most do not provide a global picture of the bridge. This is necessary to assure that the entire structure is under surveillance and to account for unknown failure mechanisms. Vibrational measurements have been successfully used in many different applications for global evaluations, including prevention of major breakdowns in manufacturing plants, maintaining continuous power generation in power plants and monitoring of aircraft parts. Vibrational data may be processed in a wide variety of ways, allowing for development of techniques which can magnify the changes which occur when the structural integrity is compromised.

A significant portion of the work at the University of Connecticut during the past decade to adapt vibrational monitoring to bridges has involved development of new equipment and adaption of evaluative techniques. This work has included: (1) A field study of the vibrations of a four span continuous steel plate girder bridge with both test vehicles and normal traffic; (2) A model study to learn how vibrational monitoring might be applied to bridges; (3) Additional field tests to define requirements for both sensors and the hardware needed for interaction with the sensors in order to apply the vibrational technique to long-term bridge monitoring; (4) Design of a prototype monitoring system for continuous, long-term installation on bridges, involving support from a State of Connecticut in conjunction with Vibra•Metrics, a Connecticut Company specializing in manufacture of vibrational equipment; (5) Installation and testing of the prototype monitoring system on a single span, new Connecticut bridge. Work with the prototype monitoring system was reported on at the Fourth International Conference on Structural Failure, Product Liability and Technical Insurance [3]. The approach involved using frequency spectra obtained from different sensors along with phase information to determine natural frequencies and mode shapes.

Two additional studies have included a field test in which a crack was introduced into a bridge [4] and use of the previous prototype vibrational monitoring system on an older bridge [5].

Preliminary results from the study in which the crack was introduced into an existing bridge [4] demonstrate that the vibrational signature does change as the crack propagates through the web. While changes in specific natural frequencies are not sufficient to fully indicate that the structural integrity has been compromised, changes in the overall vibrational signature can conclusively show that the structural behavior is different. This involves review of the lowest natural frequencies as a group and evaluation of the mode shapes. Additionally, relative acceleration levels changed and a new frequency appeared, with resulting alterations in the mode shapes. Extensive analytical studies are now underway on these field results.

The prototype monitoring system is presently mounted on an older, continuous two-span steel girder bridge [5]. The system was modified, based on findings from the first application. This study has demonstrated that the vibrational signature changed as the temperature changed, i.e. a drop in temperature resulted in increased natural frequencies. This was attributable to frozen bearings, which introduced rotational fixity in the pinned supports and axial forces in the members. This work is continuing.

These studies have demonstrated that vibrational monitoring is able to indicate changes in the overall performance, due to the introduction of a crack or frozen bearings. These changes are discernable from the normal traffic data collected over time. In application with actual traffic, it will be necessary to use statistical comparisons with multiple sets of data to demonstrate that changes in the structural integrity have occurred. As an example, frozen bearings will cause the range of values for one or more specific natural frequencies to shift up or down as the temperature changes. The key to using this information is to evaluate the behavior statistically.

2.3 Strain Monitoring

While vibrational techniques present a global picture of the bridge and thus allow for an evaluation of the overall behavior, strain monitoring and other techniques based on localized determinations allow for evaluation of specific members or connections. An example is monitoring a portion of a member subject to fatigue problems.

The research to develop vibrational monitoring techniques was also useful in research at the University of Connecticut to evaluate a series of bridges in Connecticut using strain monitoring. These studies have provided information which has been directly useful the management of the bridge infrastructure. This work has been based on the development and application of portable monitoring systems for the determination of strains and stresses. Considerable effort has been made to develop computer software to automate the monitoring and reduce the data.

These strain monitoring efforts began with development of portable monitoring hardware in 1987, and have continued with application of newer hardware to increase both the speed and capability. One study of bridges on one of the main highways in Connecticut sought information on the behavior of diaphragm connections prior to renovation. The results of this work demonstrated that repairs were needed only at the center diaphragms and not at all as recommended by the design consultant.

Another major study of one of the interstate overpasses to study the causes and potential repair of diaphragm connections [6] involved both analytical and field evaluations. In addition to detailed information on the connections, it was confirmed that diaphragms are often not needed in the bridge after construction is completed.

Recent efforts have included the development of a faster, fully portable monitoring system with multiple channels. The new system has been used on a major bridge across the Connecticut River to evaluate excessive deformations in connections which caused fatigue cracking and bolt failures [7]. Further studies [8,9] have included (1) Evaluation of the remaining strength in aged hangers for the counterweights in a draw bridge; (2) Determination of necessity for repair of cracks in connections in a steel bridge: (3) Investigation of fatigue cracking in a steel bridge with estimation of remaining life; (4) Determination of the live load rating for an existing bridge; (5) Evaluation of column cracking problems in a concrete arch bridge.

These studies have demonstrated the feasibility of using field data collected from normal traffic loading to evaluate different aspects of bridge performance. An essential requirement for continuous monitoring of bridges is the need to use normal traffic excitation for evaluation. A monitoring approach which requires use of a test vehicle, or other specialized loading, is limited to use only when sufficient manpower is available to operate the system. This type of inspection technique normally requires closing of the bridge to conduct the monitoring. This is difficult to schedule for many of the more heavily trafficked bridges on the major highways.

3 Specification for Continuous Monitoring Systems

3.1 Generic Specification

The previous efforts with both vibrational and strain monitoring systems were used to develop generic and non-site specific specifications for bridge monitoring, involving different types of bridges, different sensors and different processing. Site-specific details must be added to the generic specification. The following reviews the basis of the generic specification.

3.2 Basic Requirements for Monitoring System

There are four parts needed for a fully operational monitoring system: (1) System Control Unit including hardware and computer for overall management of monitoring and data analysis; (2) Sensors for collecting different types of data; (3) Software to control data collection and to provide operating engineers with needed information; (4) Communication capability so that control of the system and information collected may be treated in a centralized office. It is then necessary to have a computer in the centralized office for communication and management functions.

A fully functional system should allow for infrastructure management, monitoring of structural health, maintenance needs, and monitoring of environmental conditions. It is necessary to have flexibility to allow for incorporation of different types of sensors and to provide options on how data is obtained.

In general bridge monitoring systems should be expandable. Additional types of sensors should be provided for. The system software should be upgradable so that it can provide an expert diagnosis and reporting system. The monitoring system must function as an integral part of an intelligent highway management system.

For continuous monitoring, the system must operate under normal traffic conditions. It must not be a requirement that the bridge must be closed to traffic before data collection can occur.

3.3 Components of Monitoring System
The main components needed are as follows.

Control Units - The main system control unit typically incorporates a host computer and sensor excitation hardware. The host computer is used to control the excitation and data interpretation hardware, store the recorded data on the hard disk, analyze the data and provide communication capability with remote computers. The host computer is the brain for the system. The sensor excitation and data interpretation hardware provides the link between the sensors and the host computer. Signal processing, needed to activate the sensors, can be placed in either the main control unit or in satellite units. Satellite units are often useful for providing signal processing for different sets of gages. They are useful for reducing the number of wires to the main control unit, and if they are used for digitizing of the data, they can increase the distance between the sensors and the main control unit .

Signal processing hardware links the operating system directly to the sensors. It is used to excite the sensors electrically and to feed the resulting electronic voltage back to the operating system. This voltage is then converted to the appropriate engineering units of interest, i.e. strain, acceleration, temperature, etc. This hardware can be located either in the system control unit or in satellite units.

Sensors - The choice of sensor types and quantity are based on the information needed. The general categories, with some types of the desirable information, are: (1) Structural - Strains, deformations, tilt, vibrations, forces, continuity, crack development; (2) Environmental - Temperature, dew point, relative humidity, wind speed and direction, identification of precipitation type, rainfall rate, snowfall intensity, fog conditions; (3) Traffic Flow - Volume of traffic, vehicle weights, vehicle speeds; (4) Maintenance - Pavement information, corrosion and paint condition, bearing conditions, condition of cathodic protection system.

The sensor types used on bridges include strain gages, tilt meters, accelerometers, LVDT (Linear variable displacement transducers), sensors for acoustic emissions; sensors for vehicle counting, sensors for vehicle weight/speed systems; temperature sensors, icing sensors, sensors for snow/ice/rain volume, sensors for stream flow and water height and weather sensors. The considerations in choice of sensors involve the desired data range, sensitivity, durability, power requirements, signal conditioning requirements, calibration needs while on bridge, reliability of data, and compatibility with bridge monitoring system and other sensors on bridge.

Software - The software is needed to control the operating system, control the sensors, data analysis and communication. All software should be addressable from the host computer at a remote site, for alteration, replacement, additions, etc. Interfaces with all software should be readily understandable by those operating the system, using graphics and clearly worded statements or menus. The software should provide for storage so that it is not necessary to have the remote computer operating to fully operate the system. The software must provide full control of the system. The capability should be provided to allow different kinds of remote access, i.e. access to only selected sensors, access to the full sensor data or access to data after analysis.

The software for the sensors should allow for remote control of sensor characteristics, including current, levels of excitation, resolution and filter characteristics. The software must be capable of displaying the information in different formats and exporting for interaction with analysis software. It is desirable that this software be capable of testing and detecting sensor faults and alerting the operator.

All data analysis should preferably occur on the bridge, since this reduces the amount of data transmitted back to the host computer. Consideration should be given to presenting the data in different formats, including both numerical results and statements on the significance of the data, i.e. level of data and whether a critical situation exists. The software should reduce the data collected to readily usable information, with options on storing peak values only, a time history based on selected pieces of data and processing of the data into alternative values, such as natural frequencies and mode shapes for vibrational data or numerical values to indicate how much change has occurred.

Communication - Communication capabilities should be provided from the host computer to a remote site, as well as to a portable computer hooked up to the system in the field. The communication software should be cable of high speed data communication using different kinds of data.

Possible communication links include: (1) Modem with telephone line - this is desirable for large amounts of data transmission or frequent access; (2) Cellular connection - This should be considered when there is not a need for transmission of large quantities of data or connection with the remote sight at frequent intervals; (3) Data link - This type of connection requires special lines and should be justified by the quantity and types of data needed; (4) Telemetry or radio link - This is especially useful for longer distances.

4 Bridge Selection

4.1 Guidelines for Bridge Selection

Applications for monitoring include:

Structural considerations:

(1) Non-redundant - failure in one part will cause loss of stability, so that the consequences of a failure would be significant.

(2) Bridges subject to overweight vehicles or traffic volumes beyond those it was designed for.

(3) Bridges which are fatigue susceptible.

(4) Aged bridges - monitoring those that are near or have exceeded their life expectancy can extend the usable life.

(5) Damaged bridges.

(6) Bridges with new designs, when it is felt necessary to verify some of the design assumptions or necessary to review the resulting behavior.

B. Environmental factors:

(1) Source for weather information needed to schedule work, such as snow plowing.

(2) Determination of weather conditions to evaluate loading on bridge, such as from a hurricane.

(3) Stream flow, useful in determining load on a bridge and in determining if scour is likely.

(4) Determination of snow, ice build up.

C. Traffic Flow:

(1) Determination of vehicle loads and speeds.

(2) Determination of traffic volume, which effects fatigue and the aging process.

D. Maintenance Needs:

(1) Performance of members, connections, bearings, etc.

(2) Corrosion and paint soundness.

(3) Determination of when snow removal and/or chemical application is needed.

4.2 Bridges Selected for Monitoring Systems

Monitoring systems have been designed for 6 bridges, with plans to expand the system to other bridges in the future. The bridges chosen for the initial systems include different types of bridges, with different monitoring needs. Information developed with the initial monitoring systems will be sued in the to the design of subsequent systems. The goal is to develop a network of remotely monitored bridges, with different sensors types. The following bridges will be monitored first phase of this project.

(1) Curved, cast in place, three span continuous box girder bridge with post-tensioned cables. The monitoring system is designed to evaluate the structural performance and provide information on the weather conditions oat the site. The structural sensors used are 12 temperature sensors, 18 accelerometers, 24 strain gages, and 6 tiltmeters.

(2) New segmental, multispan, post tensioned, concrete box girder bridge. The system is designed to evaluate the structural performance. It will be applied to two spans, with possible expansion in the future. The sensors used are 4 temperature sensors, 8 strain gages, 8 accelerometers and 6 tiltmeters.

(3) Curved, composite, multispan, steel box section bridge. The system is designed to evaluate the influence of temperature on the bridge's structural performance. It will be applied to two spans, with possible expansion in the

future. The sensors used are 6 temperature sensors, 4 pavement temperature gages embedded in the pavement, 8 accelerometers and 6 tiltmeters.

(4) Older, multispan truss bridge. The system is intended for the center hung span, with sensors on the hung span and the cantilevers to evaluate the overall performance. Each of the hangers will also be equipped with strain monitoring sensors to monitor the forces in the hanger. The sensors include 12 accelerometers and 8 strain sensors.

(5) Scour critical bridge with high traffic volume. The system is designed to evaluate possible pier movements. The sensors include 8 accelerometers and 4 tiltmeters.

(6) Steel plate girder bridge which is regularly subjected to overweight loads. The system is designed to evaluate the structural performance of the bridge and verify that the capacity is not reduced by the overweight loads. The sensors include 2 accelerometers and 6 strain gages.

5 Summary and Conclusions

For many years, transportation agencies, universities, private firms and research organizations have temporarily instrumented individual bridges for various reasons. These temporary systems differ in complexity and functionality. Eventually these systems are abandoned or shelved with no or very little continuing effort to develop techniques for the overall monitoring and management of a network of bridge structures. The advances in technology are being used in this work to develop permanent continuous safety monitoring systems for application to a bridge network.

This paper reports on the development of the material needed to design the monitoring systems and the selection of the first bridges in the research project. This is part of a longterm effort to use modern technology to evaluate a system of bridges, supplementing the normal visual approach.

6 Acknowledgments

This research is a joint effort between the University of Connecticut and the Connecticut Department of Transportation, with funding from the Connecticut Department of Transportation and the U.S. Department of Transportation Federal Highway Administration.

7 References

1. Bordogna, J., (1994) “Intelligent Renewal Keeps Infrastructure Civil,” Civil Engineering, Vol. 64, No. 3, p. 6.

2. Chase, S.B., (1995) “Research Needs for Bridge Management,” Proceedings of American Society of Civil Engineers Conference on *Restructuring: America and Beyond*, Volume 1, Boston, Massachusetts, U.S.A., pp. 670-671.

3. DeWolf, J.T., O’Leary, P.N. and Bagdasarian, D.A., (1993) “Failure Prevention in Bridges Through Continuous Vibrational Monitoring,” Proceedings of 4th International Conference on *Structural Failure, Product Liability and Technical Insurance*, Vienna, Austria, pp. 357-364.

4. Lauzon, R.G. and DeWolf, J.T., (1993) “Full-Scale Bridge Test to Monitor Vibrational Signatures,” Proceedings of American Society of Civil Engineers Conference on *Structural Engineering in Natural Hazards Mitigation*, Volume 2, Irvine, California, U.S.A., pp. 1089-1094.

5. Conn, P.E. and DeWolf, J.T., (1995) “Bridge Signatures for Continuous Bridges,” Department of Civil and Environmental Engineering, University of Connecticut, Storrs Connecticut, U.S.A.

6. Descoteaux, T.J. and DeWolf, J.T. (1993) “Influence of Diaphragms on the Overall Performance of Bridges,” Proceedings of American Society of Civil Engineers Conference on *Structural Engineering in Natural Hazards Mitigation*, Volume 2, Irvine, California, U.S.A., pp. 963-968

7 Bernard, K.J. and DeWolf, J.T. (1994) “Fatigue Failure Investigation - Putnam Bridge,” Department of Civil and Environmental Engineering, University of Connecticut, Storrs Connecticut, U.S.A.

8. Sartor, R.R., (1995) “Strain Monitoring of Highway Bridge Structures,” M.S. Thesis, Department of Civil and Environmental Engineering, University of Connecticut, Storrs Connecticut, U.S.A.

9. O’Leary, P.N., Sartor, R. And DeWolf, J.T. (1995) “Evaluation and Retrofit of Spandrel Columns on a Historic Multi-Span Concrete Arch Bridge,” Proceedings of 12th Annual International Bridge Conference and Exhibition *A World of Bridges*, Pittsburgh, Pennsylvania, U.S.A. (In press).

34 COUPLING VIBRATION OF TURBOMACHINERY ROTOR BLADES AS A POSSIBLE CAUSE OF DISK RIM FAILURE

A.P. ZINKOVSKII
Institute for Problems of Strength, Kiev, Ukraine

Abstract
Results of numerical experiments involving the determination of possible causes of disk rim failure of a rotor wheel with blades, having the root of a dovetail joint type, are presented. The experiments were performed using the theory of coupling vibration of mechanical systems. Numerical investigations of specific features of natural vibration spectrum of the rotor design considered and of the influence of the blade frequencies mistuning upon resonant vibrations have been performed to elucidate the cause of rather high vibration stresses in the disk rim.
Keywords: Coupling vibration, failure, frequency mistuning, localization, rotor wheel.

1 Introduction

The object of the investigation is a rotor wheel of the gas turbine engine compressor with blades, having the root of a dovetail joint type. In the case of such root joints, inter-groove parts of the rim are acted upon by static and dynamic loads from adjacent blades.

The data collected during laboratory tests and operation of gas turbine engines show that one of possible defects is the formation of cracks in the disk rim and breaking off of the inter-groove part of the rim. The results of metallographic analysis of the disk rim fracture surfaces have shown that fatigue crack formation caused by the action of rather high alternating stresses can be a probable cause of the disk rim failure and always starts at an acute angle of the inter-groove part of the rim. In this connection, the analysis of possible vibration stressing of the inter-groove parts of the disk rim as the main cause of crack initiation in them is of interest.

As is known [1.2], small structural imperfections, resulting from the manufacturing

Structural Failure, Product Liability and Technical Insurance, Proceedings of Fifth International Conference. Edited by H.P. Rossmanith. Published in 1996 by Taylor & Francis, 2 Park Square, Milton Park, Abingdon, Oxon, OX14 4RN. ISBN 0 419 22080 1

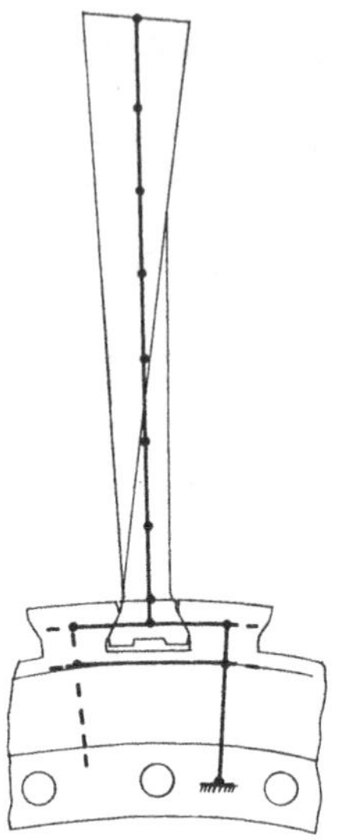

Fig. 1. Rod finite element model of a rotor wheel section

tolerances, which violate the cyclic symmetry of the rotor wheel can lead to a significal variation in the blade resonant amplitudes of vibration within the same assembly. This scatter of the blade resonant vibration amplitudes causes different dynamic stressing of the inter-groove part of the rim. Therefore, low levels of dynamic stresses observed during operational development of engines in individual inter-groove parts of the rim which did not fail give no grounds for a conclusion about low vibration stressing of other (both fractured and nonfractured) inter-groove parts.

Thus the objective of the work is to investigate numerically the influence of specific features of the natural vibration spectrum of a tuned wheel and of a possible frequency mistuning of blades upon vibration stressing of the inter-groove parts of the disk rim.

2 Modeling

2.1 Finite element model of a rotor wheel

A rod finite element model was chosen to determine the natural vibration spectrum for the rotor wheel under investigation. Figure 1 shows a model of the considered rotor wheel section as of a key element of the system with cyclic symmetry. When choosing a model it was supposed the connection of contacted inclined surfaces of the blade root and the disk rim inter-groove part occurring under the influence of centrifugal forces is such that relative displacements of these surfaces can be neglected. The validity of the choice of such a computational model proves to be true by good agreement between the results of the calculation of natural frequencies of some vibration modes of a rotor blade, obtained on the basis of the rod and shell models, and the data collected from testing a real blade clamped in the root section (Table 1). However, it is necessary to note, that for the second mode of vibration, for example, the calculated value of the

Table 1. Results of the investigation of natural vibration of a blade clamped in root section

Blade model	Values of natural frequencies of blade vibration modes investigated (hertz)			
	1st mode	2d mode	3d mode	4th mode
Experiment	340-350	1169	1550	-
Thick-walled shell: 21 elements, 168 nodes	357	1206	1587	2800
Thin shell: 84 elements,56 nodes	361	1206	1685	2714
Rod model: 7 elements, 8 nodes	354	1368	1482	3369

natural frequency, obtained using the rod model, exceeds the experimental one by 16 %, which is explained by the interaction of the bending and torsion components of the vibration mode considered. This to all appearances, explains significant distinction between the calculated frequencies of the blade fourth vibration mode .

2.2 Lumped parameter model of a rotor wheel

To determine the influence of the blade frequencies mistuning upon vibration stressing of the disk rim inter-groove parts a lumped parameter model of the rotor wheel was developed which is shown in Fig. 2. The vibration system considered is closed and has the properties of a cyclic symmetry. Each blade clamped in the root section is represented by a subsystem with a single degree of freedom. Mass $M_1^{(n)}$ and the coefficient of stiffness $k_1^{(n)}$ represent modal parameters of the blade and define natural frequency of its vibration $p^{(n)} = \sqrt{k_1^{(n)}/M_1^{(n)}}$, where n is the number of the blade vibration mode. It is assumed that there is an elastic coupling of the blade root of mass M_2 with the disk rim, which is simulated by springs with the coefficient of stiffness k_2. It should be noted, that the magnitude of the coefficient of stiffness k_2 depends on that of the centrifugal force and with an increase in the latter the elastic flexibility of the blade attachment decreases. During vibration the inter -groove parts of the disk rim ,

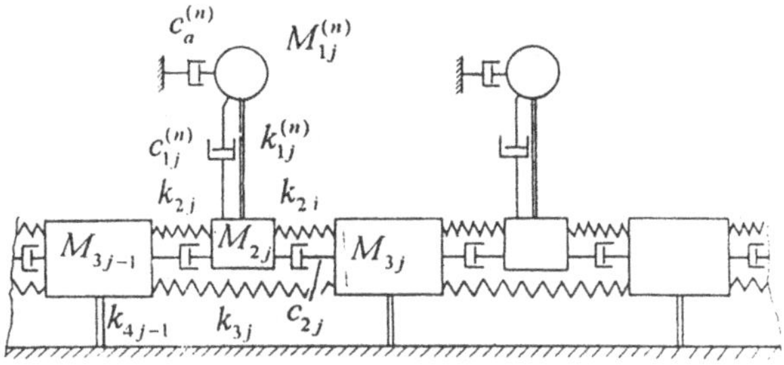

Fig. 2. Lumped parameter model of a rotor wheel

which are simulated by masses M_3, undergo cyclic deformation and can influence each other, this fact being taken into account in the proposed model through springs with the coefficient of stiffness k_3. Flexibility of a disk is simulated by springs with the coefficient of stiffness k_4 that also provides elastic coupling of the blades. Vibration damping of the rotor wheel considered is caused by inelasticity of the blade material , energy losses in the root joint and aerodynamic resistance of the flow around the blade. To take into account the sources of energy losses specified, the model involves viscous dampers with viscous damping coefficients $c_1^{(n)}, c_2, c_a^{(n)}$ respectively.

One of the main sources of the blade vibration excitation is circumferential nonuniformity of the gas flow about which the rotor wheel is rotating. Each blade rotating through circumferential nonuniformity of the gas flow is exposed to the influence of an external force of identical amplitude. The frequency ν of this force is multiple to the frequency of the rotor wheel rotation, and the phase difference is equal to the product of the excitation order m_B and the magnitude of the angle between blades. Therefore, the external force of a given excitation order, acting on the j-th subsystem, which simulates a blade, can be presented in the form

$$F_j = F_0 \cos[\nu t + (j-1)2\pi m_B / N] \tag{1}$$

Then the system of differential equations, describing forced vibrations of the rotor wheel model considered, has the following form:

$$\begin{aligned}
&M_{1j}^{(n)} \ddot{x}_{1j} + c_a^{(n)} \dot{x}_{1j} + c_{1j}^{(n)} (\dot{x}_{1j} - \dot{x}_{2j}) + k_{1j}^{(n)} (x_{1j} - x_{2j}) = \\
&= F_0 \cos[\nu t + (j-1)2\pi m_B / N]; \\
&M_{2j} \ddot{x}_{2j} + c_{1j}^{(n)} (\dot{x}_{2j} - \dot{x}_{1j}) + c_{2j} (2\dot{x}_{2j} - \dot{x}_{3j} - \dot{x}_{3,j-1}) + \\
&+ k_{1j}^{(n)} (x_{2j} - x_{1j}) + k_{2j} (2x_{2j} - x_{3j} - x_{3,j-1}) = 0; \\
&M_{3j} \ddot{x}_{3j} + c_{2j} (\dot{x}_{3j} - \dot{x}_{2j}) + c_{2,j+1} (\dot{x}_{3j} - \dot{x}_{2,j+1}) + k_{4j} x_{4j} + \\
&+ k_{2j} (x_{3j} - x_{2j}) + k_{2,j+1} (x_{3j} - x_{2,j+1}) + k_{3j} (x_{3j} - x_{3,j-1}) + k_{3,j+1} (x_{3j} - x_{3,j+1}) = 0, \\
&j = 1,2,\ldots,N, x_{k0} = x_{kN}; x_{k,N+1} = x_{k1}, k = 1,2,3.
\end{aligned} \tag{2}$$

3 Results and discussion

3.1 Natural vibration of a tuned rotor wheel

Frequency functions of the rod finite element model of the rotor wheel considered (Fig. 1) for the case of its strict symmetry are presented in Fig.3. The frequency function is understood as the dependence of the system natural frequencies $p_{mn}^{(o)}$ on the number of deformation waves m (or the number of node diameters) along the circumference, where index "o" means strict cyclic symmetry of the system, n is the number of frequency function. It should be noted, that relative values of the system natural frequencies $\overline{p}_{mn}^{(o)} = p_{mn}^{(o)} / p_{o1}$ (where p_{o1} is the first natural frequency of an individual subsystem) are plotted on the ordinate axis.

The estimation of frequency functions was made for a disk of two modifications:

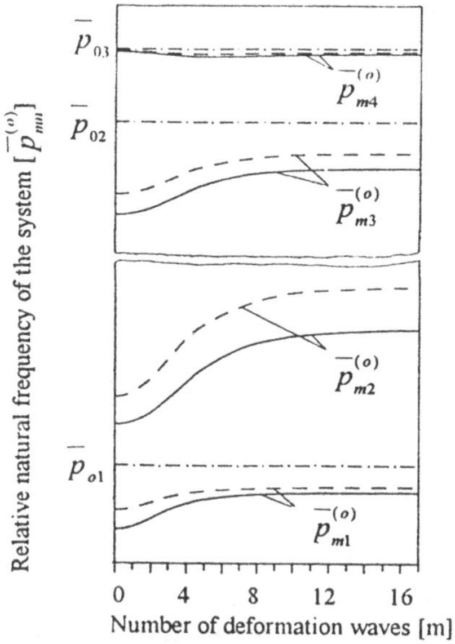

Fig. 3. Frequency functions of the rod finite element model of the tuned rotor wheel

with an initial thickness of the rim (solid lines) and with rim thickened 1.364 times (dashed lines). According to the classification presented in [1], the frequency functions shown in Fig. 3 correspond to the systems "elastic disk and elastic blades " (n=1,3,4) and " elastic disk and rigid blades " (n=2). Moreover, the Figure gives relative values $\bar{p}_{on}$(n=1,2,3) of natural frequencies of the blade clamped in the root section, which determine frequency functions of the system " rigid disk and elastic blades " (dot-and-dash line).

We shall consider some specific features of the frequency functions presented. The width of the frequency spectrum , corresponding to the given frequency function, which characterizes its density and is determined by parameter

$$\Delta p_n^{(o)} = \left|(p_{mn}^{(o)})_{m=N/2} - (p_{mn}^{(o)})_{m=0}\right|, \tag{3}$$

depends on the disk stiffness. As is evident from the values of parameter $\Delta p_n^{(o)}$ listed in Table 2, the width of the frequency spectrum, except for the frequency function n=2, decreases with higher disk rigidity owing to an increase in its thickness, and the

Table 2. The magnitudes of parameter $\Delta p_n^{(o)}$ (hertz)

Frequency function number ,n	Disk modification	
	Initial disk	Thickened disk
1	58.3	36.5
2	147.0	171.1
3	72.6	64.0
4	3.2	1.7

frequency density increases.

For the system " elastic disk and elastic blades " one mode of the blade vibration corresponds to each of the frequency functions considered. The blade mode shape is determined by an appropriate frequency function of the system " rigid disk and elastic blades ", which it is approaching asymptotically with an increase in the number of deformation waves. It should be noted, that such approach occurs faster for rotor wheels with a more rigid disk.

The second frequency function (n=2), characteristic of the system "elastic disk and rigid blades ", corresponds to the vibration of the rotor wheel accompanied by the blade displacements in the direction of the axis of their maximum stiffness as nondeformable bodies. In this case, as is pointed out in [1], dangerous vibration of disks may occur at which the displacement of blades as rigid bodies causes displacements of the disk mass in the direction of the system axis.

3.2 Resonant vibration of a mistuned rotor wheel

The violation of the symmetry of the rotor wheel as a system " elastic disk and elastic blades " induces a scatter of the resonant vibration amplitudes of blades. It is known [3], that the formation of vibrations of the mistuned rotor wheel is determined by the ratio of the blade frequency scatter to the spectrum width of the symmetric rotor wheel frequencies, corresponding to their frequency function considered. It is shown, that if the ratio of these parameters is of one order of magnitude or higher, the phenomenon of modal localization is possible. In this case , one or a few blades of the assembly vibrate on certain frequencies of the system. Moreover, it is established [4], that the degree of vibration localization may be governed by the density of only a particular portion of the frequency function of the tuned system.

With the use of the computational model developed (Fig. 2) numerical experiment was performed which involved the determination of the influence of circumferential distribution of the blade frequencies upon the formation of resonant vibrations of the wheel as a system " elastic disk and elastic blades " in the presence of the phenomenon of modal localization. As an example, the paper presents the results of computations for N=8. The analysis involved such values of stiffness coefficients of elastic elements that the first frequency function of the tuned system had the appearance shown in

Table 3. Variants of relative natural frequencies distribution of blade simulating subsystems

Subsystem number	Variant *à*	Variant *b*
1	1.000	1.000
2	1.035	1.040
3	1.010	1.010
4	1.045	1.010
5	1.015	1.045
6	1.050	1.005
7	1.005	1.050
8	1.040	1.035

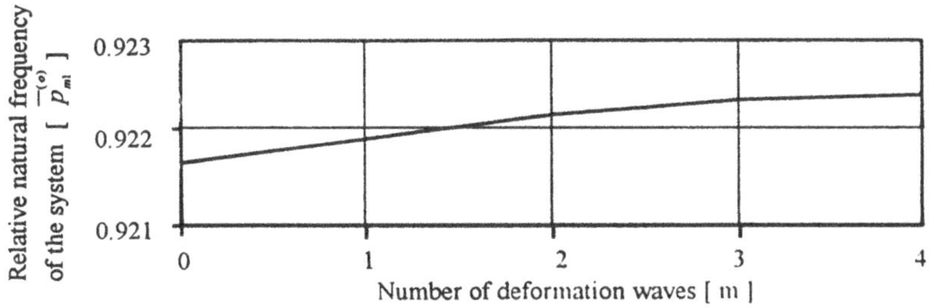

Fig. 4. First frequency function of the lumped parameter model of a tuned rotor wheel

Fig. 4. The computations were performed for two variants of subsystems frequencies distribution (Table 3), the difference between which is as follows. Variant ***a*** is close to an alternating distribution and is characterized by the fact that the neighbouring subsystems have essentially differing (up to 5 %) natural frequencies. The peculiarity of variant ***b*** is that two neighbouring subsystems (the 3d and the 4th) have identical natural vibration frequencies, which differ appreciably from those of adjacent subsystems.

The analysis of the vibration modes of the rotor wheel model considered shows that in the case of variant ***a***, when the system vibrates with a frequency of any subsystem, the adjacent subsystems will remain practically immobile owing to modal localization, thus being restrictors of the displacements of the masses, which simulate the inter-groove parts of the disk rim. For variant ***b*** owing to the excitation of such vibration modes when only two neighbouring subsystems are vibrating with identical frequencies, while others remain practically immobile , the stressing of the inter-groove parts of the disk rim between the above subsystems may increase much. To confirm these assumptions, Fig. 5 shows the distributions of relative maximum displacement amplitudes of masses , simulating a blade ($\overline{A}_1^{(max)}$) and a inter-groove part the disk rim ($\overline{A}_3^{(max)}$) achievable in a resonant zone. The analysis of the data obtained shows that in the case of subsystem frequencies distribution according to variant ***b*** the maximum vibration stressing of the disk rim inter-groove parts increases about 1.6 times in comparison with that for variant a. Thus, in this case, the highest stresses are observed in the inter-groove parts of the disk rim between blades with identical frequencies. It should also be noted, that if the scatter of the vibration amplitudes of the subsystems simulating the blades remains practically constant with the variation of their frequency distribution considered, the scatter of displacement amplitudes of masses simulating the inter-groove parts of the disk rim increases 1.5 times in comparison with variant ***a*** with the distribution of the subsystem frequencies according to variant ***b***.

4 Conclusions

To elucidate the causes of rather high dynamic stressing of a disk rim numerical investigation was performed which involved the determination of the natural vibration

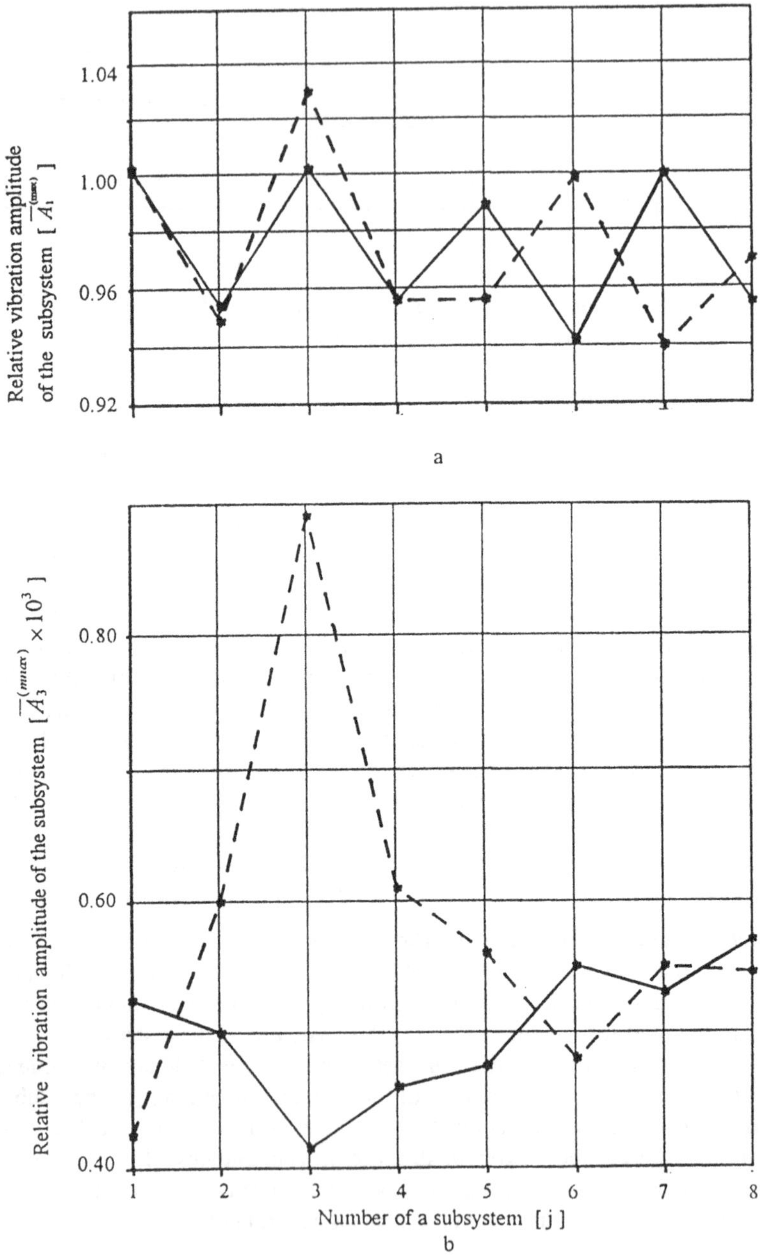

Fig. 5. **Distribution of maximum vibration amplitudes of subsystems simulating a blade (a) and disk rim inter-groove parts (b) for chosen variants of subsystem frequencies distribution: solid lines - variant *a*; dashed lines - variant *b***

spectrum for the rotor wheel design considered with the use of its rod finite element model. The presence of the vibration modes corresponding to the systems " elastic disk and rigid blades " and " elastic disk and elastic blades " is shown to be a typical feature of the natural vibration spectrum of the rotor wheel design considered.

On the basis of the given numerical estimations of natural vibrations of the system "elastic disk and elastic blades ", a model of a rotor wheel with lumped parameters has been developed which makes it possible to investigate the influence of the blade frequencies mistuning and parameters of their coupling upon vibration stressing of blades and inter-groove parts of the disk rim.

The analysis has been performed of the influence of different variants of mistuned blades distribution along the circumference of the disk rim upon resonant vibration of the rotor wheel when modal localization occurs. When the blade frequencies distribution along the circumference of the disk rim is alternating or close to it, motionless blades restrict the displacement of inter-groove parts of the rim. However, if two neighbouring blades have equal or close natural vibration frequencies, which differ appreciably (5%) from those of the neighbouring blades, an increase in vibration stressing of the inter-groove part of the disk rim and its failure may occur due to modal localization

5 References

1. Ivanov, V. P. (1983) *Vibration of Turbine Rotor Wheels*, Mashinostroenie, Moscow.
2. Ewins, D.J. (1969) The effects of detuning upon the forced vibrations of bladed disks. *J.Sound and Vibr.*, Vol. 9, No. 1. pp. 65-79.
3. Hodges, C.H. (1982) Confinement of vibration by structural irregularity. *J. Sound and Vibr.*, Vol. 82, No. 3. pp.411-24.
4. Zin'kovskii, À.P., Vasil'ev, Ì.V. and Buslenko, I.N. (1992) Distortion of the sinusoidalness of the forms of vibration of systems with a disturbed rotary symmetry and the mode localization phenomenon. Report 2, *Strength of materials*,Vol.24, No. 11. pp.678-84.

35 LIMITATIONS OF A KINEMATIC HARDENING MODEL FOR ASSESSING RATCHETTING FAILURES IN PRESSURISED PIPES

R.J. RIDER, P.J. HANCELL AND S.J. HARVEY
School of Engineering, Coventry University, UK
I.D. CHARLES
Masotech Engineering Europe, Rochford, UK

Abstract

Design procedures within the nuclear power generation industry require that the possibility of incremental collapse or ratchetting incurred in pressure vessels and large pressurised piping runs during seismic disturbances be considered. This requirement has led to a programme of experimental work to simulate component and material behaviour under such conditions and to consider theoretical models so that predictive rather than codified design analysis can be performed.

As part of this programme, the plastic deformation of thin-walled cylinders has been experimentally examined for the loading conditions of ±1% cyclic axial strain with hoop stresses of approximately 0, 1/4, 1/2 and 3/4 of the initial uniaxial yield stress.

Two materials similar to those used in the pipework of PWR nuclear plant in the U.K. have been tested, namely 304S11 stainless steel and En6 low-carbon steel. Under the loading conditions both materials incurred plastic hoop ratchet strains to varying degrees. This ratchetting behaviour was compared to the behaviour predicted by the Prager-Ziegler model of kinematic work-hardening.

The experimental results indicate that there is a clear divergence with theoretical predictions. It is shown that when the cyclic plastic strains dominate, the model predicts the limiting condition is rapidly attained within a few cycles.

Keywords: cyclic plasticity, hysteresis, kinematic hardening, ratchetting,

Nomenclature

σ_{ij} = stress tensor.
α_{ij} = translation of the yield surface in stress space
α_z = translation in axial stress direction
α_θ = translation in hoop stress direction
$\varepsilon_{\theta p}$ = plastic hoop strain
$d\mu$ = proportionality scalar

Structural Failure, Product Liability and Technical Insurance, Proceedings of Fifth International Conference. Edited by H.P. Rossmanith. Published in 1996 by Taylor & Francis, 2 Park Square, Milton Park, Abingdon, Oxon, OX14 4RN. ISBN 0 419 22080 1

1 Introduction

The risk of failure in pressure vessels and piping, due to low cycle fatigue and incremental collapse or ratchetting is of considerable concern in the nuclear power generation industry. These failure mechanisms may be induced by conditions such as thermal cycling and seismic disturbances. In order to guard against low cycle fatigue failure, pressure vessels and their associated support structures are usually designed in accordance with rules such as the ASME Code Section III [1] It has been shown however, Rangarth [2], that over-conservatism in the design rules can result in excessive piping support. This in turn can then increase the possibility of fatigue failure. Additionally, the codes are based on static collapse rather than incremental collapse and are therefore limited when cyclic plastic deformations are considered.

In the high-strain, low-cycle fatigue règime, material behaviour is complex and difficult to describe mathematically. Simple models to describe cyclic plastic deformations are therefore attractive. Most commonly, for ductile metals, the Prager-Ziegler kinematic hardening model is used in combination with the von Mises yield criterion. Whilst this model is able to qualitatively account for the Bauschinger effect observed in reversed cyclic plasticity it does not appear able to accurately represent the observed accompanying ratchetting deformations. A purely kinematic model is also unable to account for cyclic strain hardening or softening phenomena.

This paper presents some preliminary results from experimental studies into the ratchetting behaviour of steels similar to those used in PWR nuclear plant pipework. In particular, the ratchetting behaviour of a low carbon steel will be compared with the predictions obtained from the Prager-Ziegler model. The model will be applied from the beginning of the loading history and will include a simplified representation of isotropic hardening to partially take account of cyclic material hardening observed in the early stages of cyclic loading. It will be argued that whereas this material will continue to accumulate ratchet strain until fatigue failure, the Prager-Ziegler model predicts a very rapid cessation of ratchet strain accumulation.

2 Kinematic Hardening

The general expression for kinematic hardening as described by Prager [4] and applied to a bi-linear material is given by:-

$$f(\sigma_{ij} - \alpha_{ij}) = 0 \tag{1}$$

with the yield surface translating in stress space in the direction of the outward normal of the yield surface at the stress point according to :-

$$d\alpha_{ij} = c\,d\varepsilon_{ij} \tag{2}$$

where c is a function of the post yield modulus.

In a modification due to Ziegler [3], the yield surface is assumed to translate along the vector joining the current stress point to the centre of the surface, i.e. :-

$$d\alpha_{ij} = d\mu\,(\sigma_{ij} - \alpha_{ij}) \tag{3}$$

By assuming that plastic deformation is independent of hydrostatic stress it becomes convenient to consider only the deviatoric stress system. In this case, the Prager and Ziegler rules coincide. This point is shown schematically in Fig. 1 where the direction of the current load vector, $\overline{\sigma}$, is colinear with the plastic strain increment vector $d\varepsilon_p$.

Whilst the Prager-Ziegler model is most usually applied with a constant value for the post-yield modulus, this is not, as Ziegler pointed out, a requirement of the model. In this paper the model will be applied at incremental steps in the loading history , using the post yield modulus currently appertaining. The size of the yield surface will be updated at the beginning of each semi-cycle to take some account of cyclic strain hardening.

3 Materials And Test Details

The material for the results presented here was a low-carbon steel, En6, similar to those commonly used in nuclear plant pipework. The En6 steel was normalised and shot-blasted prior to machining. Thin-walled tubular push-pull specimens were manufactured from each material with a nominal gauge length of 40 mm, an outer diameter of 22 mm and bore of 20 mm.

The specimens were tested under strain-controlled cyclic push-pull conditions of ±1% axial strain at a rate of 0.1%/min. Additionally internal pressures were applied hydraulically, allowing the cyclic loading to be combined with static tensile hoop stresses of nominally 0, 1/4, 1/2 and 3/4 of the uniaxial yield stress. Tests were conducted both with the axial strain initially compressive and with the axial strain initially tensile. Tests were limited to 15 cycles of cyclic strain in order to avoid excessive amounts of buckling of the specimen gauge length. The test rig, instrumentation and experimental procedures are fully described elsewhere [5].

4 Experimental Results

The results presented here are intended to illustrate the problems of the Prager-Ziegler model and are therefore restricted to the case of the hoop stress being *nominally* 1/2 the uniaxial yield stress. This material exhibits an initial uniaxial yield strength of about 200 N/mm^2. Although 1/2 yield is therefore about 100 N/mm^2, practical limitations meant that over the period of the test, the average applied hoop stress was about 130 N/mm^2.

Hysteresis loops form under the cyclic plastic loading conditions and plastic hoop strain ratchetting occurs when a steady hoop stress is applied. The plastic strain increments shown diagrammatically in Fig. 1 illustrate how plastic hoop strain ratchetting occurs even though the magnitude of the hoop stress is of elastic order.

The development of typical hysteresis loops and the inter-cycle hoop ratchetting

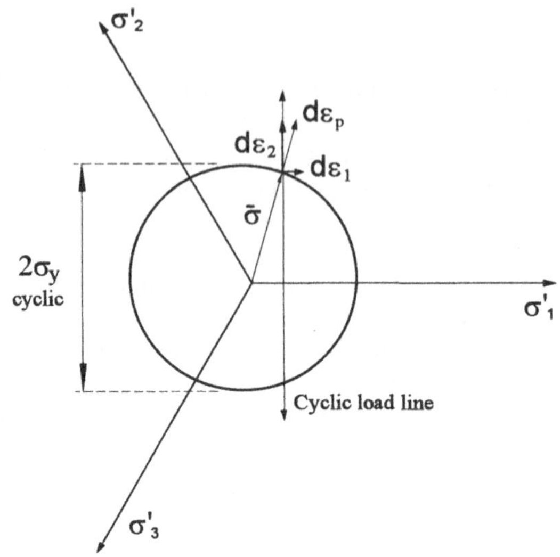

Fig. 1. Development of ratchet strains in deviatoric stress space

behaviour for the low-carbon steel En6, with an initial compressive pre-stress, are shown in Figs. 2 and 3 respectively. Fig. 2 shows that this material cyclically hardens, reaching a stable hysteresis loop by the fourth cycle.

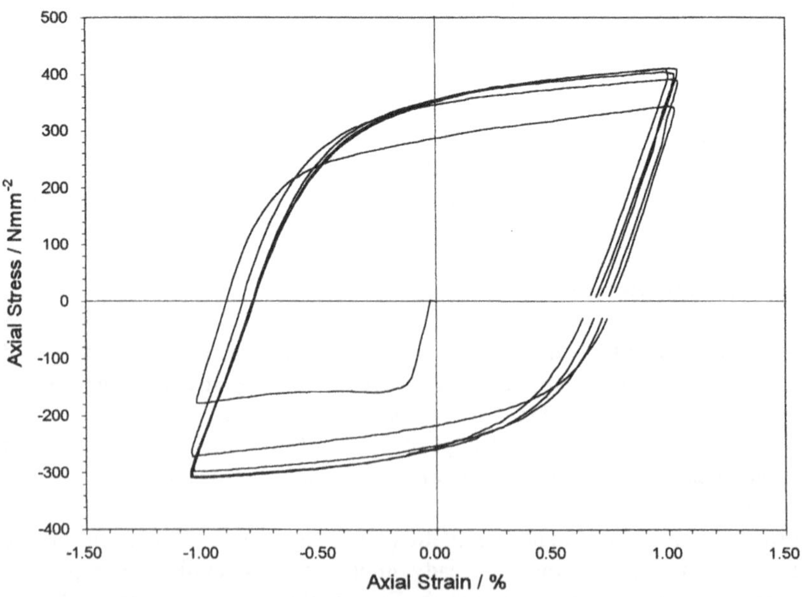

Fig. 2. Hysteresis loop for En6 low carbon steel. Compressive pre-stress. Nominal hoop stress 130 N/mm^2

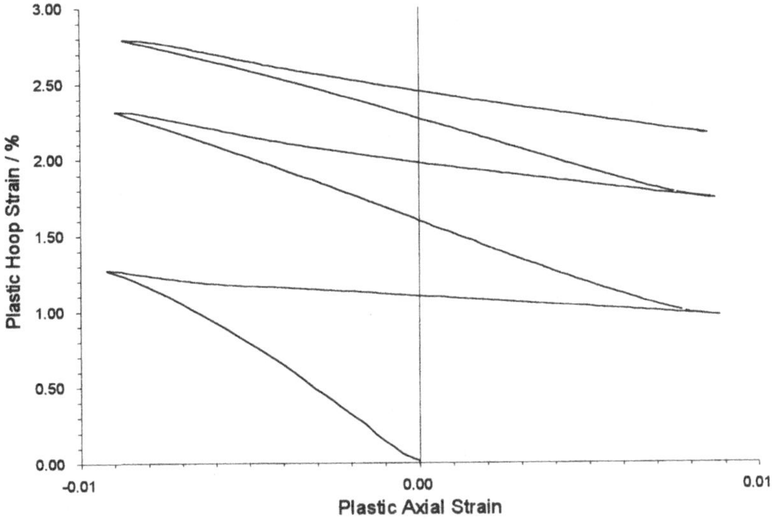

Fig. 3. Plastic hoop ratchet for En6 low carbon steel. Compressive pre-stress. Nominal hoop stress 130 N/mm^2

Fig. 3 shows that the plastic hoop ratchet strains are initially large then gradually reduce in magnitude. This pattern is reflected in Fig. 4 which shows the hoop strain accumulated at the end of each 1/2 cycle plotted against the cycle number.

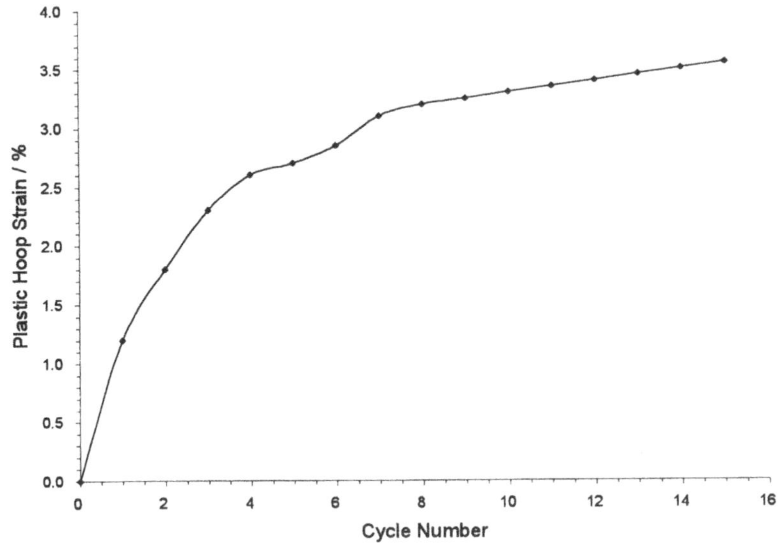

Fig. 4. Plastic hoop strain accumulation En6 low carbon steel. Compressive pre-stress.

5 Modelling

The experimental results show that even under the strictly prescribed loading conditions described here that the material response is quite complex. In attempting to apply the Prager-Ziegler model in applications such as finite element codes, it is often the case that the hysteresis loops are modelled bi-linearly and calculations based on a constant post yield modulus (H). As illustrated by Rider at al [7], a variety of bi-linear interpretations may be made which lead to a variety of assumptions for the values of the yield stress and the post yield modulus H. Predictions for the hoop ratchet strain are particularly sensitive to the assumed values of H so to avoid these problems it was decided to apply the experimental data (i.e. the hysteresis loops and hoop stress) to the incremental form of the model.

At the current stage of modelling, each 1/2 cycle is processed separately, with an estimate made from the data, of the yield stress appertaining to the beginning of that 1/2 cycle. Additionally, a constant hoop stress is assumed although in practice it varies due to limitations of the test rig equipment. Future developments in the modelling will account for the hoop stress variations and attempt to better represent the cyclic strain hardening.

The behaviour of the model under the present modelling constraints for the loading conditions described here is shown in Fig. 5. This shows the translation of the centre of the yield locus in hoop stress - axial stress stress space over four complete cycles of loading. It can be seen that more than 2/3 of the translation has occurred by the end of the first cycle and by the end of the third cycle the centre is almost coincident with the applied hoop stress. For practical purposes it can be said that in the fourth cycle the centre is coincident with the load line. At this stage, the Prager-Ziegler model predicts

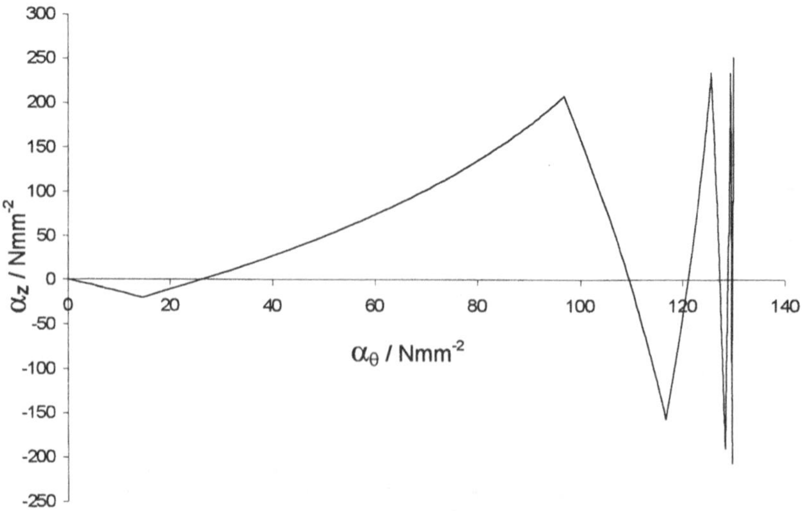

Fig. 5. Yield locus centre translation. En6 low carbon steel. Compressive pre-stress.

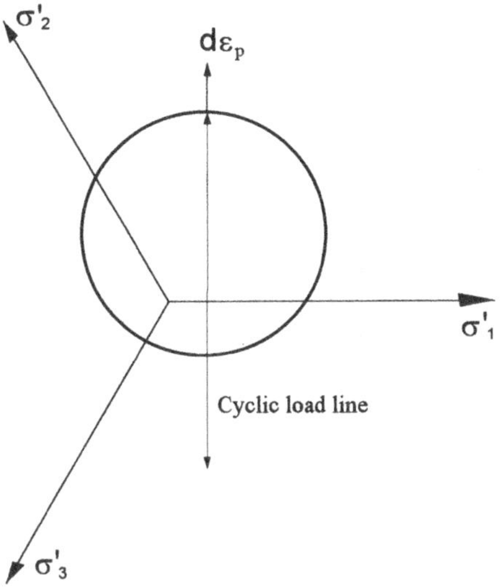

Fig. 6. Fully translated yield locus in deviatoric stress space

a complete cessation to hoop strain ratchetting. The position of the yield locus in deviatoric stress when this condition has been reached is shown schematically in Fig. 6. It can be seen from this diagram that the plastic strain increment vector contains no hoop strain ratchet component.

6 Discussion And Conclusions

Comparison of Figs. 4 and 5 clearly demonstrate the main limitation of the Prager-Ziegler kinematic hardening model. Whilst this model has predicted a limit to ratchet strain accumulation by the fourth cycle, the experimental results show that hoop strain continues to accumulate with further cycling. In this case, the tests were only conducted up to fifteen cycles. Data obtained by Bright [7] however, in cyclic plastic torsion with similar equivalent stresses, suggest that a fatigue life of about 40 cycles could be expected, producing a total hoop strain accumulation of approximately 10%.

In terms of the ASME III code nomenclature, the loading configuration effectively simulates a primary stress (the hoop stress) combined with a cyclic secondary stress (the axial stress) in which the secondary stresses are the dominant ones. It is possible that the Prager-Ziegler model may be particularly unsuited to these conditions. Further work is therefore planned to investigate the effects of different cyclic secondary stress ranges and a broader spectrum of primary stresses.

7 References

1. ASME Boiler and Pressure Vessel Code, Section III (1989).
2. S.Rangarth (1990) *Piping and Fitting Dynamic Reliability Program Vols 1-5.* GE Nuclear Energy EPRI Contract RP1543-15.
3. H.Ziegler (1959) A Modification of Prager's Hardening Rule. *Quart. Appl. Math.* **17**, pp. 55.
4. W.Prager (1955) The theory of plasticity: A survey of recent achievements. *Proc. IMechE.*, **169**, pp. 41-57.
5. I.D.Charles (1992) Ratchetting Strains in Pressurised Pipes. Ph.D. Thesis. Coventry Polytechnic CNAA.
6. R.J.Rider, S.J.Harvey, I.D.Charles (1995) The Application of a Kinematic Hardening Model for Assessing Ratchetting in Pressurised Pipes. *J. Nuclear Engng. and Design.* Topical Issue. To be published.
7. M.R.Bright. (1979) Cyclic strain induced creep and fatigue under complex stress. Ph.D. Thesis.Lanchester Polytechnic (Rugby) CNAA.

36 PROBABILISTIC PREDICTION OF FATIGUE LIFE IN THE SHORT CRACKS RANGE

D. KOCANDA, S. KOCANDA and H. TOMASZEK
Department of Mechanical Engineering, Military University of Technology, Warsaw, Poland

Abstract
Probabilistic approach to predicting fatigue life of components in the range of short cracks growth is the subject of the paper. A finite difference equation with coefficients originated from the two-parameter Weibull distribution of crack lengths has been derived to formulate a model of short cracks growth dynamics. The solution of the Fokker-Planck parabolic differential equation has been found in form of a crack length density function to calculate fatigue life in the range of short cracks growth. The power of the probabilistic method has been verified using experimental data for medium carbon steel specimens fatigued under reversed torsion. Good agreement between theoretical and real fatigue lives of specimens has been achieved.
Keywords: fatigue life, metals fatigue, probabilistic descriptions, short fatigue cracks.

1 Introduction

Predicting the fatigue life of structural components in the range of short cracks growth is one of more difficult issues within the problem of fatigue strength, which arises from the specific character of the cracks. It is assumed that short cracks are initiated and propagate within one or few grains of structure and are temporarily arrested at grains boundaries and other structural barriers. Therefore, the rate of growth of the cracks under discussion usually decreases irregularly as a number of load cycles increases. Generally, they show the step growth. The crack growth rate decreases to a certain threshold value, e.g. to the level of the threshold stress intensity factor K_{th}.

Structural Failure, Product Liability and Technical Insurance, Proceedings of Fifth International Conference. Edited by H.P. Rossmanith. Published in 1996 by Taylor & Francis, 2 Park Square, Milton Park, Abingdon, Oxon, OX14 4RN. ISBN 0 419 22080 1

Having passed this value the crack rate increases as the rate of long cracks. Such being the case, fatigue life is calculated from either the classic Paris equation or the derivative ones. It should be emphasized that the period of short crack initiation and propagation can be many times longer as the long crack growth period; therefore, it can be decisive for the lifetimes of structural members.

Irregular growth of short cracks in the presence of local plastic deformations makes analytical description of the growth extremely difficult. Laws of non-linear cracking mechanics are first and foremost used while looking for theoretical representations. Linear descriptions are not the last ones, however, to serve this aim. Most often deterministic descriptions and, to a smaller extent, probabilistic ones answer the purpose. The first class of the descriptions is represented by a relatively wide range of works, whereas the second one has rather poor literature. The studies are associated mainly with postulates for statistical descriptions of short cracks and exemplified with the works of Goto [1] and Goto et al. [2]. Goto [1] states that short cracks in the normalized steel 20 are subjected to two-and three-parameter Weibull distributions at the stages of crack initiation and propagation, respectively. The same is true for the crack lengths distributions. Microcracks in the normalized steel 45 behave in a similar way, though the character of their distribution depends on the stress level. Goto et al. in [2] give analysis of short crack rates scattering in the steel 20 being described with Weibull distributions as well, depending on the ranges of crack rates.

In general, the ways and means of searching for probabilistic descriptions of fatigue crack growth phenomena include the following groups: evolutionary models with random Markov processes; discrete models with crack growth random in its character and random values of elementary crack increments from one state of fatigue to another; models in form of stochastic differential equations. In case of the latter ones deterministic equations that describe crack growth are given probabilistic meaning with a multiplicative stochastic process taken into account. Our considerations with difference equations constituting their core are beyond the classification presented; nevertheless, they approximate stochastic models. The models have been used to describe long cracks [3] [4].

The outlined method has been verified using experimental data on short fatigue cracks growth in the medium carbon steel specimens subjected to reversed torsion.

2 Description of the method

Decrease in the growth rate of short cracks results, as already stated, from the temporary crack arrest at grains boundaries and other structural barriers. Therefore, crack growth in time t consists in passing subsequent barriers with intensity $\lambda(t)$. Taking into account the stochastic nature of both growth and density of cracks to be described with the two-parameter Weibull distribution we arrive at the following notation:

$$\lambda(t) = \tfrac{\alpha}{\theta}\, t^{(\alpha-1)} \tag{1}$$

where $\alpha > 0$ and $\theta > 0$ are the parameters of shape and scale, respectively, of this distribution. Time t in the above and subsequent equations can be expressed by numbers of loading cycles N. Let $U_{l,t}$ denote probability that the crack length in time instance t is $l = k \times h$, where k is a number of barriers, and h is related to the average distance between elementary barriers. The following difference equation has been formulated to include the assumptions made:

$$U_{l,t+\Delta t} = \left(1 - \tfrac{\alpha}{\theta}\, t^{(\alpha-1)}\, \Delta t\right) U_{l,t} + \tfrac{\alpha}{\theta}\, t^{(\alpha-1)}\, \Delta t\, U_{l-h,t} \tag{2}$$

The above-formulated equation describes crack growth dynamics in terms of the probabilistic approach. Changing over to functional notation the equation takes the form:

$$U(l, t+\Delta t) = \left(1 - \tfrac{\alpha}{\theta}\, t^{(\alpha-1)}\, \Delta t\right) U(l,t) + \tfrac{\alpha}{\theta}\, t^{(\alpha-1)}\, \Delta t\, U(l-h, t) \tag{3}$$

where $U(l,t)$ - density function of a crack length in time instance t.

With the functions $U(l, t+\Delta t)$ and $U(l-h, t)$ expanded into Taylor series as well as with transformations made the partial differential equation of the Fokker-Planck type results from formula (3):

$$\frac{\partial U(l,t)}{\partial t} = \frac{\alpha h}{\theta}\, t^{(\alpha-1)}\, \frac{\partial U(l,t)}{\partial l} + \frac{1}{2}\,\frac{\alpha h^2}{\theta}\, t^{(\alpha-1)}\, \frac{\partial^2 U(l,t)}{\partial l^2} \tag{4}$$

A specific solution of eq. (4) is to be found, i.e. one that satisfies the following conditions: for $t \to 0$ the solution shows convergence on Dirac function, i.e. $U(l,t) \to 0$ for $l \neq 0$, and $U(0,t) \to +\infty$ in such a way that integral of function $U(l,t)$ equals to unity for all $t > 0$. Such being the case, it can be shown that the solution assumes the form:

$$U(l,t) = \frac{1}{\sqrt{2\pi \frac{h^2}{\theta}\, t^{\alpha}}} \exp\left(- \frac{\left(l - \frac{h}{\theta}\, t^{\alpha}\right)^2}{2\,\frac{h^2}{\theta}\, t^{\alpha}} \right) \tag{5}$$

Function (5) is a crack length density function of Gaussian distribution with parameters of Weibull distribution. An average crack length equals to :

$$E[l] = \tfrac{h}{\theta}\, t^{\alpha} \tag{6}$$

where t^{α}/θ is closely related to a number of barriers passed by the crack. According to (5) the standard deflection is defined in the following way:

$$\sigma_t = \sqrt{\frac{h^2}{\theta} t^{\alpha}} \tag{7}$$

If function (5) is to be used, parameters h, α, and θ have to be estimated on the grounds of experimental data which can be recorded either as lengths of a crack in time instances t or in form of a number of loading cycles N:

$$(l_0, t_0), (l_1, t_1), (l_2, t_2)\ldots\ldots\ldots\ldots\ldots, (l_n, t_n) \tag{8}$$

with $l_0 = 0$, and $t_0 = 0$.

The likelihood function L for the data set (8) and function (5) has been formulated:

$$L = \prod_{k=0}^{n-1} \frac{1}{\sqrt{2\pi \frac{h^2}{\theta} (t_{k+1}^{\alpha} - t_k^{\alpha})}} \exp\left(- \frac{\left((l_{k+1} - l_k) - \frac{h}{\theta} (t_{k+1}^{\alpha} - t_k^{\alpha}) \right)^2}{2 \frac{h^2}{\theta} (t_{k+1}^{\alpha} - t_k^{\alpha})} \right) \tag{9}$$

The following expressions have been received from the condition for the maximum value of function $L(h, \alpha, \theta)$:

$$h^* = \frac{\theta l_n}{t_n^{\alpha}}, \qquad \theta^* = \frac{h^* t_n^{\alpha}}{l_n}, \qquad \alpha^* = \frac{\ln\left(\frac{\theta^* l_n}{h^*}\right)}{\ln(t_n)} \tag{10}$$

Determining the parameters h, α, and θ from formula (10) is rather inconvenient; therefore, further transformations have been made to result in the following relationships:

$$\alpha^* = \frac{\ln l_n - \ln l_1}{\ln t_n - \ln t_1}, \qquad h^* = \frac{l_n}{t_n}, \qquad \theta^* = t_n^{(\alpha^*-1)} \tag{11}$$

The estimated parameters are to be put into formula (5) that expresses crack length density function. The function gives the grounds for determining the fatigue life from the formula:

$$R^*(t) = \frac{1}{\sqrt{2\pi}} \int_{-\infty}^{l_s} e^{-l_t/2} dl \tag{12}$$

The symbols used in this formula denote: $R^*(t)$ - probability that the acceptable short crack length l_d assumed *a priori* will not be exceeded; l_t - a standardized crack length, l_s - a standardized acceptable crack length, with

$$l_t = (l_k - E_t[l]) / \sigma_t \,, \qquad l_S = (l_d - E_t[l]) / \sigma_t$$

$E_t[l]$ expresses the expected crack length in the time instance t, σ_t - standard deflection of a crack length in the time instance t, whereas l_k - actual crack length.

Finally, either time t or a number of cycles N_{KP} determined from Eq. (12) for the assumed level of probability $R^*(t)$ will constitute the fatigue life in the range of short cracks. The sum of numbers of cycles of short and long cracks growth will represent the total life of a component. The number of cycles in the range of long cracks is to be calculated in a generally known way, i.e. on the basis of the integrated Paris formula and the derivative ones.

3 Experimental verification

Some selected results of the short cracks investigations in the specimens of normalized steel 45 were used to verify the method. Specimens of geometry and dimensions shown in Fig. 1 were subjected to reversed torsion at frequency of 5 Hz. Torsional stress was calculated from the Nadai's relationship for the elasto-plastic range, using test diagram of cyclic strain: amplitude of shearing stress τ_a depending on the shear strain γ. The initiated and propagating cracks were recorded using replicas taken from the specimen surfaces after the pre-set number of cycles. Growth, population and lengths of short cracks were examined on the display screen, using the computer-aided system of the microscopic image analysis. The transmission electron microscope (TEM) was used to locate the short cracks, to reveal characteristic features and mechanism of fracture.

Short cracks in carbon normalized steels, steel 45 included, are initiated and propagate practically in ferrite grains. Slip bands are usually the source of crack initiation. Direction of crack growth under reversed torsion is determined by the planes of the highest shearing stresses, although it is not a rule. A network of the surface cracks nearly perpendicular to each other is originated with the specimen-axis-oriented cracks prevailing. This is effected by ferrite bands originated while rolling bars. Microcracks were temporarily arrested at the ferrite grains boundaries and most often completely arrested at the pearlite grains boundaries. Examples of crack initiation and growth have been shown on the micrographs (Fig. 2).

They have been taken from the sample surfaces under examination at $\tau_a = 232$ MPa after 4x10^3 cycles (Fig. 2a, b) and 10^4 cycles (Fig.2c ,d) . Figure 2a shows an individual microcrack 10 μm long, grown in the slip plane inside the ferrite grain. A system of parallel cracks shown in Fig. 2b has become visible in slip bands as well, with well-marked extrusions.

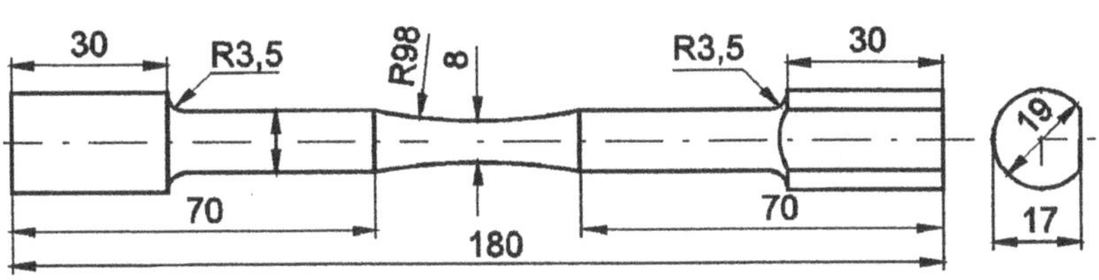

Fig.1. Geometry of specimen (dimensions in mm)

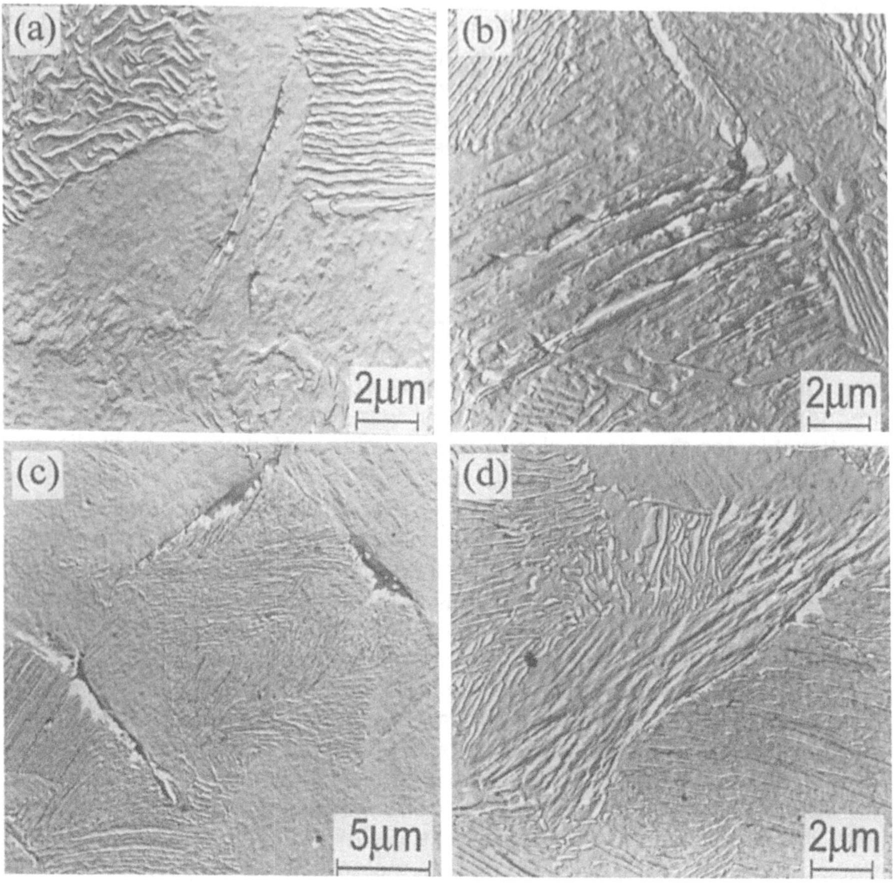

Fig. 2. Surface short cracks; see text for explanations

Three microcracks perpendicular to each other, originated from single slip bands, are shown in Fig. 2c, whereas the strongly developed slip bands even with slip lines visible are to be seen in Fig. 2d. The presented micrographs are the evidence that the cracks grow in the ferrite grains only and are arrested either at the pearlite grains boundaries or by the previously initiated cracks.

Characteristic behaviour of short crack growth have been reflected in diagrams of crack growth rates dl/dN depending on crack length l (Fig. 3a, c) and cycles ratio N_i / N_f (Fig. 3b, d) (N_i - number of current cycles, N_f - number of cycles to failure).

Exemplary diagrams have been prepared for four values of stresses τ_a. Due to high density of the diagram dots and lines only two curves per diagram have been plotted to illustrate experimental and calculated data on the expected crack rates $dE_N(l)/dN$. Continuous lines represent experimental results, whereas the dashed lines the calculated ones. Discrepancies between the results should be attributed to irregular growth of short cracks.

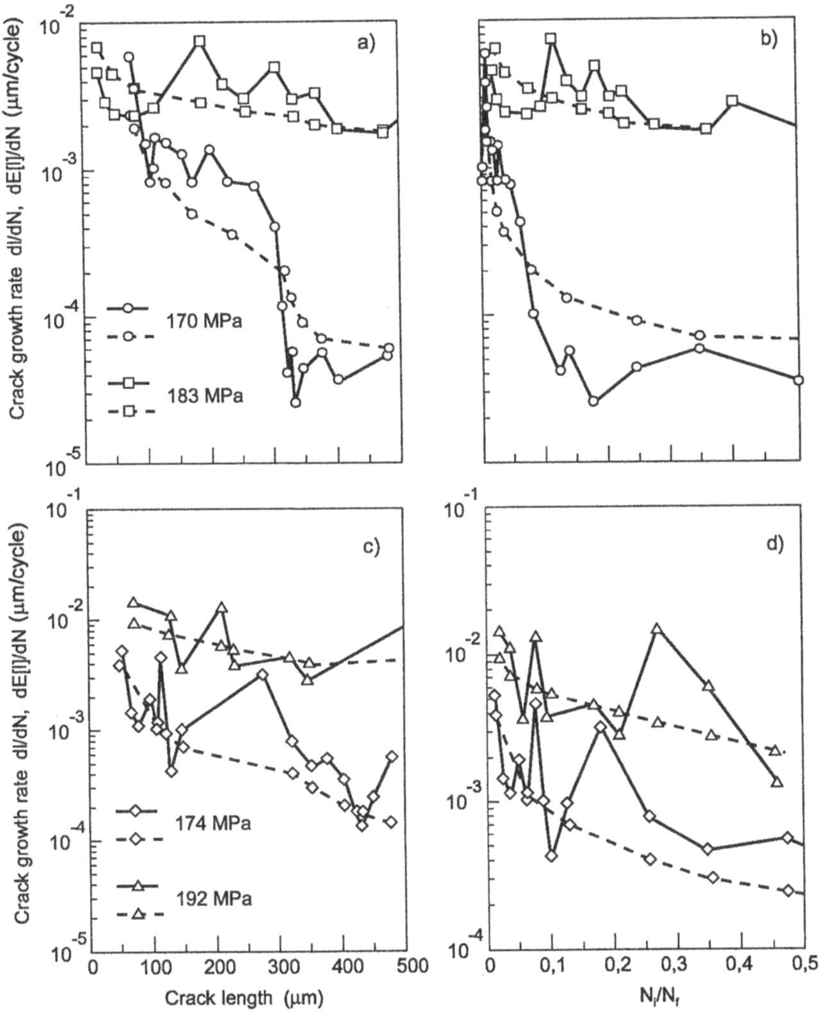

Fig. 3. Crack growth rate vs crack length (a,c) and cycle ratio N_i/N_f (b,d)

The predicted fatigue life up to the acceptable limit, the *a priori* assumed crack length l_d=0.5 mm with probability equal to unity have been shown in Fig. 4.

A number of cycles with the crack length equal to 0.5 mm has not been recorded experimentally. Lengths of cracks have been measured at numbers of cycles approximating those used in calculations. The lengths in μm have been shown on the diagram. Agreement of the experimental results with the calculated ones can be recognized pretty good. In view of the limited size of the publication rich tabular documentation showing results of calculations of the already described quantities $E_N(l)$, s_N, l_s, and R_N^* have been neglected.

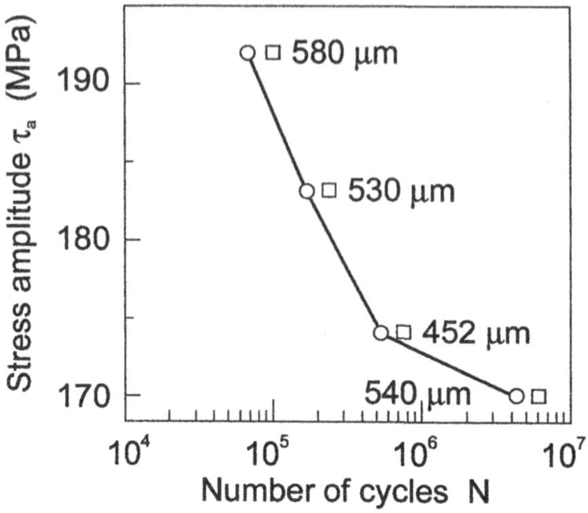

Fig. 4. Fatigue life of 0.45% C steel specimens under reversed torsion, in the range of short cracks

Index N is closely related to a number of loading cycles. It is substitute for loading time indice t. Parameters of Weibull distribution for the stress τ_a= 183 MPa : α^*= 0.6126, θ^*= 0.009637, h^*= 0.002975 have been reported purely to illustrate results of calculations.

4 Final remarks

The presented method enables simplified description of fatigue crack growth and forecasting of fatigue life in the range of short cracks. The method can be generalized to include both the short and long cracks because the curve of crack growth depends on the coefficient α from the Weibull distribution. For $\alpha < 1$ the crack rate decreases monotonically, whereas for $\alpha = 1$ the crack rate is constant in time. Both the cases refer to short cracks. For $\alpha > 1$ the crack growth rate increases, which is characteristic for long cracks. The essence of the method has been derived from earlier works of the authors, those used to determine fatigue life of aircraft undercarriages with one- and two-dimensional crack propagation phenomena [3] [4]. It is worth emphasizing that the method discussed has found confirmation in description of short cracks in components made of the laser hardened steel 45 and ones made of this steel with the surface layers hardened by means of shot peening. Experimental results have been presented in [5] [6] [7].

5 References

1. Goto, M. (1991) *Statistical investigation of the behaviour of microcracks in carbon steels,* Fatigue and Fracture of Engineering Materials and Structures, Vol.14, No. 8. pp. 833-45.
2. Goto, M., Miyagawa, H., Kawagoishi, N., Nisitani, H., *(1994) Relation between the scatter in growth rate of microcracks and small crack growth law,* Recent Advantages in Experimental Mechanics, Balkema, Rotterdam, pp. 1145-50.
3. Kocańda, S., Tomaszek, H., (1992) *Probabilistic method of evaluating fatigue life of aircraft structure components,* Low Cycle Fatique and Elasto-Plastic Behaviour of Materials. Third International Conference Berlin, Elsevier, London, pp. 485-89.
4. Kocańda, S., Tomaszek, H., (1994) *Probabilistic evaluation of fatigue life of structural components in presence of crack propagation process.* Journal of Theoretical and Applied Mechanics, No. 1. pp. 95-108.
5. Natkaniec, D., Kocańda, S., Miller. K.J., (1994) *Short fatigue crack growth in laser hardened medium carbon steel,* Journal of Theoretical and Applied Mechanics, No. 1. pp. 163-76.
6. Natkaniec, D., Kocańda, S., Miller, K.J., (1994) *Behaviour of fatigue short cracks in laser hardened medium carbon steel subjected to reversed torsion,* Proceedings of the 10th European Conference on Fracture, Berlin, EMAS, Warley U.K., Vol. 2, pp. 1259-64.
7. Natkaniec, D., Kocańda, S., Miller, K.J., (1995) *Influence of shot peening on short crack behaviour in a medium carbon steel,* 7th International conference on Mechanical Behaviour of Materials, Book of Abstracts, ESIS, Haga, pp. 705-6.

37 STOCHASTIC ANALYSIS OF THE CREEP BEHAVIOUR OF MULTI-COMPONENT SOLIDS

D.R. AXELRAD
Micromechanics Research Laboratory, McGill University,
Montreal, Canada

Abstract
The creep behaviour of a multi-component material is analyzed in terms of stochastic processes. The final state leading to fracture is treated by a simplified parabolic growth model of the inherent flaws or microcracks.
Key words: Markovian model, stochastic population process, parabolic evolution model for transition probabilities.

1 Introduction

The present study deals with the stochastic formulation of the behaviour of randomly inhomogeneous solids. It focuses in particular on the evaluation and effects of flaws or microdefects on the overall response of the material when a sample of it is subjected to a constant load at high-temperatures. This class of materials comprises the well-known high-temperature compounds or multi-component solids (M.C. systems) that has been previously investigated [1,2] with the emphasis on the transient response and the corresponding structural changes from a very brief elastic to a steady inelastic response. It has been shown that this transient and the microstructural changes of the system can be characterized by a Markov process and a mapping of the relevant field quantities from one subspace to another of a more general probabilistic function space [1] as well as by consideration of some aspects of the geometry of random fields [3].

Hence the present analysis is directed towards the development of flaws or defects, which in the sequel will be referred to as "microcracks (m.c)" in the given microstructure that ultimately leads to the fracture of the M.C. system.

Structural Failure, Product Liability and Technical Insurance, Proceedings of Fifth International Conference. Edited by H.P. Rossmanith. Published in 1996 by Taylor & Francis, 2 Park Square, Milton Park, Abingdon, Oxon, OX14 4RN.
ISBN 0 419 22080 1

Since there is no general theory of "crack nucleation and propagation" as yet available, it is evident that a statistical analysis may provide some insight into the fracture phenomenon of such systems. Since the system under consideration may be regarded as a brittle material that contains a rather large number of flaws, it is convenient to employ the theory of stochastic population processes given in [4,5] or equivalently by satisfying certain conditions, the general theory of point processes [6,7]. It is also possible in the study of evolution of microdefects during the deformation of a M.C. system to use the results of the theory of branching stochastic processes as discussed in refs. [8-16].

The development of microcracks in a given material domain can be characterized by certain "events", ie., whether they become transformed (branching) or extinct and hence depend on certain probability laws with respect to a probabilistic function space. The stochastic analysis concerned with the evolution of the m.c.'s in the material depends essentially on the material characteristic as well as on certain geometrical conditions which have to be satisfied. Although, in general, a rigorous probabilistic formulation should be based on atomistic mechanics, e.g., involving the structure (lattice) of individual elements, their interactions as well as energy considerations, it is beyond the scope of the present analysis. Hence in the stochastic formulation given in the following sections, an approximate analysis will be given, which follows closely the probabilistic and axiomatic definitions given earlier in [1]. It may be convenient to review some of these notions and definitions in the next section.

2 Basic concepts and definitions

The development of flaws or microcracks during the creep deformation is basically influenced by "local processes" on the atomic scale, whereby the material becomes organized into "clusters". The latter are nucleated due to the presence of voids in structural elements, but mainly due to grain boundary sliding (cavitations), (see also [11]). The voids in the structural elements develop into larger units which cause ultimately the failure of the material.

In the stochastic mechanics of discrete media [1], such clusters are regarded as "micro-domains", which are associated topologically with certain subspaces of a probabilistic function space X. The latter may be identified with or include a "state-space Z" and generally the material system can be represented by an "abstract dynamical systems" or $[X,\ \mathcal{F},\ \mathcal{P}]$ in which $\mathcal{F}$ denotes the σ-algebra of the events and $\mathcal{P}$ an appropriate probability measure. A finite number of microdomains forms an intermediate or mesodomain of the macroscopic material body. In this manner the transition from the microscopic formulation to the macroscopic one is carried out in two steps. Since the sites or positions (to the C.M.) of the mesodomains can be identified with the material coordinates of the continuous material body, the connection between the probabilistic and deterministic description can be established. More recently, a discussion on the thermodynamic aspects of this notion has been given by W. Muschik [12].

In accordance with the fundamental concept of the "state-space" representation [1], one can assign a "state vector ${}^{\Gamma}\mathbf{z}$" to each micro-domain or ${}^{\Gamma}\mathbf{z} \in Z \subset X$, which may generally signify an energy, the position or size of a microcrack etc. Thus, the evolution of the inherent microcracks can be regarded as a time-dependent random variable and the random phenomenon of the development of the m.c.'s as a stochastic process ${}^{\Gamma}\mathbf{z}(t) \in {}^{\Gamma}Z;\ t \geq 0$. The state of the "microcrack population" contained in Γ can be characterized by individual states, ie, $z_1, z_2, \ldots z_n$ forming the set: ${}^{\Gamma}\mathbf{z}_n = \{z_1, z_2, \ldots z_n\}$, $(n = 0, 1, \ldots)$, thus defining a "stochastic population" probability distribution or the measure $\mathcal{P}^z$ on the σ-field $\mathcal{F}^z$ of subsets of $Z \subset X$. It is often convenient to introduce a counting process by using the number $N(t)$ of microcracks within the domain Γ, which corresponds to the cardinal number of the set, ie:

$$ {}^{\Gamma}N_t\left(\left\{{}^{\Gamma}z_n\right\}\right) \equiv \overline{\overline{\{{}^{\Gamma}z_n\}}}, \qquad \Gamma \in Z \subset X \tag{1} $$

The state-space Z can be formed in two ways. Thus, if the individual m.c.'s can be distinguished from each other, they from an "ordered set" of individual states or the state of "n" members. On the other hand, if the m.c.'s are indistinguishable or unobservable, they form an "unordered set" of individual states. This distinction is important for the choice of an appropriate probability measure. A rigorous mathematical discussion on this matter is given among others in [6]. In the present analysis, it is assumed that the latter is the case and that the M.C. system consists of m-type members so that in general, one has:

$$ \left.\begin{aligned} \left\{{}^{\Gamma}z_m^{(n)}\right\}_{\nu=1}^{N} = Z^{(N)} \equiv X\ ; \quad & m = (1, 2\ \ldots),\ (n = (0, 1, 2\ \ldots) \\ & \nu = (1\ \ldots\ N) \end{aligned}\right\} \tag{2} $$

where $\left\{z_m^{(0)}\right\}$ designates an "empty" state. Since the probabilistic function space X is associated with the total population of the m.c.'s, the corresponding σ-algebra will be $\mathcal{F}_t$ of the events in X and $\mathcal{P}$ the probability measure on X. Thus for a specific micro-domain, the subspace $({}^{\Gamma}Z,\ {}^{\Gamma}\mathcal{F}_t)$ is a measurable space and T is an index set, so that the stochastic population process is also characterized by:

$$ \{\mathbf{x}_t,\ t \in T\}_{\nu=1}^{N}, \quad t \geq 0, \tag{3} $$

corresponding to the map $X \rightarrow {}^{\Gamma}Z \subset Z$ and where each mapping belongs to the equivalent classes: $X \rightarrow {}^{\Gamma}Z$ and $\mathcal{F}_t/{}^{\Gamma}\mathcal{F}_z$. By considering the development of the m.c.'s as a random phenomenon, the family of random variables signifies a stochastic process $\mathbf{z}_t \in Z$ on the subspace Z of X. This subspace is in general formed by a "state vector $\mathbf{z}$" or a set of r-parameters representing the thermomechanical states of an element of the structure without flaws or microdefects (see also [1]). From a probability point of view, the state vector $\mathbf{z}$ is also an "outcome or elementary event" in X as a result of the statistical experiment "(n)". Hence the states of the microcracks described in (2) belonging to a specific micro-domain can be characterized by:

$$ {}^{\Gamma}\mathbf{z} < {}^{\Gamma}\mathbf{z}_m^{(n)} < {}^{\Gamma}\mathbf{z} + \Delta\,{}^{\Gamma}\mathbf{z} \tag{4} $$

where ${}^{\Gamma}\mathbf{z}_m^{(n)}$ corresponds to a specific value for the m.c.'s in the set: $\left\{{}^{\Gamma}z_m^{(n)}\right\}_{\nu=1}^{N}$. Thus one obtains only subsets of the events $E \subset Z$, which may be regarded as "open spheres" and where $\bigcup_{\nu=1}^{N} E_\nu \equiv Z^{(N)} \subset X$.

An appropriate probability measure for the subsets of $Z^{(N)} \subset X$ is then of the form:

$$0 \leq \mathcal{P}\ \{E_\nu\} \leq 1; \quad \mathcal{P}\ \{E_\nu\} = 0, \text{ if } E_\nu = \phi \text{ and } \mathcal{P}\ \left\{Z^{(N)}\right\} = 1. \tag{5}$$

In accordance with probability theory, this measure represents the distribution of the relevant quantities [1]. It is important to note that a micro-domain Γ is defined by its volume, ie:

$$\left.\begin{array}{ll} {}^{\Gamma}V = {}^{m}V \bigcup {}^{1}V, & {}^{m}V \text{ in the volume of microcracks} \\ & \text{within } \Gamma \\ & {}^{1}V \text{ the volume of an individual element} \\ & \text{of the (1) - phase of the M.C. system} \end{array}\right\} \tag{6}$$

The subscript "m" used before refers to various components of the M.C. system. In the present case where considerations are given to a M.C. system with a binary structure (WC-Co system), $m = 1, 2$ with "1" referring to the WC-phase and "2" to the Co-phase of the compound material. Hence it becomes necessary to introduce the concentration of the phases within a micro-domain or:

$${}^{\Gamma}c =^{m} V/{}^{\Gamma}V \tag{7}$$

and where the probability of a "fixed number of m.c.'s" ${}^{\Gamma}N$ is given by:

$$P\ \left\{{}^{\Gamma}c\left({}^{\Gamma}N\right)\right\} = \hat{c} \quad \text{with } 0 \leq \hat{c} \leq 1. \tag{8}$$

In terms of the corresponding sets, one has:

$$\left.\begin{array}{ll} \{{}^{1}V\} = \{{}^{1}V_1,\ {}^{1}V_2\ \ldots\ {}^{1}V_k\} \equiv \{{}^{1}V_i\}_{i=1}^{N} & \text{(a)} \\ \{{}^{2}V\} = \{{}^{2}V_1,\ {}^{2}V_2\ \ldots\ {}^{2}V_l\} \equiv \{{}^{2}V_j\}_{j=1}^{l} & \text{(b)} \\ \{{}^{m}V\} = \{{}^{m}V_1,\ {}^{m}V_2\ \ldots\ {}^{m}V_n\} \equiv \{{}^{m}V_s\}_{s=1}^{{}^{\Gamma}N} & \text{(c)} \end{array}\right\} \tag{9}$$

in which the superscripts 2 in (9b) refers to the second phase or the matrix material of the WC-Co system.

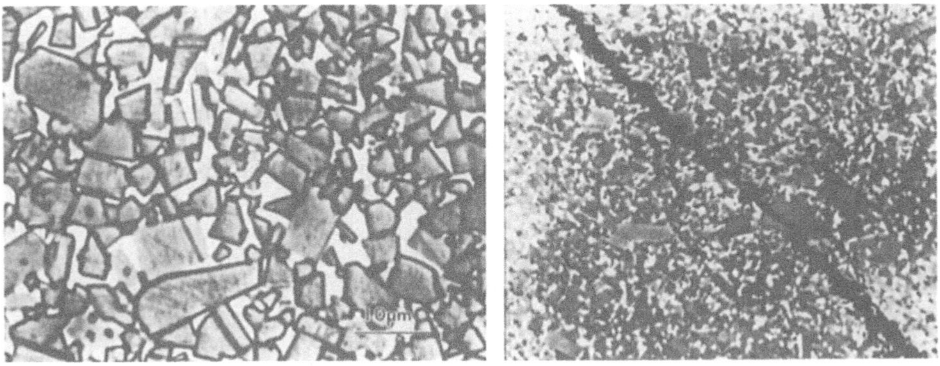

(a) undeformed configuration (WC-Co system)

(b) microcrack development during steady-state creep (WC-Co system)

Fig. 1 Micrograph (TEM) of a WC-Co specimen under creep conditions.

3 The Markov branching model for the microcrack development

Before discussing a probabilistic model of the evolution with time of the microcracks occurring in a material sample, it may be instructive to remark briefly on Markov processes and the general theory of stochastic populations. It may be seen from the observed strain-time graph of Fig. 2 representing two groups of M.C. system, ie, the Silicon-Nitride and the Tungsten-Cobalt compounds that after an initial transient (curves (OA) and (OA') respectively), the material response takes a "steady-state" form corresponding to (AB) and $(A'B')$. These transients have been analyzed in detail in ref. [2]. The present study is mainly concerned with development of the microcracks during the last transient, ie. curves (BC) and $(B'C')$ of Fig. 2 that ultimately leads to the fracture of the material specimen. Since the occurrence of microcracks is due to certain random processes on the atomic scale, the subsequent evolution or growth of these flaws can be considered in terms of a stochastic process $\{\mathbf{z}_t,\ t \in T\}$. If the joint distribution function of the random variables $\{z_{t_1},\ z_{t_2}\ \ldots\ z_{t_n}\}$ is known for all points $(n = 1, 2, \ldots)$ and all sets of values of time $(t_1, t_2\ \ldots\ t_n) \subset T$ and if these distribution functions are compatible, then there exists a probability field characterized by $[X,\ \mathcal{F}_t,\ \mathcal{P}]$. The probabilistic function space X is the set of all outcomes of the random experiments, defined previously, $\mathcal{F}_t$ the Borel field of certain subsets of X or the set of the random events $\{E_\nu\}_{\nu=1}^{N} \subset Z$ and $\mathcal{P}\ \{E\}$ the associated probability measure for $E \in \mathcal{F}_t$ (see also [1]).

The family $\{z_t,\ t \in T\}$ is defined on $Z \subset X$ for which the measure $\mathcal{P}\ \{z_{t_1} \leq x_1,\ \ldots\ z_{t_n} \leq x_n\}$ is equal to the distribution for every $(n = 1, 2\ \ldots)$ and $(t_1,\ t_2 \ldots t_n) \subset T$.

In particular, the stochastic process is Markovian, if the following conditions hold:

$$\mathcal{P}\ \{z_t \le x \mid z_{t_1} = y_1,\ z_{t_2} = y_2\ \dots\ z_{t_n} = y_n\} = \mathcal{P}\ \{z_t \le x \mid z_{t_n} = y_n\} \tag{10}$$

for all $t_1 < t_2 \dots t_n \subset T$, $(n = 1, 2 \dots)$ and all possible values of the random variables [1]. The process is uniquely determined by its initial distribution, ie.:

$$\mathcal{P}\ \{z_o \le x\} = P\ (0, x) \tag{11}$$

and the conditional distribution functions,

$$\mathcal{P}\ \{z_t \le x \mid z_s = y\} = p\ \{s, y;\ t, x\},\ s < t \tag{12}$$

which are also called "transition probabilities".

Since $\mathcal{P}\ \{z_t \le x\} = P\ (t, x)$ or:

$$P\ (t, x) = \int_{-\infty}^{\infty} p\ \{0, y;\ t, x\}\, d_y\, P\ (0, y) \tag{13}$$

one obtains by the theorem of total probability:

$$p\ \{s, y;\ t, x\} = \int p\ \{u, z;\ t, x\}\ d_z\ p\ \{s, y;\ u, z\}\ ;\ s < u < t \in T \tag{14}$$

which is the well-known Chapman-Kolmogorov functional relation [1]. Moreover, the Markov process is time homogeneous, if the probability density $p\ \{\ /\ \}$ depends only on $(t - s)$ apart from y and x. The process is additive or one with independent increments when $p\ \{s, y;\ t, x\}$ depends only on $(x - y)$ apart from t and s.

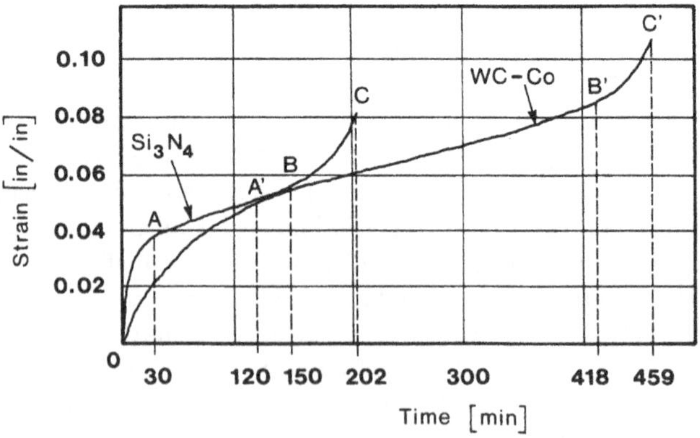

Fig. 2 Experimentally observed strains of the $Si_3\ N_4$ and WC-Co compounds during the creep response of these systems. Steady-state response for the $Si_3\ N_4$ system (30 - 150) minutes and for the WC-Co system (120 - 418) minutes during short-time creep tests in uni-axial compression.

In accordance with the formulation of the stochastic population mechanics [5,9], a continuous time parameter BD-Markov process is specified by two functions, namely $\lambda\ ({}^{\Gamma}N_t)$ and $\mu\ ({}^{\Gamma}N_t)$ in which the former refers to the probability of a "birth" or the formation of a microcrack within a time interval $[t,\ (t+\Delta t)] \in T$.

Assuming for simplification that only one m.c. exists at time t in Γ, one obtains the probability: $\lambda_N\ \Delta t + O(\Delta t)$ that it will be transformed into N microcracks between $(t+\Delta t)$, $\Delta t > 0$.

It is apparent that for a state in which a microcrack is <u>not</u> being transformed, the probability will be given by: $1 - \lambda_N\ \Delta t + O\ (\Delta t)$. Hence the evolution of the m.c.'s in a particular material domain will be characterized by the following differential equations of probabilities:

$$\left.\begin{aligned}\frac{d\,p_N(t)}{dt} &= (N+1)\ \lambda_0\ p_{N+1}(t) + N\ \lambda_1\ p_N(t) + \\ &+ (N-1)\ \lambda_2\ p_{N-1}(t) + \ \dots\ \lambda_n\ p_0(t),\ (N = 0, 1\ \dots)\end{aligned}\right\} \tag{15}$$

with the initial condition: $\lambda_N = 0,\ N \neq 1$ and $p_1(0) = 1$.

Recognizing that in the present analysis, the macroscopic strain field is given by:

$$\epsilon = \{\epsilon_1,\ \epsilon_2\ \dots\ \epsilon_N\} \in \mathcal{E} \subset Z \tag{16}$$

for the M.C. system as a whole, it can be represented also by a Markov process in which $\mathcal{E}$ designates either the strain or the "strain-rate" subspace of X ([1]). Using conditional probabilities to characterize the transition from a state z^k of the m.c.'s at time s to another state $z^j \in E$ at time $t > s$, and assuming that a "stable transition mechanism" occurs in Γ, one has in accordance with (13, 14), the following relation:

$$\left.\begin{aligned}P\ \left\{E, t \mid z^k, s\right\} = \sum_{j=0}^{N} \int_{\Gamma^j} P\ \left\{E, t\ \mid\ y^i, u\right\}\ P_j\ \left\{dy^i, u\ \mid\ z^k, s\right\};\\ (t \geq u \geq s)\end{aligned}\right\} \tag{17}$$

in which the time-axis $T \equiv R^+$ or the set of all integers in the case of a "Markov chain" (see also [1]). It is to be noted that even for a continuous time parameter process, the transitions involving the entire m.c. population will occur in form of "jumps" and then the more general theory of discontinuous Markov processes should apply (see for instance [4,8,9,13]). In this case, the probability P will satisfy the following integral equation:

$$\begin{aligned}P\ \left\{E, t \mid z^k, s\right\} = P_0\ \left\{E, t \mid z^k, s\right\} + \sum_{j=0}^{\infty} \int_s^t \int_{z^i = \Gamma^j} \\ P\ \{E, t \mid y^i, u\}\ Q^{(i)}\ (dy^i, du \mid z^k, s),\ (s, t) \in T\end{aligned} \tag{18}$$

in which P_0 is the transition probability of a transition from $z^k\ \rightarrow\ E$ during the time interval $(s, t) \in T$ without jumps, ie, by conserving the total number of individual m.c.'s or $N_t(z) \equiv N$ and Q designates the joint probability of the first jump and the corresponding population state resulting from the jumps conditional on the state z^k at time s. By taking this approach, an equivalent form of the Kolmogorov backward equation can be obtained (see also Moyal [5] and [1]). Following the discussion given in ref. [13] and recognizing that both in the "discrete time" case, where t takes on values $0, 1, 2, \ldots$ and in the continuous time case with $t \in [0, \infty)$ one can use the transition probability in (13) denoted now by $P_{s,t}\,(\epsilon_1,\ \epsilon_2)$. Here the strains $\epsilon,\ \epsilon_2$ are associated with the m.c. field taking on values in $Z \subset X$ with the understanding that each strain vector corresponds to the components of the unobservable strains induced in the micro-domains Γ_ν, $\nu = 1, \ldots N$.

Hence, the conditional probability of the process $\mathbf{z}(t) = \epsilon_2$ given that $\mathbf{z}(s) = \epsilon_1$ within the interval $0 \leq s \leq t$ becomes by the use of base vectors $\mathbf{e}_i$ in Z with components δ_i^j, ie:

$$\delta_i^i = 1 \text{ (for } i = j);\quad \delta_i^j = 0 \text{ (for } i \neq j) \tag{19}$$

and noting that $\epsilon_1 \rightarrow$ state i, $\epsilon_2 \rightarrow$ state j:

$$P_{s,t}\,(i, j) = P_{s,t}\,(e_i,\ e_j) \tag{20}$$

By assuming that the individual m.c.'s evolve independently from each other, ie. neglecting any interaction effects between them, the transition probabilities assume the following form:

$$P_{s,t}\,(\epsilon_1,\ \epsilon_2) = \sum \prod_{i=1}^{N} \prod_{j=1}^{\Gamma N} P_{s,t}\,(i,\ \epsilon_{ij}) \tag{21}$$

in which the summation is taken over all strains vectors $\epsilon_{ij}\ \epsilon \dot{\mathcal{E}} \subset Z$, $(i = 1, 2, \ldots N)$, $(j = 1, 2, \ldots {}^{\Gamma}N)$ in such a manner that the sum forms the vector ϵ_2. Moreover, if ${}^{\Gamma}N = 1$ one has the simplified form, ie:

$$\prod_{i=1}^{N} P_{s,t}\,(i, \epsilon_{ij}) = \delta\,(\epsilon_{ij}) \tag{22}$$

in which δ_i^j is defined by (19). It is seen that a Markov process in the state-space $Z \subset X$ with a transition probability satisfying (21) is a "branching process" in which the m.c.'s are of m-types. If the process $\mathbf{z}(t)$ is time homogeneous evidently (21) reduces to:

$$P_{s,t}\,(\epsilon_1,\ \epsilon_2) = P_{t-s}\,(\epsilon_1,\ \epsilon_2) \tag{23}$$

In general the m.c.'s have a "random life-time" that may be assumed to have an exponential distribution. In this case, one has for a time continuous Markov process on a "discrete state-space" ($Z \equiv R_m^+ \subset X$), an analogue to the "point-process" model (see also [6,14]). Denoting the number of m-type microcracks at time t by $N_m(t)$, then:

$$\mathbf{N}(t) = \{N_1(t),\ N_2(t)\ \ldots\ N_m(t)\}, \quad t \geq 0, \tag{24}$$

and one has a stochastic process $\{\mathbf{N}(t),\ t \geq 0\}$ on $[Z,\ \mathcal{F}^z,\ \mathcal{P}^z]$ or an one-dimensional continuous time Markov (branching) process with the following characteristics:

(i) a state-space $Z \equiv R_m^+$

(ii) it is a strong Markov process with respect to the general σ-field during the steady-state response of the system, ie.:

$$\mathcal{F}_t = \sigma \ \{N(s,x); \quad s \leq t\} \tag{25}$$

(iii) it has a transition probability:

$$P \ \left\{i, j, \ s^j, t\right\}, \quad s^j \in C_m \ \text{(complex space)}$$

satisfying:

$$\sum_{j \in R_m^+} P \ \left\{i, j \ s^j, t\right\} = \prod_{k=1}^{m} \sum_{j \in R_m^+} P \ \left\{e_k, j, \ s^j, t\right\} \tag{26}$$

in which the r.h. side of (26) holds for all $i \in R_m^+$. It has been shown by Athreya and Ney [10] that by the use of infinitesimal parameters and these transition probabilities one can obtain a unique solution of the Kolmogorov backward equation (see also [13,14]). This, however, involves an appropriate generating function as well as the concept of a "minimal process".

It is apparent from the above discussion that the stochastic population process of the microcracks in each domain Γ can be represented for the entire M.C. system by the set $\left\{{}^{\Gamma}\mathbf{z}_t, \ t \geq 0\right\}_{\Gamma=1}^{N}$ either in the state-space Z or correspondingly in the strain-rate space $\dot{\mathcal{E}} \subset Z$ such that:

$$\left.\begin{aligned} \mathbf{z}_t &= \left\{ {}^{\Gamma}z_1(t), \ {}^{\Gamma}z_2(t) \ \ldots \ {}^{\Gamma}z_N(t) \right\}_{\Gamma=1}^{N} && \text{(a)} \\ \boldsymbol{\epsilon}_t &= \left\{ {}^{\Gamma}\epsilon_1(t), \ {}^{\Gamma}\epsilon_2(t) \ \ldots \ {}^{\Gamma}\epsilon_N(t) \right\}_{\Gamma=1}^{N} && \text{(b)} \end{aligned}\right\} \tag{27}$$

in which $\boldsymbol{\epsilon}_t$ represents the observed strains during the creep response of the M.C. system (Fig. 2). Using the fixed number of m.c.'s in (24) or equivalently the set $Z^{(N)}$ and assuming that a set $Z^{(K)}$ is one without any microdefects and that these sets being disjoint, one obtains by the use of the earlier given event structure:

$$\left.\begin{aligned} & E = \sum_{N=0}^{\infty} E^{(N)}; \quad E^{(N)} = E \bigcap Z^{(N)}; \quad E \in \mathcal{F}^z \\ & \text{and} \\ & \bigcup_{t \in T} E_t = \sum_{N=0, t \in T} \bigcup E_t^{(N)}; \quad \bigcap_{t \in T} E_t = \sum_{N=0}^{\infty} \bigcap E_t^{(N)} \end{aligned}\right\} \tag{28}$$

It is seen that the stochastic process $\left\{{}^{\Gamma}z_t, \ t \in T\right\}$ in Z induces a "minimal σ-field $\mathcal{F}^{(N)}$", $(N = 0, 1, 2, \ \ldots)$ with a measure of $\mathcal{P}^{(N)}$ on $\mathcal{F}^{(N)}$ where:

$$\sum_{N=0}^{\infty} P^{(N)} \left(Z^{(N)}\right) = 1$$

and hence

$$P(E) = \sum_{N=0}^{\infty} P^{(N)} E^{(N)} \tag{29}$$

However since the total microcrack population or $N(Z)$ is finite with probability one, a set corresponding to an event E for an arbitrary micro-domain Γ constitutes a restriction of the total population size, ie.:

$$N \left\{E \mid Z^{(N)}\right\} = \sum_{i=0}^{(n)} \delta \left(E \mid Z^{(i)}\right) \tag{30}$$

It is possible to include the time homogeneous process in a more general non-homogeneous one as has been shown by Kuznetsov [15].

Another extension of the above analysis regarding the evolution of the microcracks can be obtained by considering the latter as a "random field" (see also refs. [2,14]).

4 A simplified growth model

Evidently for the comparison of the stochastic analysis with experimental observation as well as for computational purposes, a simplified model of the development of microcracks is desirable. More recently, a deterministic model has been discussed by Tvergaard [11] whereby the material failure can be ascribed to a "void growth" process and the coalescence of such voids. Earlier, a stochastic model of creep failure has been considered by Feltham [17] who considered the creep phenomenon at moderate temperatures in terms of a "thermally activated process" (dislocation kinetics).

In the present study, the transition probabilities during the final stage of the creep response (curve $B'C'$ of Fig. 2) for the chosen M.C. system can also be considered in terms of infinitesimal parameters leading to the Kolmogorov backward equation [14] in the form of:

$$\tilde{P}_{ik}(t) = -q_i\, P_{ik}(t) + \sum_{j \neq i} q_{ij}\, P_{jk}(t); \quad i,\ k \geq 0 \tag{31}$$

in which the q_{ij}'s are the infinitesimal characteristics of the N_t-process and the elements of the "transition matrix Q" of the Chapman-Kolmogorov functional relation (eqn. 13, in matrix form) (see also [1]).

It is apparent that the initial distribution: $P_0(k) = P_k^e$ will satisfy:

$$P_N^e q_k = \sum_{j \neq k} P_j\, q_{jk}; \quad k \geq 0 \tag{32}$$

in which $P_N^e = P_k(t)$ holds for all $t \geq 0$ and all k. It is therefore called an "equilibrium distribution" of the N_t-process. It certainly holds for a steady state transitions during the period $A' \rightarrow B'$ of Fig. 2. However, one may regard also the N_t-process as a BD-Markov process [1,2] in which the Q-matrix has non-zero elements, ie.:

$$q_{N,N+1} = \lambda_N; \quad q_{N+1,N} = \mu_N; \quad N \geq 0 \tag{33}$$

in which the q-elements are referred to as "intensities" of formation (birth) λ_N and a "death-rate μ_N" of the m.c.'s, so that: $\lambda_N\ \mu_{N+1} > 0$ for all $N > 0$ in an equilibrium distribution, viz.:

$$P_N^e \equiv P_N^{(K)} = K\ r_n \tag{34}$$

and where the coefficients: $K = \left(\sum\limits_{j=0}^{\infty} r_j\right)^{-1} < \infty$, $r_0 = 1$ and generally in [7]:

$$r_N = \frac{\lambda_0\ \ldots\ \lambda_N}{\mu_1\ \ldots\ \mu_N}, \quad N \geq 1 \tag{35}$$

Using the concept of a "generating function" of probability theory (see for instance [18]) for the random variable N_t taking on values $K = 0, 1, 2, \ldots$ with the probability distribution $P_N(k)$, the function $g(k)$ of $N_t < 1$ is then defined by:

$$g(N) = \sum_{k=0}^{\infty} P_N(k)\ N^k \tag{36}$$

which determines uniquely the probability distribution:

$$P_N^{(k)} = \frac{1}{k!}\ g^{(k)}(0); k = 0, 1, 2, \ldots \tag{37}$$

The probability densities given by (14) can be used to express $g(N)$ as follows:

$$g(N,t) = \sum_{(n)} N^n\ P_n(t) \tag{38}$$

for which a general solution of the process [9] can be obtained, ie.:

$$p_N^{(t)} = e^{-\lambda t}\ \left[1 - e^{-\lambda t}\right]^{n-1}; \quad (n \geq 1) \tag{39}$$

Hence an approximation of the observed transient (curve $B' \rightarrow C'$ of Fig. 2) can be obtained in a "step-wise" manner. That is from one stability level associated with the induced strain, ϵ_1 at time t_o (corresponding to the end of the steady-state stage B' in Fig. 2) to another level ϵ_2 or the point where fracture occurs. This transition is characterized by the set: $\boldsymbol{\epsilon} = \{\epsilon_1,\ \epsilon_2,\ \ldots\ \epsilon_n\}\ \epsilon\ \dot{\mathcal{E}} \subset Z$. It is to be noted however that each strain level reached represents an extreme or "maximum value" for a particular micro-domain $\Gamma\ \epsilon\ Z \subset X$ (see also [16]). Equivalently from energy consideration, one can derive an expression for the growth process in a parabolic form, ie.:

$$\frac{d \mid \epsilon_i(t) \mid}{dt} = K_i\ (\mid \epsilon_i(t) \mid)^2 \tag{40}$$

in which ϵ_i designates the observed strain (macroscopic) during $B' \rightarrow C'$ (Fig. 3, below) and K_i (ϵ_i, t) is a functional that is related to the slope of the observed strain - time function of the magnified diagram of Fig. 3. It corresponds then to the intensity parameter λ_N of the BD-Markov process mentioned previously.

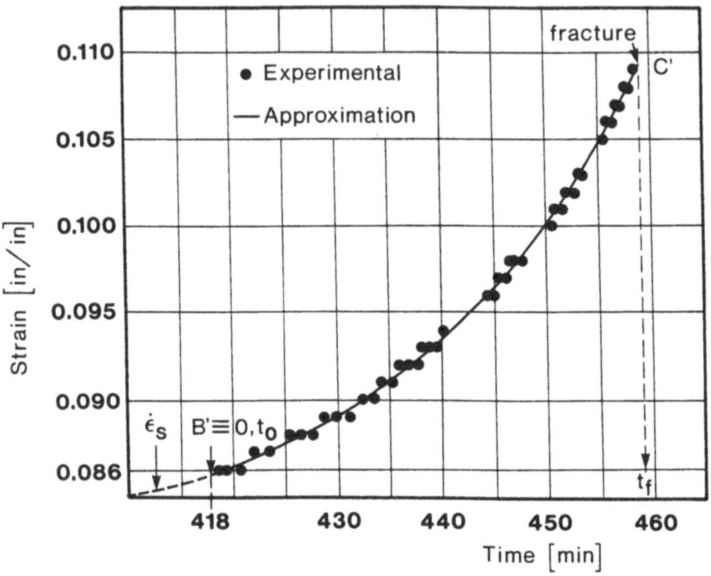

Fig. 3 Magnified diagram of the creep response curve $B' \rightarrow C'$ of Fig. 2 for a WC-Co specimen tested in uni-axial compression.

The functional $K_i \equiv K_i$ (ϵ_i, t) depends on the initial and final strain values as follows [16]:

$$K_i = \frac{|\epsilon_i^f| - |\epsilon_i^o|}{|\epsilon_i^f||\epsilon_i^o| \; (t_i^f - t_i^o)} \tag{41}$$

in which the superscripts "o and f" denote the initial and final observed macrostrains (before fracture) in Fig. 3 for the transition from the steady-state at t_o to t_f (at fracture). The solution of relation (41) is then given by:

$$|\epsilon_i(t)| = \frac{|\epsilon_i^o|}{1 - K_i \, |\epsilon_i^o| \; (t_i^f - t_i^o)}; \quad t_i \, \epsilon \, [0, t^f] \tag{42}$$

where t_i is an element of small incremental time-intervals, ie.: $\sum_{i=1}^{n} J_i = (t^f - t^o) \, \epsilon \, [0, t^f]$ in the approximation of the non-linear response of the M.C. system during the transition from $B' \rightarrow C'$.

The numerical solution of this simplified growth model for the m.c. population is compared with the experimentally assessed strains and reveals a rather close fit.

The actual tests on the WC-Co system specimen were carried out under 7.2 ton/m^2 uni-axial loading at a temperature of 1250°C. The required equipment for such tests and the procedure are described in some detail in ref. [19]. Thus the measurements of the strains in the short-time creep tests on specimens of the WC-Co compound were $|\epsilon(t_o)| = 0.086$ in/in at $t_o = 418$ minutes after the application of the uni-axial load and $|\epsilon(t^f)| = 0.11$ in/in at $t^f = 460$ minutes. It should be noted that in order to obtain an appropriate evaluation in accordance with relation (42) it becomes necessary to choose "small time steps" for the equilibrium distributions to hold. These are also related to the dimensions of each micro-domain Γ and involve the required jump-times for the growth process mentioned earlier.

5 References

1. Axelrad, D.R. (1993) *Stochastic Mechanics of Discrete Media*, Springer-Verlag, Berlin-Heidelberg.
2. Axelrad, D.R. and Frydrychowicz W. (1992) Stochastic theory of the inelastic behaviour of multi-component high temperature materials. *Math. Models and Methods in Applied Sciences*, Vol. 2, No. 2, World Scientific Publ.
3. Adler, R.J. (1981) *The Geometry of Random Fields*, John Wiley & Sons, Chichester, England.
4. Moyal, J.E. (1957) Discontinuous Markov processes. *Act. Math.*, Vol. 92.
5. Moyal, J.E. (1962) The general theory of stochastic population processes. *Act. Math.*, Vol. 2.
6. Daley, D.J. and Vere-Jones, D. (1981) An introduction to the theory of point processes. *Springer Series in Statistics*, Springer-Verlag, Berlin.
7. Brémaud, P. (1981) Point processes and ques. *Martingale Dynamics*, Springer Verlag, Berlin.
8. Bartlett, M.S. (1949) Some evolutionary stochastic processes. *J Roy. Stat. Soc., Series B*, Vol. 11.
9. Kendall, D.G. (1949) Stochastic processes and population growth. *J. Roy. Stat. Soc., Series B*, Vol. 11.
10. Athreya, K.B. and Ney, R.E. (1972) *Branching Processes*, Springer-Verlag, Berlin.
11. Tvergaard, V. (1990) Material failure by void growth to coalescence. In *Advances in Applied Mechanics*, Eds. J.W. Hutchinson and T.Y. Wu, Vol. 27, Academic Press, N.Y.
12. Muschik, W. (1993) Fundamentals of non-equilibrium thermodynamics in *Non-Equilibrium Thermodynamics with Application to Solids*, CISM Courses and Lectures No. 336, International Centre for Mechanical Sciences, ed., W. Muschik, Springer-Verlag, N.Y.
13. Gihmann, I.I. and Skorohod, A.V. (1975) *The Theory of Stochastic Processes*, Vol. I, II, Springer-Verlag, Berlin.

14. Axelrad, D.R., Frydrychowicz, W. and Kamal, M.R. (1992) Stochastic analysis of polymer melts with a binary structure. *Z. angew. Math. Phys. (ZAMP)*, Vol. 43, Birkhäuser Verlag, Basel.
15. Kuznetsov, S.E. (1984). *Non-Homogeneous Markov Processes*, (English translation), Plenum Press, N.Y.
16. Axelrad, D.R. and Frydrychowicz, W. (1995) Stochastic analysis of the fracture of solids with microcrack. *Z. angew. Math. Phys. (ZAMP)*, Vol. 46, Birkhäuser, Verlag, Basel.
17. Feltham, P. (1974) Collective dislocation models in crystal plasticity in *Reviews on the Deformation Behaviour of Materials*, Vol. I, No. 1, Freund Publ. Tel-Aviv.
18. Prohorov Y.V. and Rosanov Y.A. (1969) *Probability Theory*, Springer-Verlag, Berlin.
19. Axelrad, D.R. (1965) A new high-temperature materials testing apparatus in *Int. Symposium on Machines for Materials and Environmental Testing, Proc. Appl. Mechs. Group, Inst. Mech. Eng.*, Vol. 180, Part 3A, London, England.

38 RELIABILITY EVALUATION OF COMPONENTS USING FRACTURE MECHANICS IN SPITE OF UNCERTAIN DATA

T. WINKLER and E. KAULFERSCH
Chemnitzer Werkstoffmechanik GmbH, Chemnitz, Germany
B. MICHEL
Fraunhofer-Institution of Reliability and Micro-Integration, Berlin, Germany

Abstract
The reliability evaluation of components commonly includes a comparison of stresses acting in the component with stresses tolerably well by the material. Since both sides are randomly influenced in many cases also probability methods have to be applied. They lead to probability terms or safety margins, respectively. It is not trivial to extend this probability concept to complicated problems of fracture mechanics. In the paper, a method to take into account random input parameters in Finite Element analyses is described and illustrated. This so-called *Stochastic FEM* is based on perturbation theory and especially applicable to problems with multiple random parameters showing small scattering. The randomly influenced results like displacement or stress fields can be derived from a series expansion in the small deviations. This enables to compute mean and variance of stresses and derived terms of fracture mechanics as well as - under certain assumptions - non-exceeding probabilities.
The method is applied to the reliability evaluation of thermo-mechanically stressed micromechanical components, e.g. chip-on-board and glob top/packaging. Parameters with uncertain scattering are the geometry of the joints or packages as well as material parameters like heat transfer coefficients and Young's moduli. The Stochastic FEM is used first of all for sensitivity analyses with respect to the level of influence of the different uncertain parameters.
Keywords: finite element method, fracture mechanics, microsystem technology, reliability, stochastic FEM, uncertain data.

Structural Failure, Product Liability and Technical Insurance, Proceedings of Fifth International Conference. Edited by H.P. Rossmanith. Published in 1996 by Taylor & Francis, 2 Park Square, Milton Park, Abingdon, Oxon, OX14 4RN. ISBN 0 419 22080 1

1 Introduction

Investigations in reliability evaluation of components commonly include a comparison of reactions in the component due to working conditions with those reactions tolerably well by the material. A resulting quantity of the component is compared with a corresponding critical quantity of the component. Depending on the problem this quantity can be, for instance, load, stress, strain, displacement, defect size, stress intensity factor, CTOD, J-Integral or more advanced fracture mechanic quantities. Since both sides of the comparison are randomly influenced in many cases also probability methods have to be applied. They lead to probability terms or safety margins, respectively. It must be noted that the probability approach does not substitute the common deterministic mechanical analysis. The approach is based on the mechanical model but generalises its assumptions. It is not trivial to extend this probability concept to complicated problems of fracture mechanics where the resulting quantity in the component cannot be obtained analytically but has to be computed. Additionally the quantity has to be determined not only in a single point but as a field.

The finite element method has found a wide application for problems in fracture mechanics. It is -from the viewpoint of our context- a purely deterministic method. The random scatter of materials could be taken into account by means of statistical simulation. In the paper an other method is explained which can -under some restrictions- lead to practicable solutions under reduced efforts. The method will be illustrated by applying it to problems of microcomponents.

Components of microsystem technology often exhibit a large scattering of local material parameters, e.g. between thin layers and bulk material, in inhomogeneous material regions or in material interconnects. The scatter leads not infrequently to big uncertainties in the reliability evaluation and lifetime prediction for these components. For the material strength, for instance, statistical assumptions derived from generalised Weibull models are made to take into account the scatter of parameters [1]. Also in the field of fatigue crack growth stochastic approaches are partially integrated in the evaluation of mechanical reliability and lifetime prediction [2]. For the determination of stress and strain fields taking into account random input quantities a so-called method of stochastic finite element analysis will be applied.

For small scatter around the mean it is possible to combine the classical finite element method with the perturbation theory aiming at results concerning the variance and higher moments of the probability distribution of mechanical and thermal field quantities [3]. Under certain conditions the numerical effort can be reduced in this way in comparison with commonly used Monte-Carlo simulations which require numerous deterministic computation runs, since only one "stochastic FE computation" is necessary. This advantage has its effect especially on the complex three-dimensional thermoelastic and elastoplastic computations of microcomponents.

2 Fundamentals of the stochastic finite element method

Before going to apply FEM to microcomponents let us give a short survey about the stochastic FE approach. The finite element method has found a wide application for problems in fracture mechanics [4]. Unfortunately, in certain practical applications not all input variables are deterministic. For example the crack length can be uncertain. The same is valid for the Young's modulus of certain structural components. In the same way the loads can be stochastically distributed. In these cases the conventional finite element computer codes cannot directly be applied. To get probability evaluations regarding the displacements, strains, stresses, stress intensity factors etc., and related fatigue and fracture quantities a large number of computations with various realisations of stochastic input variables such as the co-ordinates of nodal points, material properties or loading parameters is necessary. An advanced finite element approach avoiding this problem is a stochastic FEM for stochastic material and structural behaviour.

The stochastic FEM can be applied to problems of linear and non-linear fracture mechanics as well. We here outline one possible approach only (for details see e.g. [5]). If we have, for example, some co-ordinates x_i and x_j which are assumed to be uncertain, then these stochastic co-ordinates $\underline{x}_j$ are represented by the formula

$$\underline{x}_j = x_j(1+\underline{\tilde{x}}_j) \qquad (1)$$

where $\underline{\tilde{x}}_j$ are small stochastically distributed displacement corrections with a small value of variance (the underline denotes stochastic quantities instead of deterministic ones). The global stiffness matrix K of the system can be constructed as a series of uncertain values $\underline{\tilde{x}}_j$ by means of equ. (1):

$$K = K^0 + K^i \underline{\tilde{x}}_i + K^j \underline{\tilde{x}}_j + K^{ii} \underline{\tilde{x}}_j^{\,2} + K^{ij}\, \underline{\tilde{x}}_i \underline{\tilde{x}}_j + \ldots \qquad (2)$$

The matrix K^0 corresponds to the stiffness matrix of the deterministic problem, when all $\underline{\tilde{x}}_j$ are equal to zero. The matrix $\underline{K}$ is obtained from the finite element discretization with consideration of the stochastic values. It must be solved as a stochastic system of equations

$$\underline{K} \cdot \underline{u} = f \qquad (3)$$

where $\underline{u}$ is the stochastic displacement vector and f takes into consideration special kind of boundary conditions.

The stochastic displacements $\underline{u}$ are also described as a series of the uncertain values $\underline{\tilde{x}}_j$:

$$\underline{u} = u^0 + u^i \underline{\tilde{x}}_i + u^j \underline{\tilde{x}}_j + \ldots + u^{ii} \underline{\tilde{x}}_i^{\,2} + u^{ij} \underline{\tilde{x}}_i \underline{\tilde{x}}_j + \ldots \qquad (4)$$

The coefficients (u^i and u^{ij}) of this series are determined by the help of perturbation theory. The vector u^0 is the solution of the corresponding deterministic problem. Its derivatives and related field quantities as strains, stresses and the fracture quantities are also represented in terms of the stochastic quantities.

The mean value, the variance and the moments of higher order can be computed using the stochastic series above. If the density functions of the stochastically distributed uncertain displacements are known, then also the density and distribution functions of the results can be determined from equ. (4).

3 Examples of application

3.1 Demonstration at standard fracture specimens

In order to test the method at first there were computed some standard fracture specimens.

As an example for the stochastic FEM Fig. 1 shows a comparison between stochastic and deterministic calculations of the J-integral of a ceramic CT-specimen with varying crack length (see e.g. [3]). The comparison is performed in the following manner:

(1) We compute $u = u(X_j)$ for a given value X_j by conventional FEM.

(2) We write $X_j = x_j(1+\tilde{x}_j)$ following (1) with small and fixed $\tilde{x}_j$. Then equ. (2) can be applied which gives an approximation $\tilde{u}$ for $u = u(X_j)$.
A linear as well as a second order approximation of equ. (2) have been computed.

(3) The comparison of u (obtained from FEM) and $\tilde{u}$ (from STOFEM) for different values $\tilde{x}_j$ gives hints for the applicability of the series (2).

The results in Fig.1 show that in the example the second order approximation is in good agreement with the deterministic values whereas the linear approximation of the series (2) should be sufficient only for a scatter less than 5%.

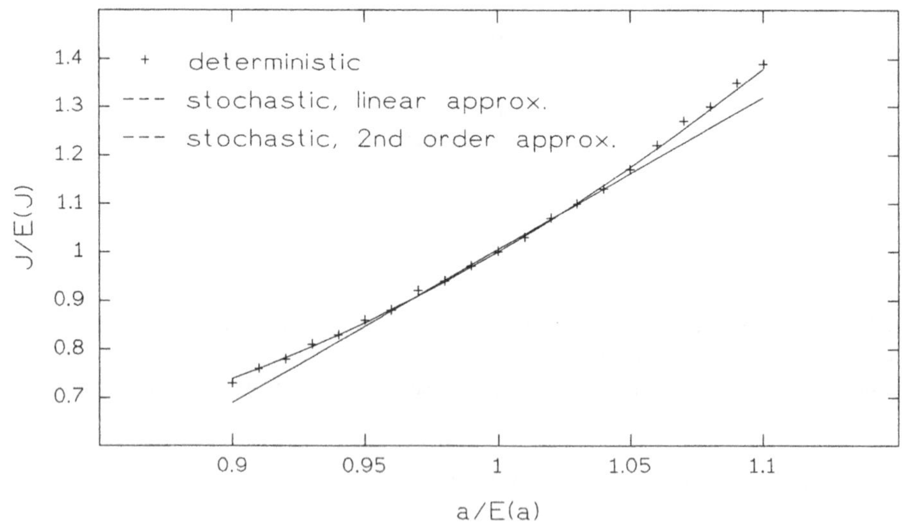

Fig. 1: Deterministic versus stochastic FE computation of J integral for varying crack length a.

3.2 Application to a surface mounted device with random material parameters

We consider a printed circuit board (PCB) with several small ceramic capacitors mounted on the board. The technology is called SMT (Surface Mounted Technology) and used in numerous modern technical applications. The result is a complicated material transition region between three or more different materials (solder, ceramic capacitor and polymeric board). Small fatigue cracks due to the thermal "misfit" between board and ceramic may be a dangerous failure reason of the PCB in thermal fatigue loading [6].

To study the deformation and fracture behaviour of the PCB under bending tests a theoretical method is applied which is related the finite element approach. A simple FE model of the PCB (plane model) is shown in Fig. 2 which was generated automatically from the video image of a metallographic section. The mesh was adaptively refined due to the results of FE computations. In the given form it is well suited for an investigation of the crack near region in the gap between solder and board.

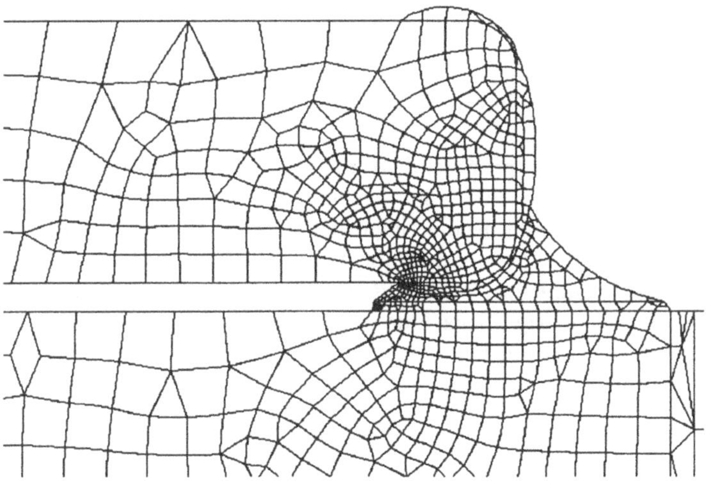

Fig. 2: Mesh used in FE computation of PCB.

In Fig. 3 the FEM result for the printed circuit board (PCB) mentioned above is shown. It was the task of this stochastic FE calculation to study the influence of a scatter in Young's moduli within different highly stressed regions of the PCB to get some more detailed information on the confidence range as the moduli are not very well known in practical applications. The uncertain values in this example are the Young's moduli of the capacitor, the solder and the board, respectively. A variation coefficient of 10 % was prescribed. Fig. 3 shows the contours of the stress component σ_{yy} which is common practice in conventional FEM. In addition the contours of so-called confidence regions obtained from stochastic FEM are presented. The stresses depend on the random Young's moduli and lie with the probability 0.9 in the range from $\sigma_{yy} - d$ to $\sigma_{yy} + d$. For the Young's moduli a normal distribution was assumed. From Fig. 2 we can draw the conclusion that a high sensitivity is observed within the highly stressed regions near the "gap" and at the outer surface of the solder.

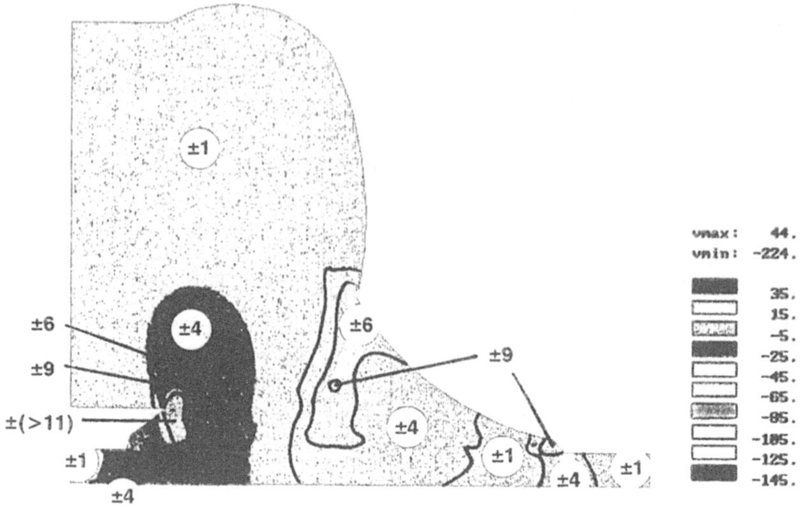

Fig. 3: Confidence regions for the stresses in every point of PCB [6].

3.3 Application to a spider lead with fuzzy Young's modulus under thermal load
Another approach to take into account the uncertainty of the elastic material parameters which might become important in microelectronics and microengineeering is a fuzzy approach. If we assume the Young's moduli to be "fuzzy numbers" we also get interesting results for the same PCB which may be a good help for the evaluation of the fatigue and damage behaviour of materials and components. For details of this method we refer to the research report [7], which also includes other examples of technical application.

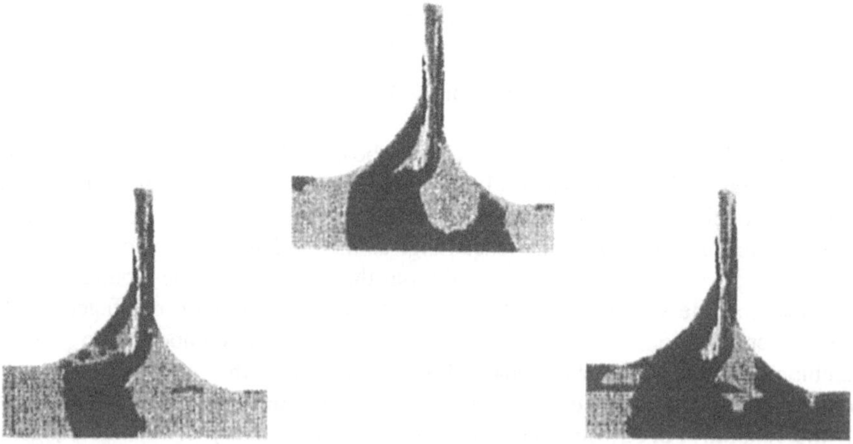

Fig. 4: Stress in a spider lead [7]. Top: mean; bottom left: lower bound; bottom right: upper bound.

Fig. 4 shows results obtained by SKURT [7] for another kind of interconnections in microelectronics, the so-called spider leads. We take into account the Young's modulus E of all materials to a "fuzzy" number. The special purpose computer code ASTOR was applied which contains a special software tool for "fuzzy"-FEM applications. Fig. 4 gives an impression of the spread of the result due to the spread of the input. This approach can also generalise "worst case" considerations.

3.4 Application to a glob top with random material parameters and geometry

Another characteristic group of microelectronic assemblies comes from the packaging. For example in many automotive applications the microelectronic devices have to be protected against environmental influences which can be done by an encapsulation. The quiet inhomogeneous epoxy encapsulant as well as the adhesive show a large scatter both in the thermal expansion coefficients and the Young's moduli. Additionally the shape of the so-called glob tops can slightly differ due to the production process. The question was: which scattering quantity has the greatest influence?

Going from top to bottom in Fig. 5 we see a dominating influence of the variance of the thermal expansion coefficient of the encapsulant (e_1). Further investigations have shown that this variance due to the thermal expansion coefficient is much more pronounced than those due to the Young's modulus and the geometry imperfection.

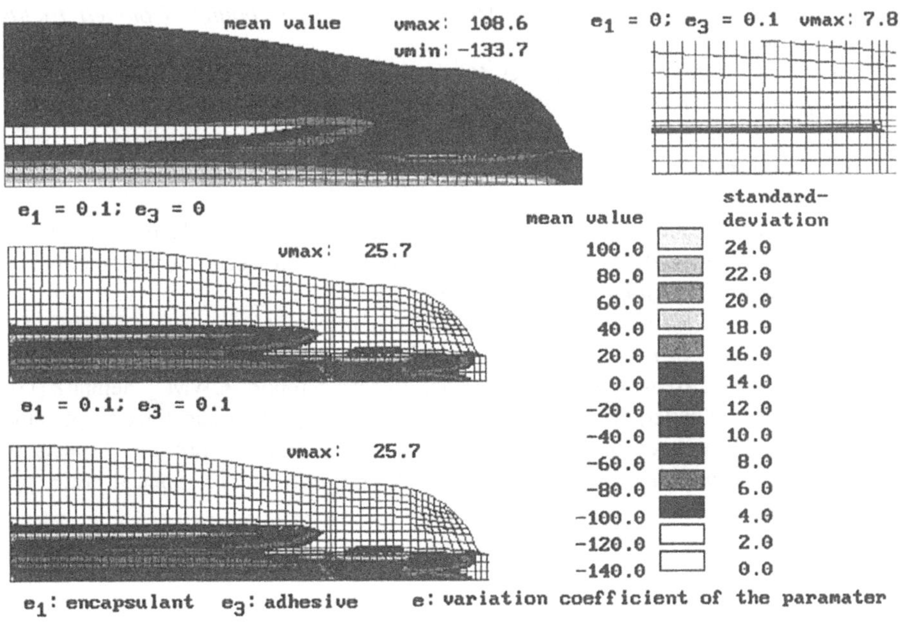

Fig. 5: Variation of σ_{yy}-stress at an glob top for different variations of input quantities. Top, left: mean value; top, right: standard deviation due to random adhesive only; middle: random encapsulant only; bottom: random adhesive and encapsulant.

4 Summary

A special numerical method of solid and fracture mechanics was presented. Under certain conditions it is possible to obtain results regarding also the scatter of reliability parameters used in the mechanical reliability and quality evaluation. The so-called Stochastic FEM enables to consider small imperfections, small load scattering or small parameter uncertainties in linear and nonlinear finite element analyses and to estimate the resulting scatter of derived quantities. Also "fuzzy FE methods" can complete the evaluation of mechanical-thermal reliability for microtechnology applications.

References

1. Lau, J.H. (ed.) (1991) *Solder Joint Reliability - Theory and Applications.* Van Nostrand Reinhold, New York.
2. Michel, B., Winkler, T., Skurt, L. (1993) Physical and Micromechanical Aspects of Stochastic Fatigue Crack Growth., in *Stochastic Approach to Fatigue* (ed. K. Sobczyk). CISM Courses and Lectures No. 334, Springer Verlag Wien-New York, pp. 79-120.
3. Skurt, L.(1986) Anwendung einer Stochastischen Finite Elemente Methode in der Bruchmechanik, *Series Fracture Mechanics, Micromechanics, Coupled Fields (FMC)*, No. 19, Chemnitz, pp. 45-54.
4. Liu, W.K., Belytschko, T. (ed.s) (1989) *Computational Mechanics of Probabilistic and Reliability Analysis.* Elmepress, Lausanne-Washington.
5. Skurt, L. and Michel, B. (1992) Stochastic finite element method for solid mechanic problems with uncertain values. *Mathematical Research,* vol.86, pp.28-33.
6. Michel, B., Schubert, A., Dudek, R. and Grosser, V. (1994) Experimental and numerical investigations of thermo-mechanically stressed micro-components. *Microsystem Technologies,* vol.1, no.1, pp. 14-22
7. Chemnitzer Werkstoffmechanik GmbH (1992) *Untersuchungen zur mechanischen Zuverlässigkeit von Komponenten der Mikrosystemtechnik mittels numerischer Simulation und experimenteller Methoden.* Research report. CWM, Chemnitz (D).
8. Michel, B., Winkler, T. and Sommer, J.-P. (1992) Theoretical Concepts and Experiments in Fracture Mechanics - Macroscopic and Microscopic Aspects of Modelling. *European Journal of Mechanics, A/Solids,* vol 11, Special issue, pp. 115-133.

39 A PROPOSAL OF A QUALITY CONVENTION USED FOR DIMENSIONING OF STEEL STRUCTURES AGAINST FATIGUE FAILURE

W. SOBCZYKIEWICZ
Institute of Heavy Machinery Engineering, Warsaw University of Technology, Warsaw, Poland

Abstract
The presented proposal is an attempt of including the problem of dimensioning of steel structures against fatigue failures into the process of achieving required product quality. The necessity of such an approach arises due to the growing significance of product quality in any economic undertaking. An assumption was made that the product durability and reliability belong to the basic properties of quality. These features are measured with the probability of no failure survival, the failure rate and percentile fatigue life. The cause-effect ROBUST method was used for evaluation of the product quality.
Keywords: Carrying steel structures, designing, durability, fatigue, quality.

1 Introduction

Four main factors play an essential role in development of society:
- natural resources
- resources of physical and intellectual labour
- capital equipment, and
- quality.

Historically, one can observe the continuous growth of the importance of the last factor. Actually, it has became a decisive element in the economic activity.

In the contemporary world, a competition takes place in the field of widely understood quality. Those who can design and manufacture products of higher quality and who subsequently can exploit them at the equally high quality level, they gain additional economical advantages in relation to producers and users of products of standard and low quality.

Structural Failure, Product Liability and Technical Insurance, Proceedings of Fifth International Conference. Edited by H.P. Rossmanith. Published in 1996 by Taylor & Francis, 2 Park Square, Milton Park, Abingdon, Oxon, OX14 4RN. ISBN 0 419 22080 1

These advantages are created in the subsequent phases of existence of the product: designing, manufacturing and exploitation, and concern practically all aspects of its existence: material and energy consumption, productivity, ecological effects etc.

2 Quality assessment

Generally, the quality is a set of features of a product, estimating the degree to which the product corresponds to the requirements of user in a given period of time.

The quality level of a product is evaluated according to: features of quality, their properties, measures and target values and is presented in Fig. 1.

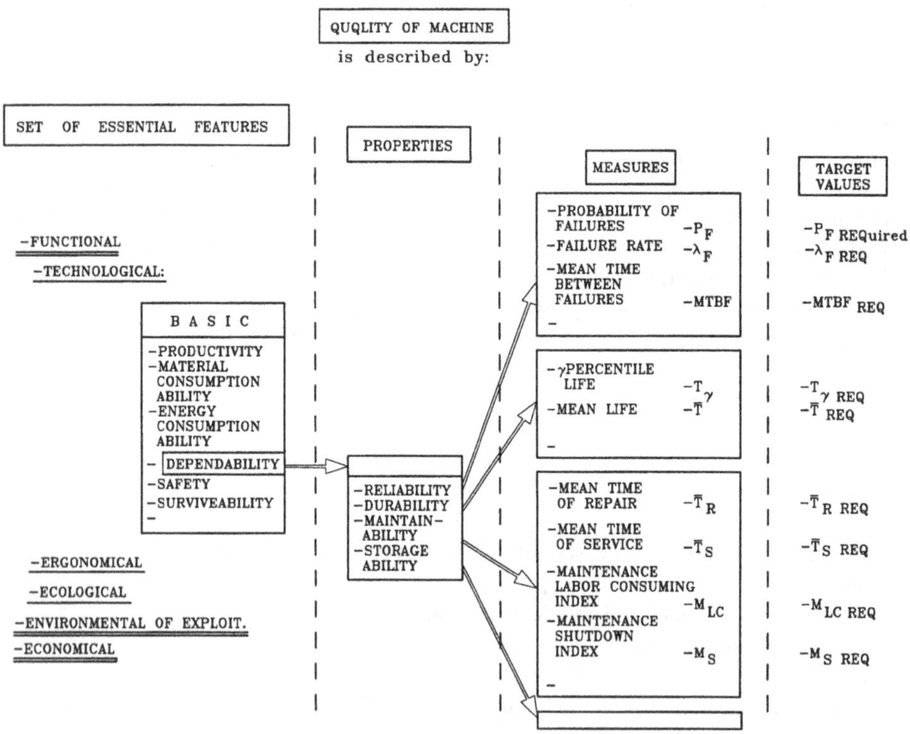

Fig. 1. Flow chart for description of the quality of product.

Contemporarily, dependability is, because of economic reasons, considered more and more as a dominant component of quality. Dependability is also a complex feature

which can be described by four properties:
- reliability (failure free operation)
- durability
- maintainability
- storageability.

In every case each property is described by measures, compared in the evaluation process to the target values. The evaluation chart is illustrated in Fig. 1.

The conception of dimensioning of an object, in this case the steel structure, using the theory of dependability is based on the occurrence of failure. The main task of the theory and engineering of dependability is to create methods, procedures and instructions, in order to enable the most cost effective realization of objects, taking fully into account the results caused by their failure. For this purpose it is necessary to posses a set of dependability properties measures that enable the dependability level in scope of above mentioned four basic properties to be estimated and controlled.

In the case of load carrying steel structures, mainly damaged due to fatigue process, the storageability is not considered at all. Maintainability can also be neglected, because it plays a relatively minor role. The assumption was made, that for steel structures durability and reliability are the basic properties of quality. These properties can be expressed by the following measures: the probability of no failure (survival), the failure rate, the mean time between failures, and the γ-percentile fatigue life.

3 Robust conception

The cause - effect ROBUST conception [1] seems to be the most suitable for evaluation of the product quality. According to that concept, results of dimensioning against the fatigue failure, i.e. meeting the required quality measures mentioned above, depend on the causative factors that can be grouped as follows:

1. signal factors - SF
2. control factors - CF
3. noise factors - NF
4. tolerance factors - TF

According to the ROBUST conception the designer should identify all these causative factors that are important for the product, a welded load carrying steel structure, from the fatigue life and reliability points of view, using the scheme shown in Fig. 2.
The causative factors can be described as follows:

- Signal Factors - SF, quantities which can be varied by the user of the product to obtain the intended values of the quality measures
- Control Factors - CF, quantities which can be varied freely by the designer without essential changes of the manufacturing costs
- Noise Factors - NF, quantities which are too difficult or too expensive to control in service, because they change from unit to unit, from environment to environment, and in function of time. They are described by means of statistical parameters.
- Tolerance Factors - TF, quantities which can be varied by the designer. Their changes effect strongly on the costs of manufacturing.

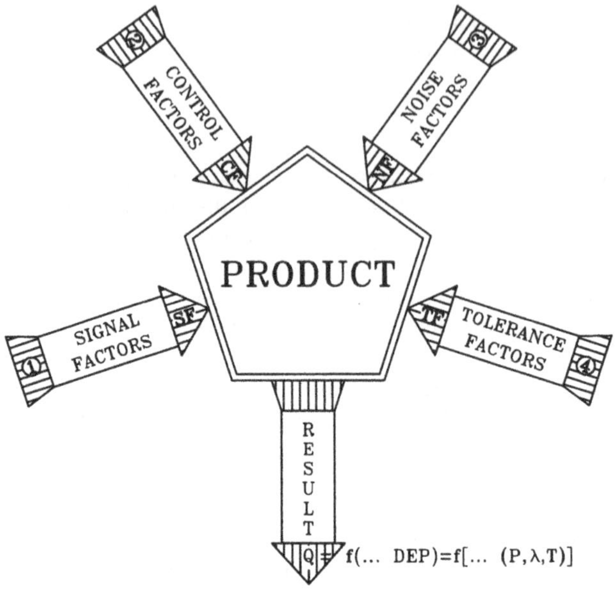

Fig. 2. Quality as a result of causative factors.
Q - quality, DEP - dependability.

The proposition of the creation of the quality properties mentioned above: reliability and durability, estimated by means of their adequate essential measures (probability of no failure - P_{NF}, failure rate - γ_F, γ percentile life - T_γ) in approaches of three methods of fatigue dimensioning, called: FALSN, FALIN, FALPRO [2] are presented in Fig. 3,4,5. These methods, belong to the subsequent stages of the Design & Development process .

The stress description key is shown in Fig. 6. The cause-effect layout, proper for ROBUST conception, generally illustrated in Fig. 2, has been carried out by means of the methods FALSN, FALIN, FALPRO. In this convention the values of measures describing the quality of product are results of acting of properly choosen causative factors, by means of which it is possible to controll the quality level. The estimated quality level, obtained in the subsequent stages of the D&D process called: preliminary design - by means of FALSN method, research prototype - by means of FALIN method, industrial prototype - by means of FALPR method, ought to be verified with the same measures of the quality, by the tests carried out during the realization of the next stage of the D&D process-field testing and finally during the service. The methods mentioned above are based on the high cycle (σ - N curve), low cycle (notch-strain curve) and fracture mechanics (da/dN - ΔK curve) approaches, respectively [2].

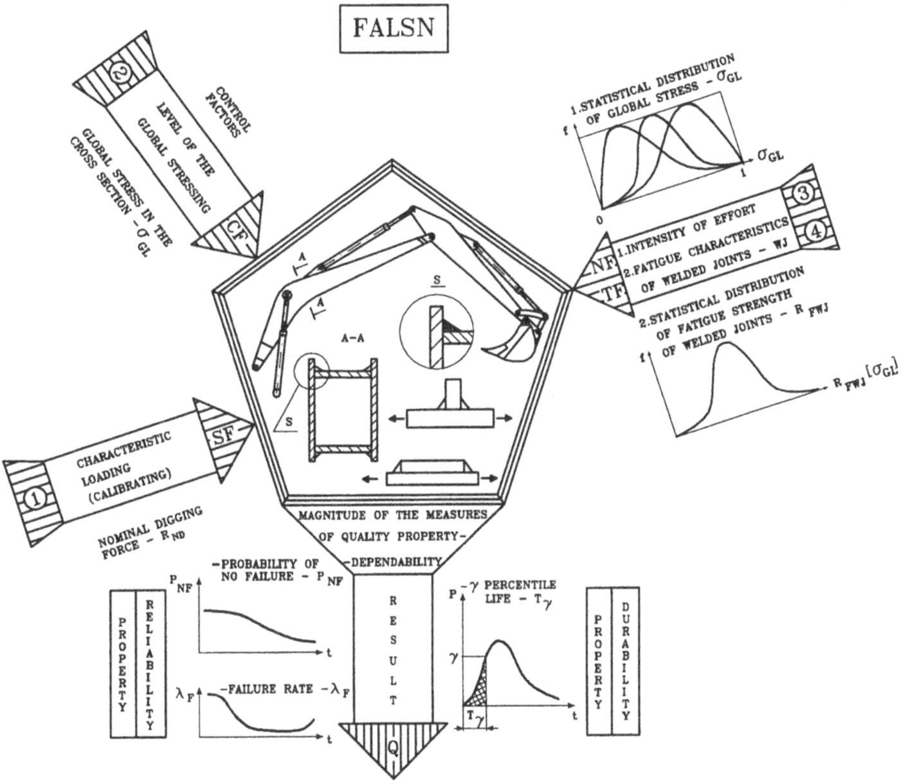

Fig. 3. Factors creating the magnitudes of the quality measures according to FALSN method. Property dependability. Preliminary design stage - PDS. R_{FN} - fatigue strength of welded joint.

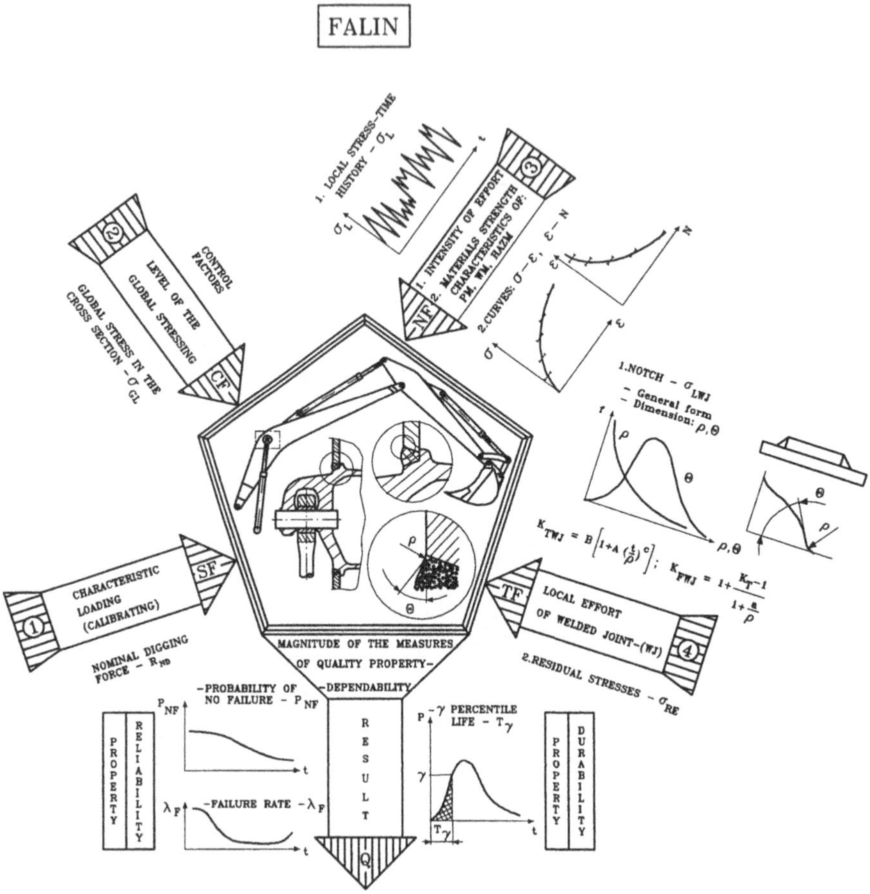

Fig. 4. Factors creating the magnitudes of the quality measures according to FALIN method. Property dependability. Research prototype stage - RPS, t - thickness, a - material constant, ρ, Θ - dimensions of the notch bottom, K_{TWJ} - stress concentration factor of welded joint, K_{FWJ} - fatigue factor of welded joint, PM - parent metal, WM - weld metal, HAZM - heat affected zone metal.

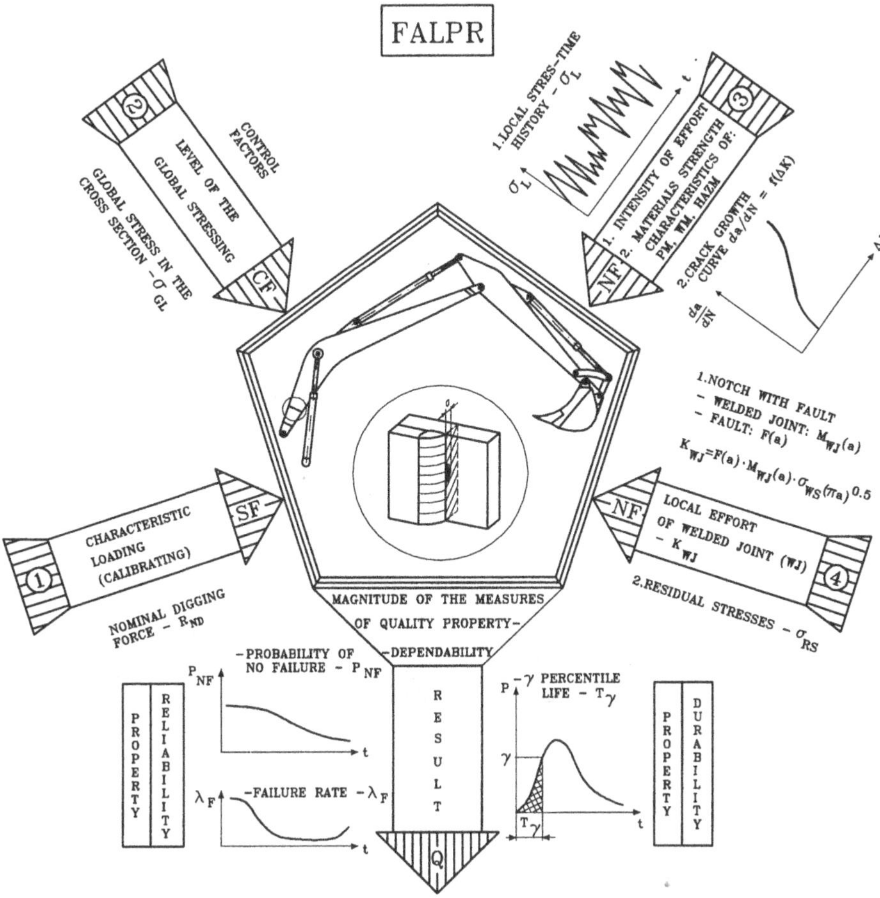

Fig. 5. Factors creating the magnitudes of the quality measures according to FALPR method. Property dependability. Industrial prototype stage - IPS, K_{WJ} - stress intensity factor of welded joint, F(a) - correction function of the crack, $M_{WJ}(a)$ - correction function of the welded joint, PM - parent metal, WM - weld metal, HAZM - heat affected zone metal.

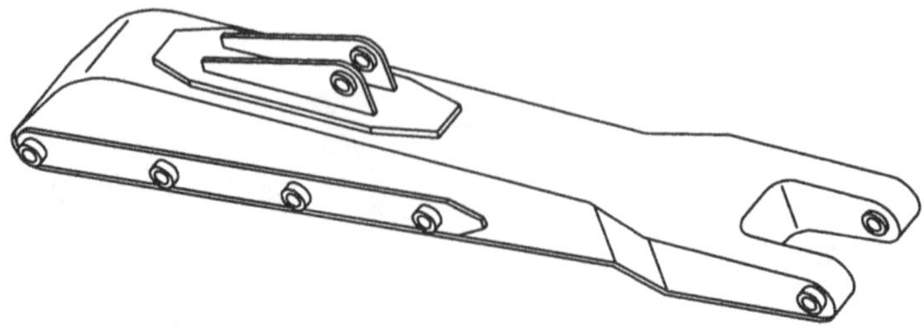

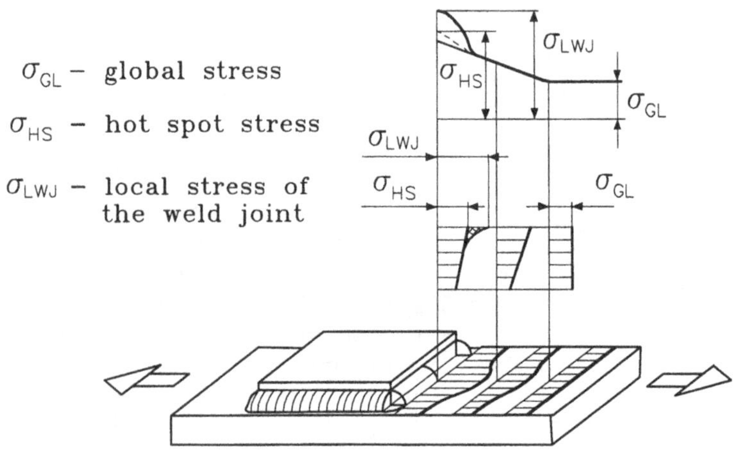

Fig. 6. Description of the stresses in the region of the welded joint.

4 Conclusion

The presented procedure makes it possible to design of the steel structures against fatigue failures in the quality terms. This way the quality can be quantitatively controlled, by means of quality measures created by values of the properly choosen causative factors.

5 References

1. Phadhke, M.S.(1989) Quality Engineering Using Robust Design. Prentice Hall ed.
2. Sobczykiewicz, W., Jakubczak,H., Glinka,G. (1992) Fatigue Design of Earthmoving Machines, Fatigue Design 1992, Helsinki, May 1992 VTT Symposium 131.

40 QUANTITATIVE EVALUATION OF EFFECTS OF SMALL DEFECTS AND NONMETALLIC INCLUSIONS ON FATIGUE STRENGTH OF METALS AND APPLICATION TO QUALITY CONTROL OF MACHINE PARTS

Y. MURAKAMI
Department of Mechanical Science and Engineering, Kyushu University, Fukuoka, Japan

Abstract
Specimens and machine components containing small cracks, defects and nonmetallic inclusions having the same value of the square root of projection area, $\sqrt{area}$, have the same fatigue strength regardless of different shapes, different stress concentration factors. The reason is explained on the basis of both microscopic observation and numerical analysis of cracking from small cracks, defects and inclusions.

The fatigue limit σ_w of components containing surface defects can be predicted by the equation, using two parameters, the geometrical parameter $\sqrt{area}$ and the Vickers hardness H_V.

$$\sigma_w = 1.43\ (H_V + 120)\ /\ (\sqrt{area}\ )^{1/6}$$

where the units are σ_w in MPa, H_V in kgf/mm^2 and $\sqrt{area}$ in μm. The above equation is valid over a range that is dependent on the material.

Applications based on fatigue strength prediction equation and quality control of materials based on statistics of extreme of material defects and inclusions are also outlined.
Keywords: Crack, defects, fatigue, mass production, nonmetallic inclusions, product liability, quality control.

1 Introduction

The importance of the influence of small defects and nonmetallic inclusions on fatigue of metals has been recognized for a long time.

In many early reports [1-4], the correlation between the cleanliness of steels and

Structural Failure, Product Liability and Technical Insurance, Proceedings of Fifth International Conference. Edited by H.P. Rossmanith. Published in 1996 by Taylor & Francis, 2 Park Square, Milton Park, Abingdon, Oxon, OX14 4RN. ISBN 0 419 22080 1

fatigue strength was investigated, but the results were not satisfactory. Many efforts have been made to evaluate the stress concentration factors of inclusions by assuming that their shapes were spherical or ellipsoidal cavity [5-9]. These assumptions are, however, just rough estimations, because a slight deviation from the geometry or small protrusions could greatly affect the stress concentration factors.

The problems of small cracks have recently attracted attention in the field of the fracture mechanics approach to fatigue. The threshold stress intensity factor range, ΔK_{th}, is useful for determining the maximum allowable stress when cracks or defects are detected in machine parts and structures under service loading. However, recent experimental studies [10-15] show that the value of ΔK_{th} is dependent on the crack size, i.e. the smaller the crack, the smaller the value of ΔK_{th}. Kitagawa and Takahashi [12] pointed out that ΔK_{th} decreases with decreasing crack size. However, an expression for ΔK_{th} as a function of crack size was not given explicitly.

In previous papers [16-19], Murakami and Endo have discussed the same subjects by using a new geometrical parameter, $\sqrt{area}$, which is defined as the square root of the area of a defect or crack projected onto a plane perpendicular to the maximum tensile stress. The essence of their discussion will be introduced and reviewed in the present paper in order to understand the framework of their study. The usefulness of *the* $\sqrt{area}$ *parameter model* of Murakami and Endo will be reconfirmed with the additional experimental data on various materials. Finally, the extended applications of *the* $\sqrt{area}$ *parameter model* will be shown.

2 Background of fatigue threshold for cracks emanating from small defects and nonmetallic inclusions

2.1 Small Defects and Cracks

In order to study the effects of small surface defects, Murakami and his colleagues [16, 20-23] have conducted systematic experiments using specimens containing artificial small holes with diameters ranging from 40 μm to 500 μm.

Figure 1 shows the crack emanating from an artificial hole observed at the fatigue limit [24]. This crack is a so-called nonpropagating crack which started from the hole edge in the early stages of stress cycle and afterward stopped propagating under the same stress level. Many similar observations have been given in the literature [16, 20-22, 25, 26]. These facts are extremely important for the quantitative and systematic evaluation of the effect of small defects on fatigue strength. Namely, the condition of the nonpropagation of cracks from the defect at the fatigue limit can be regarded as a crack problem rather than as a stress concentration problem of a notch in fatigue, because the final stress field in the vicinity of a nonpropagating crack from a defect is not affected by the three-dimensional volume of the defect, but is prescribed by the magnitude of the projection area of the defect. This is the reason why $\sqrt{area}$ must be introduced as the geometrical parameter [16]. This parameters, $\sqrt{area}$, has an intimate correlation with the maximum stress intensity factor K_{Imax} along crack front [27].

2.2 Nonmetallic Inclusions

It is known that there is a good correlation between the rotating bending or uniaxial fatigue strength σ_{w0} of unnotched specimens and the ultimate tensile strength σ_B or the

Vickers hardness H_V for low- and medium-strength steels. This correlation can be empirically formulated as follows.

$$\sigma_{w0} = 1.6H_V \pm 0.1H_V \quad (H_V < 400) \tag{1}$$

where σ_{w0} is in MPa and H_V is in kgf/mm^2.

However in the higher range of the Vickers hardness or ultimate tensile strength, this linear correlation no longer holds and there is more scatter in the fatigue strength. It is quite common that a white circular mark named *fish-eye* is observed on the fatigue fracture surface of high-strength steels. Usually a nonmetallic inclusion exists at the center of a fish-eye but it is sometimes not found when it is dropped out of the fracture surface. Figure 2 shows the typical nonmetallic inclusion at fish-eye. This implies that the fatigue fracture origin of high-strength steels is mostly at the nonmetallic inclusion such as MnS, Al_2O_3, SiO_2 and TiN.

3 The $\sqrt{area}$ parameter model for defects and inclusions

3.1 Correlation between K_{Imax} along Crack Front and $\sqrt{area}$ of Crack

It is reasonable to seek the geometrical parameter from the standpoint that the effects of shapes and sizes of cracks on fatigue strength may be correlated with stress intensity factors, especially with the maximum stress intensity factor K_{Imax} along the three-dimensional crack front. Previous numerical studies [16, 27-29] on this problem can be summarized by the following approximate equations:

$$K_{Imax} = 0.65\ \sigma_0 \sqrt{\pi\sqrt{area_s}} \quad \text{; for surface crack} \tag{2}$$

$$K_{Imax} = 0.50\ \sigma_0 \sqrt{\pi\sqrt{area_i}} \quad \text{; for internal crack} \tag{3}$$

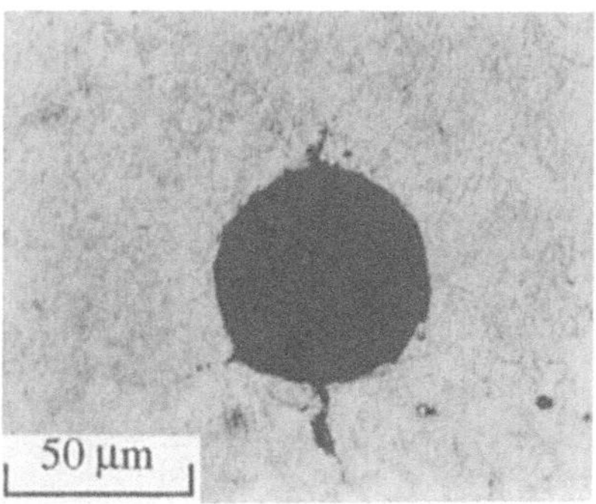

↔ Axial direction

Fig. 1. Nonpropagating crack observed at fatigue limit of Maraging steel (H_V = 740) [24]. σ_w = 666 MPa, σ_m = –255 MPa, Hole diameter = 50 μm, Hole depth = 70 μm.

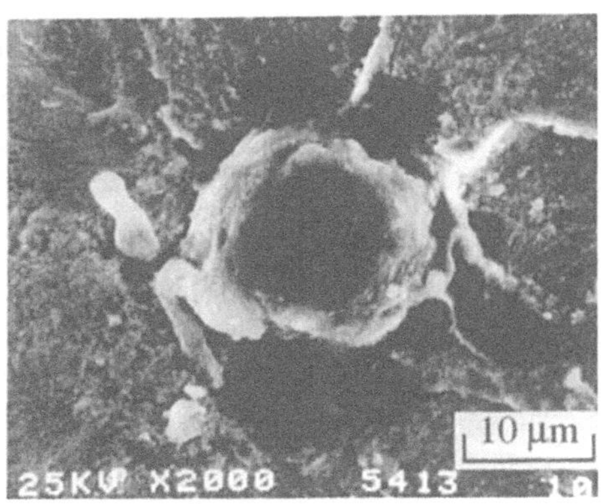

Fig. 2. Inclusion at the center of a fish-eye. SAE9254 (H_V = 641), σ = 980 MPa, $N_f = 1.69 \times 10^6$, $\sqrt{area}$ = 17.9 μm, Stress at the inclusion σ' = 927 MPa.

where σ_0 : the maximum principal stress, $area_s$: the area of a surface crack and $area_i$: the area of an internal crack.

3.2 Stress Intensity Factor for a Crack Emanating from Defects or Inclusions

When a specimen contains a three-dimensional defect other than a planar crack, the fatigue limit is determined by the threshold condition of the crack emanating from the defect [16, 20-22, 25, 26]. In this case, the initial three-dimensional shape of the defect is not directly correlated with the stress intensity factor. Rather, the planar domain (area) which is occupied by projecting the defect onto the plane perpendicular to the maximum principal stress should be regarded as the equivalent crack, and the stress intensity factor should be evaluated from the equivalent crack (see Fig. 3).

Figure 4 [30] shows the stress intensity factors for a crack emanating from ellipsoidal cavities. This figure indicates that with crack extension, a cavity with a crack readily becomes mechanically equivalent to a planar penny shaped crack. Considering this results together with similar numerical analyses on various nonmetallic inclusions [30], we can understand that $\sqrt{area}$ is the rational geometrical parameter for various defects and inclusions.

3.3 Modified Rule for the Estimation of $\sqrt{area}$ for Two-dimensional Cracks and Defects

With regard to an estimation of $\sqrt{area}$ for irregularly shaped, very slender and very deep cracks and defects, some simple corrections in the estimation of $\sqrt{area}$ are necessary [16, 17, 23, 27] for the application of Eqs. (2) and (3) (see Ref. [18]).

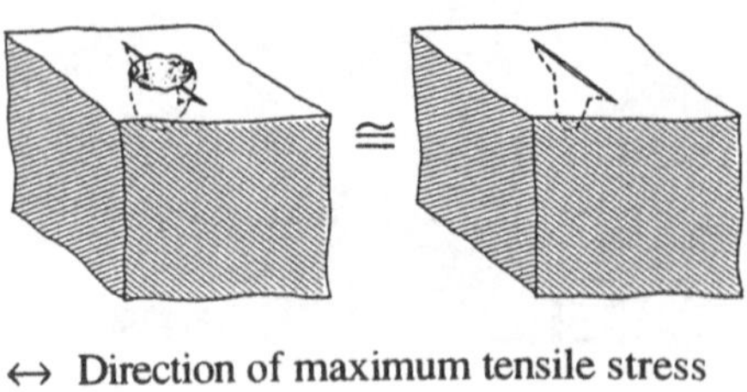

Fig. 3. Defect with cracks and its equivalent crack [17].

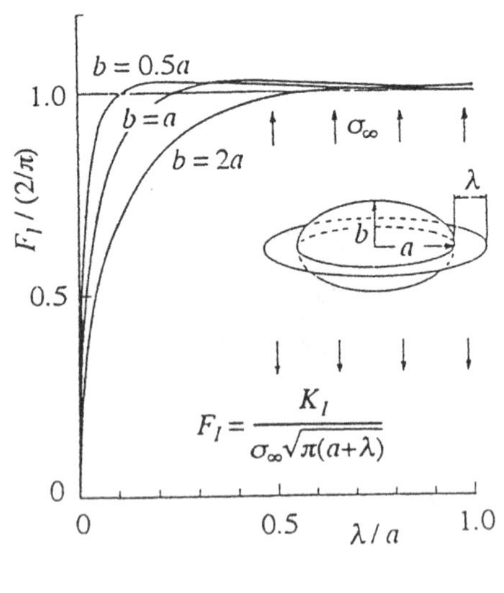

Fig. 4. Stress intensity factors for a crack emanating from ellipsoidal cavities [30].

4 Fatigue limit σ_w as the function of $\sqrt{area}$ and H_V

Considering more than 100 fatigue data on more than 15 materials containing various artificial defects and having various Vickers hardness H_V ranging from 70 to 700, Murakami and Endo obtained the equation for predicting fatigue limit for stress ratio R=-1 as follows [17]:

$$\sigma_w = 1.43\,(H_V + 120) \,/\, (\sqrt{area}\,)^{1/6} \text{ ; for surface defects and inclusions} \qquad (4)$$

where the units are σ_w : MPa, H_V : kgf/mm^2 and $\sqrt{area}$: μm.

Considering the comparison between Eqs. (2) and (3), Murakami *et al.* derived fatigue limit prediction equation for internal defects and inclusions as follows [31, 32]:

$$\sigma_w = 1.56\,(H_V + 120) \,/\, (\sqrt{area}\,)^{1/6} \text{ ; for internal defects and inclusions} \qquad (5)$$

Murakami *et al.* [24,33] extended the equations (4) and (5) for the case R≠-1 as follows. These equations are useful for predicting fatigue strength under mean stress and residual stress.

$$\sigma_w = C\,(H_V + 120) \,/\, (\sqrt{area}\,)^{1/6} \cdot [\,(1 - R)\,/\,2\,]^{\alpha} \qquad (6)$$

$$\alpha = 0.226 + H_V \times 10^{-4}, \quad C : \text{constant in Eq. (4) or (5)}$$

5 Application of the prediction equation

5.1 Small Defects and Cracks

From the viewpoint of examining *the $\sqrt{area}$ parameter model,* the comparison of fatigue limit of specimens containing an artificial drilled hole and an artificial surface crack having the same value of $\sqrt{area}$ is interesting. Table 1 shows the shape of the initial hole and crack [34]. The crack was introduced by fatigue test using a specimen

Table 1. Comparison of rotating bending fatigue limit of specimens containing a small artificial hole and that of specimens containing a small crack (Annealed 0.46% C steel) [34]

Shape of initial hole and crack		$\sqrt{area}$ μm	Experimental value of fatigue limit, σ_w MPa
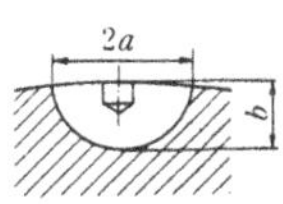	$2a = 213\ \mu$m Aspect ratio: $b/a = 0.7$	112	201
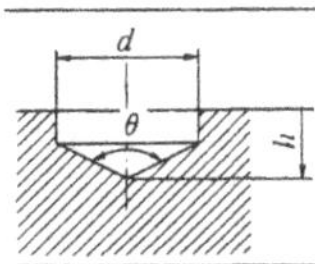	$d = 200\ \mu$m $h = 95\ \mu$m $\theta = 120°$	115	201

containing a smaller artificial hole with the diameter of 40 μm. The specimen was annealed after preliminary fatigue test to relieve the prior fatigue history. As shown in Table 1, the fatigue limits measured with these two kinds of specimens are completely the same regardless of the big difference in stress concentration factor of hole and crack. Additional fatigue tests results [16, 21, 25, 35] which are in good agreement with the predictions are shown in Fig. 5.

To examine more in detail the validity of *the $\sqrt{area}$ parameter model*, specimens of a hard steel (maraging steel: $H_V = 510$) were tested by tension-compression fatigue [34]. As shown in Fig. 6, the fatigue limits of the specimen having an initial crack (C) and a two-hole defect (B) with fatigue crack introduced by preliminary fatigue test are the same (402 MPa) for the same value of $\sqrt{area}$, though that for the specimen containing one hole (A) is 421 MPa which is only 4.7% higher than (B) and (C).

5.2 Nonmetallic Inclusions

5.2.1 Nonmetallic inclusions considered to be mechanically equivalent to small defects
When nonmetallic inclusions cause fatigue failure, it is very difficult to predict the fatigue strength in advance of fatigue test. We can identify the size and location of the fatal inclusion appeared on the fracture surface of specimen after fatigue fracture. In such cases, the stress applied at the fatal inclusion should be larger than the value which is calculated by the fatigue strength prediction Eqs. (4)-(6). In this manner, we can examine the applicability of *the $\sqrt{area}$ parameter model* to inclusion problems. Murakami and his colleagues [31, 32, 34, 36-42] investigated many data of fish-eye marks and verified the validity of the prediction model. Table 2 shows an example of the comparison of such investigation [31]. We can see that the stresses (σ') applied at the fatal inclusions are actually larger than the estimated fatigue limits (σ_w').

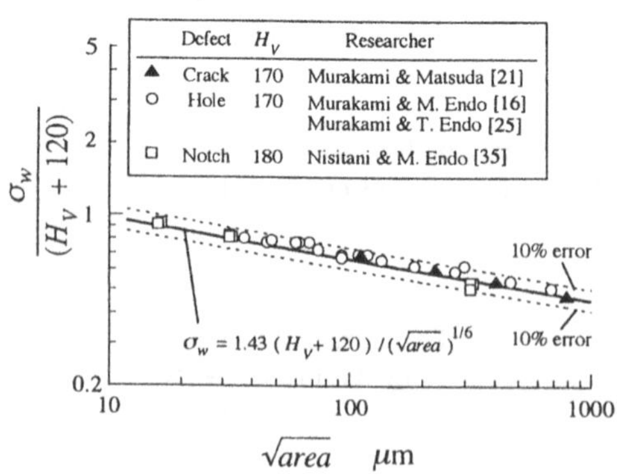

Fig. 5. Relationship between $\sigma_w / (H_V + 120)$ and $\sqrt{area}$ for annealed medium-carbon steel specimens containing a crack or small defect [34].

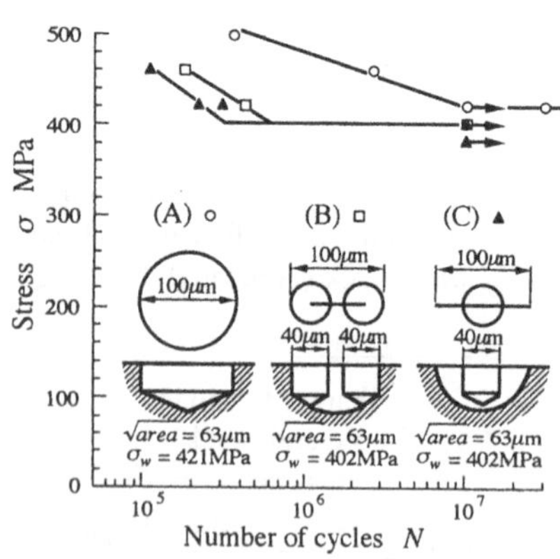

Fig. 6. S–N curve for maraging steel specimens containing a hole or a crack with the equal initial value of $\sqrt{area}$ (63 μm). $H_V = 510$ [34].

Table 2 Size and location of inclusions and fatigue limit predicted by Eq. (5) (Bearing steel ; $H_V = 758$) [31]

Nominal stress at surface, σ (MPa)	Cycles to failure, N_f	Inclusion size, $\sqrt{area}$ (μm)	Distance from surface, h (μm)	Shape of inclusions	Nominal stress at inclusion, σ' (MPa)	Fatigue limit predicted by Eq.(5), σ_w' (MPa)	σ'/σ_w'
1030	125.25×10^4	24.0	310		954	806	1.18
1030	556.77	15.9	140		995	863	1.15
981	422.95	15.2	28		973	870	1.12
981	898.01	9.9	74		963	934	1.03
981	175.51	12.5	10		978	899	1.09
1030	224.64	18.6	74		1011	841	1.20
1030	30.52	22.2	24		1023	817	1.25
1030	26.48	28.4	110		1003	784	1.28
971	735.45	28.9	350		886	782	1.14

The concept of *the* $\sqrt{area}$ *parameter model* suggests that there should be no difference in the influence of inclusion shape if the values of $\sqrt{area}$ of two inclusions are the same. This prediction is completely agreement with the results of the fatigue tests conducted by Duckworth and Ineson [8] who artificially introduced spherical and angular alumina particles into ingots (see Fig. 7) and found no definite difference of effect of inclusion shape. Their experimental data can be arranged clearly by *the* $\sqrt{area}$ *parameter model* [43].

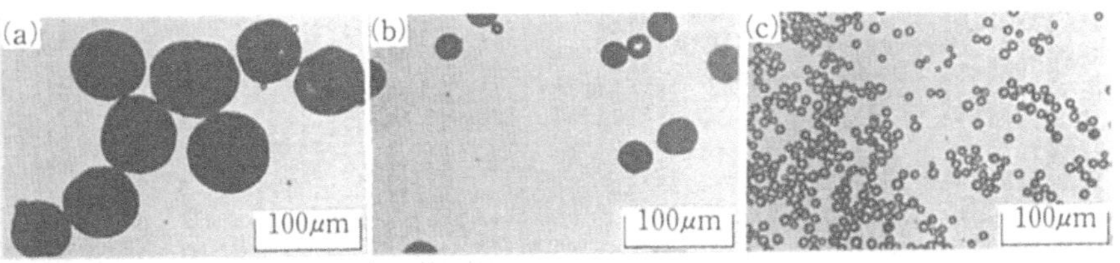

(A) Typical spherical particles ;
(a) 73 μm nominal size, (b) 40 μm nominal size, (c) 10 μm nominal size.

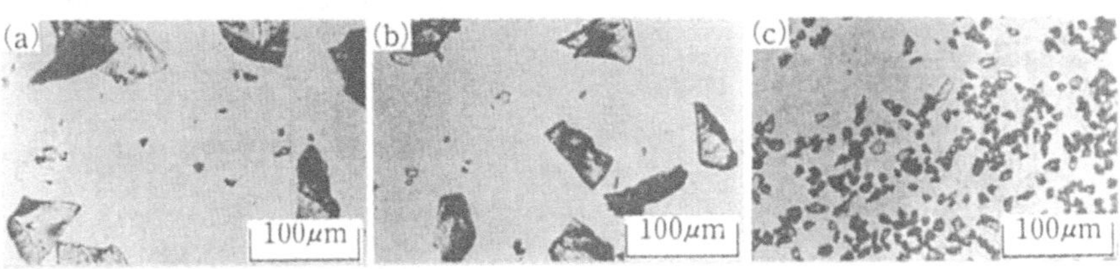

(B) Typical angular particles ;
(a) 73 μm nominal size, (b) 40 μm nominal size, (c) 10 μm nominal size.

Fig. 7. Typical examples of alumina particles artificially added into the ingots [8, 43].

5.2.2 Prediction of fatigue strength scatter caused by nonmetallic inclusions

Murakami *et al.* [31] and Murakami and Usuki [32] proposed a method of predicting the upper and the lower bound of fatigue strength. The upper bound of fatigue strength is obtained when fatigue fracture is not affected by defects or inclusions, and the value can be estimated by Eq. (1).

The lower bound σ_{wl} of fatigue strength can be obtained when the largest inclusion is located in contact with the surface of a specimen [32]. The prediction equation for σ_{wl} is given by Eq. (7).

$$\sigma_w = 1.41\,(H_V + 120)\,/\,(\sqrt{area}_{max})^{1/6} \tag{7}$$

where $\sqrt{area}_{max}$ is the expected maximum size of inclusions contained in a definite volume. In order to estimate the expected maximum size of inclusions contained in a definite number of specimens, Murakami and his colleagues [32, 36-42] applied the statistics of extreme values [44] to the distribution of inclusions. Fig. 8 shows the distribution of the maximum size $\sqrt{area}_{max}$ of inclusions observed in the unit area S_0 (= 0.0309 mm^2) of two kinds of bearing steel (SAE52100, Japanese Industrial Standard SUJ2). Estimating the return period T for N specimens, the expected maximum size of inclusion $\sqrt{area}_{max}$ can be determined by the procedure indicated in the figure.

Introducing H_V and $\sqrt{area}_{max}$ to Eq. (7), the lower bound σ_{wl} of fatigue strength for N specimens is predicted as in Fig. 9. The lower bound of experimental data is well predicted by the equation. Similar applications to various materials have been

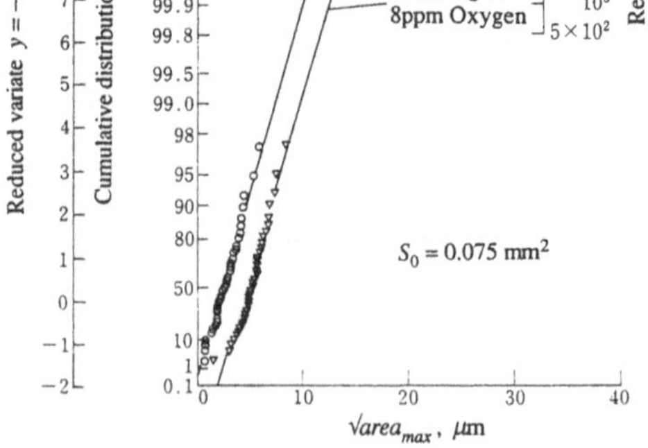

Fig. 8. Statistical distribution of the maximum size $\sqrt{area}_{max}$ of inclusions for two kinds of bearing steel (JIS SUJ2).

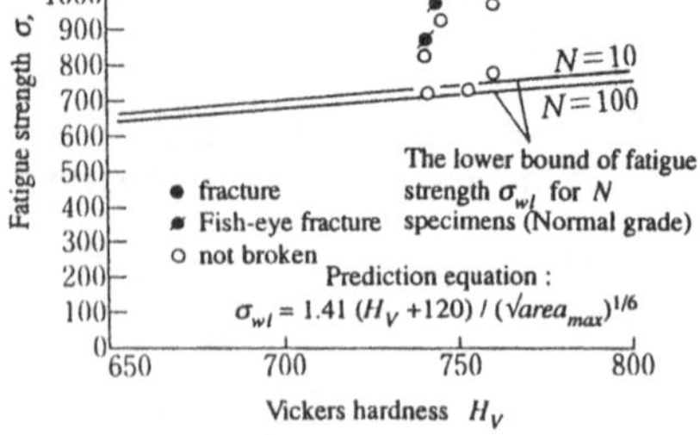

Fig. 9. Comparison of rotating bending fatigue test data with predicted upper- and lower-bounds of fatigue strength.

reported in references [32, 36-42]. The accurate prediction of the lower bound of fatigue strength will be very useful not only for the design of machine elements or structures but also for the improvement of the grade and quality of materials. The line of the lower bound of fatigue strength in Fig. 9 is more helpful for the determination of reliable allowable stress than taking an ambiguous safety factor. Figure 8 can provide a guide for the control of nonmetallic inclusions in the steel-making process.

5.3 Carbides

Carbides sometimes become fatigue fracture origins. Tool steels which contain various carbides are usually formed at high temperature to avoid carbide cracking. However, if carbides behave like nonmetallic inclusions, it was expected by *the* $\sqrt{area}$ *parameter model* and was verified by the experiment [45] that it is not substantial in fatigue strength if carbides are cracked or not cracked before fatigue test.

5.4 Spheroidal Graphites in Cast Irons

Nodular cast iron contains numerous small defects as graphite nodules in the structure (see Fig. 10). Since the fatigue strength is strongly affected by the nodules or casting defects, a lot of successful attempts [46-53] has been recently made to apply *the* $\sqrt{area}$ *parameter model* to the fatigue strength evaluation of this material.

5.5 Second Phases in Al-Si Eutectic Alloy

The inhomogeneity phases such as Si and Al-Fe-Mn in Al-12%Si eutectic alloys behave like hard nonmetallic inclusions in fatigue, because the hardness of these phases is $H_V \cong 700$ while that of aluminium matrix ranges from 100 to 170. Fatigue cracks initiate at the interface between these hard second phases and the soft matrix. Thus, the same concept of fatigue strength prediction for nonmetallic inclusions has been successfully applied to this material [54].

6 Definition of defects in engineering and product liability

Materials inherently contain various defects such as point defects, dislocations, twins, non-metallic inclusions and pores. However, these defects are not necessarily forensic

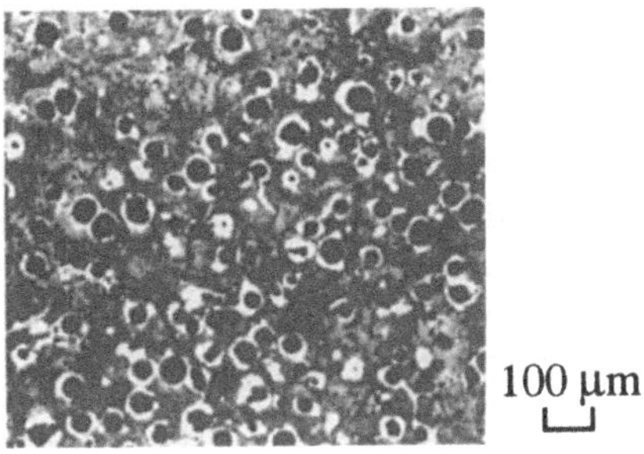

Fig. 10. Microstructures of nodular cast irons: JIS FCD700 (pearlite + graphite nodules surrounded by ferrite) [50, 52].

defects which are discussed in PL cases.

Even if a nonmetallic inclusion is found on the fracture surface of a failured component, it cannot be said that the component contained a defect in forensic meaning, because the component might be designed considering the existence and size of nonmetallic inclusions and the failure might be caused by over load higher than designed stress.

Thus, the forensic definition of material and structural defects in PL cases are not so simple.

7 Conclusion

The √area parameter model was introduced for the prediction of fatigue strength of metallic materials containing small defects, small cracks, nonmetallic inclusions and inhomogeneities such as carbides and eutectic silicon. The conclusions are summarized as follows.

1. The fatigue limit of specimens or machine components containing small defects, small cracks, nonmetallic inclusions, graphites in cast irons and other inhomogeneities in metals is controlled by the Vickers hardness H_V of the microstructure and the square root of the projection area of the defect in these inhomogeneities, $\sqrt{area}$.

   The prediction equations are given in the following.

$$\sigma_w = C\,(H_V + 120) / (\sqrt{area})^{1/6} \cdot [\,(1 - R)/2\,]^{\alpha}$$

$$\alpha = 0.226 + H_V \times 10^{-4}$$

   C = 1.43 ; for surface defects and inclusions
   C = 1.56 ; for internal defects and inclusions

   R : stress ratio $(= \sigma_{min} / \sigma_{max})$

   where the units are σ_w : MPa, H_V : kgf/mm^2 and $\sqrt{area}$: μm.
2. The scatter of fatigue strength in high hardness region is caused by the fracture from nonmetallic inclusions. Since the size ($\sqrt{area}$) and location of the nonmetallic inclusion which leads a specimen to final fracture are not constant, it is almost impossible to determine the fatigue strength of individual specimens. Only the upper and lower bound of fatigue strength σ_{wu} and σ_{wl} can be estimated. σ_{wu} is empirically proportional to H_V. σ_{wl} can be estimated using the statistics of extreme values which enables one to find the expected maximum size $\sqrt{area}_{max}$ of inclusion to be contained in many specimens and machine components.
3. Inclusion rating by the statistics of extreme can be applied to the quality control of materials from the viewpoint of cleanliness.
4. Various defects in engineering materials can not necessarily be regarded as forensic defects in PL cases.

8 Reference

1. Ineson, E., Clayton-Cave, J. and Taylor, R. J. (1958) Variation in fatigue properties over individual casts of steel - part II. *J. Iron Steel Inst.*, Vol.190, pp.277-283.
2. Atkinson, M. J. (1960) The influence of nonmetallic inclusions on the fatigue properties of ultra-high-tensile steels. *J. Iron Steel Inst.*, Vol.195, pp.64-75.
3. Uhrus, L. O. (1963) Through-hardening steels for ball bearings-effect of inclusions on endurance. *Clean Steel, Iron Steel Inst., Sp. Rep.*, Vol.77, pp.104-109.
4. de Kazinczy, F. (1970) Effect of small defects on the fatigue properties of medium-strength cast steel. *J. Iron Steel Inst.*, Vol.208, pp.851-855.
5. Frith, P. H. (1955) Fatigue tests on rolled alloy steels made in electric and open-hearth furnaces. *J. Iron Steel Inst.*, Vol.180, pp.26-38.
6. Hyler, W. S., Tarasov, L. P. and Favor, R. J. (1958) Distribution of fatigue failures in flat hardened steel test bars. *Proc. ASTM*, Vol.58, pp.540-551.
7. Johnson, R. F. and Swell, J. F. (1960) The bearing properties of 1%C-Cr steel as influenced by steelmaking practice. *J. Iron Steel Inst.*, Vol.196, pp.414-444.
8. Duckworth, W. E. and Ineson, E. (1963) The effects of externally introduced alumina particles on the fatigue life of En24 steel. *Clean Steel, Iron Steel Inst., Sp. Rep.*, Vol.77, pp.87-103.
9. Kawada, Y., Nakazawa, H. and Kodama, S. (1963) The effects of the shapes and the distributions of inclusions on the fatigue strength of bearing steel in rotary bending. *Trans. Japan Soc. Mech. Eng.*, Vol.29, No.206, pp.1674-1683.
10. Frost, N. E., Pook, L. P. and Denton, K. (1971) A fracture mechanics analysis of fatigue crack growth data for various materials. *Engng. Fract. Mech.*, Vol.3, pp.109-126.
11. Kobayashi, H. and Nakazawa, H. (1972) A stress criterion for fatigue crack propagation in metals. *Proc. 1st Int. Conf. Mech. Behavior Mater.*, Kyoto, pp.199-208.
12. Kitagawa, H. and Takahashi, S. (1976) Applicability of fracture mechanics to very small cracks or the cracks in the early stage. *Proc. 2nd Int. Conf. Mech. Behavior Mater.*, Boston, pp.627-631.
13. Romanov, O. N., Siminkovich, V. N. and Tkack, A. N. (1981) Near-threshold short fatigue crack growth. *Fatigue Thresholds: Fundamentals and Eng. Appl.*, Vol.II, Edited by Bäcklund, J., Blom, A. F. and Beevers, C. J., EMAS Publ., West Midlands, pp.799-807.
14. Leis, B. N., Kanninen, M. F., Hopper, A. T., Ahmad, J. and Broek, D. (1983) A critical review of the short fatigue crack in fatigue. Battelle's Columbus Labo., Rep. No.AFWAL-TR-83-4019, pp.1-135.
15. Suresh, S. and Ritchie, R. O. (1983) The propagation of short fatigue cracks. Rep. No.UCB/RP/83/1014, pp.1-70.
16. Murakami, Y. and Endo, M. (1983) Quantitative evaluation of fatigue strength of metals containing various small defects or cracks. *Engng. Fract. Mech.*, Vol.17, pp.1-15.
17. Murakami, Y. and Endo, M. (1986) Effects of hardness and crack geometries on ΔK_{th} of small cracks emanating from small defects. *The Behaviour of Short*

Fatigue Cracks, EGF Publ.1, Edited by Miller, K. J. and de los Rios, E. R., Mech. Engng. Publ., London, pp.275-293.

18. Murakami, Y. and Endo, M. (1992) The $\sqrt{area}$ parameter model for small defects and nonmetallic inclusions in fatigue strength: experimental evidences and applications. *Proc. Conf. Theoretical Concepts and Numerical Analysis of Fatigue*, Birmingham, pp.51-71.
19. Murakami, Y. and Endo, M. (1994) Effects of defects, inclusions and inhomogeneities on fatigue strength. *Fatigue*, Vol.16, pp.163-182.
20. Murakami, Y., Tazunoki, Y. and Endo T, (1984) Existence of the coaxing effect and effects of small artificial holes on fatigue strength of an aluminum alloy and 70-30 brass. *Metall. Trans.*, Vol.15A, pp.2029-2038.
21. Murakami, Y. and Matsuda, K. (1987) Threshold stress for fatigue crack initiation from a small crack. *Trans. Japan Soc. Mech. Eng.*, Ser.A, Vol.53, No.489, pp.871-877.
22. Endo, M. and Murakami, Y. (1987) Effects of an artificial small defect on torsional fatigue strength of steels. *Trans. ASME, J. Engng. Mater. Tech.*, Vol.109, pp.124-129.
23. Murakami, Y. and Endo, M. (1991) Prediction equation for ΔK_{th} of various metals containing small defects in terms of the vickers hardness (H_V) and square root of the projected area of defects ($\sqrt{area}$). *Current Japanese Materials Research*, Vol.8, Soc. Mater. Sci., Japan, Elsevier Applied Science, pp.105-124.
24. Murakami, Y., Uemura, Y., Natsume, Y. and Miyakawa, S. (1990) Effect of mean stress on the fatigue strength of high-strength steels containing small defects or nonmetallic inclusions. *Trans. Japan Soc. Mech. Eng.*, Ser. A, Vol.56, No.525, pp.1074-1081.
25. Murakami, Y. and Endo, T. (1980) Effects of small defects on fatigue strength of metals. *Int. J. Fatigue*, Vol.2, pp.23-30.
26. Murakami, Y. and Endo, T. (1981) The effects of small defects on the fatigue strength of hard steels. *Proc. Fatigue 81*, Warwick Univ., UK, pp.431-440.
27. Murakami, Y. (1985) Analysis of stress intensity factors of modes I, II and III for inclined surface cracks of arbitrary shape. *Engng. Fract. Mech.*, Vol.22, No.1, pp.101-114.
28. Murakami, Y. and Nemat-Nasser, S. (1983) Growth and stability of interacting surface flows of arbitrary shape. *Engng. Fract. Mech.*, Vol.17, No.3, pp.193-210.
29. *Stress Intensity Factors Handbook*, Edited by Murakami, Y. *et al.* (1987) Committee on Fracture Mechanics, Soc. Mater. Sci., Japan & Pergamon Press, Vol.2, p.822.
30. *Stress Intensity Factors Handbook*, Edited by Murakami, Y. *et al.* (1992) Committee on Fracture Mechanics, Soc. Mater. Sci., Japan & Pergamon Press, Vol.3, pp.559-562.
31. Murakami, Y., Kodama, S. and Konuma, S. (1989) Quantitative evaluation of effects of non-metallic inclusions on fatigue strength of high strength steels. I: Basic fatigue mechanism and evaluation of correlation between the fatigue fracture stress and the size and location of non-metallic inclusions. *Int. J. Fatigue*, Vol.11, No.5, pp.291-298.
32. Murakami, Y. and Usuki, H. (1989) Quantitative evaluation of effects of non-

metallic inclusions on fatigue strength of high strength steels. II: Fatigue limit evaluation based on statistics for extreme values of inclusion size. *ibid.*, pp.299-307.

33. Murakami, Y. (1993) *Metal Fatigue: Effects of Small Defects and Nonmetallic Inclusions*, Yokendo Ltd., Tokyo.
34. Toriyama, T. and Murakami, Y. (1993) The $\sqrt{area}$ parameter model for evaluation of effects of various artificial small defects and mutual interaction of small defects on fatigue limit. *J. Soc. Mater. Sci., Japan*, Vol.42, No.481, pp.1160-1166.
35. Nisitani, H. and Endo, M. (1985) Fatigue strength of carbon steel specimens having an extremely shallow notch. *Engng. Fract. Mech.*, Vol.21, No.1, pp.215-227.
36. Murakami, Y. (1988) Fundamental aspects of fatigue threshold of metals containing small defects. *Basic Mechanisms in Fatigue of Metals, Materials Science* , Edited by P. Lukás and J. Polák, Elsevier, New York, pp.343-350.
37. Murakami, Y. (1989) Effects of small defects and nonmetallic inclusions on the fatigue strength of metals. *JSME Int. J., Ser. I*, Vol.32, No.2, pp.167-180.
38. Murakami, Y. (1990) Effects of nonmetallic inclusions on the fatigue strength of metals. *Micromechanics and Inhomogeneity,* The Toshio Mura 65th Anniversary Volume, Edited by G. J. Weng, M. Taya and H. Abé, Springer-Verlag, New York, pp.283-295.
39. Natsume, Y., *et al.* (1990) Prediction of fatigue strength of high speed steel based on estimation of the maximum size of inclusions. *Proc. Fatigue 90*, Honolulu, Vol.I, pp.349-354.
40. Murakami, Y. (1990) Effects of small defects, small cracks and nonmetallic inclusions on fatigue strength of high strength steels. *ibid.*, pp.377-382.
41. Murakami, Y (1991) Effects of nonmetallic inclusions on fatigue strength of high strength steels: Historical overview and new method of quantitative evaluation. *Impact of Improved Materials Quality on Properties, Product Performance, and Design*, MD-Vol.28, ASME, pp.89-102.
42. Kobayashi, M. *et al.* (1991) Fatigue strength prediction of automobile suspension spring steels: Explicit analysis of nonmetallic inclusions, shot peening, decarburized layer, surface roughness and corrosion pits. *ibid.*, pp.171-183.
43. Murakami, Y., Kawakami, K. and Duckworth W. E. (1991) Quantitative evaluation of effects of shape and size of artificially introduced alumina particles on the fatigue strength of 1.5Ni-Cr-Mo (En24) steel. *Int. J. Fatigue*, Vol.13, No.6, pp.489-499.
44. Gumbel, E. J. (1957) *Statistics of Extremes*, Columbia University Press, New York.
45. Natsume,Y., Muramatsu, T. and Miyamoto, T, (1990) Effect of carbide crack on fatigue strength of alloy-tool steel under cold working. *Proc. JSME Meeting*, No.900-86, pp.323-325, and *private communication.*
46. Endo, M. (1989) Effects of graphite shape, size and distribution on the fatigue strength of spheroidal graphite cast irons. *J. Soc. Mater. Sci., Japan*, Vol.38, No.433, pp.1139-1144.
47. Sakai, T., Kogo, S. and Suzuki, M. (1989) Statistical fatigue properties of nodular cast iron in rotating bending. *J. Soc. Mater. Sci., Japan*, Vol.38, No.434, pp.1268-1274.

48. Endo, M. (1990) Fatigue threshold for small cracks in spheroidal graphite cast iron. *Proc. Fatigue 90*, Honolulu, Vol.III, pp.1357-1362.
49. Doi, S. *et al.* (1990) Fatigue behavior of spheroidal graphite ductile iron (2nd report, Effect of small defect on high cycle fatigue). *Proc. JSME Meeting*, No.900-86, pp.613-615.
50. Endo, M. (1991) Fatigue strength prediction of nodular cast irons containing small defects. *Impact of Improved Materials Quality on Properties, Product Performance, and Design*, MD-Vol.28, ASME, pp.125-137.
51. Sugiyama, Y., Asami, K. and Matsuoka, S. (1992) The quantitative evaluation method on fatigue limit of materials with defects by using ductile iron and its application to the fatigue limit improvement. *69th JSME Spring Ann. Meeting*, Vol.A, No.920-17, pp.476-478.
52. Endo, M. (1992) Fatigue strength prediction of spheroidal graphite cast irons based on the root-area parameter model. *ibid.*, pp.479-481.
53. Kato, Y. and Hasegawa, N. (1992) Fatigue limits and thresholds in fatigue crack growth of a spheroidal cast iron FCD450. *ibid.*, pp.494-496.
54. Murakami, Y., Ikeda, H. and Toriyama, T. (1991) Quantitative evaluation of effects of imhomogeneity phases on the fatigue strength of Al-Si new alloys. *ICM6*, pp.433-438.

41 REPRODUCIBILITY OF THE FATIGUE CRACK GROWTH RESISTANCE OF MATERIALS AND ITS CONSEQUENCES

L. TOTH and J. LUKACS
Department of Mechanical Engineering, University of Miskolc, Hungary

Abstract

The reproducibility of different mechanical testing methods is analysed referring specially to the determination of cyclic crack growth resistance of materials. This is important in reliability assessment of quasistatic or cyclic loaded structural elements having crack like defects. Numerical results for several welded joints are given. The definition of a crack propagation sensitivity index of structural elements is proposed. Application of the sensitivity index concept assists in joining the roles of the uncertainties of crack growth resistance testing and crack detection methods in reliability assessment.

Keywords: crack growth, design curve, fatigue, reliability, reproducibility, resistance, sensitivity, welded joint.

1 Introduction

One of the main goals of fracture mechanics testing is to provide data on crack growth resistance of materials which can be used in the reliability assessment of quasistatically or cyclically loaded structural components having crack like defects. However each one of the decisions taken in reliability assessment has its own risk. The task of the expert is to minimize the risk of false decision. This can only be realised when the reliability of crack length measurement and the reproducibility of crack growth resistance properties are known and the validity of the applied fracture mechanics model is verified. Therefore, the strategy of periodical inspection of the working structural elements includes the fundamental questions:

- Which is the most dangerous part of the element?
- What kind of non-destructive method shall be employed?

Structural Failure, Product Liability and Technical Insurance, Proceedings of Fifth International Conference. Edited by H.P. Rossmanith. Published in 1996 by Taylor & Francis, 2 Park Square, Milton Park, Abingdon, Oxon, OX14 4RN. ISBN 0 419 22080 1

- What is the required reliability in flaw detection?
- How can the measure of danger of different flaws be assessed and compared to each other?

These questions can be supplemented by specific problems:

- consequences of a local or global failure,
- role of the structural element in the whole system,
- repairing possibilities, etc.

This paper present a definition for CRACK PROPAGATION SENSITIVITY INDEX OF CYCLICALLY loaded structural elements. The practical applicability of this definition is illustrated by worked examples. The role of scatter of crack growth resistance of materials in reliability estimation will also be discussed.

2 Crack propagation sensitivity index of cyclically loaded structural element

Def: The *crack propagation sensitivity index* of a *cyclically loaded* structural element is defined as the slope of the logarithm of residual life time vs. crack length, lgN_R vs. a curves as shown in Fig. 1.

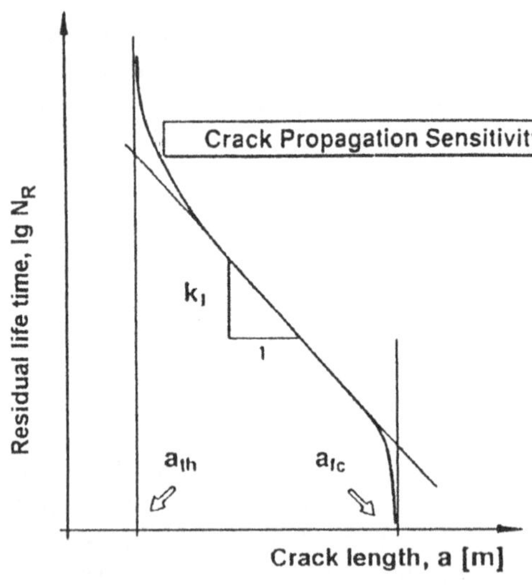

Fig. 1. Definition of crack propagation sensitivity index of cyclically loaded structural element

Considering that the residual life time vs. crack length curve can be calculated by integration of the crack propagation law within the range of the given crack length (a) and the critical crack length (a_{crit})

$$N_R=\int(da/dN)^{-1}da=\int[f(\Delta K, R, \text{crack growth resistance})]^{-1}da, \quad (1)$$

the crack propagation sensitivity index will depend on the

* type of construction element,
* crack geometry, position and length,

* loading mode (I, II. or III. type),
* load level and asymmetry factor,
* local fatigue crack growth resistance of material.

This definition combines the required reliability of the estimated residual life and the crack length measurements, as will be presented in the next paragraph. Therefore this concept is well applicable in developing an inspection strategy of structural components.

3 Practical use of the crack propagation sensitivity index concept

The scheme of practical applicability of the crack propagation index concept for cyclically loaded structural components is illustrated in Fig. 2. The logarithm of the residual life times *vs.* crack lengths, i.e. lgN_R *vs.* a curves for **element N°1** and **N°2** are different. If the levels of required reliability in the residual lifetimes are the same, then they can only be satisfied by different required reliability in crack length measuring.

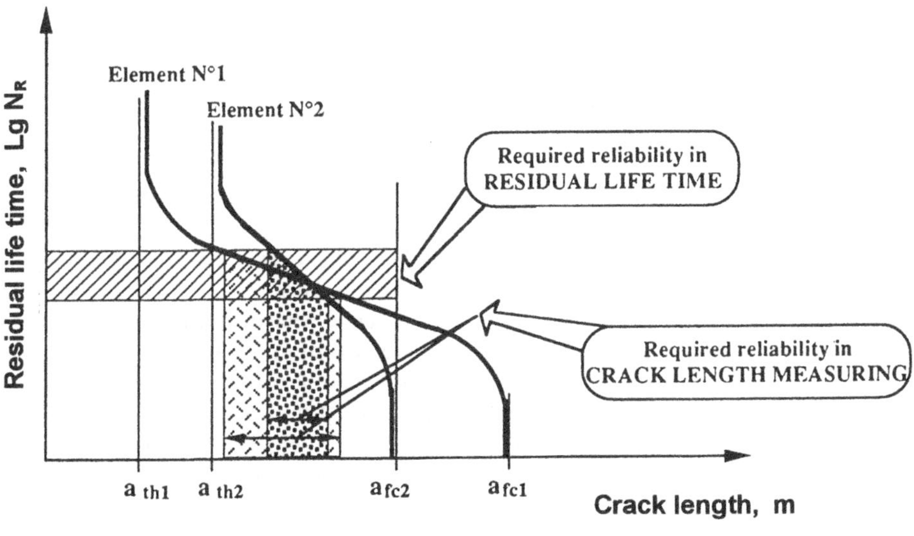

Fig. 2. Practical use of crack propagation sensitivity index concept for comparing the dangers of cyclically loaded structural elements

Exactly the same scheme can be drawn when the scatter of the fatigue crack growth resistance is considered. The *materials* **N°1** and **N°2** in Fig. 3 represent the upper and lower bounds of crack growth resistance which belong to minimum and maximum life times. The interrelation between the required reliability in life time estimations and crack length measuring is the same as was detailed previously.

Concerning Fig. 2. and Fig. 3. it can be concluded, that an exact system can be developed for the required levels of reliability defendet on material properties, crack length measuring, safety factors relating to load level, lifetime or crack length, by using the crack propagation sensitivity index concept.

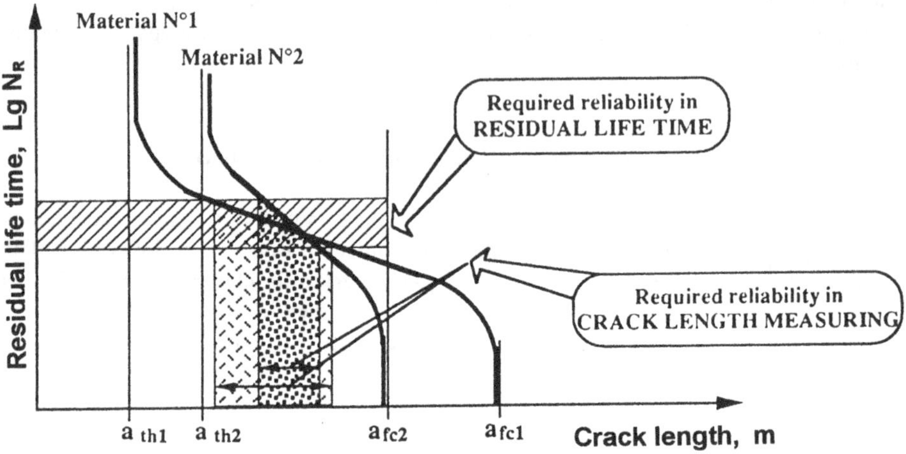

Fig. 3. Practical use of crack propagation sensitivity index concept for cyclically loaded structural element considering the scatter of crack growth resistance

4 Role of the scatter of fatigue crack growth resistance in lifetime estimations

The procedures for determination of the mechanical properties of materials are well defined in standards, handbooks and textbooks. Even when tests are well performed, a scatter band of results has to be considered. Its range depends on test type, local and global inhomogenity (structural and residual stresses) of material and test accuracy. It is always difficult to compare the reproducibility and reliability of different mechanical testing methods and results. The standard deviation of the results is not a satisfactory parameter because the mean values could have different order as well. For the purpose of comparison it seems that one of the best parameters is the relative deviation, i.e. variation coefficient defined by:

$$v = s/x \; 100 \quad \text{in } \% \qquad (2)$$

where the average value is denoted by x and the standard deviation by s. Of course, each type of tests has its own upper and lower limit.

These limits experienced by us are summarised in Fig. 4 for hardness testing and mechanical properties determined by tensile testing (yield stress, ultimate strength, percentage reduction of area, elongation, absorbed specific fracture energy) and for fracture mechanics parameters measured at quasistatic or cyclic loading conditions. The ranges of the acceptable reproducibility are shaded. Below the shaded limits the fair performance of a test series could be the topic of a further discussion. Fig. 3. shows exactly that the variation coefficient for the fracture mechanics parameters is minimum

5%, but a test series with v=10÷15% has to be regarded as a well performing one, as well. This value for welded joints is remarkably greater.

In reliability assessment of the cyclically loaded structural elements having crack like defects the reproducibility of the ΔK_{th} and the slope of the Paris-law regions (***m***) have a well defined role. The scatter of ΔK_{fc} results does not have a significant influence on the estimated residual life time. Using the Weibull function with three parameters for characterisation of different crack growth resistance properties for 37 C and E 420 types construction steels with yield stress R_y=270 MPa and R_y=450 MPa, respectively, the distribution of ΔK_{th} is plotted in Fig. 5. The distribution of Paris law exponent (***m***) in Fig. 6 and the scatter of ΔK_{fc} values in Fig. 7.

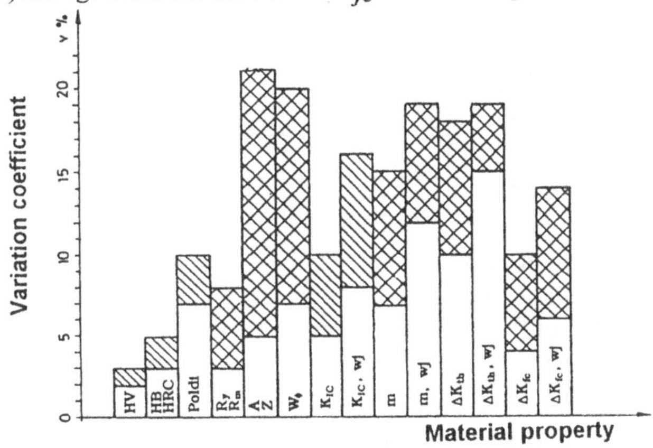

Fig. 4. The reproducibility of hardness, tensile and fracture mechanics testing methods.

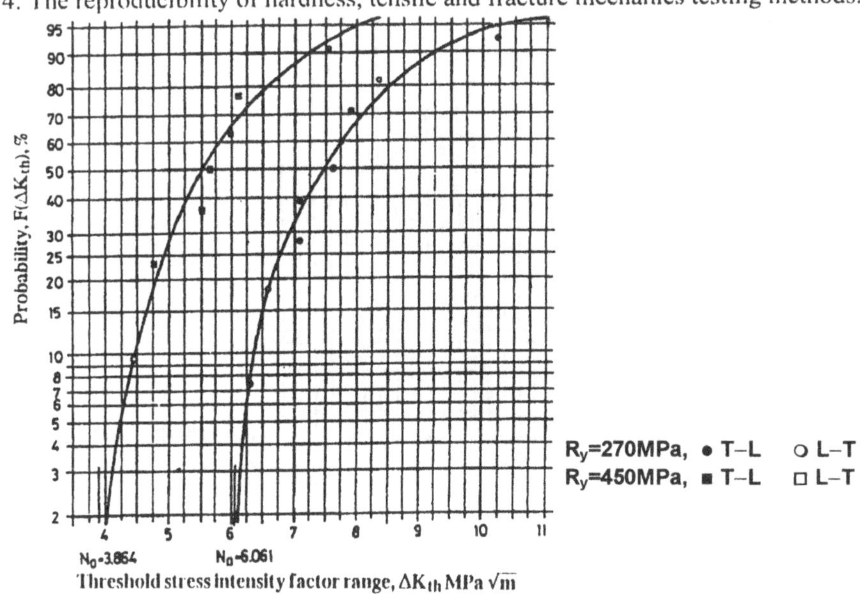

Fig. 5. The distribution of ΔK_{th} values for construction steels with a yield stress of 270 MPa and 450 MPa.

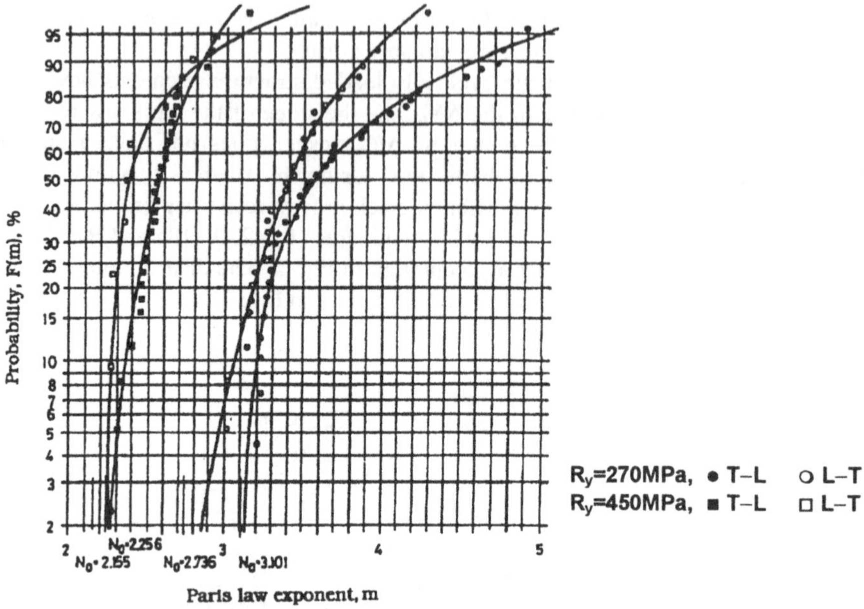

Fig. 6. The distribution of Paris law exponents, ***m*** values for construction steels with a yield stress of 270 MPa and 450 MPa.

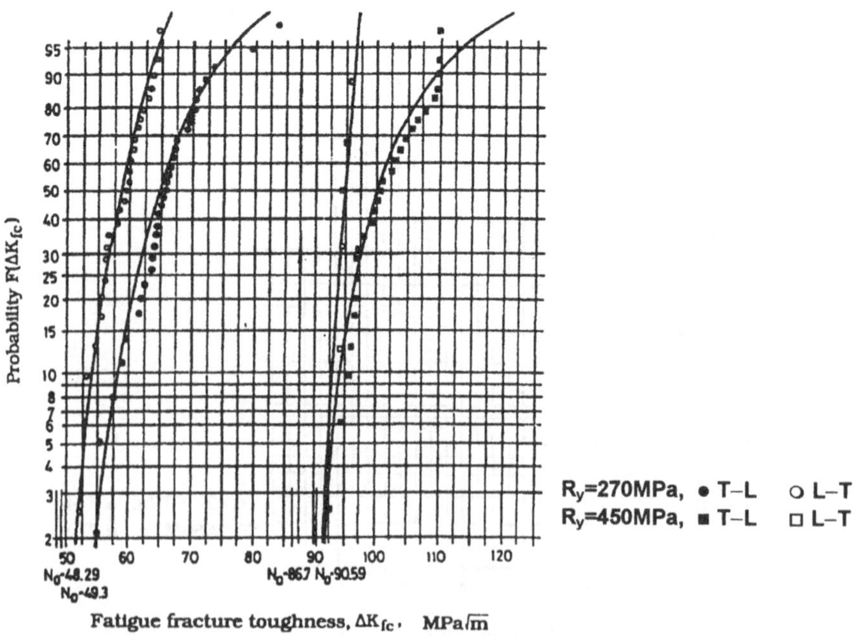

Fig. 7. The distribution of ΔK_{fc} values for construction steels with a yield stress of 270 MPa and 450 MPa.

The scatter bands of these properties for welded joints are significantly greater.

Following from the statistical nature of different fatigue crack growth resistance parameters the "*design curves*" for different classes of materials and for their welded joints are widely used in reliability assessment [1-3]. On the basis of the results presented in Fig. 5-7. *design curves with different probabilities* can be constructed.

Considering the decision risk minimisation it seems satisfactory to give the *"design curve"* with probabilities of 95 % for ΔK_{th} and of 95 % for Paris exponent, respectively. This decision gives the most acceptable risk during residual lifetime calculations. Such *"design curves"* are illustrated in Fig. 8.

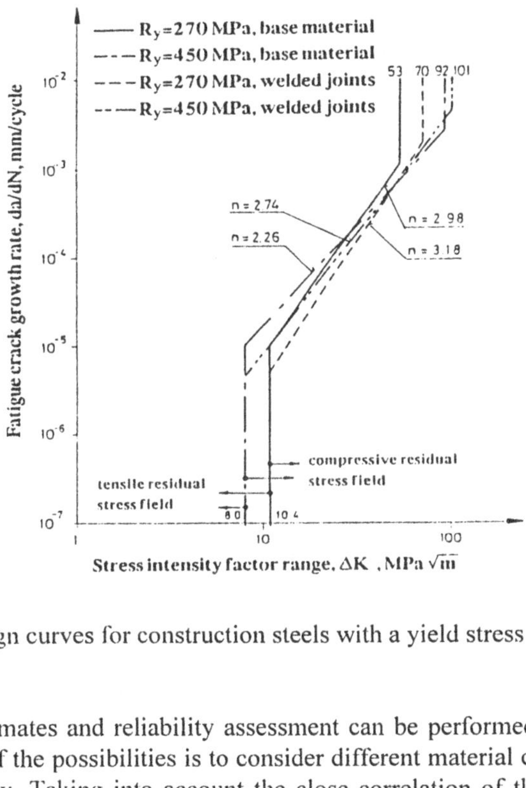

Fig. 8. The design curves for construction steels with a yield stress of 270 MPa and 450 MPa.

Lifetime estimates and reliability assessment can be performed using probabilistic methods. One of the possibilities is to consider different material characteristics with a given probability. Taking into account the close correlation of the parameters of the Paris law, residual lifetime calculations can be performed by varying only the exponent value. These estimations give us information about the effect of this single parameter on the reliability of the structural element. On the basis of ΔK_{th} distributions the propagating crack lengths are estimated, and the applicable non-destructive test methods can then be discussed.

The presentedlife time estimates do no consider such important factors as the problem of short cracks, the residual stress effect (which could also be explained by the importance of crack closure effect) the random character and the multiaxility of the real loading condition. Therefore, the necessity of the cooperation of specialist working in the field of *NDT - Mechanical Testing and Loading Analysis* should be emphasised.

Without this cooperation there is no chance for a satisfactory estimation of the real risk. Using the crack propagation sensitivity index concept of the structural elements, risk analysis can be performed on the basis of reproducibility of the mechanical and NDT test results.

5 Concluding remarks

Considering the aim of this paper and the results presented, the following conclusions can be drawn:

1 The crack propagation sensitivity index has been proposed for cyclically loaded construction elements. It is defined by the slope of the logarithm of the residual lifetime vs. crack length.

2. The application of the crack propagation sensitivity index of the construction elements provides the possibility to combine the reliability assessment calculation and the reproducibility of the NDT or crack growth resistance test results.

REFERENCES

1. Ohta, A. at al., (1986) Statistical evaluation of fatigue crack propagation properties including threshold stress intensity factor. *Eng. Fract. Mech.* **24**, pp. 789-802 .
2. Ohta, A. at al. (1989) Fatigue crack propagation curve for design of welded structures. *Trans. Japan Weld. Soc.* **20**, pp. 17-23.
3. Lukács, J. (1992) Reliability of cyclically loaded welded joints having crack like defects. (in Hungarian). University of Miskolc, Department of Mechanical Engineering. *PhD. Thesis.*

42 THE ADVANTAGES OF VERY FINE GRAIN SIZE STEELS

T. VARGA
Technical University Vienna, TVFA, Vienna, Austria

Abstract

The steels developed nowadays need enhanced strength, toughness and purity to become competitive. The price must be kept, however, low. Grain refining is one of the most effective methods, to obtain adequate improvements. Main characteristics of very fine grain size steels, which are thermomechanically treated, will be therefore discussed.
Keywords: low alloy, high strength, steel, toughness, fine grain

1 Introduction

As usual with most metals, there exist coupled temperature and deformation ranges, which result in very fine grain during and after rolling.

Grain refining was known in steel at least since World War II: VOeST Linz introduced silicon and aluminium killed steels, the Aldur brand. The aluminium did further refine the grain because of forming of Al_2O_3 and AlN microparticles, which did act als grain nuclei.

A few years ago, the very strong desulphurizing and deoxidizing effect of calcium became applied in industrial scale. Calcium containing coatings of welding rods, and fluxes for submerged arc welding did reduce the sulphur and oxygen content drastically, increasing the toughness to values, which were unknown up to this time even for unalloyed steels at higher temperatures. A similar effect was achieved by using Titanium and Boron containing coatings for hand electrodes, whose weld metal did represent excellent toughness down to – 100 deg. C (if cooling time remained in

Structural Failure, Product Liability and Technical Insurance, Proceedings of Fifth International Conference. Edited by H.P. Rossmanith. Published in 1996 by Taylor & Francis, 2 Park Square, Milton Park, Abingdon, Oxon, OX14 4RN. ISBN 0 419 22080 1

acceptable ranges). The basis for this improvement were fine Titanium and Boron carbides, which served as grain growth nuclei too. The advantages of fine grain materials were therefore known all over the years.

However, very fine grain may be produced by applying optimum deformation and temperature combinations during rolling alone. By subsequent quenching after rolling the very fine grain structure is well stabilized and excellent yield strength and toughness are achieved.

This quenching cannot be compared to quenching and subsequent tempering. In QT steels the material hardens by usual martensitic transformation. During the subsequent annealing process hardness will be reduced and toughness restored. Hardening, however, requires sufficient carbon content. In thermomechanically treated steels the carbon content is reduced to an extent that hardening becomes nearly impossible.

Because of the very fine grain the limit of elasticity, i.e. the yield strength, is increased so much, that less than half of the usual carbon content is sufficient to guarantee the same mechanical properties as before. The very fine grain size increases toughness essentially even at low temperatures. The fine grain steel is, however, not to be applied for high temperature use.

In large sections the rolling at optimum temperature and deformation ranges are supported by microalloying, which develops fine precipitates. They act as additional grain refiners. Therefore the excellent yield strength values and the low temperature toughness are possible up to about 90 mm wall thickness.

2 Main characteristics of the TM-treatment

The main features of the thermomechanical heat treatment is shown in Fig. 1a [1]. This diagram contains typical temperature-time diagrams in a schematic form. The basic normalizing heat treatment and rolling ("N") are compared to thermomechanical (TM) treatments. The main characteristics of the thermomechanical heat treatment during rolling are:

The aim is a microstructure with a very fine effective grain size which yields advantages in application.

The composition of the steel is adequately chosen.

The following steps are adepted according to time and temperature:

- Heating of the slab is to be applied by using a range of appropriate temperature
- Rolling is performed according a fixed sequence at final rolling in the non recristallizing austenitic or ($\alpha + \gamma$)-two phase region.
- Cooling down: either on air or in a staple or exposed in a cooling region to a defined cooling rate and temperature.

In some cases additional heat treatment (annealing) is needed.

As it can be seen from the operations listed above, the rolling in a multiple phase state is characterized by preliminary deformation in the recristallizing austenite and

- final rolling in a state of non recristallizing austenite or in
- ($\alpha + \gamma$)-two phase region.

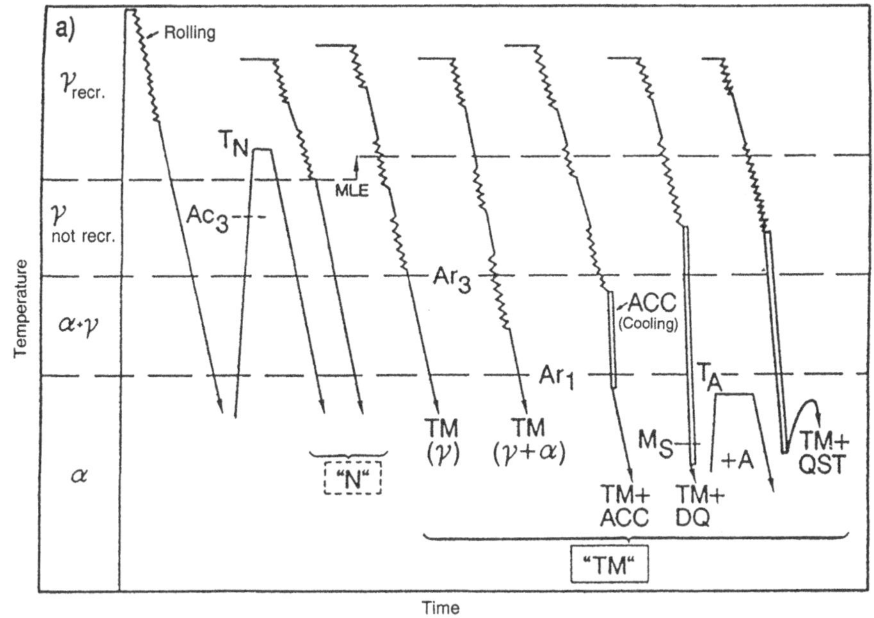

Fig. 1a. Schematic temperature – time graphs are represented for different TM rolling methods. The three diagrams shown apply accelerated cooling.

The rolling is followed by cooling down in air or by interrupted water cooling into the ferritic/perlitic or ferritic/bainitic region (by ACC, accelerated cooling). Cooling with the highest possible chill effect leads to martensite (direct quenching). An anneal can follow after this chilling of the material. Another procedure employs self–annealing, where the surface is cooled very quickly and shortly and the residual heat is used for annealing of the full volume.

Very important is the exact temperature and deformation control, because they are essential for the resulting combination of toughness and strength.

In Fig. 1 b [1] the resulting microstructures and limit grain size coupled with the charpy impact values is shown. The steel was of the CMnNb–type with carbon equivalent of about 0.35 and a plate thickness of 15 mm.

Using flat tension test specimens and Charpy V transversal probes the values have been measured at −40°C. The gain in yield strength, caused by TM–rolling in comparison to the normalizing heat treatment is shown. Further, other possibilities to increase the yield strength by applying accelerated cooling methods were used.

In Fig. 1 b [1] the analysis and plate thickness yields the results as given.

In Fig. 2 [1] the parameters which have to be defined before TM–rolling are listed [1, 2].

The heating of the slabs depends on the geometry, the initial temperature, the time of heating, the overheating, the through–thickness temperature and the temperature of furnace door opening.

Thermomechanically controlled rolling depends on both temperature and time. Important parameters are the geometry of the plate to be rolled, the number of the rolling passes, start and end temperature of those passes, the shape factor during the passes, the rolling velocity, the reduction of the section, the intermediate cooling, the rolling steps and the control of the geometry.

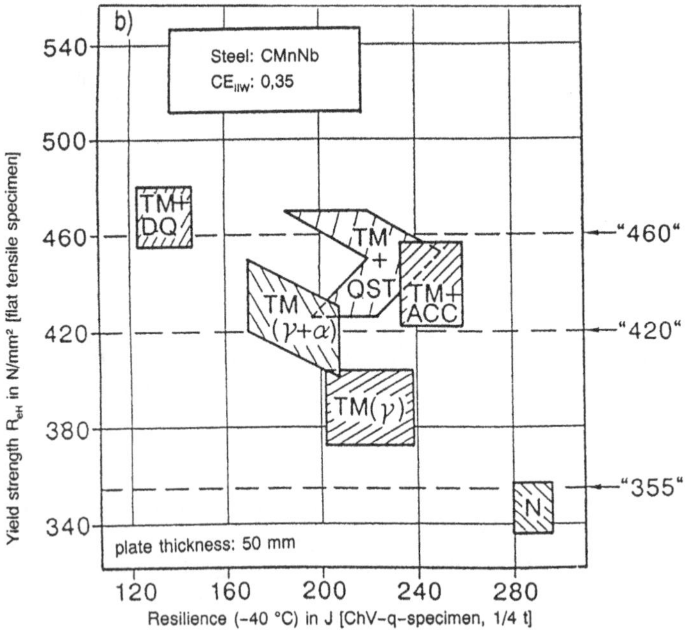

Fig. 1b. Comparison of properties as achieved by different TM rolling conditions. The steel composition analysis and plate thickness are kept constant.

The cooling rate is governed by the transfer time, the start temperature, the final temperature, the maximum gradient, the distribution of the water quantity, the rolling speed, the cooling method (continous or oscillating). As an additional operation shape forming and staple system, sometimes annealing conditions have to be fixed, too.

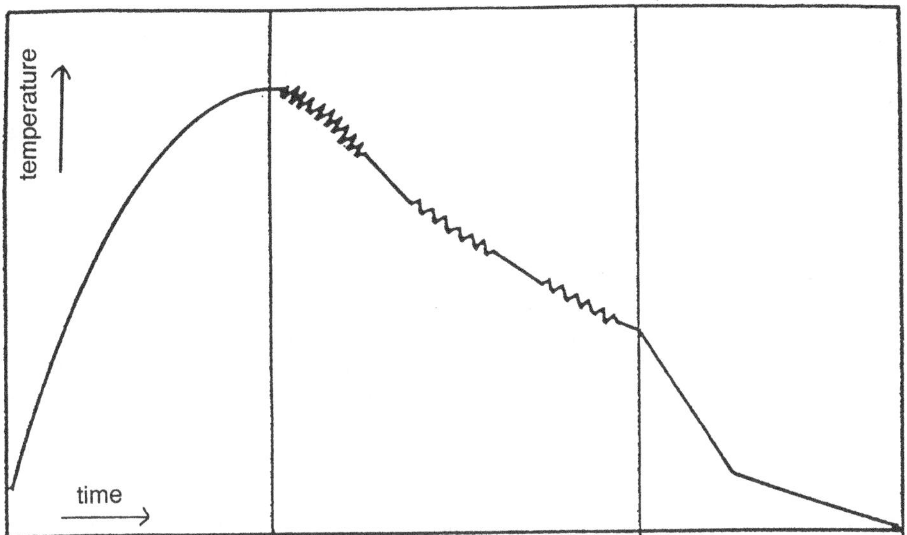

slab preheat	TM–rolling	cooling down
– slab geometry ------------------ – initial temperature – time interval in the oven – overheat – through thickness warming – finish temperature	– geometry of the plate ------------------ – number of rollings – start and finish temperature of the phases – shape factor – rolling speed – impact passes – intermediate cooling – rolling scheme – control of geometry and plane geometry	– transfer time – start temperature – finish temperature – maximum gradient – water distribution – velocity – cooling (continous or oszillating) ------------------ – warm straightening – stapling – annealing (if needed)

Fig. 2. Parameters which have to be taken into account for TM rolling operations.

The heating up of the slabs should be reproducible and well defined, therefore the heat up has to be programmed based on the physical conditions of the heating of the slab.

A considerable influence on the deformation process has the technical equipment and especially the output of the rolling equipment. Especially rolling with the TM–methods need low final rolling temperatures and high rolling forces. The reproduction is possible only if a quick and exact procedure is at disposal which is based on a rather exact knowledge of the plate temperature, plate thickness and of the rolling pressure force. Cool down of the finalized plate occurs on free air, on a cooling bed, in a staple or, if its disponible and necessary, accelerated in a cooling line [1, 3].

3 Results of the TM–process

A homogenous value of the yield strength and the ultimate tensile strength can be achieved only by a well regulated cooling line. For instance, the surface at the upper and lower end of the plate have to be cooled less intensively, because these ends are already cooler when they come to the rolls. Therefore an end masking, that means a slower cooling combined with a premature stop of the water will help to achieve a homogenous strength and toughness in the section.

The scatter of the yield strength is about 2 s = 25 N/m² similar to the TM steel with accelerated cooling.

Concerning toughness, which should yield very high values, there is very important to employ an optimized proportionality of all processing parameters. For instance by reducing the initiation temperature of the slabs from 1180°C to 1080°C yield on a TM + ACC treated offshore steel a decrease of the transition temperature curve by about 25 K.

For crack arrest the very stringent conditions of the Pellini–drop–weight test may be used. The nil ductility transition (NDT) temperature is shown in Fig. 3 [1]. The difference in transition temperature between a steel St 52–3 normalized and a steel StE 460 TM is about 40°C, i.e. the TM–steel has an approximate crack arrest temperature which is about –65°C, the conventional steel of about –25°C.

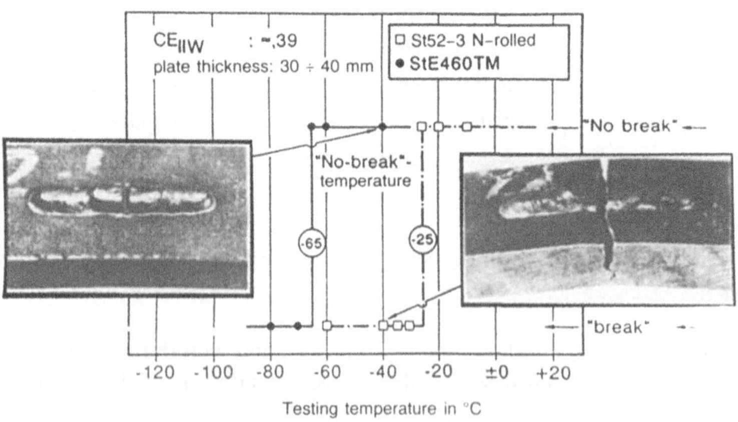

Fig. 3. Comparison of the NDT–temperatures and the no–break conditions for a steel □ St 52–3 (conventional) and ● StE 460 TM (thermo–mechanically treated)

4 Properties of TM–steels

For the user of a TM–steel it is very important to know whether he has to change the welding and heat treatment conditions or not. Because of the much lower content of carbon the welding procedure is less expensive, because there is less or no need at all of preheating. Especially for accelerated cooling after rolling the strength properties are maintained even with higher wall thicknesses and with strongly decreased contents of carbon and of alloying and microalloying elements. In Fig. 4 [1] there is the time $t_{8/5}$ shown and also the hardness of the heat affected zone for the steel types of the yield strength class of 355 N/mm². The hardness values were measured on one layer weldments very near to the fusion line. The different cooling times were simulated by variation of the heat input.

According to the lower content of carbon of the TM–steels the martensite hardness is decreased. This factor defines the hardness of the heat affected zone. With higher heat inputs the influence of alloying elements are becoming stronger. Carbon, but also Molybdenum, Chromium, Manganese and Titanium and Boron have to be mentioned.

In relation to the hardness the delayed cold cracking susceptibility due to hydrogen is of influence. Because of the lower carbon content and the increasing toughness in the heat affected zone the formation of cracks will become more less pronounced. Cracking can be described by using the P_{cM}–formula, which describes the relation between chemical composition of the steels, above all the carbon content and their susceptibility to cracking.

Investigations using Tekken–type crack susceptibility specimens showed the advantage of the low carbon containg TM–steels, see Fig. 4 b [1]. A conventional steel St 52–3 required a preheat of 150°C, but there was no preheat necessary at all even in welding of the root pass of TM–steel E 355 of a plate thickness of 40 to 60 mm.

Of course, to realize welding without preheat electrodes with low content of hydrogen are needed. Otherwise hydrogen cracking (cold cracking) can not be avoided.

It is valid for all steels, that higher purity of the steel, low content of carbon, some nitrogen, vanadium and alloying of a low amount of titanium will be advantageous with normalized steels as well as with TM–steels. This is valid both for the fracture mechanics characteristics as for the impact strength values of the heat affected zone [4, 5].

On the other hand it could be proven that TM–treated steels even in case of high heat inputs during welding show better toughness values in the heat affected zone as usual [6].

Therefore the increased application of TM–treated steels increases the safety of a welded component and allows for higher deposit rate during welding. Further, the stress relief heat treatment of the welded TM–steels can be omitted because of the very high toughness in the heat affected zone in the as welded condition. Rules as they exist, however, have to be regarded in every case.

The toughness of the HAZ, as described by CTOD over temperature can be seen in Fig. 5 [1]. On Gleeble simulated specimens of 10 x 10 mm², a/W = 0,3 the CTOD is demonstrated after a 1350°C treatment, $C_{8/5}$ being 40 s. The TM–steel St E 355 TM is by far the best concerning CTOD, followed by St E 460 TM and StE

420 TM. The steel St E 355 N, normalized, is the weakest material of those compared.

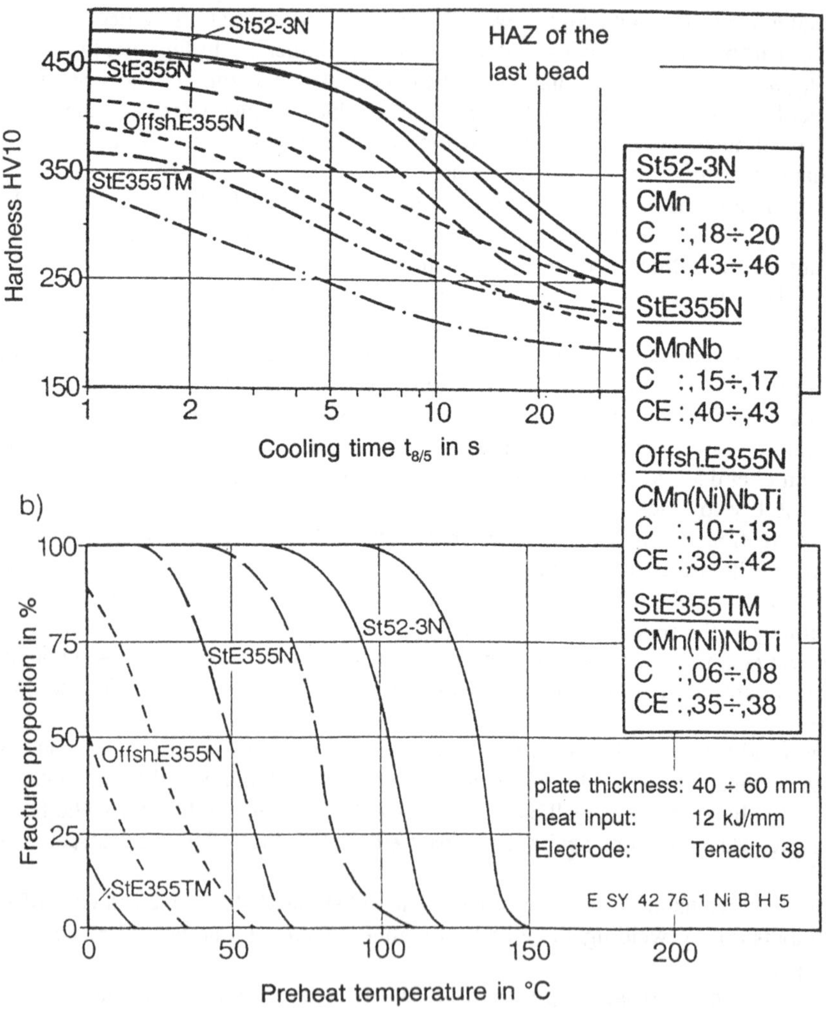

Fig. 4. Influence of the C–content and of the C–aequivalent on the weldability of the steels of a yield strength of ≥ 355 N/mm²

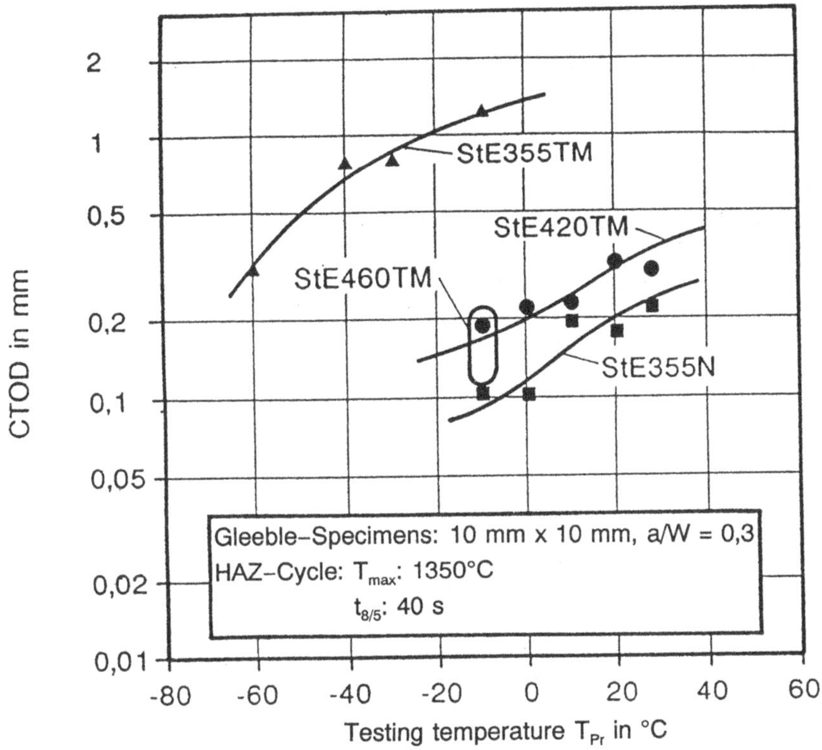

Fig. 5. Comparison of the transition curves as measured by using CTOD values on simulated specimens, normalized (N) and TM–treated steels.

5 Conclusions

Main characteristics of TM–steels as of those with the finest grain size have a very small grained microstructure with a composition which is adequately chosen, heat treated by using the optimum deformation temperature and time combinations. Also cooling is optimized for fine grain results. There is a high impact strength and fracture toughness, CTOD included, resulting.

Because of less than half of the usual carbon content it is in most cases sufficient to weld with reduced or no preheat, but in most cases also stress relief heat treatment is not needed any more.

6 Literature

1. Streißelberger A., Oswald W., Bauer J. and Hanus F (1991) *Verfahren der TM–Behandlung mit beschleunigter Kühlung zur Herstellung von Grobblech aus Baustählen*, Sonderdruck aus "Stahl und Eisen" 111, Heft 5, Seite 65 - 73

2. Hubo R. and Hanus F. E. (1994) *Thermomechanisch gewalzte Grobbleche für den Stahlbau*, Sonderdruck aus "Stahlbau" 63, Heft 3, Seite 84 - 89

3. Hubo R. *Entwicklungen auf dem Gebiet hochfester schweißgeeigneter Stähle*, Conference "Schweißen '95, Probleme und Erkenntnisse in der Schweißtechnik, Technologien und Geräte", Seite 61 - 68, Mai - Juni 1995, St. Petersburg

4. EN 10113 *Warmgewalzte Erzeugnisse aus schweißgeeigneten Feinkornbaustählen, Teil 3: Lieferbedingungen für thermomechanisch gewalzte Stähle*, DIN, April 1993

5. R. Hubo et al., (1993) *Manufacturing and fabrication of thermomechanically rolled heavy plates*, Steel Research 64, No. 8/9

6. Nakao H. and Suzuki S. (1991) *The state–of–the art of steel developments for offshore use*, presented at the Symposium of Norway Japan Cooperative Project, June 28, 1991 Stavanger (Norway)

43 DETERMINING THE RELIABILITY OF TRANSPORT FACILITY ELEMENTS USING FATIGUE TESTING EQUIPMENT TO SERVICE

M. KOPECKY
Department of Mechanics and Strength, University of Transport and Communications, Zilina, Slovak Republic

Abstract
Mechanical systems have a number of very specific features, particularly relating to character and mode of failures. The material must be thoroughly investigated for mechanical properties such as fatigue life or maximum strength to permit efficient and economical, as well as safe design of component structures.

The method described in this paper has been conceived in order to over come the lack of rapid data recording and processing equipment and of fatigue testing equipment capable of applying random loadings for service simulation testing of components.

The results of its application are restriced to transport facility elements.
Keywords: Fatigue, statistical reliability, random loads, laboratory tests, probability density, statistical moment, equivalent frequency.

1 Introduction

Modern technology has advanced the development of many new materials and products which in turn have created the need for new and advanced test methods. As far as fatigue is concerned, it is required to find the loading equivalent sinusoid-constant amplitude and frequency, for a stationary random loading of any subassembly of metallic construction.

This particular method of solution necessitate the determination of the equivalent sinusoid of random stationary loading and of Gaussian

Structural Failure, Product Liability and Technical Insurance, Proceedings of Fifth International Conference. Edited by H.P. Rossmanith. Published in 1996 by Taylor & Francis, 2 Park Square, Milton Park, Abingdon, Oxon, OX14 4RN. ISBN 0 419 22080 1

distribution.This is performed by the use of the concept of power spectral density. This method plots the energy as a function of frequency so that all variables are then taken into account: frequency, amplitude and the number of occurrences. This method usually requires a computer for analysis. The results of application are restricted to load-carrying motor-cycles components.

2 Theory

For a random stationary and Gaussian distribution time series several parameters describing the time evolution of the quantity can be expressed by the function of the statistical moments of the power spectral density in relation to the zero frequency axis is defined as follows:

$$m_n = \int_{-\infty}^{\infty} f^n . S(f)\, df \tag{1}$$

The mean number of crossings in time unit $\mathbf{N_k}$ through k-level is

$$N_k = \sqrt{(m_2 / m_0). \exp(-k / 2m_0)} \tag{2}$$

An equivalent frequency of the phenomenon can be obtained taking k=0 :

$$f_s = \sqrt{(m_2 / m_0)} \tag{3}$$

In this case one should consider the equivalent amplitude sinusoid of a random loading equal to the mean square value of the random evolution with zero mean value

$$\sigma^2_{amp} = 1/T \int_0^T (\sigma)^2\, (t)\, dt \tag{4}$$

and

$$\sigma_{red} = \sigma_{amp} + \sigma_s \tag{5}$$

where

$$\sigma_s = 1/n \sum_{i=1}^{n} \sigma_i \tag{6}$$

3 Application of statistical reliability

The starting point of theoretical solutions reliability is the Weibull model [3].

The dependence between random loads and life, N_f, of components must be completed by a variable, $R(N_f)$, which expressed digital guarantee in the probability form.

A parameters distribution may be expressed as

$$R(N_f) = \exp\left(-((N_f - a) / b^{1/k})^k\right) \tag{7}$$

where: a, b, k are a parameters of distribution.

The probability density function related to **(7)** is of the form

$$f(N_f) = k/b^{1/k} \,.\, ((N_f - a) / b^{1/k})^{k-1} \,.\, \exp\left(-((N_f - a) / b^{1/k})^k\right) \tag{8}$$

The determination of the parameters of this distributinn **a, b, k**, are achieved by the moment of function **(8)** numerically.

Common value of n-th moment for variables $(N_f - a) / b^{1/k}$ is:

$$m_n = \Gamma(1 + n/k)$$

The first and second central moment of basic distribution eq.**(8)** are:

$$m_1 = \Gamma(1 + 1/k)\,; \qquad m_2 = \Gamma(1 + 2/k) - \Gamma^2(1 + 1/k)$$

The coeficient of obliquity is definited with second and third central of moment:

$$m_3 / m_2^{3/2} = [\,\Gamma(1+3/k) - 3\Gamma(1+2/k) \,.\, \Gamma(1+1/k) + 2\Gamma^3(1+1/k)\,] \,/\, /\,[(\Gamma(1+2/k) - \Gamma^2(1+1/k))^{3/2}\,] \tag{9}$$

If we return to original variable N_f, the first moment is a reply to estimate the moment $m_1(N_f)$ for basic random selection:

$$m_1(N_f) \approx N_s \tag{10}$$

where: N_s is a middle value of longevity.

The dispersion of original variable N_f may be expressed in second moment and the estimate of moment is:

$$m_2(N_f) \approx S_N^2 \tag{11}$$

where: S_N is the standard deviation.

The coeficient of obliquity of variable N_f may be expressed in second and third moment and the estimate is:

$$m_3(N_f) / m_2(N_f)^{3/2} \approx (n^2 / (n-1) . (n-2)) .$$

$$. ((N_S^3 - (3N_S^2) . N_S + 2N_S^3)) / S_N^3 = B(k) \tag{12}$$

From the second moment and eq.**(11)** is:

$$S_N . \mathbf{D}(k) = b^{1/k} \tag{13}$$

From the first moment and eq.**(10)** is:

$$N_S - (S_N . \mathbf{D}(k)) . \mathbf{C}(k) = a \tag{14}$$

The values of functions **B**(k), **C**(k) and **D**(k) which achieves eq.**(9)**, are introduced for the practical application in eq.**(7)** for variables **1/k** .

With parameters of distribution, we may define the result by the statistical curve of longevity, which in a form of probability characterized the longevity form eq.**(7)**

$$\ln(-\ln \mathbf{R}(N_f)) = k [\ln(N_f - a) - \ln b + \ln\ln e] \tag{15}$$

4 Experiment and results

The applications of this method in this paper are restricted to load-carrying parts of motor-cycles of small capacity.

For illustration are shown the results of laboratory simulating test for a several parts of construction:

4.1 The frame construction

The frame is the most important load-carrying part of a motor-cycle. The test of simulation regime for a frame construction has been made upon the special purpose machine in a laboratory, as shown the scheme in Fig.**1**.

The laboratory test has been made for 3 alternative frame constructions. Experimental results of the test are shown in Tab.**1**.

Alternative **A** is seen to be the most reliable. The results of random excitation in operating state are to be seen in Tab. **2**.

Table 1. Frame constructions

Reliability: Number of cycles to failure	Alternative: A:	B: x 10^6	C:
N_{f1}	22,4656	6,8759	12,1511
N_{f2}	25,4133	1,2145	3,8624
N_{f3}	22,4967	4,4231	1,8765
N_{f4}	48,2970	4,0824	2,6661

Table 2. Frame constructions

Statistical moment	Loading equivalent to random loading (MPa)	Equivalent frequency (s^{-1})
m_0 = 922,36	σ_s = 15,56	
m_1 = 1690,24	σ_{amp} = 30,3644	
m_2 = 1346291,7	σ_{red} = 45,92144	
		38,21

The statistical curve of longevity for **A** alternative frame construction from eq.**(7)** is shown in Fig.**2**.

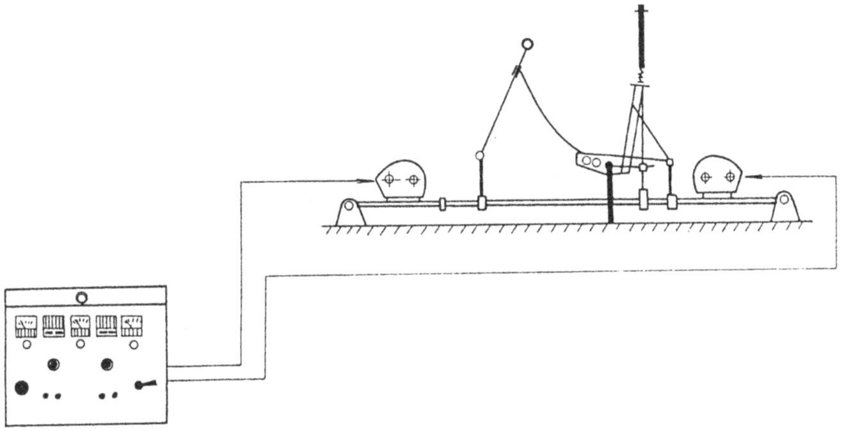

Fig. 1. Scheme of the special-purpose testing for frames.

Equation **(8)** is an answer to the curves of probability density, from Fig.**3** in our case for frames.

4.2 The luggage carrier constructions

The test of simulation regime for a luggage carrier has been made upon the special-purpose machine. The laboratory test has been made for 3 alternative luggage carrier constructions. Experimental results from the laboratory tests are shown in Tab.**3**.

Alternative **B** is seen to be the most reliable. The results of random excitation in operating state are to be seen in Tab.**4**.

The statistical curve of longevity for **B** alternative from eq.**(15)** is shown in Fig.**4**. Equation **(16)** is an answer to the curves of probability density from Fig.5. in our case.

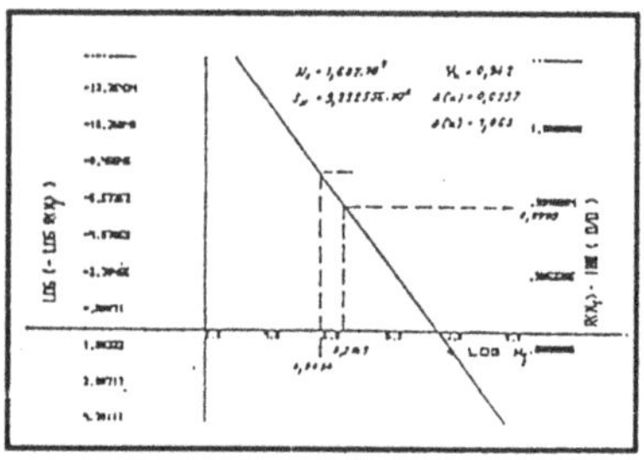

Fig. 2. Statistical curve of the longevity A-alternative of frame

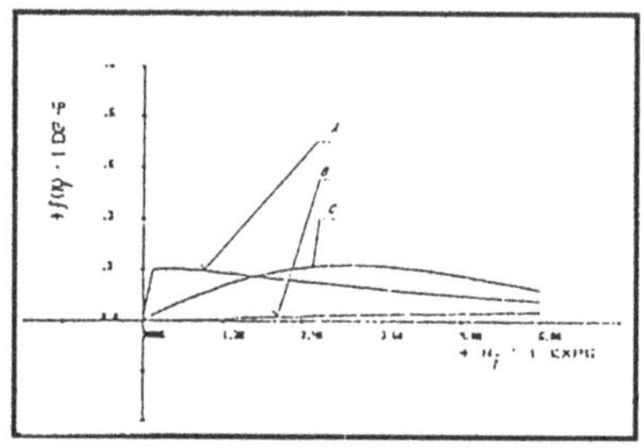

Fig. 3. Curve of probability density distribution of damage for frames

Table 3. Luggage carrier constructions

Reliability: Number of cycles to failure		Alternative:	
	A:	B: x 10^6	C:
N_{f1}	1,79	3,10	0,62
N_{f2}	1,33	0,904	1,53
N_{f3}	0,436	4,59	1,26
N_{f4}	2,2	14,35	1,811

Table 4. Luggage carrier constructions

Statistical moment	Loading equivalent to random loading (MPa)	Equivalent frequency (s^{-1})
m_0 = 594,01	σ_s = 51,6	
m_1 = 574,48	σ_{amp} = 24,37	
m_3 = 457261,21	σ_{red} = 75,97	
		27,24

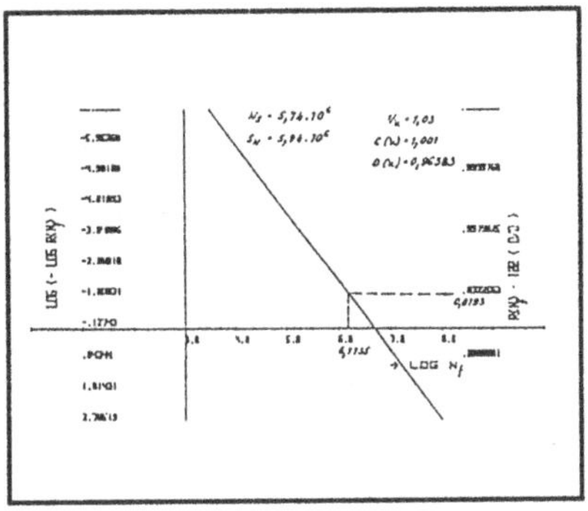

Fig.4. Statistical curve of the longevity for B-alternative of luggage carrier.

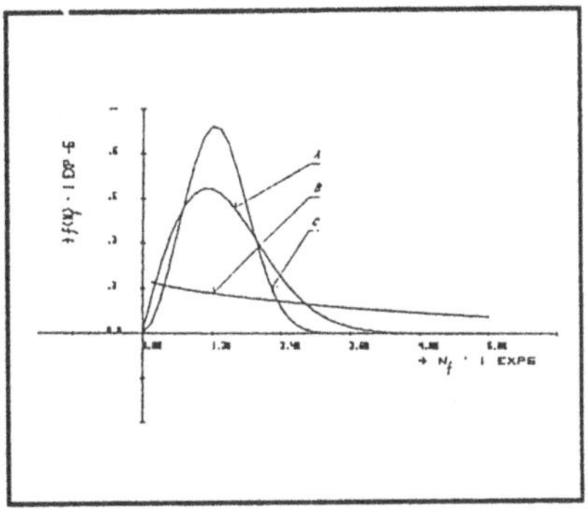

Fig.5. Curves of probability density distribution of damage for luggage carrier.

5 Conclusions

Tests are frequently completed on construction subassemblies, as they are shown in Fig.**6**

Tests of the longevity of load-carrying parts of motor-cycles at laboratory make variable extreme conditions of random excitation in operating state possibles. Prototype vehicles or components of vehicles can be more closely evaluated.

The application of this method shorten knowledge of the time to failure of machine components for transportation and contribute to the safety and economy of mechanical systems.

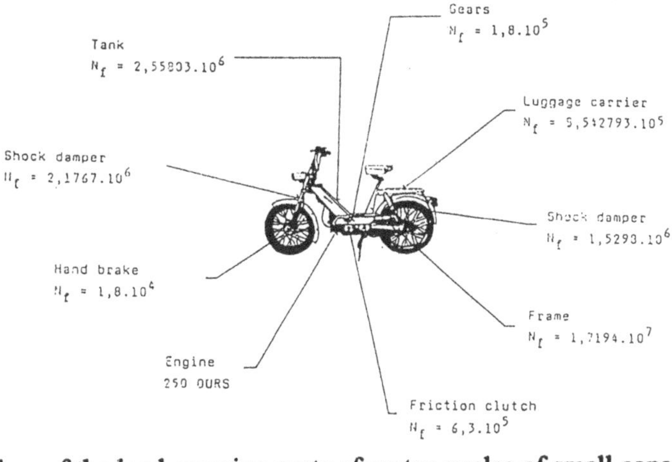

Fig.6. View of the load-carrying parts of motor-cycles of small capacity.

6 References

1 Kopecky,M. (1990) Experimental-numerical method of random loading analysis. *9th Inter. Conference on Exprimental Mechanics,* Vol. 3, pp. 1006-1012, Copenhagen, Denmark.

2 Kopecky,M. (1993) Life assessment of some load-carrying machine components for transportation. *Structural Safety,* Vol. 12, pp. 145-149, Elvesier Science Publishers, The Netherlands.

3 Weibull,W. (1951) A statistical distribution functions of wide applicability. *Journal of Applied Mech.,* No. 3.

44 PREDICTION OF DYNAMIC LOAD INPUTS FOR FINITE ELEMENT MODELS

M.V. BLUNDELL, J.C. WILLCOX and A. MACKIE
School of Engineering, Coventry University, Coventry, UK

Abstract

This paper describes the use of multibody systems analysis for the modelling and simulation of a vehicle towing operation. The work carried out was used to demonstrate the use of software to predict typical in service dynamic loads which were then used as inputs to a finite element stress analysis. Dynamic simulations were carried out to represent driving on flat or inclined road surfaces and the effects of road bumps and potholes. The results of these studies were used to predict the peak longitudinal and vertical components of load acting on the tow hitch. An overview is provided of the ADAMS® dynamic analysis program chosen to perform the simulations and the modelling elements used to represent the system under consideration. The use of finite element stress models is well established and is not discussed here.
Keywords: ADAMS®, Multibody Systems Analysis, Predictive Software

1 Introduction

The design of a commercially available towing bracket is often based on a traditional approach. The new design will evolve from an existing design which has been proven in service. It is then customary to produce a prototype, test it, and strengthen it where necessary. Commercially available towing brackets can be fitted to cars, minibuses, vans or small trucks. They can be subjected to a wide range of usage towing caravans, horse-boxes, or trailers. In terms of strength the problems are similar to those faced by the automotive designer when dealing with suspension components where two separate factors can influence the design. The first of these is that during the life of the towing bracket it will be subjected to repeated loading due to accelerating, braking and cornering. These loads may not be severe but may pose a fatigue problem. A more

Structural Failure, Product Liability and Technical Insurance, Proceedings of Fifth International Conference. Edited by H.P. Rossmanith. Published in 1996 by Taylor & Francis, 2 Park Square, Milton Park, Abingdon, Oxon, OX14 4RN. ISBN 0 419 22080 1

difficult problem is the possibility of failure due to a sudden unexpected load due to a large obstacle in the road. In order to design a towing bracket which is strong enough for these situations it is necessary to be able to estimate the magnitude and direction of these loads. Although this can also be achieved by testing, this paper concentrates on the modelling and simulation approach using the ADAMS® program, the usage of which is widely established within the automotive industry.

The ADAMS® *(Automatic Dynamic Analysis of Mechanical Systems)* software is used to study the behaviour of systems consisting of rigid or flexible parts connected by joints and moving through large displacements. ADAMS® [1] is supplied as a main analysis code and a fully interactive graphical pre- and post-processor. The software is installed at Coventry University on a suite of Sun workstations. The program may be used to model mechanical systems comprising a number of rigid or flexible bodies. The bodies may be connected to each other elastically or rigidly using a comprehensive library of joints and constraints allowing for large relative displacements. The system automatically identifies and eliminates redundant constraints and there is no restriction on the topological configuration of a system.

Users prepare a model which provides a physical description of the multibody system together with any time dependant force and displacement inputs. On running a simulation the software assembles the equations of motion and solves them automatically. It is also possible to include differential equations directly in the model formulation. The output will be calculated at selected points in time during the simulation. Results include displacements, velocities, accelerations and forces. These results can be resolved globally or relative to any other part in the system. It is also possible to formulate customised output using any mathematical combination of the normal requested output. The output can be presented as tabular data or as two-dimensional plots where results can be plotted in the time or frequency domain. It is also possible to visualise the results either as still frames or more usefully as continuous graphic animation.

Software of this type is used by companies and institutions throughout North America, Europe and Asia involved in the process of designing and analysing complicated mechanisms. The main users are to be found in the automotive, aerospace, general machinery, electro-mechanical, and construction equipment industries. The program may be used during concept or feasibility studies to predict the response and motion of a new design. This stage may involve sensitivity studies, investigating certain trends due to successive variations in a system parameter. In this phase the initial load predictions may also be extracted and the process repeated after more detailed design of individual components.

At a later stage it is possible to evaluate the performance of existing designs or in parallel with the development and testing of full prototypes. In some cases multibody systems analysis may be used to investigate extreme operating conditions which could lead to a dangerous or expensive failure if attempted with a prototype. In some cases simulation may also be used to reconstruct or investigate the behaviour of a system which is not operating correctly or is experiencing repeated failures. The most dramatic of these investigations can involve determining the reasons for an accident involving the operation of a system or vehicle. The main users in industry are automotive where the program is well established for the support of suspension design or for full vehicle ride and handling studies [2-4].

The procedure adopted for this study was to use the program to model the dynamic behaviour of a light truck towing a caravan and to determine the maximum forces acting on a component within the system during the simulation. These forces were then used as inputs to a finite element model of the component. The project is a realistic representation of the use of the software in industry. The automotive industry use ADAMS® to predict suspension loads under braking, cornering and hitting obstacles [3].

2 Modelling

The first step in this project was to establish the level of modelling required for the analysis in ADAMS®. For this project it was important to keep the level of detail within sensible bounds in order for the work to be completed within the time scales of the study. Another factor was the difficulty in obtaining detailed data such as bush rates or mass properties of suspension components. It was therefore necessary to make simplifying assumptions in the modelling.

The model of the vehicle and caravan includes a description of the rigid parts, connecting joints, applied forces, spring dampers and tyre properties. In the model discussed here the major components are as follows.

2.1 Parts and Markers

For each rigid body part in the system it is necessary to define the mass, centre of mass location, and mass moments of inertia. In this model parts were used to model the main body of the vehicle and the main body of the caravan. These were treated as rigid, therefore ignoring the effects of structural flexibility. The front and rear suspensions of the vehicle were modelled as single lumped masses as were the road wheels for the vehicle and caravan. It is also necessary to include one non-moving part which is referred to as the ground part. A schematic of the model is provided in Figure 1.

It was only necessary to model a half of the vehicle and caravan as it was decided the simulations would not consider turning and the use of symmetry could be exploited to reduce the problem size. The joints and forces shown in Figure 1 are explained later in this section. Obtaining the mass properties data for a complicated model can be very difficult but the task can be made easier by considering the nature of the analysis to be performed. In this case the vehicle and caravan were constrained so that lateral or turning motion would not be considered. The bodies had only three degrees of freedom with translational motion available along the global X-axis, global Z-axis, and pitching rotation about the global Y-axis. It was therefore only necessary to estimate the mass moment of inertia associated with pitching since body roll motion and yaw motion were constrained.

Markers are geometric points defined in global or local co-ordinate systems and considered to belong to and move with a part during the simulation. Markers are used to define centre of mass locations, joint locations and orientations, force locations and directions. In this model markers were also included for the various outline graphics used to represent the vehicles.

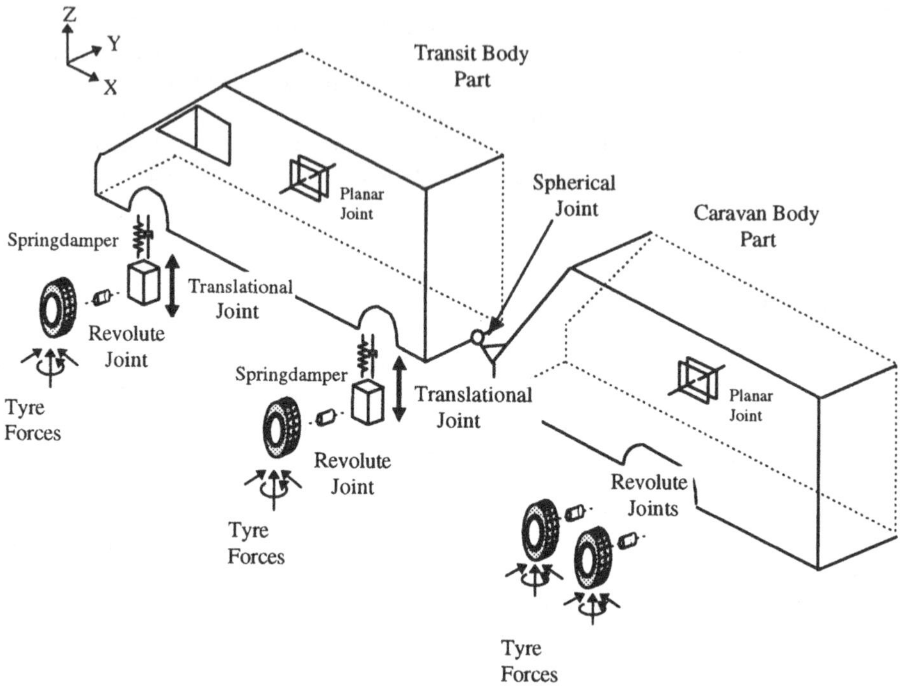

Figure 1 System Schematic

2.2 Constraints

The relative motion between different parts in the system can be constrained using joints, joint primitives, couplers, gears and user defined constraints. In this manner any of the six relative rigid body degrees of freedom can be constrained between two parts. The joints used in this model are indicated in the system schematic provided in Figure 1. A planar joint was located on the centre plane of the vehicle and the caravan. These joints connected the bodies to the ground part. The points on each of the bodies were free to move only in the ZX plane of the ground. The planar joint removes three degrees of freedom for each body namely, translational motion in the global Y-direction and rotation about the global X-axis and Z-axis. The planar joints effectively act as a symmetry constraint. The front and rear suspensions were connected to the vehicle body by translational joints. The translational joint removes five relative degrees of freedom allowing only the translation along the axis of the joints which in this case were initially aligned with the global Z-axis. The road wheels were connected using revolute joints which also remove five degrees of freedom allowing only rotation about the axes of the joints which were set parallel to the global Y-axis. The caravan body was connected to the vehicle by a spherical joint located at the coupling ball location. The spherical joint removes the three relative translational degrees of freedom between the two connected bodies. In this case both bodies were already constrained from translational motion in the Y-direction by the planar joints. The Y-constraint from the spherical joint was therefore identified as a redundant constraint so that the joint only removed two degrees of freedom instead of the normal three. The total degree of freedom audit for the system is calculated in Table 1.

Table 1 Summary of Degrees of Freedom in the System

Parts	+8 x 6	=	+48
Translationals	-2 x 5	=	-10
Revolutes	-4 x 5	=	-20
Planars	-2 x 3	=	-6
Spherical	-1 x 3	=	-3
Redundants	+1 x 1	=	+1
TOTAL DOF		=	10

2.3 Single Component Forces

External forces can be constant, time histories or functionally dependent on any state variable. These forces can also be defined to be translational or rotational. They can act in the global system or can act in the local system of the part so that they effectively 'follow' the part during the simulation. In ADAMS® terminology external forces are referred to as Action-Only forces. For this model torques were applied about the revolute joints connecting the wheels to the suspension. These torques were based on functional equations and were used to either drive or brake the vehicle.

Internal force elements were used to define the springs and the dampers acting between a point on the suspension part and a point on the vehicle body. The springs were modelled as linear force elements based on the stiffness and free length of the spring. The force generated in the springs was dependent on the magnitude of the displacement between the two ends. The dampers were non-linear and asymmetric with different properties in bump and rebound. Spline functions were used to define the damper characteristics which were dependent on the relative velocities between the suspension and vehicle body parts.

The access to system variables can be a particularly powerful modelling tool. The user can effectively access any displacement, velocity, acceleration or other force in the system when defining the force equation. Forces can also be defined as a function of time to vary or switch on and off as the simulation progresses.

2.3 Tyre Properties

The modelling of forces at the tyre-road surface contact patch was achieved using the ADAMS/Tire auxiliary program which was linked in with the main ADAMS® code. The data required to define the properties of the tyres and the road profile was set up in separate data files. For each tyre in the model the program will calculate the forces and torques acting at the wheel centre as a result of the conditions at the tyre contact patch. For this to be achieved there is a flow of information between ADAMS® and ADAMS/Tire regarding the position, velocity and orientation of the wheel centre marker and any changes in the topography of the road surface. Road surface definitions were set up in different files based on the simulation to be considered. These were flat or inclined road surfaces or surface events including bumps and potholes. Once the forces have been calculated at the contact patch they are resolved back to the wheel centre. ADAMS® will then integrate through time to find the new position and orientation of the vehicle. In order to obtain the data required to physically describe the properties of the tyre it is necessary to test the tyre in a test rig which can measure the forces and moments acting at the wheel centre for varying loads at the tyre/road contact patch. This data was obtained using the flat bed tyre test machine installed within the School of Engineering at Coventry University.

3 Running the Analysis

The types of analysis simulation which can be performed may be summarised as follows.

3.1 Kinematic Analysis

Kinematic analysis is applicable to systems possessing zero rigid body degrees of freedom. Any movement in this type of system will be due to prescribed motions at joints. The program uncouples the equations of motion and force and then solves separately and algebraically for displacements, velocities accelerations, and forces. No numerical integration is necessary. A simple example of a kinematic model would be the hand of a clock which is attached to the clock face (ground) by a revolute joint. A motion is applied at the joint to rotate the hand through one revolution every 60 seconds. If we remove the motion the system has one degree of freedom and becomes essentially a pendulum. This problem is now dynamic and cannot be solved by kinematic analysis. The model discussed here had ten degrees of freedom and could not be analysed kinematically.

3.2 Static Equilibrium Analysis

For static analysis the velocities and accelerations are set to zero and the applied loads are balanced against the reaction forces until an equilibrium position is found. This may involve the system moving through large displacements between the initial configuration and the equilibrium position and will require a number of iterations before convergence on the solution closest to the initial configuration. For this project a static analysis was performed as a preliminary to a dynamic analysis. The static analysis establishes the configuration of the vehicle at 'kerb height' before the vehicle moves forward in the dynamic phase of the simulation. Quasi-static analysis is a series of static equilibrium solutions at selected time steps. Although the system can be considered to be moving the dynamic response is not required. An example would be to perform a quasi-static analysis on a vehicle mounted on a tilting surface. As the surface rotates to different angles with time the static equilibrium of the vehicle can be calculated at selected angles.

3.3 Dynamic Analysis

Dynamic analysis is performed on systems with one or more degrees of freedom. The differential equations representing the system are automatically formulated and then numerically integrated to provide the position, velocities, accelerations and forces at successively later times. Although the required output is specified at various points in time the program will often compute solutions at many intermediate points in time. The interval between each of these solution points is known as an integration time step. In ADAMS® the size of the integration time step is constantly adjusted using internal logic although the user may override the system defaults if so desired. For this project the following simulations were performed:

(i) Acceleration on a flat road
(ii) Braking on a flat road
(iii) Constant velocity on a slope
(iv) Hitting a road bump at constant velocity
(v) Hitting a pothole at constant velocity

The first three cases allowed for hand calculations as a check against the simulation output. These results showed good agreement permitting progress to the more difficult bump and pothole cases. The arrangement for the road and vehicle graphics for these last two cases is shown in Figure 2.

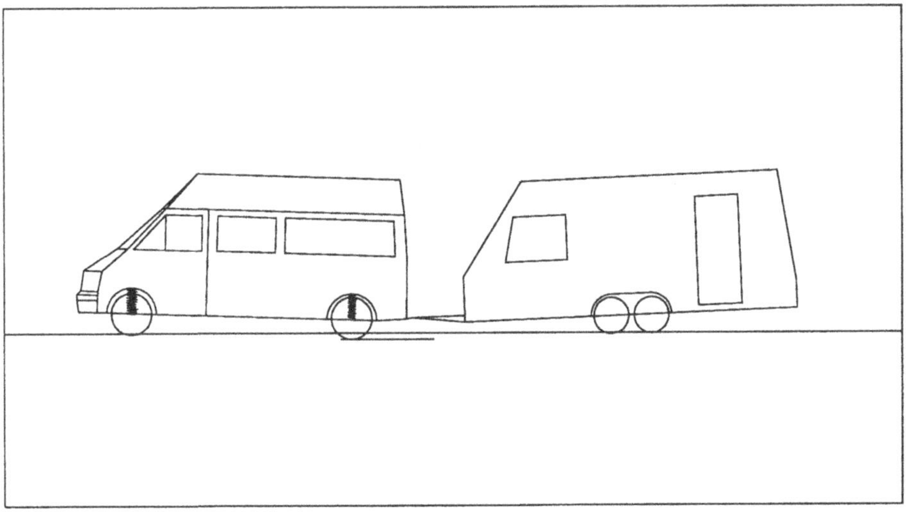

Figure 2 Road and Vehicle Graphics

The most severe loading occurred during the pothole simulation. The pothole was modelled as 1500mm long with a depth of 100mm. The vehicle velocity was fixed at a constant 10m/s. The simulation lasted 7 seconds with peak forces occurring at 4.3s when the rear wheels of the vehicle entered the pothole. The vehicle graphics at this time point are shown in Figure 3. A time history plot for the vertical and horizontal forces is shown in Figure 4.

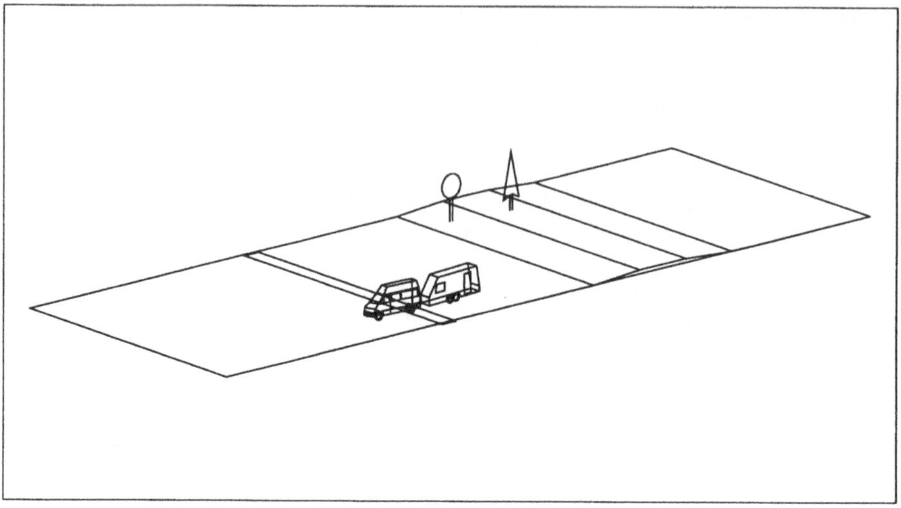

Figure 3 Point of Maximum Load during Pothole Simulation

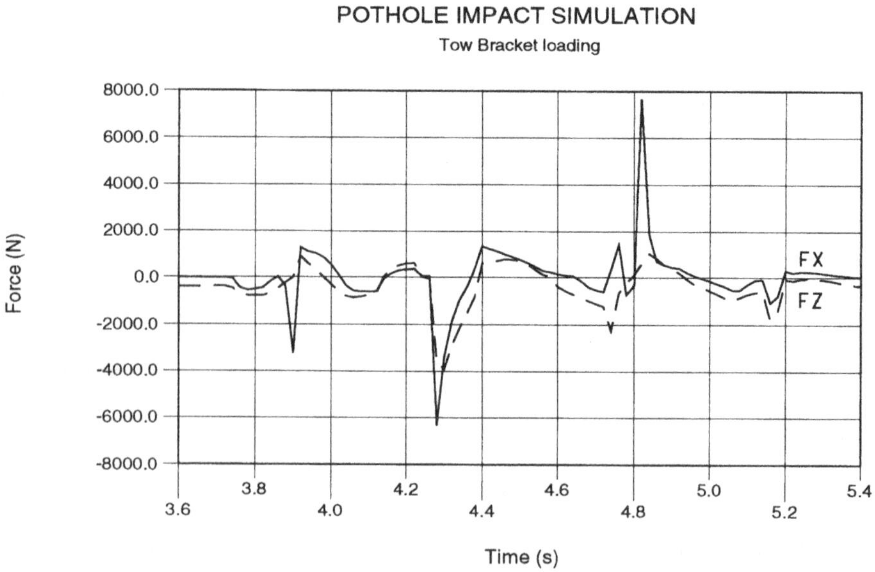

Figure 4 Force Time History

4 Conclusions

The study described here has demonstrated the application of modelling and simulation using a multibody systems analysis program. Although it has not been possible to carry out physical testing to validate the models the exercise has been successful in applying a technology which could have potential within the design process of commercial towing brackets. More work is needed to investigate the influence of model refinement upon the accuracy of simulation outputs. This work only investigated forces occurring during straight-line motion. Future work could extend the model to consider lateral forces due to cornering or single wheel impacts.

Acknowledgements

The authors wish to thank the staff of York Towbars Limited, Ford Motor Company, Avondale Coachcraft, and MDI Limited for providing data and information in support of this project. The authors would also like to acknowledge Stefan Huppertz who contributed greatly to this project at Coventry University whilst on exchange from Fachochschule Aachen.

References

[1] Ryan, R.(1990) *ADAMS® Multibody Systems Analysis Software*, Multibody Systems Handbook/Werner Schielen (Editor),.

[2] Orlandea,N. and Chase,M. *Simulation of a Vehicle Suspension with the ADAMS® computer program*, SAE paper 770053.

[3] Rai, N.S and Soloman, A.R. *Computer simulation of suspension abuse tests using ADAMS®*, SAE paper 820079.

[4] Antoun, R.J.,Hackert, P.B., O'Leary, M.C., Sitchen, A. *Vehicle dynamic handling computer simulation - Model development, correlation, and application using ADAMS®*, SAE paper 860574.

45 ASSESSMENT OF FATIGUE FAILURES - FINITE ELEMENT ANALYSIS OF STRUCTURAL JOINTS UNDER VIBRATION LOAD

P. ZEMAN
GEC-Marconi Dynamics, Inc., West Lake, California, USA
A. PURI
GEC-Marconi GMROS, Stanmore, UK

Abstract

This presentation discusses a method of utilizing Finite Element Analysis (FEA) to determine mean and alternating stresses within bolted joints. This method is particularly useful for evaluating unconventionally designed joints. On preloaded joints, the external load (in this case an alternating load) is shared by both bolts and flanges. Only through FEA can the mean and alternating stress be determined accurately for each component. In order to determine stresses in the joint components, this unique FEA method can be applied. This method involves a forced closed "gap element" to preload the joint.

In the first step of the method, the bolt is loaded and the flange is restrained to simulate actual stress conditions within the joint introduced by the bolt torque, without external load.

Then, the bolt and the flange are coupled together through a gap element, so that when a gap is closed by arbitrarily high forces, the same stress conditions are achieved as in the first step.

Now, the alternating load is added, and from resulting stresses, the mean and alternating stress can be determined.

Finally, the fatigue risk can be assessed using conventional cumulative fatigue methodology.

Keywords: Bolted joints, fatigue, finite element analysis, force closed gap elements.

Structural Failure, Product Liability and Technical Insurance, Proceedings of Fifth International Conference. Edited by H.P. Rossmanith. Published in 1996 by Taylor & Francis, 2 Park Square, Milton Park, Abingdon, Oxon, OX14 4RN. ISBN 0 419 22080 1

1 Introduction

This presentation discusses a method utilizing Finite Element Analysis (FEA) to determine mean and alternating stresses within bolted joints. This method is particularly useful for evaluating unconventionally designed joints. On preloaded joints, the external load (in this case an alternating load) is shared by both the bolts and flanges. Through FEA, the mean and alternating stress can be determined accurately for each component.

In order to correctly determine stresses in the joint components, this unique FEA method can be applied. This method involves a forced closed "gap element" to preload the joint.

In the first step, the bolt is loaded and the flange is restrained to simulate actual stress conditions within the joint introduced by the bolt torque, without external load.

Then, the bolt and the flange are coupled together through a gap element, so that when the gap is closed by an arbitrarily high forces, the same stress conditions within the joint are achieved, as in the first step.

Now, the alternating load is added, and from the resulting stresses, the mean and alternating stress can be determined.

Finally, the fatigue risk can be assessed using conventional cumulative fatigue methodology.

2 Joint preload introduced by known forces from the bolt torque

(Note: No gap elements are used, and this method will not yield correct results when combined with external loads.)

The preload is introduced between the bolt head and the mating surface. This simulates the actual stress conditions within the joint that are introduced by the bolt torque, in the absence of an external load.

The bolt preload can be calculated from the following formula:

Preload = Bolt Torque/ (Nominal bolt diameter X Nut factor)

Nut factor = (0.10 - 0.20) depending on the thread pitch, lubricant, material.

A simple bolted joint model was generated (Figure 1), to illustrate the stress conditions in the joint components (Figure 2).

Note, that only the first few engaged threads are loaded, with the maximum load on the first engaged thread. Also, there is a stress concentration in the first bolt thread from the shank.

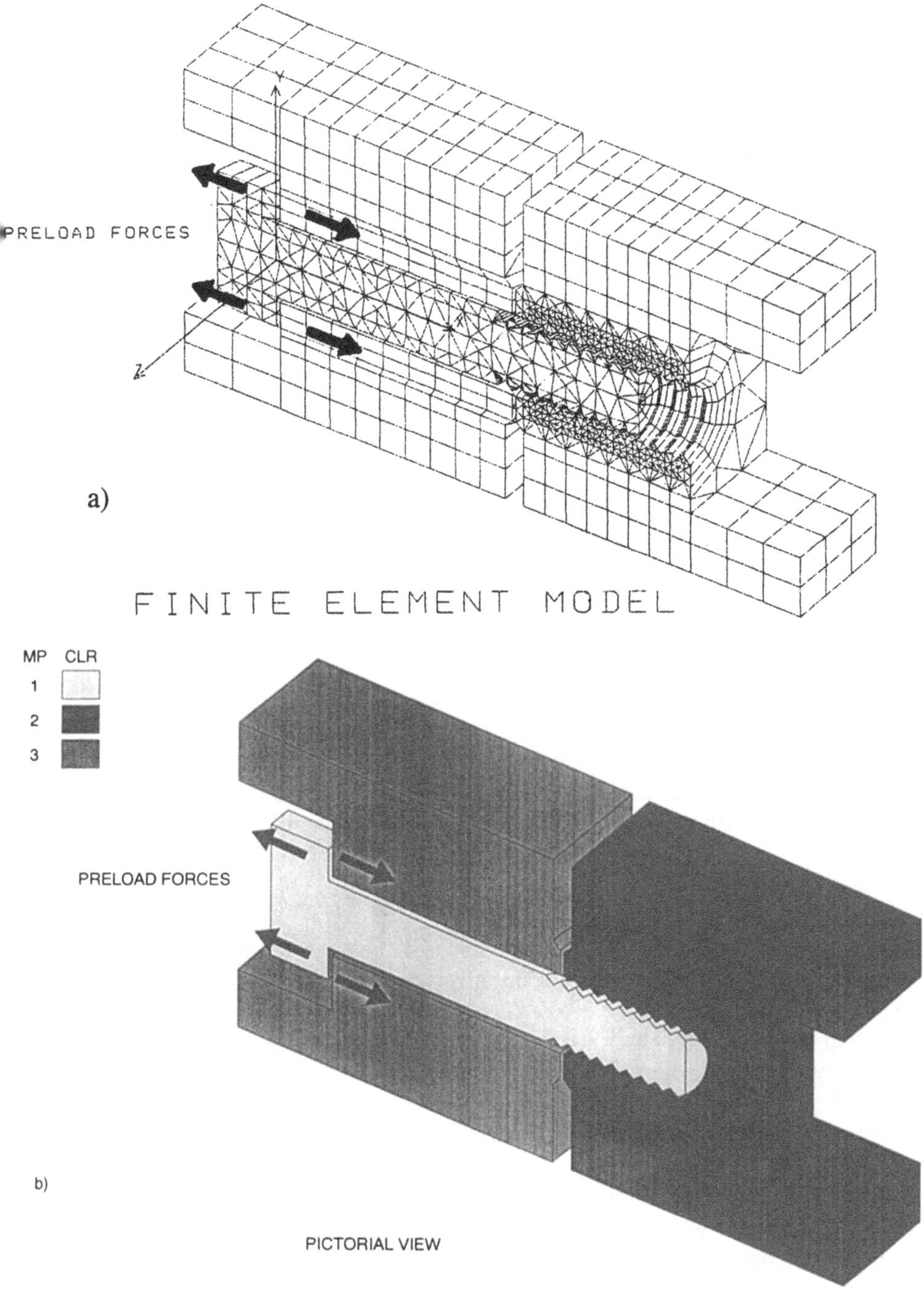

Figure 1 Finite element analysis of a bolted structural joint with joint preload only (preload force applied to bolt head)
a) finite element model; b) pictorial view

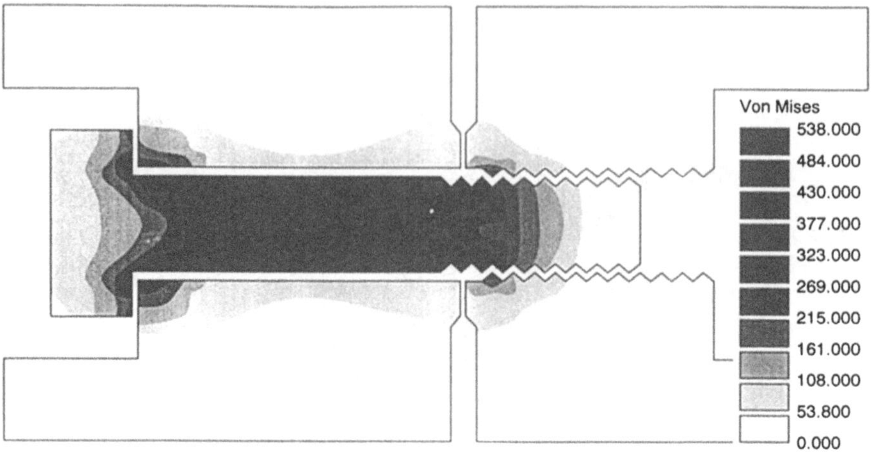

Figure 2 The result of the finite element analysis: joint preload only

3 Problems encountered while applying an external load

As it can be seen from Figure 3, the entire external load is supported by the flange and the stress in the bolt is not affected. This is not a correct result. On preloaded joints, the external load is shared by both the bolts and flanges. In this case the external load is not shared, because the bolt and its mating surface with the flange can move relative to each other.

4 Force closed gap

In this application, "tensile gap elements" will be used to obtain correct results (Figure 4). Tensile gaps are initially open, and they are closed by a tensile load.
The gap dimension (G dist) is specified by the gap's input data, and it is very easy to change without changing the model.

5 Bolt preload introduction by a force closed gap

The bolt head and the flange are coupled together through gap elements (Figure 5), so that when the gaps are closed by arbitrarily high forces, the same stress conditions within the joint are achieved as in the first step (bolt preloaded by the preload force). The bolt preload does not depend on the magnitude of the applied forces, but rather on the gap value (G dist). It is recommended that forces of at least one order of magnitude greater than the bolt preload be applied to prevent the gaps from opening, when the external

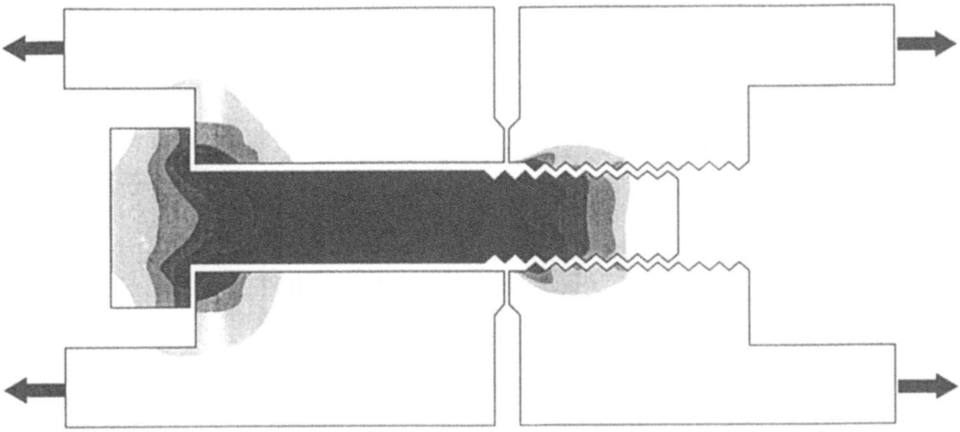

Figure 3a The result of the finite element analysis: joint preload and external load

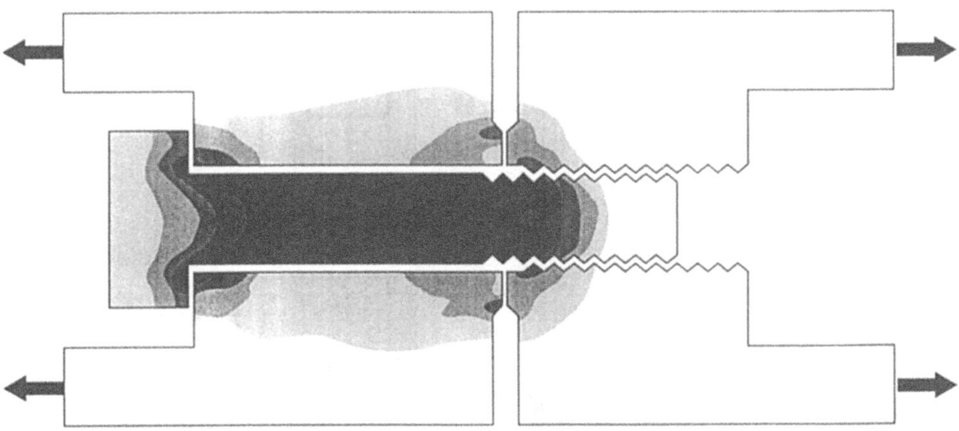

Figure 3b The result of the finite element analysis: Preload force applied to bolt head

load is applied. Gaps under the bolt head resemble a washer which expands under the head to a predetermined thickness, and freezes in this expanded position. Stresses in the bolt thread and in the flanges (Figure 6) are the same as the results from the first case (Figure 2), where the bolt preload was applied directly to the bolt head.

6 External load applied to the joint

Now, the alternating load is added. One at the time, 1 Sigma, 2 Sigma and 3 Sigma alternating loads are applied, and the corresponding stresses are obtained. From the resulting stresses, the mean and alternating stresses can be determined.

When force closed gap elements are used, the external (alternating) load is correctly shared between the flanges and the bolt. This can be seen in the stresses in the bolts shown in Figure 7.

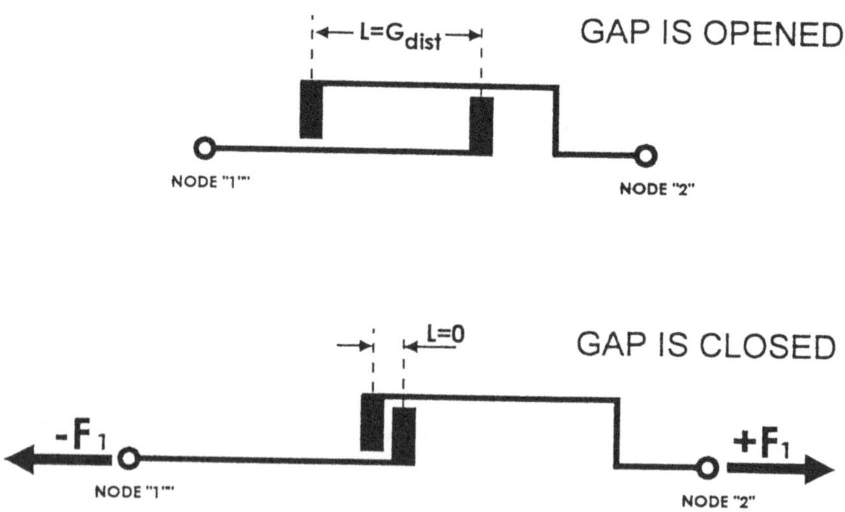

Figure 4 Force closed gap

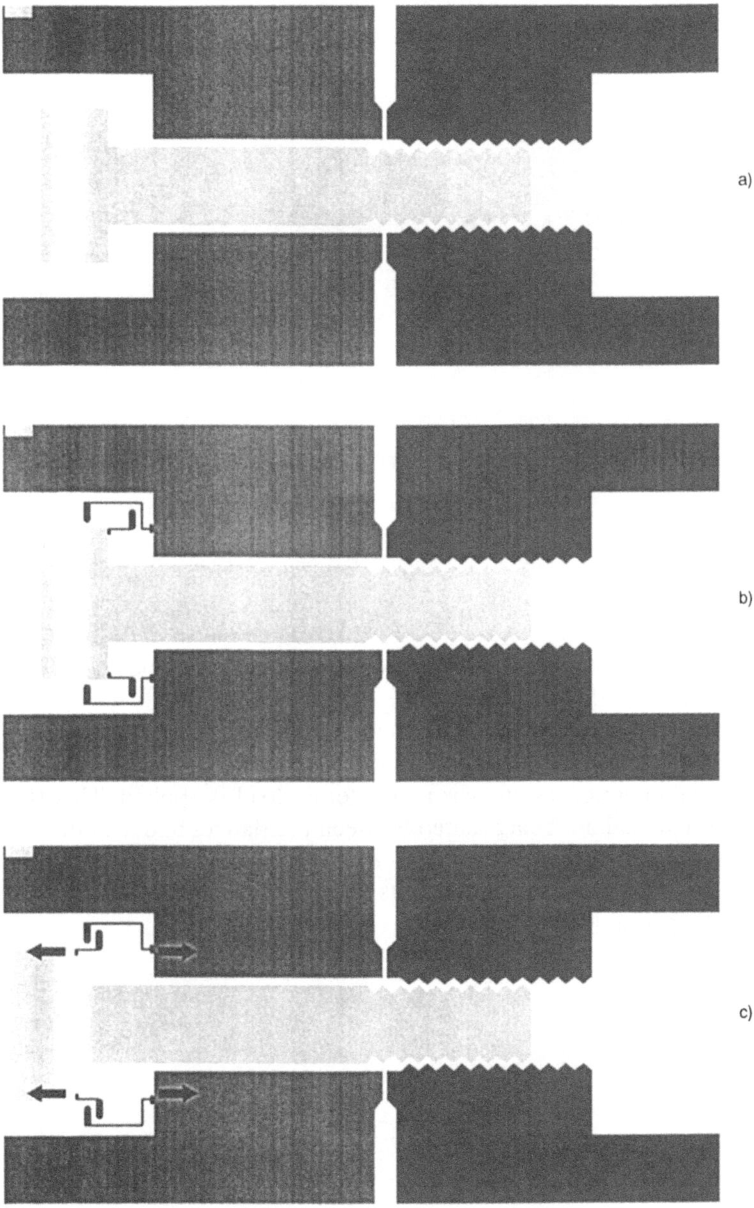

Figure 5 Bolt and gap loading in the finite element analysis: a) the bolt is not preloaded and no gap elements are located under the bolt head; b) the bolt is not preloaded but the gap elements are opened; and c) the bolt is preloaded and the gap elements are closed.

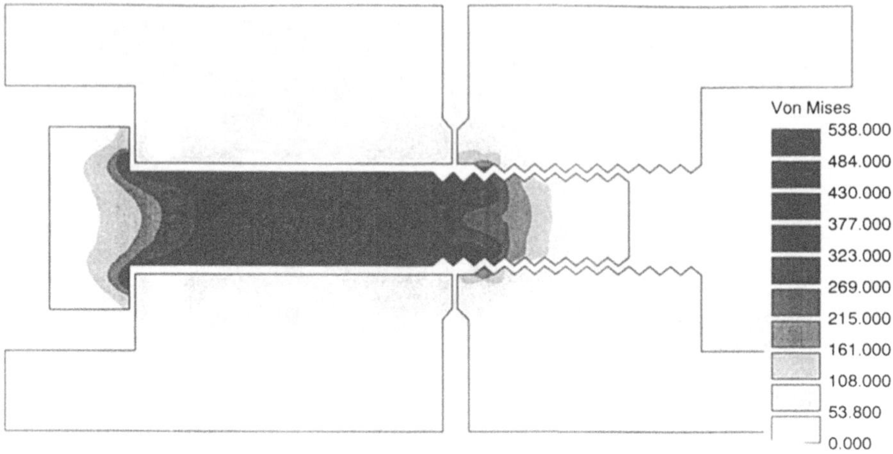

Figure 6 The result of the finite element analysis: joint preload only

7 Conclusion

The purpose of this presentation was to introduce a method to preload a joint by force closed "gap elements".
This presentation described two different methods to preload a joint. In the first method, the joint preload is introduced directly to the bolt head and does not yield correct results when an external load is applied. This is due to the external load not being shared between the flanges and the bolt.
The second method, involves a force closed "gap elements" to preload the joint. This method yields correct results, because the external load is shared by all participating joint components.

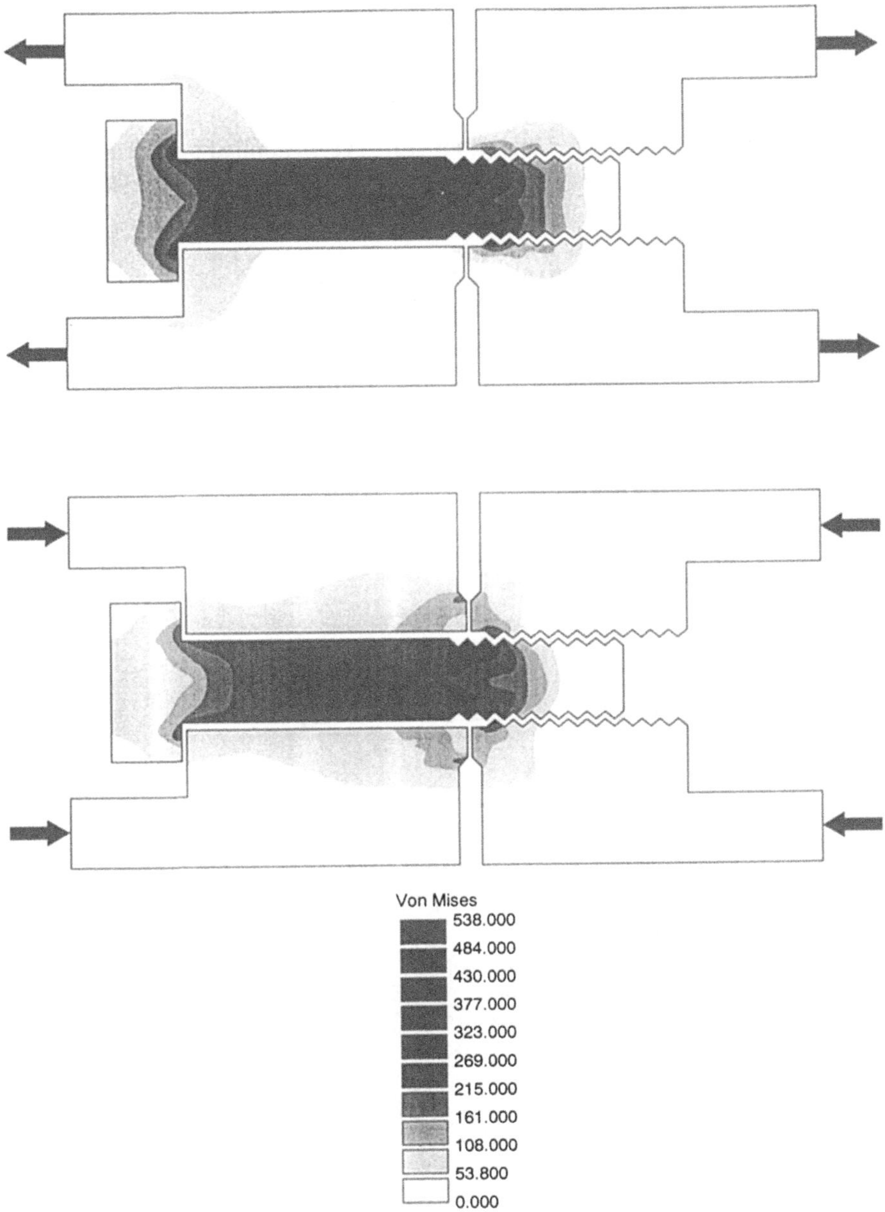

Figure 7 The result of the finite element analysis:
a) joint preload + tensile load;
b) joint preload + compressive load

ASSESSMENT OF THE FATIGUE DAMAGE

The fatigue risk can be assessed using conventional cumulative fatigue methodology(Miner's rule).

1) Determine alternating load and number of cycles applied to the joint. In the case of random vibration, determine 1 Sigma, 2 Sigma, 3 Sigma load and corresponding number of cycles (n1, n2, n3).
2) Determine critical locations.
3) Determine mean and alternating stresses (1 Sigma, 2 Sigma, 3 Sigma) in critical locations.
4) Determine the number of cycles (N1, N2, N3) required to produce failures for corresponding stress levels (1 Sigma, 2 Sigma, 3 Sigma) and corresponding mean stress (Goodman line, Soderberg line, Gerber line or measured line). The displayed diagram (Figure 8) is from MIL-HDBK-5C and illustrates a sample problem.
5) Cumulative fatigue damage can be calculated using Minor's rule:

Rn=n1/N1 + n2/N2 + n3/N3 (for random vibration)
Rn $\geq$1 constitutes fatigue failure

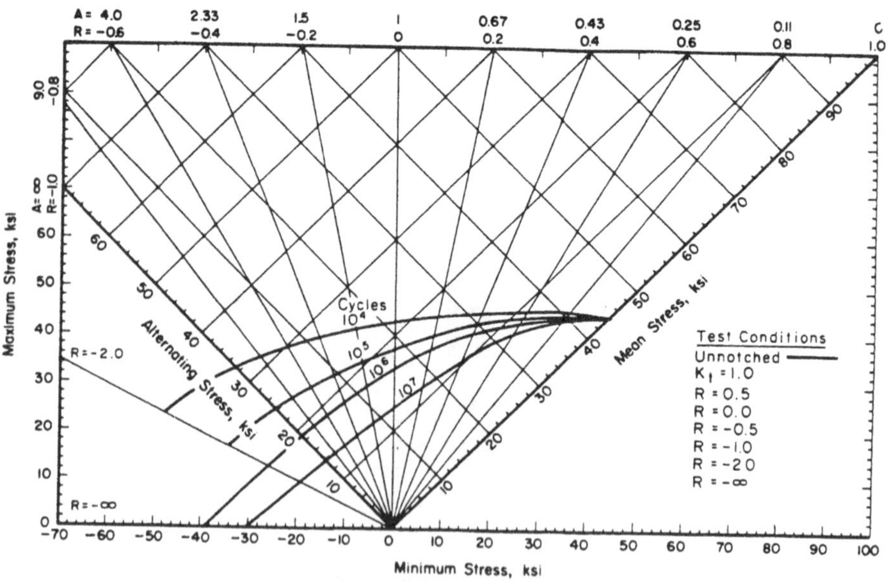

Figure 8 Typical constant life fatigue diagram

46 MEASURING METHOD OF STRESS INTENSITY FACTOR OF ORTHOTROPIC MATERIAL USING SPATE

J.S. HAWONG and J.G. SUH
Department of Mechanical Engineering, Yeungnam University, Gyung-San City, Korea
R.E. ROWLANDS
Department of Engineering Mechanics and Astronautics, University of Wisconsin, Madison, Wisconsin, USA

Abstract
SPATE(Stress Pattern Analysis by Thermal Emission) can be effectively used to analyzed the stress distribution of the orthotropic structure under the repeated load by non-contact. In this research, the measuring conception and method of stress intensity factor of orthotropic material using SPATE are suggested. The validity of the method and conceptions were certified through SPATE experiment.

The relationships between the maximum values of SPATE signal and $1/\sqrt{x'}$(or $1/\sqrt{y'}$) are theoretically established in the vicinity of crack tip of the orthotropic material. It is certified through SPATE experiment that their linear quality is very excellent.
Keywords: Orthotropic material, stress intensity factor, SPATE(stress pattern analysis by thermal emission), thermoelastic effect.

1 Introduction

Though the conceptions of thermoelastic effect which is the basic theory of SPATE were derived in 1853 by Lord Kelvin[1], until 1978, SPATE has not been nearly used and studied because of the nondevelopment of the measuring device associated with the conceptions of thermoelastic effect. Since it has been developed in 1978, SPATE has been applied to the various researches of isotropic material, the thermoelastic theory of composite[2,3,4,5], the damage of composite[6] the thermoelastic theory of micro-crack in the composite[7], the

Structural Failure, Product Liability and Technical Insurance, Proceedings of Fifth International Conference. Edited by H.P. Rossmanith. Published in 1996 by Taylor & Francis, 2 Park Square, Milton Park, Abingdon, Oxon, OX14 4RN. ISBN 0 419 22080 1

analysis of micro-crack in the composite[8] and the separating method of stress components of composite by hybrid method[9] etc..

Though authors and another many researchers have studied the measuring method of stress intensity factor of isotropic material using SPATE[10, 11, 12, 13, 14], but until now, as the researches of fracture mechanics using SPATE have not completely been established in the isotropic material, the various fields of fracture mechanics using SPATE should be studied. Moreover, in the orthotropic material, the researches of fracture mechanics using SPATE have not been initiated by any one. Therefore, to establish the measuring method of stress intensity factor using SPATE in the orthotropic material is the purpose of this research.

2 Basic theory

2.1 Relationships between SPATE signals and stress components

As linear relationships are established between the temperature changes and the sum changes of the principal stresses in the isotropic materials under adibatic conditions, eqt.(1) which is thermoelastic effect is also established in the orthotropic material under adibatic conditions.

$$dT=-\frac{T}{\rho C_\sigma}\frac{\partial \varepsilon_{ij}}{\partial T}d\sigma_{ij} \tag{1}$$

where dT is the adibatic temperature changes, T is absolute temperature, ρ is density (kgm^{-3}), C_σ is the specific heat at a constant stress and σ_{ij} and $\varepsilon_{ij}$are strains and stresses respectively.

The thermal expansion coefficients α_{11} and α_{22} are defined as follows

$$\frac{\partial \varepsilon_{11}}{\partial T}=\alpha_{11}\ ,\quad \frac{\partial \varepsilon_{22}}{\partial T}=\alpha_{22}\ ,\quad \frac{\partial \varepsilon_{12}}{\partial T}=\alpha_{12} \tag{2}$$

Substituting eqt.(2) for eqt.(1) under principal plane, We can obtain a eqt.(3) of the thermoelastic effect in the isotropic material.

$$dT=-\frac{T}{\rho C_\sigma}(\alpha_{11}d\sigma_{11}+\alpha_{22}d\sigma_{22}) \tag{3}$$

Applying eqt.(3) to L-T plane of orthotropic material, we can obtain a eqt.(4) of the thermoelastic effects in the orthotropic material

$$dT=-\frac{T}{\rho C_\sigma}(\alpha_L\sigma_L+\alpha_T\sigma_T) \tag{4}$$

Where, L is fiber direction and T is transverse direction to the fiber direction. α_L and α_T are respectively thermal expansion coefficients of fiber direction and of normal direction to the fiber direction. In the orthotropic material, principal axes of stress do not necessarily coincided with the principal axes of material(material axes).

Note that there is no shear stress σ_{12} in the principal plane and there is no

α_{LT} in the orthotropic material. Therefore, equation(3) can be transformed into eqt.(4). SPATE(=thermoelastic) signals(= S_d) are proportional to the change in surface temperature of specimen as eqt.(5)

$$S_d = 4eBRT^3 dT \tag{5}$$

where, e is surface emissivity, B is Stefan-Boltzmann constant and R is detector response factor. Substituting eqt. (4) for eqt. (5), eqt. (6) is obtained.

$$S_d = f_L d\sigma_L + f_T d\sigma_T \tag{6}$$

where, $f_L = -\frac{4eBRT^4}{\rho C_p}\alpha_L$, $f_T = -\frac{4eBRT^4}{\rho C_p}\alpha_T$, f_L and f_T are respectively thermo- mechanical calibration coefficients of fiber direction and of transverse direction to the fiber direction. Eqt.(6) shows the relationships between SPATE signals S_d and stress components and between SPATE signals S_d and thermomechanical calibration coefficients(f_L and f_T). Thermomechanical calibration coefficients f_L and f_T are measured by experiment or calculated from formula given in the SPATE manual.

2.2 Relationships between stress components in the vicinity of crack tip and SPATE signals

Stress components in the vicinity of crack tip are[15]

$$\begin{aligned}
\sigma_{xx} &= \frac{K_{\mathrm{I}}}{\sqrt{2\pi r}} Re\left[\frac{\mu_1\mu_2}{\mu_1-\mu_2}\left(\frac{\mu_2}{\sqrt{(\cos\theta+\mu_2\sin\theta)}} - \frac{\mu_1}{\sqrt{(\cos\theta+\mu_1\sin\theta)}}\right)\right] + \\
&\quad \frac{K_{\mathrm{II}}}{\sqrt{2\pi r}} Re\left[\frac{1}{\mu_1-\mu_2}\left(\frac{\mu_2^2}{\sqrt{(\cos\theta+\mu_2\sin\theta)}} - \frac{\mu_1^2}{\sqrt{(\cos\theta+\mu_1\sin\theta)}}\right)\right] + \sigma_{ox} \\
\sigma_{yy} &= \frac{K_{\mathrm{I}}}{\sqrt{2\pi r}} Re\left[\frac{1}{\mu_1-\mu_2}\left(\frac{\mu_1}{\sqrt{(\cos\theta+\mu_2\sin\theta)}} - \frac{\mu_2}{\sqrt{(\cos\theta+\mu_1\sin\theta)}}\right)\right] + \\
&\quad \frac{K_{\mathrm{II}}}{\sqrt{2\pi r}} Re\left[\frac{1}{\mu_1-\mu_2}\left(\frac{1}{\sqrt{(\cos\theta+\mu_2\sin\theta)}} - \frac{1}{\sqrt{(\cos\theta+\mu_1\sin\theta)}}\right)\right] \\
\tau_{yy} &= \frac{K_{\mathrm{I}}}{\sqrt{2\pi r}} Re\left[\frac{\mu_1\mu_2}{\mu_1-\mu_2}\left(\frac{1}{\sqrt{(\cos\theta+\mu_1\sin\theta)}} - \frac{1}{\sqrt{(\cos\theta+\mu_2\sin\theta)}}\right)\right] + \\
&\quad \frac{K_{\mathrm{II}}}{\sqrt{2\pi r}} Re\left[\frac{1}{\mu_1-\mu_2}\left(\frac{\mu_1}{\sqrt{(\cos\theta+\mu_2\sin\theta)}} - \frac{\mu_2}{\sqrt{(\cos\theta+\mu_1\sin\theta)}}\right)\right]
\end{aligned} \tag{7}$$

μ_1 and μ_2 of eqt.(7) are determined from the characteristic equation associated with compatibility equation of orthotropic material. Stress components of eqt.(7) are transformed into stress components of fiber direction and transverse direction to the fiber. They are

$$\begin{aligned}
\sigma_L &= \frac{1}{2}(\sigma_{xx}+\sigma_{yy}) + \frac{1}{2}(\sigma_{xx}-\sigma_{yy})\cos 2\phi + \tau_{xy}\sin 2\phi \\
\sigma_T &= \frac{1}{2}(\sigma_{xx}+\sigma_{yy}) - \frac{1}{2}(\sigma_{xx}-\sigma_{yy})\cos 2\phi - \tau_{xy}\sin 2\phi \\
\tau_{LT} &= -\frac{1}{2}(\sigma_{xx}-\sigma_{yy}) + \tau_{xy}\cos 2\phi
\end{aligned} \tag{8}$$

Where ϕ is the angle of fiber direction with respect to x axis of reference coordinates. Eqt.(6) can be rewritten by eqt.(9)

$$S_d = \delta_{ij} f_i d\sigma_j \qquad (i, j = L, T) \tag{9}$$

where, δ_{ij} is Kronecker delta, f_i is the thermomechanical calibration coefficient, σ_j is stress components. Substituting eqt.(7) for eqt. (8), stress component σ_j is obtained.

$$\sigma_j = \frac{K_{\mathrm{I}}}{2\sqrt{2\pi r}} Re\{Ap_j + Br_j\} + \frac{K_{\mathrm{II}}}{2\sqrt{2\pi r}} Re\{Aq_j + Bs_j\} + \frac{1}{2}\sigma_{ox}(1 + \cos 2\phi_j) \qquad (j = L,\ T) \tag{10}$$

where ,

$$p_j = -(\mu_1^2\mu_2 + \mu_2) - (\mu_1^2\mu_2 - \mu_2)\cos 2\phi_j + 2\mu_1\mu_2 \sin 2\phi_j$$
$$q_j = -(\mu_1^2 + 1) - (\mu_1^2 - 1)\cos 2\phi_j + 2\mu_1 \sin 2\phi_j$$
$$r_j = \mu_1\mu_2^2 + \mu_1 + (\mu_1\mu_2^2 - \mu_1)\cos 2\phi_j - 2\mu_1\mu_2 \sin 2\phi_j$$
$$s_j = \mu_2^2 + 1 + (\mu_2^2 - 1)\cos 2\phi_j - 2\mu_2 \sin 2\phi_j\ ,\ \phi_L = \phi\ , \quad \phi_T = \frac{\pi}{2} + \phi\ ,$$
$$A = \frac{1}{\mu_1 - \mu_2}\frac{1}{\sqrt{\cos\theta + \mu_1 \sin\theta}}\ ,\ B = \frac{1}{\mu_1 - \mu_2}\frac{1}{\sqrt{\cos\theta + \mu_2 \sin\theta}}$$

$d\sigma_j$ of eqt.(9) can be replaced with σ_j in the elastic region. Therefore eqt.(11) is obtained as follows

$$S_d = \frac{1}{2\sqrt{2\pi r}} Re[\ A_0 K_{\mathrm{I}} + B_0 K_{\mathrm{II}}\ ] + C_0 \tag{11}$$

where,

$$A_0 = A\sum_{j=L}^{T} p_j f_j + B\sum_{j=L}^{T} r_j f_j \quad (j = L, T),\quad B_0 = A\sum_{j=L}^{T} f_j q_j + B\sum_{j=L}^{T} f_j s_j \quad (j = L, T),$$
$$C_0 = \frac{1}{2}\sigma_{0x}(f_L + f_T) + \frac{1}{2}\sigma_{0x}(f_L - f_T)$$

If the geometrical conditions of specimen, materials and loading conditions etc. are given, A_0, B_0, C_0, K_I and K_{II} are determined. Therefore, eqt.(11) is the function of $1/\sqrt{r}$ and θ used in the A and B. Eqt.(11) with two variables can be transformed into the equation with one variable as following procedures.

In general, SPATE data are recorded with x′-y′coordinates, but stress components are represented with the crack coordinates. In the Fig.1, x′ coordinates of arbitrary point along AB-line are constant and y′ coordinates are only variable. y′ coordinates of arbitrary point along CD-line are constant and x′ coordinates are only variable. Their conceptions are respectively established on the arbitrary lines parallelled AB-line or CD-line. Therefore eqt.(12) is established.

$$x_j' = r\sin(\theta_j + \beta_j) \qquad (j = L, T) \tag{12}$$

Where, $x_L' = y'$, $x_T' = x'$, $\beta_L = \beta$, $\beta_T = \beta + \frac{\pi}{2}$

Substituting r of eqt.(12) for r of eqt.(11), eqt.(11) becomes only function of θ. Therefore, from eqt.(13), we can obtain $\theta_j (= \theta)$ which the maximum of SPATE signals are occurred

$$\frac{\partial S_d}{\partial \theta}=0 \tag{13}$$

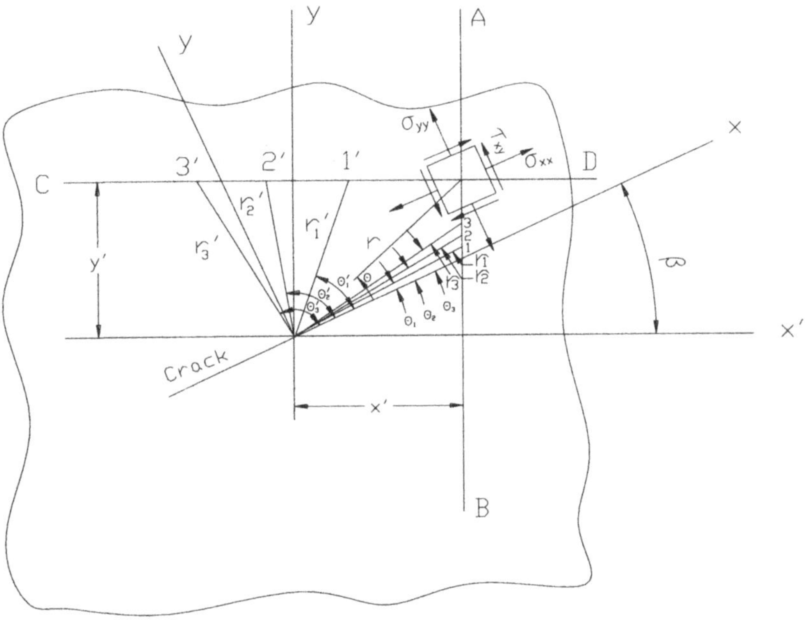

Fig.1 Relationships between SPATE coordinates(x'-y'coordinates) and crack coordinate(x-y coordinates)

2.3 x'=constant (y'=constant)

Substituting $\frac{x_j}{\sin(\theta+\beta_j)}$ for r of eqt.(11) and applying $\frac{\partial S_d(\theta)}{\partial \theta}=0$ to the eqt.(11). Eqt.(14) is produced.

$$\frac{\partial}{\partial \theta} Re[\ A_o\sqrt{\frac{\sin(\theta+\beta_j)}{\cos\theta+\mu_1\sin\theta}}+B_o\sqrt{\frac{\sin(\theta+\beta_j)}{\cos\theta+\mu_2\sin\theta}}\ ]=0 \tag{14}$$

Eqt.(14) is rectified into eqt.(15)

$$A_j'\cos^4\theta+B_j'\cos^2\theta+C_j'=0 \tag{15}$$

Where, $A_j=\mu_2^2-a_j'\mu_1^2+a_j'-1$, $B_j=2(\mu_2-\mu_1 a_j')$, $C_j=a_j'\mu_1^2-\mu_2^2$,

$A_j'=Re\ (A_j^2+B_j^2)$, $B_j'=Re\ (2A_jC_j-B_j^2)$ and $C_j'=Re\ C_j^2$

Eqt. (16) is obtained from eqt. (15), θ_j of eqt. (16) is the angle($\theta=\theta_j$) which the maximum of SPATE signals is occurred.

$$\theta_j = \cos^{-1} Re\left[\pm\sqrt{\frac{B_j' \pm \sqrt{B_j'^2 - 4a_j C_j'}}{2A_j'}} \right] \quad (j = L,\ T) \tag{16}$$

Where, θ_L or θ_T are respectively angles which the maximum of SPATE signals is occurred when y′ is constant or x′ is constant. Substituting θ_j of eqt.(16), μ_1, μ_2, f_L, f_T, $r = \frac{x_j}{\sin(\theta_j + \beta_j)}$ and ϕ_j etc. for equation (11), eqt. (17) is obtained.

$$S_{dmax} = \frac{1}{\sqrt{x_j'}}(a_j K_{\text{I}} + b_j K_{\text{II}}) + C_j \tag{17}$$

where, $a_j = Re\frac{\sqrt{\sin(\theta_j + \beta_j)}}{2\sqrt{2\pi}} A_0$, $b_j = Re\frac{\sqrt{\cos(\theta_j + \beta_j)}}{2\sqrt{2\pi}} B_0$ and $C_j = C_0$

In eqt.(17), the relationship between S_{dmax} and $1/\sqrt{x_j'}$ is linear, $a_1 K_I + b_1 K_{II}$ is slope of eqt.(17) and C_j is the intercept of eqt.(17). If θ_j is experimentally measured and is substituted for eqt.(17), slope of eqt.(17)(= $a_1 K_I + b_1 K_{II}$) is only function K_I and K_{II}, that is, K_I and K_{II} are only unknown variables in the slope of eqt. (17). Slopes(= $a_1 K_I + b_1 K_{II}$) of eqt.(17) which experimental values θ_j are substituted are equal to the slopes(= T_j) determined from the relationship between maximum SPATE signals and $1/\sqrt{x_j'}$. Therefore, the following equation is obtained

$$a_j K_I + b_j K_{II} = T_j \tag{18}$$

K_I and K_{II} are determined from two equations of eqt.(18) and σ_{0x} from C_j.
The slopes between maximum SPATE signals and $1/\sqrt{x_j'}$ had better be determined from the curve-fitting equation. When β_j is 0 and load is mixed mode, the equation between S_{dmax} and $1/\sqrt{x_j'}$ is

$$S_{dmax} = \frac{1}{\sqrt{x_j'}}(a_j K_{\text{I}} + b_j K_{\text{II}}) + C_j \tag{19}$$

where, $a_j = Re\frac{\sqrt{\sin\theta_j}}{2\sqrt{2\pi}} A_0$, $b_j = Re\frac{\sqrt{\sin\theta_j}}{2\sqrt{2\pi}} B_0$ and $C_j = C_0$
In this case, we can obtain K_I, K_{II} and σ_{0x} from the previous method. When β_j is 0 and load is mode I or load is mode II, the equations between S_{dmax} and $1/\sqrt{x_j'}$ are respectively

$$S_{dmax} = \frac{1}{\sqrt{x_j'}} a_j K_{\text{I}} + C_j, \quad S_{dmax} = \frac{1}{\sqrt{x_j'}} b_j K_{\text{II}} + C_j \tag{20}$$

where, $a_j = Re\frac{\sqrt{\sin\theta_j}}{2\sqrt{2\pi}} A_0$, $b_j = Re\frac{\sqrt{\sin\theta_j}}{2\sqrt{2\pi}} B_0$ and $C_j = C_0$
When load is mode I or mode II, K_I and σ_{0x} or K_{II} and σ_{0x} are respectively determined from the above approaches.

3 Experiment and experimental method

3.1 Experimental device and experimental method

Experimental device used in this research is SPATE-8000. Temperature changes can be transformed into stress changes by SPATE device. Therefore, SPATE device is used to measure stress changes of structure under cyclic load.

Surface temperature of structure increases when structure is under compression and the temperature decreases when structure under tension. The material employed in this research is glass fiber epoxy composite. its laminate code is [0/90/ 0_{10}/90/0]. Specimen size(w×ℓ×t) is approximately 40×300×3 mm^3. The basic physical properties(E_L, E_T, G_{LT}, ν_{LT} (ν_{TL})) of the glass fiber epoxy composite employed in this research are measured using the measuring method of basic physical properties of orthotropic material[16]. thermomechanical calibration coefficients of the glass fiber epoxy composite are measured by calibration against a calculated stress or calibration against measured stress[17]. They are shown in Table.1

Table 1. Thermomechanical calibration coefficient and the basic physical properties

	f_L (A/D/MPa)	f_T (A/D/MPa)	E_L (GPa)	E_T (GPa)	G_{LT} (GPa)	ν_{LT} (ν_{TL})
Glass-fiber epoxy composite [0/90/0_{10}/90/0]	61.763* 59.395**	156.210* 151.686**	32.5	14.1	3.73	0.227 (0.095)

L : Fiber direction T : Normal direction to fiber direction

*(**) : Thermomechanical calibration coefficient measured by calibration against a calculated stress (calibration against a measured stress)[17]

3.2 The measuring method of stress intensity factor using SPATE data

The measuring method of stress intensity factor using SPATE data are

1. SPATE signals(maximum values in this research) are selected from the stored SPATE signals.
2. When x′ or y′ is constant in SPATE signals, θ_j(θ_L or θ_T) is measured through the curved fitting. θ_j is the direction which the maximum values of SPATE signals are occurred.
3. When x′ or y′ is constant, we obtain the slope of linear equation obtained in the relationships between S_{dmax}(maximum SPATE signals) and $1/\sqrt{x_j'}$ ($1/\sqrt{y_j'}$). The linear equation is obtained using curve fitting method.
4. Substituting θ_j(measured by experiment), $r=\frac{x_j'}{\sin(\theta_j+\beta_j)}$, f_j and β_j for

eqt.(17), we obtain the slopes and intercepts of eqt.(17) associated with the experimental conditions.

5. Slopes and intercepts determined from experimental data are respectively equal to those of the fourth clause. There are three conditions in the determinating method of the slopes and the intercept of the linear equation. They are x′ = constant, y′<0 (y′=const.) and y′>0 (y′=const.). Therefore, there are three cases in the determinating method of stress intensity factor and non-singular term from the three linear equations.

Case Ⅰ : The determinating method of stress intensity factor from the the relationships between S_{dmax} and $1/\sqrt{x'}$ and between S_{dmax} and $1/\sqrt{y'}$. When x′ is constant, y′ is constant (y′>0).

Case Ⅱ : The determinating method of stress intensity factor from the the relationships between S_{dmax} and $1/\sqrt{x'}$ and between S_{dmax} and $1/\sqrt{|y'|}$. When x′ is constant, y′ is constant(y′<0).

Case Ⅲ : The determinating method of stress intensity factor from the the relationships between S_{dmax} and $1/\sqrt{y'}$ and between S_{dmax} and $1/\sqrt{|y'|}$. When y′ is constant (y′>0 and y′<0) .

4 Experimental results and discussions

Fig.2 shows SPATE stress patterns which α is 0°. Here α is the angle between ϕ loading direction and fiber direction. 2a is crack length(mm), w the width of specimen(mm), t specimen thickness(mm), ℓ specimen hight(mm), N loading cyclic number(Hz), $d\sigma$(MPa) is difference between σ_{max} and σ_{min}.

Fig.3 and Fig.4 respectively show the first biggest values and the second biggest values of SPATE signals in the Fig. 2 when x′ is constant and y′ constant.

θ_T and θ_L are respectively measured from Fig.3 and from Fig.4. θ_T's are approximately -27° in the Fig.3. It is measured from Fig.3. When y′ is greater than 0, θ_L are approximately 52° in the Fig.4. When y′ is less than 0, θ_L are approximately −71° in the Fig.4. they are respectively measured from Fig.4. Fig.5 shows the relationships between maximum SPATE signals and $1/\sqrt{x'}$, between maximum SPATE signals and $1/\sqrt{y'}$(y′>0) and maximum SPATE signals and $1/\sqrt{|y'|}$(y′<0) in the Fig.2. Fig.6 shows the relationships between maximum SPATE signals and $1/\sqrt{x'}$, between maximum SPATE signals and $1/\sqrt{y'}$(y′>0) and between maximum SPATE signals and $1/\sqrt{|y'|}$(y′<0) in the another specimens.

In the Fig.5 and Fig.6 , rectangular symbol(-□-□-□-), triangular symbol(-△-△-△-) and star symbol(-☆-☆-☆-) respectively represent experimental values when x′ is constant, those when y′ is constant(y′>0) and those when y′ is constant(y′<0).

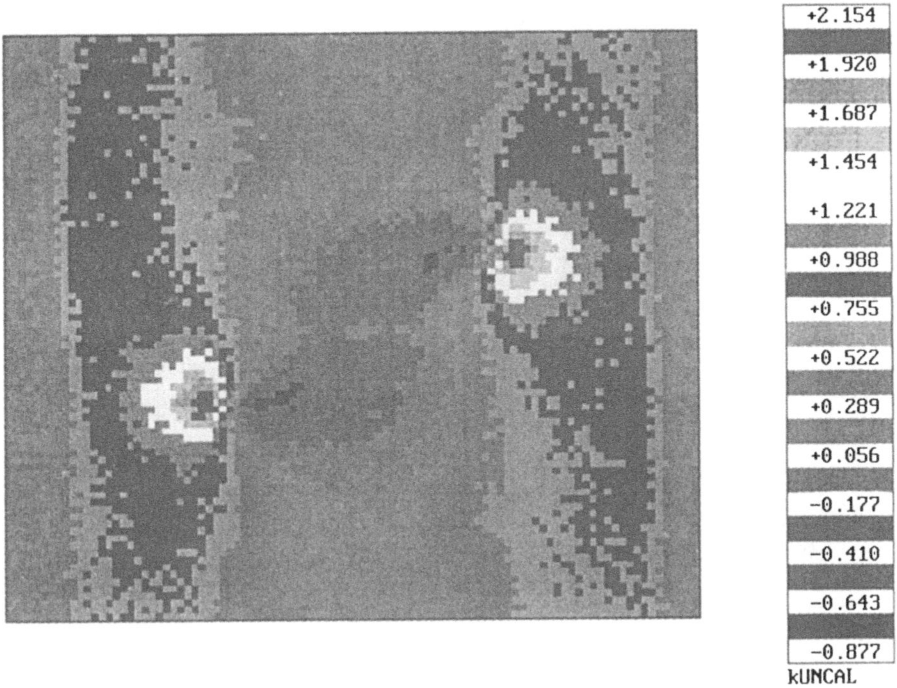

w=37.95, $2a/w=0.542$, $\beta=27°$, N=23, $d\sigma=11.945$ MPa, t=3.21, ℓ=293

Fig.2 SPATE stress patterns ($\alpha=0°$)

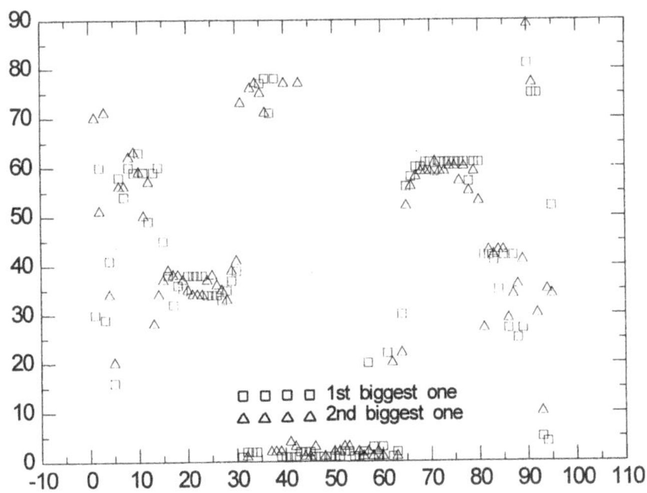

Fig.3 Positions of the first and second biggest values of SPATE signals
(x' = constant, $\beta=27°$, $\alpha=0°$)

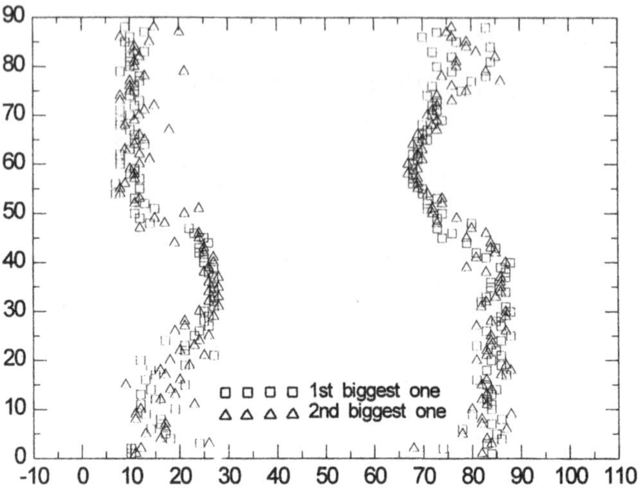

Fig.4 Positions of the first and second biggest values of SPATE signals
(y' = constant, $\beta = 27^{\circ}$, $\alpha = 0^{\circ}$)

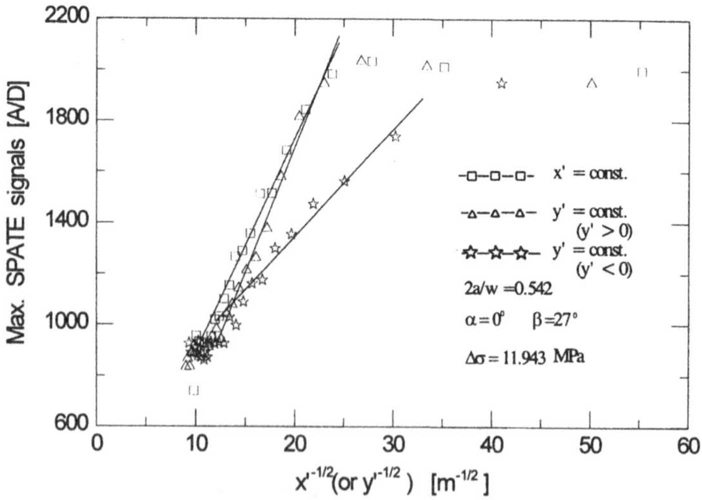

Fig.5 The relationships between $1/\sqrt{x_j'}$ and maximum SPATE signals
($\beta = 27^{\circ}$, $\alpha = 0^{\circ}$, $2a/w = 0.542$)

We can know from Fig.5 to Fig.7 that linear qualities between $1/\sqrt{x_j'}$ and maximum SPATE signals are very excellent. In each Figure, linear equation are obtained using curve-fitting. Slopes of each linear equation are more and more bigger as the fictitious crack tip is more and more far away from the actual crack tip within the some region. Therefore, it is very important to determine crack-tip exactly in the analysis of fracture mechanics using SPATE.

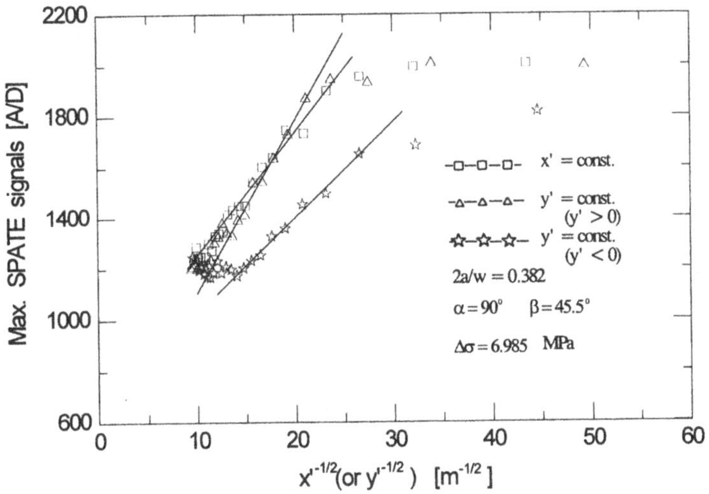

Fig.6 The relationships between $1/\sqrt{x_j'}$ and maximum SPATE signals ($\beta=45.5°$, $\alpha=90°$, $2a/w=0.382$)

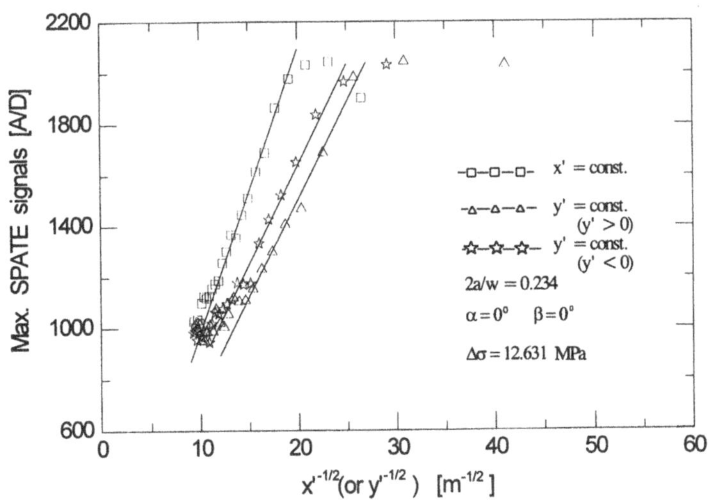

Fig.7 The relationships between $1/\sqrt{x_j'}$ and max. SPATE signals

Table 2. Comparison photoelastic experimental values(theoretical values) with SPATE experimental values
(experimental model material : glass-fiber epoxy composite
(0/90/ 0_9/90/0))

Specimen, loading conditions		$\frac{K_I}{K_0}$		$\frac{K_{II}}{K_0}$	
		Exp. val. (Photo.)	Exp. val. (SPATE)	Exp. val. (Photo.)	Exp. val. (SPATE)
$\alpha=0°$ $\beta=0°$, $\frac{2a}{w}=0.2343$	case 1	1.07*	0.9807	0	2.953×10^{-2}
w=37.643	case 2	1.07*	0.9807	0	8.643×10^{-2}
$d\sigma$=12.631, σ_{mean}=8.559	case 3	None	None	None	None
$\alpha=0°$ $\beta=27°$, $\frac{2a}{w}=0.542$	case 1	0.9379**	0.9236	0.4693**	0.4859
w=37.95	case 2	0.9379**	0.9309	0.4693**	0.4707
$d\sigma$=11.943, σ_{mean}=7.962	case 3	0.9379**	0.9254	0.4693**	0.4775
$\alpha=90°$ $\beta=45.5°$, $\frac{2a}{w}=0.382$	case 1	0.535**	0.5569	0.55**	0.5882
w=37.95 , ℓ=305	case 2	0.535**	0.5259	0.55**	0.5277
$d\sigma$=6.985, σ_{mean}=5.734	case 3	0.535**	0.5299	0.55**	0.5591

case 1 : x'=const., y'=const. and y'>0 , case 2 : x'=const., y'=const. and y'<0
case 3 : y'=const., y'>0 and y'<0
w : Width specimen (mm), 2a : Crack length, β : Crack angle
α : Angle of fiber direction with loading direction, $d\sigma : \sigma_{max} - \sigma_{min}$ (MPa),
$\sigma_{mean} = (\sigma_{max} + \sigma_{min})/2$ (MPa), * : Theorectical value of Bowie[19]
** : Photoelastic experimental value[18], Photo : Photoelastic experiment

Experimental values of SPATE, photoelastic experimental values [16] and Bowie's theoretical values [19] are shown in the Table 2. It is known from Table 2. that they are very similar to each other. Therefore, we can know that the measuring method and conception of stress intensity factor using SPATE are very effective. There are three cases in the determinating method of stress intensity factor using SPATE signals. Among them, arbitrary case can be used to obtain stress intensity factor using SPATE signal.

5 Conclusions

The following results are obtained through the basic theory and the previous experiments.

1. As shown in Table 2, it is certified through experiment that the determinating method of stress intensity factor using SPATE signals are very useful and its precisions are very good.
2. There are two kinds of error in the determinating processes of stress intensity factor using SPATE signals. One is the error occurred in the measurement of θ_j, another is the error occurred in the slope measurement of relative equation between maximum SPATE signals and $1/\sqrt{x_j'}$. Errors occurred in the determination of stress intensity factor using SPATE signals are mainly resulted from the error of the latter, and it is resulted from errors produced in the determination of crack tip position.
3. There are three cases in the determinating method of stress intensity factor from the slope of relative equation between the maximum SPATE signals and $1/\sqrt{x_j'}$ and from that of eqt.(17). Among them, arbitrary case can be used to obtain stress intensity factor using SPATE signals.

6 References

1. Thomson W.(Lord Kelvin) (1853), On Dynamic Theory of Heat, Trans, Roy, Soc, Edinburgh, Vol.20, PP.261-283.
2. Potter R. T. and Greavers L.J.(1987), The Application of Thermoelastic Stress Analysis Technique to Fiber Composite, Proceedings of the Society of Photo-Optical Instrumentation Engineers, pp. 134-146.
3. Potter R. T.(1987), Stress Analysis in Laminated Fiber Composites by Thermoelastic Emission, Second International Conf. on Stress Analysis by Thermoelastic Techniques, pp.20-1SA-20-10SA.
4. Kageyama K., Ueki K. and Kikuchi M.(1988), Thermolelastic Technique Applied to Stress Analysis Carbon Fiber Reinforced Composite Materials, Proc. of the VI International Congress on Experimental Mechanics, pp.931-936.
5. Wong A. K.(1991), A Non-adibatic Thermoelastic Theory for Composite Laminates, Journal Phys. Chem. Solids Vol. 52, No.3. pp. 483-494.
6. Jones R., Heller M., Lombardo D., Dunn S., Paul J. and Saunders D.(1989), Thermoelastic Assessment of Damage Growth in Composites, Composite Structures 12 , pp.291-314.
7. Bakis Charles E. and Reifsnider Kenneth L. (1986), Application of Micromechanics in the Themoelasticity of Laminated Fibrous Composites, Department of Engineering Science and Mechanics, Virginia Polytechnical Institute and State University, Blacksburg, Virginia, 24061~4899 pp. 10~20
8. Reifsnider Kenneth L. and Giacco Alexander(1990), The Use of Microcrack

Analysis in Performance Simulation for Composite Material Systems, The American Society of Mechanical Engineers, AMD-Vol. III, MD-Vol. 22, pp.19~27.
9. Lin Shun-Tien , Quantitative(1993), Thermoelastic Stress Analysis of Orthotropic Composite Structures, Ph D. Prelim. Document, Department of Engineering Mechanics and Astronautics, University of Wisconsin Madison.
10. Hawong, J. S., Suh, J. G., Choi S. H. and Rowlands R.E.(1994) Measurement of Stress Intensity Factor of Isotropic Material Using SPATE, Proceedings of the KSME Fall Annual Meeting '94, Vol. 1, The Korean Society of Mechanical Engineers(KSME), pp. 56-62.
11. Stanley P. and Chan W.K.(1986), Mode II Crack Studies Using the "SPATE" Technique, Proc.SEM Spring Conference on Experimental Mechanics, New Orleans, pp. 916-923
12. Rukas Stephan R.(1987), Theoretical Considerations for Determining Stress Intensity Factors via Thermoelastic Stress Analysis, Second International Conference of Stress Analysis by Thermoelastic Techniques, pp. 9-1SA~9-12SA
13. Leaity G. P. and Smith R. A. (1989), The Use of SPATE to Measure Residual Stress and Fatigue Crack Growth, Fatigue Fracture Engineering Material Structure, Vol. 12, No. 4, pp. 271-282.
14. Stanley P. and Dulieu J. M.-Smith(1993), Progress in the Thermoelastic Evaluation of Mixed-Mode Stress Intensity Factors, Proceedings of the 1993 Society for Experimental Mechanics Conference, pp. 617~626.
15. Sih S. G., Paris G. R. and Irwin G. R.(1965), On Cracks in Rectlinearly Anisotropic Bodies, Inter. Journal of Fracture Mechanics, pp 189-203.
16. Hawong J. S., Shimamoto A., Takahashi S. and Choi S. H.(1990), A Method of Measuring Physical Properties from Stress Fringe Values, JSME International Journal, the Japan Society of mechanical Engineers, Vol. 33, No. 4, pp 497-505.
17. Dulieu-Smith J. M.(1995), Alternative Calibration Techniques for Quantitative Thermoelastic Stress Analysis, Strain, February, pp. 9-16.
18. Hawong J. S., Shimamoto A., Takahashi S. and Choi S.H.(1990), A Study on the Development of the Composite Material for Photoelastic Experiment (3rd Report Analysis of the Fracture Mechanics of Orthotropic Material by Photoelastic Experiment), the Japan Society of Mechanical Engineers (A), Vol. 56, No. 527, pp. 96-104.
19. Bowie O. L. and Freese L. E.(1972), Central Crack in Plane Orthortropic Rectangular Sheet, International J. of Fracture Mechanics, Vol. 8, pp. 49-58.

Author Index

Subject Index

For Product Safety Concerns and Information please contact our
EU representative GPSR@taylorandfrancis.com Taylor & Francis
Verlag GmbH, Kaufingerstraße 24, 80331 München, Germany